SPRINGER PROCEEDINGS IN PHYSICS 98

SPRINGER PROCEEDINGS IN PHYSICS

74 **Time-Resolved Vibrational Spectroscopy VI**
Editors: A. Lau, F. Siebert, and W. Werncke

75 **Computer Simulation Studies in Condensed-Matter Physics V**
Editors: D.P. Landau, K.K. Mon, and H.-B. Schüttler

76 **Computer Simulation Studies in Condensed-Matter Physics VI**
Editors: D.P. Landau, K.K. Mon, and H.-B. Schüttler

77 **Quantum Optics VI**
Editors: D.F. Walls and J.D. Harvey

78 **Computer Simulation Studies in Condensed-Matter Physics VII**
Editors: D.P. Landau, K.K. Mon, and H.-B. Schüttler

79 **Nonlinear Dynamics and Pattern Formation in Semiconductors and Devices**
Editor: F.-J. Niedernostheide

80 **Computer Simulation Studies in Condensed-Matter Physics VIII**
Editors: D.P. Landau, K.K. Mon, and H.-B. Schüttler

81 **Materials and Measurements in Molecular Electronics**
Editors: K. Kajimura and S. Kuroda

82 **Computer Simulation Studies in Condensed-Matter Physics IX**
Editors: D.P. Landau, K.K. Mon, and H.-B. Schüttler

83 **Computer Simulation Studies in Condensed-Matter Physics X**
Editors: D.P. Landau, K.K. Mon, and H.-B. Schüttler

84 **Computer Simulation Studies in Condensed-Matter Physics XI**
Editors: D.P. Landau and H.-B. Schüttler

85 **Computer Simulation Studies in Condensed-Matter Physics XII**
Editors: D.P. Landau, S.P. Lewis, and H.-B. Schüttler

86 **Computer Simulation Studies in Condensed-Matter Physics XIII**
Editors: D.P. Landau, S.P. Lewis, and H.-B. Schüttler

87 **Proceedings of the 25th International Conference on the Physics of Semiconductors**
Editors: N. Miura and T. Ando

88 **Starburst Galaxies**
Near and Far
Editors: L. Tacconi and D. Lutz

89 **Computer Simulation Studies in Condensed-Matter Physics XIV**
Editors: D.P. Landau, S.P. Lewis, and H.-B. Schüttler

90 **Computer Simulation Studies in Condensed-Matter Physics XV**
Editors: D.P. Landau, S.P. Lewis, and H.-B. Schüttler

91 **The Dense Interstellar Medium in Galaxies**
Editors: S. Pfalzner, C. Kramer, C. Straubmeier, and A. Heithausen

92 **Beyond the Standard Model 2003**
Editor: H.V. Klapdor-Kleingrothaus

93 **ISSMGE**
Experimental Studies
Editor: T. Schanz

94 **ISSMGE**
Numerical and Theoretical Approaches
Editor: T. Schanz

95 **Computer Simulation Studies in Condensed-Matter Physics XVI**
Editors: D.P. Landau, S.P. Lewis, and H.-B. Schüttler

96 **Electromagnetics in a Complex World**
Editors: I.M. Pinto, V. Galdi, and L.B. Felsen

97 **Fields, Networks and Computations**
A Modern View of Electrodynamics
Editor: P. Russer

98 **Particle Physics and the Universe**
Proceedings of the 9th Adriatic Meeting, Sept. 2003, Dubrovnik
Editors: J. Trampetić and J. Wess

Homepage: springeronline.com

Volumes 46–73 are listed at the end of the book.

J. Trampetić J. Wess (Eds.)

Particle Physics and the Universe

Proceedings of the 9th Adriatic Meeting, Sept. 2003, Dubrovnik

RUDJER BOŠKOVIĆ INSTITUTE

Max-Planck-Gesellschaft
zur Förderung der Wissenschaften e.V.

Springer

Professor Josip Trampetić
Rudjer Boskovic Institute
Theoretical Physics Division
P.O.Box 180
10 002 Zagreb
Croatia

Professor Julius Wess
Sektion Physik
der Ludwig-Maximilians-Universität
Theresienstr. 37
80333 München
and
Max-Planck-Institut für Physik
(Werner-Heisenberg-Institut)
Föhringer Ring 6
80805 München
Germany

ISSN 0930-8989

ISBN 3-540-22803-9 Springer Berlin Heidelberg New York

Library of Congress Control Number: 2004109784

Springer is a part of Springer Science+Business Media

springeronline.com

Printed in Germany

Cover concept: eStudio Calamar Steinen
Cover production: *design & production* GmbH, Heidelberg

Printed on acid-free paper 62/3141/ts 5 4 3 2 1 0

For Prof. Dubravko Tadić
31 October 1934–6 March 2003

Dubravko Tadić was born in Zagreb, Croatia, and graduated from the University of Zagreb with his B.Sc. in 1958. He completed his Ph.D. in 1961, during the time of Vladimir Glaser and Borivoj Jakšić under the supervision of Gaja Alaga. His thesis dealt with nuclear beta decay and the structure of the weak interaction, interests which he continued to pursue thereafter as a member of the Rudjer Bošković Institute and later at the University of Zagreb. He was a leader of a theory research group at the Rudjer Bošković Institute, and later became head of the theory division of the Faculty of Sciences (PMF-Zagreb) at the University of Zagreb. He was honored for his many contributions to physics by being elected as an extraordinary member of the Yugoslav Academy of Sciences and Arts in 1981, and as a full member of the Croatian Academy of Sciences and Arts in 1991.

Professor Tadić was well known in international circles, having spent time in Birmingham while Rudolf Peierels was present, and later at Brookhaven National Laboratory. We met while Dubravko was at Brookhaven and we started a lifelong collaboration and friendship. Among our papers was the first major review of parity-violating nuclear interactions, which incorporated the then newly-developed techniques of current algebra to study models of the weak Hamiltonian.

Dubravko maintained a lifelong interest in nuclear physics, but moved later in his career into elementary particle physics, particularly weak interactions and quark models. His research was characterized by deep insight and clarity of thought along with great attention to detail. By example he served as a role model for a generation of younger physicists including the Ph.D. students he supervised in Zagreb which include B. Eman, B. Guberina, H. Galić, I. Picek, J. Trampetić, P. Colić, D. Horvat, A. Ilakovac, Z. Narančić, S. Žganec, G. Omanović, and B. Podobnik. Along with his students and other collaborators he authored or coauthored 127 publications in scientific journals whose impact on physics will be felt for many years to come.

Dubravko had a broad range of interests outside of physics which included military history and the history of Croatia. He was an avid hiker and enjoyed entertaining his visitors on hikes with details of local history. Although serious when working, he had a wonderful sense of humor when relaxing with family and friends.

It is appropriate that we remember Dubravko Tadić in these Proceedings because he was one of the prime organizers of the Adriatic meetings, and other international events which have served to showcase the work of students and younger researchers in the Central European region. He will be deeply missed not only by his family and his lovely wife Gordana, but by the whole physics community.

Ephraim Fischbach
West Lafayette, Indiana, April 2004

Preface

The Adriatic Meetings have traditionally been conferences on the most advanced status of science. They are one of the very few conferences in physics aiming at a very broad participation of young and experienced researchers with different backgrounds in particle physics.

Particle physics has grown into a highly multi-faceted discipline over the sixty years of its existence, mainly because of two reasons: Particle physics as an experimental science is in need of large-scale laboratory set-ups, involving typically collaborations of several hundreds or even thousands of researchers and technicians with the most diverse expertise.

This forces particle physics, being one of the most fundamental disciplines of physics, to maintain a constant interchange and contact with other disciplines, notably solid-state physics and laser physics, cosmology and astrophysics, mathematical physics and mathematics.

Since the expertise necessary in doing research in particle physics has become tremendously demanding in the last years, the field tends to organize purely expert conferences, meetings and summer schools, such as for detector development, for astroparticle physics or for string theory.

The Adriatic Meeting through its entire history has been a place for establishing exchange between theory and experiment. The 9th Adriatic Meeting successfully continued this tradition and even intensified the cross-discipline communication by establishing new contacts between the community of cosmologists and of particle physicists. The exchange between theorists and experimentalists was impressively intensive and will certainly have a lasting effect on several research projects of the European and world-wide physics community.

As the title of the conference suggests, cosmology and astroparticle physics and their relation to particle physics was one of the main topics of the conference. The reason for this choice is the overwhelming quality of the results obtained in cosmology throughout the recent years.

Another reason for intensifying the contact with cosmology is that the laboratory experiments at the Large Hadron Collider (LHC) in CERN are due to come into operation only in about four to five years from now. These experiments are expected to deliver for the first time sound data about physics beyond the Standard Model. It is quite unclear when or even whether there will be experiments going beyond the LHC energy scale, simply because of the large financial and organizational problems for building such projects. Therefore particle physics may be forced to look elsewhere for potential tests of its models, and extraterrestrial sources are the only conceivable alternative.

On the theoretical side, the currently intensively discussed topic of Lorentz symmetry violation was presented as a potential window into quantum gravity phenomenology. It was emphasized how stringently current astrophysical results already constrain potential extensions of the Standard Model.

Neutrino physics was discussed as a newly discovered hot topic during the 8th Adriatic Meeting. It has now firmly established its results obtained two years ago.

There is a common goal underlying theoretical research in particle physics– a unified description of all forces in Nature. Part of the research effort in this direction is known as Grand Unified Theories.

A crucial question in theoretical physics is the unification of quantum field theory (as the basis of the Standard Model) and the theory of general relativity (as the basis for the theory of gravity). The most prominent candidate for achieving this unification of two quite differently structured theories is string theory. String field theory is an attempt to use quantum field theory tools for solving string theory. The real excitement in the last two years came from the theoretical proposal that our 3+1 dimensional world might be a cosmic defect (brane-world) within higher-dimensional spacetime, with Standard Model fields and gravity localized on such a brane. This proposal also exhibits an exponential hierarchy of the Planck mass scale, an induced de Sitter metric on the brane and a phenomenologically acceptable value of the cosmological constant.

The concept of noncommutative spacetime has a long history, both in mathematics and physics, but recently it attracted a lot of attention since it was shown that noncommutativity provides an effective description of physics of strings in an external background field. The research of the last few years provides a solid mathematical basis for constructing gauge field theories on noncommutative spacetime.

The Standard Model of electroweak and strong interactions has been in place for nearly thirty years, but experimental tests of these theories today have reached a level of precision that permits glimpses of physics beyond this impressive structure. Such glimpses appear to be largely associated with the yet-to-be discovered Higgs boson. A crucial theoretical input for any such prediction are precision calculations in the theoretical models, even more, precision calculations enter into the design of the experimental setup itself.

Experiments in the K and B sectors (mixings etc.) of meson physics are achieving an impressive accuracy as well today and could yield cracks in the Standard Model at any time. Theoretical predictions were presented for possible new physics in this sector.

The weak and rare heavy quark decays together with CP violation are studied through the energy, forward-backward and CP asymmetries by using methods like pQCD, QCD sum rules, relativistic quark models, QCD on the lattice, etc.

We would like to thank young members of the Theory Division of the Rudjer Bošković Institute for their help during the Conference: A. Babić, G. Duplančić, D. Jurman and K. Passek-Kumerički. We would especially like to thank: L. Jonke, H. Nikolić and H. Štefančić for a substantial help during the organization of the Conference. We would also like to thank L. Jonke for preparing this book of Proceedings.

Zagreb,
August 2004

Josip Trampetić
Julius Wess

Contents

Part I

Neutrinos, Astroparticle Physics, Cosmology and Gravity

matrix has 7 possible limits: 3 have the normal hierarchy, 2 have the inverted hierarchy, and 2 have 3 nearly degenerate masses.

In neutrinoless double beta decay, the effective mass is $m_0 = |a+2b+2c|$. In the 2 cases of inverted hierarchy, we have

$$m_0 \simeq \sqrt{\Delta m^2_{atm}} \simeq 0.05\,\mathrm{eV}\ , \tag{8}$$

$$m_0 \simeq \cos 2\theta_{sol} \sqrt{\Delta m^2_{atm}}\ , \tag{9}$$

respectively for $m_1/m_2 = \pm 1$, i.e. for their relative CP being even or odd. In the 2 degenerate cases,

$$m_0 \simeq |m_{1,2,3}|\ , \tag{10}$$

$$m_0 \simeq \cos 2\theta_{sol} |m_{1,2,3}|\ . \tag{11}$$

With $\mathcal{M}_\nu$ of (7), U_{e3} is zero necessarily, in which case there can be no CP violation in neutrino oscillations. However, suppose we consider instead [5, 6]

$$\mathcal{M}_\nu = \begin{pmatrix} a+2b+2c & d & d^* \\ d & b & a+b \\ d^* & a+b & b \end{pmatrix}\ , \tag{12}$$

where d is now complex, then the Z_2 symmetry of (7) is broken and U_{e3} becomes nonzero. In fact, it is proportional to $iImd$, thus predicting maximal CP violation in neutrino oscillations.

3 Nearly Degenerate Majorana Neutrino Masses

Suppose that at some high energy scale, the charged lepton mass matrix and the Majorana neutrino mass matrix are such that after diagonalizing the former, i.e.

$$\mathcal{M}_l = \begin{pmatrix} m_e & 0 & 0 \\ 0 & m_\mu & 0 \\ 0 & 0 & m_\tau \end{pmatrix}\ , \tag{13}$$

the latter is of the form

$$\mathcal{M}_\nu = \begin{pmatrix} m_0 & 0 & 0 \\ 0 & 0 & m_0 \\ 0 & m_0 & 0 \end{pmatrix}\ . \tag{14}$$

From the high scale to the electroweak scale, one-loop radiative corrections will change $\mathcal{M}_\nu$ as follows:

$$(\mathcal{M}_\nu)_{ij} \to (\mathcal{M}_\nu)_{ij} + R_{ik}(\mathcal{M}_\nu)_{kj} + (\mathcal{M}_\nu)_{ik} R^T_{kj}\ , \tag{15}$$

where the radiative correction matrix is assumed to be of the most general form, i.e.

$$R = \begin{pmatrix} r_{ee} & r_{e\mu} & r_{e\tau} \\ r^*_{e\mu} & r_{\mu\mu} & r_{\mu\tau} \\ r^*_{e\tau} & r^*_{\mu\tau} & r_{\tau\tau} \end{pmatrix} . \qquad (16)$$

Thus the observed neutrino mass matrix is given by

$$\mathcal{M}_\nu = m_0 \begin{pmatrix} 1+2r_{ee} & r_{e\tau}+r^*_{e\mu} & r_{e\mu}+r^*_{e\tau} \\ r^*_{e\mu}+r_{e\tau} & 2r_{\mu\tau} & 1+r_{\mu\mu}+r_{\tau\tau} \\ r^*_{e\tau}+r_{e\mu} & 1+r_{\mu\mu}+r_{\tau\tau} & 2r^*_{\mu\tau} \end{pmatrix} . \qquad (17)$$

Let us rephase ν_μ and ν_τ to make $r_{\mu\tau}$ real, then the above $\mathcal{M}_\nu$ is exactly in the form of (12), with of course a as the dominant term. In other words, we have obtained a desirable description of all present data on neutrino oscillations including CP violation, starting from almost nothing.

4 Plato's Fire

The successful derivation of (17) depends on having (13) and (14). To be sensible theoretically, they should be maintained by a symmetry. At first sight, it appears impossible that there can be a symmetry which allows them to coexist. The solution turns out to be the non-Abelian discrete symmetry A_4 [7, 8]. What is A_4 and why is it special?

Around the year 390 BCE, the Greek mathematician Theaetetus proved that there are five and only five perfect geometric solids. The Greeks already knew that there are four basic elements: fire, air, water, and earth. Plato could not resist matching them to the five perfect geometric solids and for that to work, he invented the fifth element, i.e. quintessence, which is supposed to hold the cosmos together. His assignments are shown in Table 1.

Table 1. Properties of Perfect Geometric Solids

Solid	Faces	Vertices	Plato	Group
tetrahedron	4	4	fire	A_4
octahedron	8	6	air	S_4
icosahedron	20	12	water	A_5
hexahedron	6	8	earth	S_4
dodecahedron	12	20	?	A_5

The group theory of these solids was established in the early 19th century. Since a cube (hexahedron) can be imbedded perfectly inside an octahedron and the latter inside the former, they have the same symmetry group. The same holds for the icosahedron and dodecahedron. The tetrahedron (Plato's "fire") is special because it is self-dual. It has the symmetry group A_4, i.e. the finite group of the even permutation of 4 objects. The reason that it is special for the neutrino mass matrix is because it has three inequivalent one-dimensional irreducible representations and one three-dimensional irreducible representation exactly. Its character table is given below.

Table 2. Character Table of A_4

Class	n	h	χ_1	χ_2	χ_3	χ_4
C_1	1	1	1	1	1	3
C_2	4	3	1	ω	ω^2	0
C_3	4	3	1	ω^2	ω	0
C_4	3	2	1	1	1	-1

In the above, n is the number of elements, h is the order of each element, and

$$\omega = e^{2\pi i/3} \tag{18}$$

is the cube root of unity. The group multiplication rule is

$$\underline{3} \times \underline{3} = \underline{1} + \underline{1}' + \underline{1}'' + \underline{3} + \underline{3}\,. \tag{19}$$

5 Details of the A_4 Model

The fact that A_4 has three inequivalent one-dimensional representations $\underline{1}$, $\underline{1}'$, $\underline{1}''$, and one three-dimensional reprsentation $\underline{3}$, with the decomposition given by (19) leads naturally to the following assignments of quarks and leptons:

$$(u_i, d_i)_L, \quad (\nu_i, e_i)_L \sim \underline{3}\,, \tag{20}$$

$$u_{1R}, \quad d_{1R}, \quad e_{1R} \sim \underline{1}\,, \tag{21}$$

$$u_{2R}, \quad d_{2R}, \quad e_{2R} \sim \underline{1}'\,, \tag{22}$$

$$u_{3R}, \quad d_{3R}, \quad e_{3R} \sim \underline{1}''\,. \tag{23}$$

Heavy fermion singlets are then added:

$$U_{iL(R)}, \quad D_{iL(R)}, \quad E_{iL(R)}, \quad N_{iR} \sim \underline{3}\,, \tag{24}$$

together with the usual Higgs doublet and new heavy singlets:

$$(\phi^+, \phi^0) \sim \underline{1}, \qquad \chi_i^0 \sim \underline{3}\,. \tag{25}$$

With this structure, charged leptons acquire an effective Yukawa coupling matrix $\bar{e}_{iL} e_{jR} \phi^0$ which has 3 arbitrary eigenvalues (because of the 3 independent couplings to the 3 inequivalent one-dimensional representations) and for the case of equal vacuum expectation values of χ_i, i.e.

$$\langle\chi_1\rangle = \langle\chi_2\rangle = \langle\chi_3\rangle = u \ , \tag{26}$$

which occurs naturally in the supersymmetric version of this model [8], the unitary transformation U_L which diagonalizes $\mathcal{M}_l$ is given by

$$U_L = \frac{1}{\sqrt{3}} \begin{pmatrix} 1 & 1 & 1 \\ 1 & \omega & \omega^2 \\ 1 & \omega^2 & \omega \end{pmatrix} \ . \tag{27}$$

This implies that the effective neutrino mass operator, i.e. $\nu_i\nu_j\phi^0\phi^0$, is proportional to

$$U_L^T U_L = \begin{pmatrix} 1 & 0 & 0 \\ 0 & 0 & 1 \\ 0 & 1 & 0 \end{pmatrix} \ , \tag{28}$$

exactly as desired.

6 New Flavor-Changing Radiative Mechanism

The original A_4 model [7] was conceived to be a symmetry at the electroweak scale, in which case the splitting of the neutrino mass degeneracy is put in by hand and any mixing matrix is possible. Subsequently, it was proposed [8] as a symmetry at a high scale, in which case the mixing matrix is determined completely by flavor-changing radiative corrections and the only possible result happens to be (17). This is a remarkable convergence in that (17) is in the form of (12), i.e. the phenomenologically preferred neutrino mixing matrix based on the most recent data from neutrino oscillations.

We should now consider the new physics responsible for the r_{ij}'s of (16). Previously [8], arbitrary soft supersymmetry breaking in the scalar sector was invoked. It is certainly a phenomenologically viable scenario, but lacks theoretical motivation and is somewhat complicated. Here a new and much simpler mechanism is proposed [9], using a triplet of charged scalars under A_4, i.e. $\eta_i^+ \sim \underline{3}$. Their relevant contributions to the Lagrangian of this model is then

$$\mathcal{L} = f\epsilon_{ijk}(\nu_i e_j - e_i\nu_j)\eta_k^+ + m_{ij}^2\eta_i^+\eta_j^- \ . \tag{29}$$

Whereas the first term is invariant under A_4 as it should be, the second term is a soft term which is allowed to break A_4, from which the flavor-changing radiative corrections will be calculated.

Let

$$\begin{pmatrix} \eta_e \\ \eta_\mu \\ \eta_\tau \end{pmatrix} = \begin{pmatrix} U_{e1} & U_{e2} & U_{e3} \\ U_{\mu 1} & U_{\mu 2} & U_{\mu 3} \\ U_{\tau 1} & U_{\tau 2} & U_{\tau 3} \end{pmatrix} \begin{pmatrix} \eta_1 \\ \eta_2 \\ \eta_3 \end{pmatrix} \ , \tag{30}$$

where $\eta_{1,2,3}$ are mass eigenstates with masses $m_{1,2,3}$. The resulting radiative corrections are given by

$$r_{\alpha\beta} = -\frac{f^2}{8\pi^2} \sum_{i=1}^{3} U^*_{\alpha i} U_{\beta i} \ln m_i^2 \, . \tag{31}$$

To the extent that $r_{\mu\tau}$ should not be larger than about 10^{-2}, the common mass $m1_0$ of the three degenerate neutrinos should not be less than about 0.2 eV in this model. This is consistent with the recent WMAP upper bound [22] of 0.23 eV and the range 0.11 to 0.56 eV indicated by neutrinoless double beta decay [11].

7 Models Based on S_3 and D_4

Two other examples of the application of non-Abelian discrete symmetries to the neutrino mass matrix have recently been proposed. One [12] is based on the symmetry group of the equilateral triangle S_3, which has 6 elements and the irreducible representations $\underline{1}$, $\underline{1}'$, and $\underline{2}$. The 3 families of leptons as well as 3 Higgs doublets transform as $\underline{1} + \underline{2}$ under S_3. An additional Z_2 is introduced where $\nu_R(\underline{1})$ and $H(\underline{2})$ are odd, while all other fields are even. After a detailed analysis, the mixing matrix of (6) is obtained with $U_{e3} \simeq -3.4 \times 10^{-3}$ and $0.4 < \tan\theta_{sol} < 0.8$. The neutrino masses are predicted to have an inverted hierarchy satisfying (8).

Another example [13] is based on the symmetry group of the square D_4, which has 8 elements and the irreducible representations $\underline{1}^{++}$, $\underline{1}^{+-}$, $\underline{1}^{-+}$, $\underline{1}^{--}$, and $\underline{2}$. The 3 families of leptons transform as $\underline{1}^{++} + \underline{2}$. The Higgs sector has 3 doublets with $\phi_3 \sim \underline{1}^{+-}$ and 2 singlets $\chi \sim \underline{2}$. Under an extra Z_2, ν_R, e_R, ϕ_1 are odd, while all other fields are even, including ϕ_2. This results in the neutrino mass matrix of (7) with an additional constraint, i.e. $m_1 < m_2 < m_3$ such that the m_0 of neutrinoless double beta decay is equal to $m_1 m_2/m_3$.

8 Form Invariance of the Neutrino Mass Matrix

Consider a specific 3×3 unitary matrix U and impose the condition [14]

$$U \mathcal{M}_\nu U^T = \mathcal{M}_\nu \tag{32}$$

on the neutrino mass matrix $\mathcal{M}_\nu$ in the $(\nu_e, \nu_\mu, \nu_\tau)$ basis. Iteration of the above yields

$$U^n \mathcal{M}_\nu (U^T)^n = \mathcal{M}_\nu \, . \tag{33}$$

Therefore, unless $U^{\bar{n}} = 1$ for some finite $\bar{n}$, the only solution for $\mathcal{M}_\nu$ would be a multiple of the identity matrix. Take for example $\bar{n} = 2$, then the choice

$$U = \begin{pmatrix} 1 & 0 & 0 \\ 0 & 0 & 1 \\ 0 & 1 & 0 \end{pmatrix} \tag{34}$$

leads to (7). In other words, the present neutrino oscillation data may be understood as a manifestation of the discrete symmetry $\nu_e \to \nu_e$ and $\nu_\mu \leftrightarrow \nu_\tau$.

Suppose instead that $\bar{n} = 4$, with U^2 given by (34), then one possible solution for its square root is

$$U_1 = \begin{pmatrix} 1 & 0 & 0 \\ 0 & (1-i)/\sqrt{2} & (1+i)/\sqrt{2} \\ 0 & (1+i)/\sqrt{2} & (1-i)/\sqrt{2} \end{pmatrix}, \tag{35}$$

which leads to

$$\mathcal{M}_1 = \begin{pmatrix} 2b+2c & d & d \\ d & b & b \\ d & b & b \end{pmatrix}, \tag{36}$$

i.e. the 4 parameters of (7) have been reduced to 3 by setting $a = 0$.

Another solution is

$$U_2 = \frac{1}{\sqrt{3}} \begin{pmatrix} 1 & 1 & 1 \\ 1 & \omega & \omega^2 \\ 1 & \omega^2 & \omega \end{pmatrix}, \tag{37}$$

which leads to

$$\mathcal{M}_2 = \begin{pmatrix} 2b+2d & d & d \\ d & b & b \\ d & b & b \end{pmatrix}, \tag{38}$$

i.e. $\mathcal{M}_1$ has been reduced by setting $c = d$. The 3 mass eigenvalues are then $m_{1,2} = 2b \mp \sqrt{2}d$ and $m_3 = 0$, i.e. an inverted hierarchy, with $\tan^2 \theta_{sol}$ predicted to be $2 - \sqrt{3} = 0.27$, as compared to the allowed range [15] 0.29 to 0.86 from fitting all present data.

9 New Z_3 Model of Neutrino Masses

Very recently, two new complete models of lepton masses have been obtained, one based on Z_4 [16] and the other on Z_3 [17]. The former does not fix the solar mixing angle, whereas the latter predicts $\tan^2 \theta_{sol} = 0.5$. Here I will discuss only the Z_3 case. Let $\mathcal{M}_\nu$ be given by

$$\mathcal{M}_\nu = \mathcal{M}_A + \mathcal{M}_B + \mathcal{M}_C , \tag{39}$$

where

$$\mathcal{M}_A = A \begin{pmatrix} 1 & 0 & 0 \\ 0 & 1 & 0 \\ 0 & 0 & 1 \end{pmatrix}, \quad \mathcal{M}_B = B \begin{pmatrix} -1 & 0 & 0 \\ 0 & 0 & -1 \\ 0 & -1 & 0 \end{pmatrix}, \quad \mathcal{M}_C = C \begin{pmatrix} 1 & 1 & 1 \\ 1 & 1 & 1 \\ 1 & 1 & 1 \end{pmatrix}. \tag{40}$$

Since the invariance of $\mathcal{M}_A$ requires only $U_A U_A^T = 1$, U_A can be any orthogonal matrix. As for $\mathcal{M}_B$ and $\mathcal{M}_C$, they are both invariant under the Z_2 transformation of (34) and each is invariant under a Z_3 transformation, i.e. $U_B^3 = 1$ and $U_C^3 = 1$, but $U_B \neq U_C$. Specifically,

$$U_B = \begin{pmatrix} -1/2 & -\sqrt{3/8} & -\sqrt{3/8} \\ \sqrt{3/8} & 1/4 & -3/4 \\ \sqrt{3/8} & -3/4 & 1/4 \end{pmatrix}, \quad U_C = \begin{pmatrix} 0 & 1 & 0 \\ 0 & 0 & 1 \\ 1 & 0 & 0 \end{pmatrix}. \tag{41}$$

Note that U_B commutes with U_2, but U_C does not. If U_C is combined with U_2, then the non-Abelian discrete symmetry S_3 is generated.

First consider $C = 0$. Then $\mathcal{M}_\nu = \mathcal{M}_A + \mathcal{M}_B$ is the most general solution of

$$U_B \mathcal{M}_\nu U_B^T = \mathcal{M}_\nu \,, \tag{42}$$

and the eigenvectors of $\mathcal{M}_\nu$ are ν_e, $(\nu_\mu + \nu_\tau)/\sqrt{2}$, and $(\nu_\mu - \nu_\tau)/\sqrt{2}$ with eigenvalues $A-B$, $A-B$, and $A+B$ respectively. This explains atmospheric neutrino oscillations with $\sin^2 2\theta_{atm} = 1$ and

$$(\Delta m^2)_{atm} = (A+B)^2 - (A-B)^2 = 4BA \,. \tag{43}$$

Now consider $C \neq 0$. Then in the basis spanned by ν_e, $(\nu_\mu + \nu_\tau)/\sqrt{2}$, and $(\nu_\mu - \nu_\tau)/\sqrt{2}$,

$$\mathcal{M}_\nu = \begin{pmatrix} A-B+C & \sqrt{2}C & 0 \\ \sqrt{2}C & A-B+2C & 0 \\ 0 & 0 & A+B \end{pmatrix}. \tag{44}$$

The eigenvectors and eigenvalues become

$$\nu_1 = \frac{1}{\sqrt{6}}(2\nu_e - \nu_\mu - \nu_\tau), \;\; m_1 = A - B \,, \tag{45}$$

$$\nu_2 = \frac{1}{\sqrt{3}}(\nu_e + \nu_\mu + \nu_\tau), \;\; m_2 = A - B + 3C \,, \tag{46}$$

$$\nu_3 = \frac{1}{\sqrt{2}}(\nu_\mu - \nu_\tau), \;\; m_3 = A + B \,. \tag{47}$$

This explains solar neutrino oscillations as well with $\tan^2 \theta_{sol} = 1/2$ and

$$(\Delta m^2)_{sol} = (A-B+3C)^2 - (A-B)^2 = 3C(2A-2B+3C) \,. \tag{48}$$

Whereas the mixing angles are fixed, the proposed $\mathcal{M}_\nu$ has the flexibility to accommodate the three patterns of neutrino masses often mentioned, i.e.

(I) the hierarchical solution, e.g. $B = A$ and $C \ll A$;
(II) the inverted hierarchical solution, e.g. $B = -A$ and $C \ll A$;
(III) the degenerate solution, e.g. $C \ll B \ll A$.

In all cases, C must be small. Therefore $\mathcal{M}_\nu$ of (39) satisfies (42) to a very good approximation, and $Z_2 \times Z_3$ as generated by U_2 and U_B should be taken as the underlying symmetry of this model.

Since $\mathcal{M}_C$ is small and breaks the symmetry of $\mathcal{M}_A + \mathcal{M}_B$, it is natural to think of its origin in terms of the well-known dimension-five operator [18]

$$\mathcal{L}_{eff} = \frac{f_{ij}}{2\Lambda}(\nu_i \phi^0 - l_i \phi^+)(\nu_j \phi^0 - l_j \phi^+) + H.c. \,, \tag{49}$$

where (ϕ^+, ϕ^0) is the usual Higgs doublet of the Standard Model and Λ is a very high scale. As ϕ^0 picks up a nonzero vacuum expectation value v, neutrino masses are generated, and if $f_{ij}v^2/\Lambda = C$ for all i, j, $\mathcal{M}_C$ is obtained. Since Λ is presumably of order 10^{16} to 10^{18} GeV, C is of order 10^{-3} to 10^{-5} eV, and $A - B + 3C/2$ is of order 10^{-2} to 1 eV. This range of values is just right to encompass all three solutions mentioned above.

To justify the assumption that U_B operates in the basis $(\nu_e, \nu_\mu, \nu_\tau)$, the complete theory of leptons must be discussed. Under the assumed Z_3 symmetry, the leptons transform as follows:

$$(\nu, l)_i \to (U_B)_{ij}(\nu, l)_j, \quad l_k^c \to l_k^c \,, \tag{50}$$

implemented by 3 Higgs doublets and 1 Higgs triplet:

$$(\phi^0, \phi^-)_i \to (U_B)_{ij}(\phi^0, \phi^-)_j, \quad (\xi^{++}, \xi^+, \xi^0) \to (\xi^{++}, \xi^+, \xi^0) \,. \tag{51}$$

The Yukawa interactions of this model are then given by

$$\begin{aligned} \mathcal{L}_Y = \; & h_{ij}[\xi^0 \nu_i \nu_j - \xi^+(\nu_i l_j + l_i \nu_j)/\sqrt{2} + \xi^{++} l_i l_j] \\ & + f_{ij}^k (l_i \phi_j^0 - \nu_i \phi_j^-) l_k^c + H.c. \end{aligned} \tag{52}$$

with

$$h = \begin{pmatrix} a-b & 0 & 0 \\ 0 & a & -b \\ 0 & -b & a \end{pmatrix}, \quad \mathcal{M}_\nu = 2h\langle \xi^0 \rangle \,, \tag{53}$$

and

$$f^k = \begin{pmatrix} a_k - b_k & d_k & d_k \\ -d_k & a_k & -b_k \\ -d_k & -b_k & a_k \end{pmatrix} . \tag{54}$$

Note that the d terms are absent in h because it has to be symmetric. Assume $v_{1,2} \ll v_3$, and $d_k \ll b_k \ll a_k$, then $V_L \mathcal{M}_l \mathcal{M}_l^\dagger V_L^\dagger$ = diagonal implies that V_L is nearly diagonal. This justifies the original choice of basis for $\mathcal{M}_\nu$.

Any model of neutrino mixing implies the presence of lepton flavor violation at some level. In this case, ϕ_1^0 couples dominantly to $e\tau^c$ and ϕ_2^0 to $\mu\tau^c$. Taking into account also the other couplings, the branching fractions for $\mu \to eee$ and $\mu \to e\gamma$ are estimated to be of order 10^{-12} and 10^{-11} respectively for a Higgs mass of 100 GeV. Both are at the level of present experimental upper bounds.

10 Conclusions

The correct form of $\mathcal{M}_\nu$ is now approximately known. In the $(\nu_e, \nu_\mu, \nu_\tau)$ basis, it obeys the discrete symmetry of (34). Using (32), the phenomenologically successful (7) is obtained, which has 7 possible limits for $\mathcal{M}_\nu$.

Assuming some additional symmetry, such as A_4 or Z_3, with possible flavor changing radiative corrections, complete theories of leptons (and quarks) may be constructed with the prediction of specific neutrino mass patterns and other experimentally verifiable consequences.

Acknowledgements

I thank Josip Trampetic, Silvio Pallua, and the other organizers of Adriatic 2003 for their great hospitality at Dubrovnik. This work was supported in part by the U. S. Department of Energy under Grant No. DE-FG03-94ER40837.

Appendix

It is amusing to note the parallel between the 5 perfect geometric solids and the 5 anomaly-free superstring theories in 10 dimensions. Whereas the former are related among themselves by geometric dualities, the latter are related by S, T, U dualities: Type I $\leftrightarrow$ SO(32), Type IIa $\leftrightarrow$ $E_8 \times E_8$, and Type IIb is self-dual. Whereas the 5 geometric solids may be embedded in a sphere, the 5 superstring theories are believed to be different limits of a single underlying M Theory.

Afterword

This talk was given on September 11, 2003. Exactly one year ago, I gave a talk at TAU 2002 in Santa Cruz, California, and exactly two years ago, I gave a talk at TAUP 2001 in Assergi, Italy.

References

1. K. Eguchi et al., KamLAND Collaboration, Phys. Rev. Lett. **90**, 021802 (2003).
2. Q. R. Ahmad et al., SNO Collaboration, Phys. Rev. Lett. **89**, 011301, 011302 (2002).
3. For a recent review, see for example C. K. Jung, C. McGrew, T. Kajita, and T. Mann, Ann. Rev. Nucl. Part. Sci. **51**, 451 (2001).
4. M. Apollonio et al., Phys. Lett. B 466, 415 (1999); F. Boehm et al., Phys. Rev. D 64, 112001 (2001).

5. E. Ma, Phys. Rev. **D66**, 117301 (2002).
6. W. Grimus and L. Lavoura, hep-ph/0305309.
7. E. Ma and G. Rajasekaran, Phys. Rev. D 64, 113012 (2001); E. Ma, Mod. Phys. Lett. A 17, 289; 627 (2002).
8. K. S. Babu, E. Ma, and J. W. F. Valle, Phys. Lett. **B552**, 207 (2003).
9. E. Ma, Mod. Phys. Lett. **A17**, 2361 (2002).
10. D. N. Spergel et al., Astrophys. J. Suppl. **148**, 175 (2003).
11. H. V. Klapdor-Kleingrothaus et al., Mod. Phys. Lett. A 16, 2409 (2001).
12. J. Kubo, A. Mondragon, M. Mondragon, and E. Rodriguez-Jauregui, Prog. Theor. Phys. **109**, 795 (2003).
13. W. Grimus and L. Lavoura, Phys. Lett. **B572**, 189 (2003).
14. E. Ma, Phys. Rev. Lett. **90**, 221802 (2003).
15. M. Maltoni, T. Schwetz, and J. W. F. Valle, Phys. Rev. **D67**, 0903003 (2003).
16. E. Ma and G. Rajasekaran, Phys. Rev. **D68**, 071302(R) (2003).
17. E. Ma, hep-ph/0308282.
18. S. Weinberg, Phys. Rev. Lett. **43**, 1566 (1979).

Neutrinos – Inner Properties and Role as Astrophysical Messengers

Georg G. Raffelt

Max-Planck-Institut für Physik (Werner-Heiseberg-Institut), Föhringer Ring 6, 80805 München, Germany
raffelt@mppmu.mpg.de

1 Introduction

The observed flavor oscillations of solar and atmospheric neutrinos determine several elements of the leptonic mixing matrix, but leave open the small mixing angle Θ_{13}, a possible CP-violating phase, the mass ordering, the absolute mass scale m_ν, and the Dirac vs. Majorana property. Many attempts are in progress to determine these missing elements, notably in the area of long-baseline, tritium endpoint, and 2β decay experiments. In addition, astrophysics and cosmology are considerably contributing to this effort. The best constraint on the overall neutrino mass scale m_ν obtains from cosmological precision observables, implying that neutrinos contribute very little to the dark matter. On the other hand, if neutrinos are Majorana particles, they may well be responsible for ordinary matter by virtue of the leptogenesis mechanism for creating the baryon asymmetry of the universe. Independently of the details of the intrinsic neutrino properties, neutrinos are expected to play an important role as "astrophysical messengers" if point sources are discovered in high-energy neutrino telescopes such as Amanda II or the future Antares or IceCube. In low-energy neutrino astronomy, a high-statistics observation of a galactic supernova would allow one to observe directly the dynamics of stellar collapse and perhaps to discriminate between certain mixing scenarios. Even the observation of the tiny flux of all relic neutrinos from all past supernovae in the universe has come within reach. In the following we sketch the status of some of these developments.

2 Status of Neutrino Flavor Oscillations

Neutrino oscillations are now firmly established by measurements of solar and atmospheric neutrinos and the KamLAND and K2K long-baseline experiments [1, 2, 3, 4, 5, 10]. Evidently the weak interaction eigenstates ν_e, ν_μ and ν_τ are non-trivial superpositions of three mass eigenstates ν_1, ν_2 and ν_3,

$$\begin{pmatrix}\nu_e\\ \nu_\mu\\ \nu_\tau\end{pmatrix} = U \begin{pmatrix}\nu_1\\ \nu_2\\ \nu_3\end{pmatrix} . \tag{1}$$

The leptonic mixing matrix can be written in the canonical form

$$U = \begin{pmatrix} 1 & 0 & 0 \\ 0 & c_{23} & s_{23} \\ 0 & -s_{23} & c_{23} \end{pmatrix} \begin{pmatrix} c_{13} & 0 & e^{i\delta} s_{13} \\ 0 & 1 & 0 \\ -e^{-i\delta} s_{13} & 0 & c_{13} \end{pmatrix} \begin{pmatrix} c_{12} & s_{12} & 0 \\ -s_{12} & c_{12} & 0 \\ 0 & 0 & 1 \end{pmatrix} , \tag{2}$$

where $c_{12} = \cos\Theta_{12}$ and $s_{12} = \sin\Theta_{12}$ with Θ_{12} the 12-mixing angle, and so forth. One peculiarity of 3-flavor mixing beyond the 2-flavor case is a non-trivial phase δ that can lead to CP-violating effects, i.e. the 3-flavor oscillation pattern of neutrinos can differ from that of anti-neutrinos.

The atmospheric neutrino oscillations essentially decouple from the solar ones and are controlled by the 23-mixing that may well be maximal (45°). The solar case is dominated by 12-mixing that is large but not maximal. The CHOOZ reactor experiment provides an upper limit on the small 13-mixing. From a global 3-flavor analysis of all data one finds the 1σ ranges for the mass differences and mixing angles summarized in Table 1.

Table 1. Neutrino mixing parameters from a global analysis of all experiment [5] (1σ ranges)

Combination	Mixing angle Θ	Δm^2 [meV^2]
12	32°–36°	67–77
23	41°–49°	2200–3000
13	$< 8°$	$\approx \Delta m^2_{23}$

The only evidence for flavor conversions that is inconsistent with this picture comes from LSND, a short-baseline accelerator experiment. If the excess $\bar\nu_e$ counts are interpreted in terms of $\bar\nu_\mu$-$\bar\nu_e$-oscillations, the allowed mixing parameters populate two islands within $\Delta m^2 = 0.2$–$7\,\mathrm{eV}^2$ and $\sin^2 2\Theta = (0.3$–$5) \times 10^{-2}$ [7]. One possibility to accommodate this Δm^2 with the atmospheric and solar values is a fourth sterile neutrino appearing as an intermediate state to account for the LSND measurements, although this scheme is now almost certainly ruled out [8]. Another solution is the radical conjecture that the masses of neutrinos differ from those of anti-neutrinos, implying a violation of the CPT symmetry [9], although this interpretation does not fare very well in the light of recent data either [10]. In any case, if LSND is confirmed by the ongoing MiniBooNE project [11] the observed flavor conversions imply something far more fundamental than neutrino mixing.

Assuming MiniBooNE will refute LSND so that there is no new revolution, the mass and mixing parameters given in Table 1 still leave many questions open. Is the 23-mixing truly maximal while the 13-mixing is not? How large is the small 13-mixing angle? Is there a CP-violating phase? Moreover, it is possible that two mass eigenstates separated by the small "solar" mass difference could form a doublet separated by the large "atmospheric" difference from a lower-lying single state ("inverted hierarchy").

These issues will be addressed by long-baseline experiments involving reactor and accelerator neutrinos. KamLAND and K2K in Japan are already taking data, while the Fermilab to Soudan and CERN to Gran Sasso projects, each with a baseline of 730 km, are under construction. Future projects involving novel technologies (superbeams, neutrino factories, beta-beams) [12, 13] and their physics potential [14, 15, 16] are being discussed. The "holy grail" of these efforts is finding leptonic CP violation. It is noteworthy that the elusive 13-mixing angle can be measured at a realistic new $\sim$1 km baseline reactor experiment if it is not too far below the current CHOOZ limit [17, 18].

3 Cosmic Structure Formation and Neutrino Masses

The most direct limit on the overall mass scale m_ν derives from tritium experiments searching for a deformation of the β end-point spectrum. The final limit from Mainz and Troitsk is [19]

$$m_\nu < 2.2\,\mathrm{eV} \qquad (95\%\ \mathrm{CL})\,. \tag{3}$$

This number is much larger than the mass splittings, obviating the need for a detailed interpretation in terms of mixed neutrinos. In future, the KATRIN experiment [19] is expected to reach a sensitivity of about 0.3 eV.

Traditionally cosmology provides the most restrictive m_ν limits. Standard big bang cosmology predicts a present-day density of

$$n_{\nu\bar\nu} = \frac{3}{11}\, n_\gamma \approx 112\ \mathrm{cm}^{-3} \quad \text{per flavor}\,. \tag{4}$$

This translates into a cosmic neutrino mass fraction of

$$\Omega_\nu h^2 = \sum_{i=1}^{3} \frac{m_i}{92.5\,\mathrm{eV}}\,, \tag{5}$$

where h is the Hubble parameter in units of 100 km s^{-1} Mpc^{-1}. The oscillation experiments imply $m_\nu > 40\,\mathrm{meV}$ for the largest neutrino mass eigenstate so that $\Omega_\nu > 0.8 \times 10^{-3}$ if $h = 0.72$. On the other hand, the tritium limit (3) implies $\Omega_\nu < 0.14$ so that neutrinos could still contribute significantly to the dark matter.

This possibility is severely constrained by large-scale structure observations. Neutrino free streaming in the early universe erases small scale density fluctuations so that the hot dark matter fraction is most effectively constrained by the small-scale power of the cosmic matter density fluctuations. The recent 2dF Galaxy Redshift Survey data imply [20, 21, 22, 23, 24, 25]

$$\sum m_\nu < 0.7\text{–}1.1\,\mathrm{eV} \qquad (95\%\ \mathrm{CL})\,. \tag{6}$$

To arrive at this limit other cosmological data were used, notably the angular power spectrum of cosmic microwave background radiation as measured by WMAP as well as reasonable priors on other parameters such as the Hubble constant. The range of nominal 95% CL limits depends on the exact data sets used and the assumed priors. The rather narrow range of limits found by different authors suggests that an upper limit of about 1 eV is quite robust. The dependence of such limits on priors and other assumptions is discussed in [21, 22].

In future the Sloan Digital Sky Survey [26] will improve the galaxy correlation function while additional CMBR data from WMAP and later from Planck will improve the matter power spectrum, enhancing the cosmological m_ν sensitivity [27, 28]. Especially promising are future weak lensing data [29, 30] that may come surprisingly close to the lower limit $\sum m_\nu >$ 40 meV implied by the atmospheric neutrino oscillations.

While the progress in precision cosmology has been impressive one should keep worrying about systematic effects that do not show up in statistical confidence levels. Even when the cosmological limits are nominally superior to near-future experimental sensitivities, there remains a paramount need for independent laboratory experiments.

4 Cosmic Neutrino Density

To translate a laboratory m_ν measurement or limit into a hot dark matter fraction Ω_ν and the reverse one usually assumes the standard cosmic neutrino density (4). However, thermal neutrinos in the early universe are characterized by unknown chemical potentials μ_ν or degeneracy parameters $\xi_\nu = \mu_\nu/T$ for each flavor. While the small baryon-to-photon ratio $\sim 10^{-9}$ suggests that all degeneracy parameters are small, large asymmetries between neutrinos and anti-neutrinos could exist and vastly enhance the overall density.

The recent WMAP measurement of the CMBR angular power spectrum provides new limits on the cosmic radiation density [20, 21, 31, 32]. However, the most restrictive limits on neutrino degeneracy parameters still obtain from big-bang nucleosynthesis (BBN) that is affected in two ways. First, a larger neutrino density increases the primordial expansion rate, thereby increasing the neutron-to-proton freeze-out ratio n/p and thus the cosmic helium abundance. Second, electron neutrinos modify $n/p \propto \exp(-\xi_{\nu_e})$. Depending on the sign of ξ_{ν_e} this effect can compensate for the expansion-rate effect of ν_μ or ν_τ so that no significant BBN limit on the overall neutrino density obtains [33]. If ξ_{ν_e} is the only chemical potential, the observed helium abundance yields $-0.01 < \xi_{\nu_e} < 0.07$.

However, neutrino oscillations imply that the individual flavor lepton numbers are not conserved so that in thermal equilibrium there is only one chemical potential for all flavors. If equilibrium is achieved before n/p freeze-out, the restrictive BBN limit on ξ_{ν_e} applies to all flavors, $|\xi_\nu| < 0.07$, fixing the cosmic neutrino density to within about 1%. The approach to flavor equilibrium by neutrino oscillations and collisions was recently studied [34, 35, 36, 37]. The details are subtle due to the large weak potential caused by the neutrinos themselves, causing the intriguing phenomenon of synchronized flavor oscillations [38, 39, 40].

The bottom line is that effective flavor equilibrium before n/p freeze-out is reliably achieved only if the solar oscillation parameters are in the favored LMA region. Now that KamLAND has confirmed LMA, for the first time BBN provides a reliable handle on the cosmic neutrino density. As a consequence, for the first time the relation between Ω_ν and m_ν is uniquely given by the standard expression (5).

5 Majorana Masses and Leptogenesis

The neutrino contribution to the dark matter density is negligible. Intriguingly, however, they may play a crucial role for the baryon asymmetry of the universe (BAU) and thus the presence of ordinary matter [41]. The main ingredients of this leptogenesis scenario are those of the usual see-saw mechanism for small neutrino masses. The ordinary light neutrinos have right-handed partners with large Majorana masses. The left- and right-handed states are coupled by Dirac mass terms that obtain from Yukawa interactions with the Higgs field. The heavy Majorana neutrinos will be in thermal equilibrium in the early universe. When the temperature falls below their mass, their equilibrium density becomes exponentially suppressed. However, if at that time they are no longer in thermal equilibrium, their abundance will exceed the equilibrium distribution. The subsequent out-of-equilibrium decays can lead to the net generation of lepton number. CP-violation is possible by the usual interference of tree-level with one-loop diagrams. The generated lepton number excess will be re-processed by standard-model sphaleron effects which respect $B - L$ but violate $B + L$. It is straightforward to generate the observed BAU by this mechanism.

The requirement that the heavy Majorana neutrinos freeze out before they get Boltzmann suppressed implies an upper limit on the same parameter combination of Yukawa couplings and heavy Majorana masses that determines the observed small neutrino masses [42]. Most recently, a robust upper limit on all neutrino masses of

$$m_\nu < 120\,\text{meV} \tag{7}$$

was claimed [43]. Degenerate neutrinos with a "large" common mass scale of, e.g., 400 meV require a very precise degeneracy of the heavy Majorana masses to better than 10^{-3}.

A necessary ingredient for this mechanism is the Majorana nature of neutrino masses that can be tested in the laboratory by searching for neutrinoless 2β decay. This process is sensitive to

$$\langle m_{ee}\rangle = \left|\sum_{i=1}^{3} \lambda_i \left|U_{ei}\right|^2 m_i\right| \tag{8}$$

with λ_i a Majorana CP phase. Therefore, we have two additional physically relevant phases beyond the Dirac phase δ of the previously discussed mixing matrix. If neutrinos have Majorana masses their mixing involves three mass eigenstates, three mixing angles, and three physical phases.

Actually, several members of the Heidelberg-Moscow collaboration have claimed first evidence for this process [44, 45], implying a 95% CL range of $\langle m_{ee}\rangle = 110$–$560\,\mathrm{meV}$. Uncertainties of the nuclear matrix element can widen this range by up to a factor of 2 in either direction. The significance of this discovery has been fiercely critiqued by many experimentalists working on other 2β projects [46]. Even when taking the claimed evidence at face value the statistical significance is only about 97%, too weak for definitive conclusions. More sensitive experiments are needed and developed to explore this range of Majorana masses [47].

6 Astrophysical High-Energy Neutrinos

The observed sources of astrophysical neutrinos remain limited to the Sun and Supernova 1987A, apart from cosmic-ray secondaries in the form of atmospheric neutrinos. This situation could radically change in the near future if the high-energy neutrino telescopes that are currently being developed begin to discover astrophysical point sources.

The spectrum of cosmic rays reaches to energies of at least $3\times10^{20}\,\mathrm{eV}$, proving the existence of cosmic sources for particles with enormous energies [48, 49]. Most of the cosmic rays appear to be protons or nuclei so that there must be hadronic accelerators both within our galaxy and beyond. Wherever high-energy hadrons interact with matter or radiation, the decay of secondary pions produces a large flux of neutrinos At the source one expects a flavor composition of $\nu_e:\nu_\mu:\nu_\tau = 1:2:0$, but the observed oscillations imply equal fluxes of all flavors at Earth. High-energy neutrino astronomy offers a unique opportunity to detect the enigmatic sources of high-energy cosmic rays because neutrinos are neither absorbed by the cosmic photon backgrounds nor deflected by magnetic fields.

While there are many different models for possible neutrino sources [49, 50], the required size for a detector is generically 1 km^3. The largest existing neutrino telescope, the AMANDA ice Cherenkov detector at the South Pole, is about 1/10 this size. It has not yet observed a point source, but the detection of atmospheric neutrinos shows that this approach to measuring high-energy neutrinos works well [51]. It is expected that this instrument is upgraded to the full 1 km^3 size within the next few years under the name of IceCube [52]. Similar instruments are being developed in the Mediterranean [53]. Moreover, air-shower arrays for ordinary cosmic rays may detect very high-energy neutrinos by virtue of horizontal air showers [54]. Although this field is in its infancy, it holds the promise of exciting astrophysical discoveries in the foreseeable future.

7 Supernova Neutrinos

The observation of neutrinos from the supernova (SN) 1987A in the Large Magellanic Cloud was a milestone for neutrino astronomy, but the total of about 20 events in the Kamiokande and IMB detectors was frustratingly sparse. The chances of observing a galactic SN are small because SNe are thought to occur with a rate of at most a few per century. On the other hand, many neutrino detectors and especially Super-Kamiokande have a rich physics program for perhaps decades to come, notably in the area of long-baseline oscillation experiments and proton decay searches. Likewise, the south-pole detectors may be active for many decades and would be powerful SN observatories [55, 56, 57]. Therefore, it remains worthwhile to study what can be learned from a high-statistics SN observation.

The explosion mechanism for core-collapse SNe remains unsettled as long as numerical simulations fail to reproduce robust explosions. A high-statistics neutrino observation is probably the only chance to watch the collapse and explosion dynamics directly and would allow one to verify the standard delayed explosion scenario [58]. The neutrinos arrive a few hours before the optical explosion so that a neutrino observation can serve to alert the astronomical community, a task pursued by the Supernova Early Warning System (SNEWS) [59]. For particle physics, many of the limits based on the SN 1987A neutrino signal [60] would improve and achieve a firm experimental and statistical basis. On the other hand, the time-of-flight sensitivity to the neutrino mass is in the range of a few eV [61], not good enough as the "m_ν frontier" has moved to the sub-eV range.

Can we learn something useful about neutrino mixing from a galactic SN observation? This issue has been addressed in many recent studies [57, 62, 63, 64, 65, 66, 67, 68] and the answer is probably yes, depending on the detectors operating at the time of the SN, their geographic location, and the neutrino mixing scenario, i.e. the magnitude of the small mixing angle Θ_{13} and the ordering of the masses. Any observable oscillation effects depend on the spectral and flux differences between the different flavors. We have recently shown that previous studies overestimated these differences [69, 70, 71] because traditional numerical simulations used a schematic treatment of ν_μ and ν_τ transport. Distinguishing, say, between the normal and inverted mass ordering remains a daunting task at long-baseline experiments. Therefore, a future galactic SN observation may still offer a unique opportunity to settling this question.

The relic flux from all past SNe in the universe is observable because it exceeds the atmospheric neutrino flux for energies below 30–40 meV. Recently

Super-Kamiokande has reported a limit that already caps the more optimistic predictions [72]. Significant progress depends on suppressing the background caused by the decay of sub-Cherenkov muons from low-energy atmospheric neutrinos. One possibility is to include an efficient neutron absorber such as gadolinium in the detector that would tag the reactions $\bar{\nu}_e + p \to n + e^+$ [73]. If this approach works in practice the detection of relic SN neutrinos has come within experimental reach.

8 Conclusions

After neutrino oscillations have been established, the next challenge is to pin down the as yet undetermined elements of the mixing matrix and the absolute masses and mass ordering. Long-baseline experiments can address many of these questions and may even discover leptonic CP-violation. The Majorana nature of neutrinos can be established in 2β experiments if the 0ν decay mode is convincingly observed. Majorana neutrinos with masses $<120\,\mathrm{meV}$ fit nicely into the leptogenesis scenario for creating the baryon asymmetry of the universe so that neutrinos may well be responsible for the ordinary matter in the universe. Their contribution to the dark matter is non-zero but negligible. Still, precision observations of cosmological large-scale structure remain the most powerful tool to constrain the absolute mass scale, although independent laboratory confirmation remains crucial. In the past, neutrino oscillation physics was dominated by solar and atmospheric neutrinos, but long-baseline experiments are about to "take over." In future, neutrinos from natural sources are likely more important as "astrophysical messengers" than they are for probing intrinsic neutrino properties. The search for astrophysical point sources with high-energy neutrino telescopes may soon open a new window to the universe. Observing a high-statistics neutrino signal from a future galactic supernova remains perhaps the most cherished prize for low-energy neutrino observatories. Meanwhile, the search for the cosmic relic neutrinos from all past supernovae has become a realistic possibility. Neutrinos will remain fascinating for a long time to come!

Acknowledgements

This work was supported, in part, by the Deutsche Forschungsgemeinschaft under grant No. SFB-375 and by the European Science Foundation (ESF) under the Network Grant No. 86 Neutrino Astrophysics.

References

1. M. C. Gonzalez-Garcia and Y. Nir "Developments in neutrino physics," Rev. Mod. Phys. **75**, 345 (2003) [hep-ph/0202058].

2. G. L. Fogli, E. Lisi, A. Marrone, D. Montanino, A. Palazzo and A. M. Rotunno, "Solar neutrino oscillation parameters after first KamLAND results," Phys. Rev. D **67**, 073002 (2003) [hep-ph/0212127]. Addendum hep-ph/0308055.
3. J. N. Bahcall, M. C. Gonzalez-Garcia and C. Pena-Garay, "Solar neutrinos before and after KamLAND," *JHEP* **02**, 009 (2003) [hep-ph/0212147].
4. P. C. de Holanda and A. Y. Smirnov, "LMA MSW solution of the solar neutrino problem and first KamLAND results," *JCAP* **02**, 001 (2003) [hep-ph/0212270].
5. M. C. Gonzalez-Garcia and C. Pena-Garay, "Three-neutrino mixing after the first results from K2K and KamLAND," Phys. Rev. D **68**, 093003 (2003) [hep-ph/0306001].
6. M. Maltoni, T. Schwetz, M. A. Tortola and J. W. F. Valle, "Status of three-neutrino oscillations after the SNO-salt data," Phys. Rev. D **68**, 113010 (2003) [hep-ph/0309130].
7. E. D. Church, K. Eitel, G. B. Mills and M. Steidl, "Statistical analysis of different $\bar{\nu}_\mu \to \bar{\nu}_e$ searches," *Phys. Rev. D* **66**, 013001 (2002) [hep-ex/0203023].
8. M. Maltoni, T. Schwetz, M. A. Tortola and J. W. Valle, "Global analysis of neutrino oscillation data in four-neutrino schemes," *Nucl. Phys. Proc. Suppl.* **114**, 203 (2003) [hep-ph/0209368].
9. G. Barenboim and J. Lykken, "A model of CPT violation for neutrinos," Phys. Lett. B **554**, 73 (2003) [hep-ph/0210411].
10. M. C. Gonzalez-Garcia, M. Maltoni and T. Schwetz, "Status of the CPT violating interpretations of the LSND signal," Phys. Rev. D **68**, 053007 (2003) [hep-ph/0306226].
11. R. Tayloe [MiniBooNE Collaboration], "The Miniboone Experiment: Status And Plans," Nucl. Phys. Proc. Suppl. **118**, 157 (2003). See also http://www-boone.fnal.gov
12. Neutrino Factory and Muon Storage Rings at CERN http://muonstoragerings.web.cern.ch
13. Neutrino Factory Working Group http://nfwg.home.cern.ch/nfwg
14. V. D. Barger, S. Geer, R. Raja and K. Whisnant, "Exploring neutrino oscillations with superbeams," *Phys. Rev. D* **63**, 113011 (2001) [hep-ph/0012017].
15. A. Cervera, A. Donini, M. B. Gavela, J. J. Gomez Cadenas, P. Hernandez, O. Mena and S. Rigolin, "Golden measurements at a neutrino factory," *Nucl. Phys. B* **579**, 17 (2000); Erratum *ibid.* **593**, 731 (2001) [hep-ph/0002108].
16. M. Freund, P. Huber and M. Lindner, "Systematic exploration of the neutrino factory parameter space including errors and correlations," *Nucl. Phys. B*, **615**, 331 (2001) [hep-ph/0105071].
17. P. Huber, M. Lindner, T. Schwetz and W. Winter, "Reactor neutrino experiments compared to superbeams," Nucl. Phys. B **665**, 487 (2003) [hep-ph/0303232].
18. H. Minakata and H. Sugiyama, "Exploring leptonic CP violation by reactor and neutrino superbeam experiments," Phys. Lett. B **580**, 216 (2004) [hep-ph/0309323].
19. C. Weinheimer, "Direct neutrino mass search," hep-ex/0210050.
20. S. Hannestad and G. Raffelt, "Cosmological mass limits on neutrinos, axions, and other light particles," hep-ph/0312154.
21. S. Hannestad, "Neutrino masses and the number of neutrino species from WMAP and 2dFGRS," JCAP **0305**, 004 (2003) [astro-ph/0303076].

22. O. Elgaroy and O. Lahav, "The role of priors in deriving upper limits on neutrino masses from the 2dFGRS and WMAP," JCAP **0304**, 004 (2003) [astro-ph/0303089].
23. D. N. Spergel et al., "First Year Wilkinson Microwave Anisotropy Probe (WMAP) Observations: Determination of Cosmological Parameters," Astrophys. J. Suppl. **148**, 175 (2003) [astro-ph/0302209].
24. S. W. Allen, R. W. Schmidt and S. L. Bridle, "A preference for a non-zero neutrino mass from cosmological data," Mon. Not. Roy. Astron. Soc. **346**, 593 (2003) [astro-ph/0306386].
25. V. Barger, D. Marfatia and A. Tregre, "Neutrino mass limits from SDSS, 2dF-GRS and WMAP," hep-ph/0312065.
26. K. Abazajian et al. [SDSS Collaboration], "The First Data Release of the Sloan Digital Sky Survey," Astron. J. **126**, 2081 (2003) [astro-ph/0305492].
27. W. Hu, D. J. Eisenstein and M. Tegmark, "Weighing neutrinos with galaxy surveys," *Phys. Rev. Lett.* **80**, 5255 (1998) [astro-ph/9712057].
28. S. Hannestad, "Can cosmology detect hierarchical neutrino masses?," Phys. Rev. D **67**, 085017 (2003) [astro-ph/0211106].
29. W. Hu, "Dark energy and matter evolution from lensing tomography," *Phys. Rev. D* **66**, 083515 (2002) [astro-ph/0208093].
30. K. N. Abazajian and S. Dodelson, "Neutrino mass and dark energy from weak lensing," Phys. Rev. Lett. **91**, 041301 (2003) [astro-ph/0212216].
31. P. Crotty, J. Lesgourgues and S. Pastor, "Measuring the cosmological background of relativistic particles with WMAP," Phys. Rev. D **67**, 123005 (2003) [astro-ph/0302337].
32. E. Pierpaoli, "Constraints on the cosmic neutrino background," Mon. Not. Roy. Astron. Soc. **342**, L63 (2003) [astro-ph/0302465].
33. H. S. Kang and G. Steigman, "Cosmological constraints on neutrino degeneracy," *Nucl. Phys. B* **372**, 494 (1992).
34. C. Lunardini and A. Y. Smirnov, "High-energy neutrino conversion and the lepton asymmetry in the universe," *Phys. Rev. D* **64**, 073006 (2001) [hep-ph/0012056].
35. A. D. Dolgov, S. H. Hansen, S. Pastor, S. T. Petcov, G. G. Raffelt and D. V. Semikoz, "Cosmological bounds on neutrino degeneracy improved by flavor oscillations," *Nucl. Phys. B* **632**, 363 (2002) [hep-ph/0201287].
36. Y. Y. Wong, "Analytical treatment of neutrino asymmetry equilibration from flavour oscillations in the early universe," *Phys. Rev. D* **66**, 025015 (2002) [hep-ph/0203180].
37. K. N. Abazajian, J. F. Beacom and N. F. Bell, "Stringent constraints on cosmological neutrino antineutrino asymmetries from synchronized flavor transformation," *Phys. Rev. D* **66**, 013008 (2002) [astro-ph/0203442].
38. S. Samuel, "Neutrino oscillations in dense neutrino gases," *Phys. Rev. D* **48**, 1462 (1993).
39. S. Pastor, G. G. Raffelt and D. V. Semikoz, "Physics of synchronized neutrino oscillations caused by self-interactions," *Phys. Rev. D* **65**, 053011 (2002) [hep-ph/0109035].
40. S. Pastor and G. Raffelt, "Flavor oscillations in the supernova hot bubble region: Nonlinear effects of neutrino background," *Phys. Rev. Lett.* **89**, 191101 (2002) [astro-ph/0207281].
41. M. Fukugita and T. Yanagida, "Baryogenesis without grand unification," *Phys. Lett. B* **174**, 45 (1986).

42. W. Buchmüller and M. Plümacher, "Neutrino masses and the baryon asymmetry," *Int. J. Mod. Phys. A* **15**, 5047 (2000) [hep-ph/0007176].
43. W. Buchmüller, P. Di Bari and M. Plümacher, "The neutrino mass window for baryogenesis," Nucl. Phys. B **665**, 445 (2003) [hep-ph/0302092].
44. H. V. Klapdor-Kleingrothaus, A. Dietz, H. L. Harney and I. V. Krivosheina, "Evidence for neutrinoless double beta decay," *Mod. Phys. Lett. A* **16**, 2409 (2001) [hep-ph/0201231].
45. H. V. Klapdor-Kleingrothaus, "Neutrinoless double-beta decay: Status of evidence and future," *Phys. Atom. Nucl.* **65**, 2135 (2002) [*Yad. Fiz.* **65**, 2198 (2002)].
46. C. E. Aalseth et al., "Comment on 'Evidence for neutrinoless double beta decay'," *Mod. Phys. Lett. A* **17**, 1475 (2002) [hep-ex/0202018].
47. O. Cremonesi, "Neutrinoless double beta decay: Present and future," Nucl. Phys. Proc. Suppl. **118**, 287 (2003) [hep-ex/0210007].
48. G. Sigl, "Ultrahigh-energy cosmic rays: A probe of physics and astrophysics at extreme energies," *Science* **291**, 73 (2001) [astro-ph/0104291].
49. F. Halzen and D. Hooper, "High-energy neutrino astronomy: The cosmic ray connection," *Rept. Prog. Phys.* **65**, 1025 (2002) [astro-ph/0204527].
50. E. Waxman, "Astrophysical sources of high energy neutrinos," Nucl. Phys. Proc. Suppl. **118**, 353 (2003) [astro-ph/0211358].
51. J. Ahrens et al. (AMANDA Collaboration), "Search for point sources of high energy neutrinos with AMANDA," *Astrophys. J.* **583**, 1040 (2003) [astro-ph/0208006].
52. F. Halzen, "High-energy neutrino astronomy: From AMANDA to IceCube," astro-ph/0311004.
53. J. Carr, "Neutrino telescopes," *Nucl. Phys. Proc. Suppl.* **113**, 26 (2002).
54. A. Letessier-Selvon, "Auger: A large air shower array and neutrino telescope," Nucl. Phys. Proc. Suppl. **118**, 399 (2003) [astro-ph/0208526].
55. F. Halzen, J. E. Jacobsen and E. Zas, "Ultra-Transparent Antarctic Ice as a Supernova Detector," Phys. Rev. D **53**, 7359 (1996) [astro-ph/9512080].
56. J. Ahrens et al. (AMANDA Collaboration), "Search for supernova neutrino-bursts with the AMANDA detector," *Astropart. Phys.* **16**, 345 (2002) [astro-ph/0105460].
57. A. S. Dighe, M. T. Keil and G. G. Raffelt, "Detecting the neutrino mass hierarchy with a supernova at IceCube," JCAP **0306**, 005 (2003) [hep-ph/0303210].
58. T. Totani, K. Sato, H. E. Dalhed and J. R. Wilson, "Future detection of supernova neutrino burst and explosion mechanism," *Astrophys. J.* **496**, 216 (1998) [astro-ph/9710203].
59. The Supernova Early Warning System (SNEWS) http://hep.bu.edu/~snnet/
60. G. G. Raffelt, "Particle physics from stars," *Ann. Rev. Nucl. Part. Sci.* **49**, 163 (1999) [hep-ph/9903472].
61. J. F. Beacom, R. N. Boyd and A. Mezzacappa, "Black hole formation in core collapse supernovae and time-of-flight measurements of the neutrino masses," *Phys. Rev. D* **63**, 073011 (2001) [astro-ph/0010398].
62. A. S. Dighe and A. Y. Smirnov, "Identifying the neutrino mass spectrum from the neutrino burst from a supernova," *Phys. Rev. D* **62**, 033007 (2000) [hep-ph/9907423].
63. K. Takahashi and K. Sato, "Earth effects on supernova neutrinos and their implications for neutrino parameters," *Phys. Rev. D* **66**, 033006 (2002) [hep-ph/0110105].

64. K. Takahashi and K. Sato, "Effects of neutrino oscillation on supernova neutrino: Inverted mass hierarchy," hep-ph/0205070.
65. C. Lunardini and A. Y. Smirnov, "Probing the neutrino mass hierarchy and the 13-mixing with supernovae," hep-ph/0302033.
66. A. S. Dighe, M. T. Keil and G. G. Raffelt, "Identifying earth matter effects on supernova neutrinos at a single detector," JCAP **0306**, 006 (2003) [hep-ph/0304150].
67. R. Tomas, D. Semikoz, G. G. Raffelt, M. Kachelriess and A. S. Dighe, "Supernova pointing with low- and high-energy neutrino detectors," Phys. Rev. D **68**, 093013 (2003) [hep-ph/0307050].
68. A. S. Dighe, M. Kachelriess, G. G. Raffelt and R. Tomas, "Signatures of supernova neutrino oscillations in the earth mantle and core," JCAP **0401**, 004 (2004) [hep-ph/0311172].
69. G. G. Raffelt, "Mu- and tau-neutrino spectra formation in supernovae," *Astrophys. J.* **561**, 890 (2001) [astro-ph/0105250].
70. R. Buras, H. T. Janka, M. T. Keil, G. G. Raffelt and M. Rampp, "Electron-neutrino pair annihilation: A new source for muon and tau neutrinos in supernovae," Astrophys. J. **587**, 320 (2003) [astro-ph/0205006].
71. M. T. Keil, G. G. Raffelt and H. T. Janka, "Monte Carlo study of supernova neutrino spectra formation," Astrophys. J. **590**, 971 (2003) [astro-ph/0208035].
72. M. Malek et al. [Super-Kamiokande Collaboration], "Search for supernova relic neutrinos at Super-Kamiokande," Phys. Rev. Lett. **90**, 061101 (2003) [hep-ex/0209028].
73. J. F. Beacom and M. R. Vagins, "GADZOOKS! Antineutrino spectroscopy with large water Cherenkov detectors," hep-ph/0309300.

Lepton Flavor Violation in the SUSY Seesaw Model: An Update

Frank Deppisch, Heinrich Päs, Andreas Redelbach and Reinhold Rückl

Institut für Theoretische Physik und Astrophysik
Universität Würzburg, 97074 Würzburg, Germany

1 Introduction

The evidence for flavor violating neutrino oscillations provides strong motivation to search for LFV also in the charged lepton sector. While charged lepton-flavor violating processes are suppressed in the Standard Model with right-handed neutrinos [1] due to the light neutrino masses, in supersymmetric models new sources of lepton flavor violation exist. For example, the massive neutrinos affect the renormalization group equations of the slepton masses and the trilinear couplings, and give rise to non-diagonal matrix elements inducing LFV [2]. This finding has initiated a prospering research activity both on LFV in rare decays [3, 4, 5] as well as on lepton-flavor violating processes at future colliders [6]. Recently, we performed a comparative study of the sensitivity of rare radiative decays on the right-handed Majorana mass scale M_R [7] and the reach in slepton-pair production at a future linear collider [8] in the context of the SUSY seesaw model. We used mSUGRA benchmark scenarios [12] designed for linear collider studies and paid particular attention to the uncertainties in the neutrino input parameters. Here we present an update of these works, implementing the most recent neutrino data fits and refined post WMAP mSUGRA benchmark scenarios.

2 Neutrino Parameters

In the last decade a rather unique picture of neutrino mixing has emerged. As can be seen in Fig. 1, large to maximal mixing has been established for solar and atmospheric neutrinos, while the third angle is strongly constrained by reactor measurements. Recently, this picture of neutrino mixing has been refined further. The results of the KamLAND reactor experiment confirmed the disappearance of solar electron neutrinos, while the SNO experiment allowed for the first time to study both solar neutrino appearance and disappearance via the separate measurement of charged current, elastic scattering and neutral current interactions. Moreover, the first data of the K2K long baseline accelerator experiment indicate a confirmation of atmospheric neutrino oscillations. At the time when a linear collider will be in operation, even more

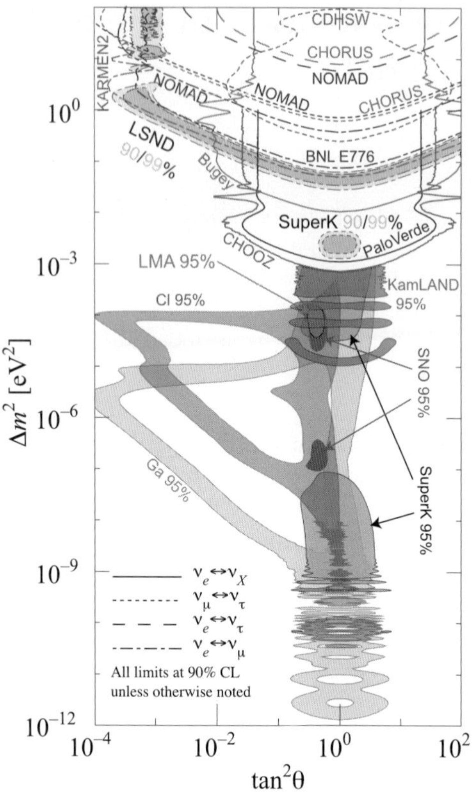

Fig. 1. Summary of evidences for neutrino oscillations [9]

precise measurements of the neutrino parameters will be available than today. In order to simulate the expected improvement, we take the central values of the mass squared differences $\Delta m_{ij}^2 = |m_i^2 - m_j^2|$ and mixing angles θ_{ij} from the most recent global fit to existing data [10] with errors that indicate the anticipated 2σ intervals of running and proposed experiments as further explained in [7]:

$$\tan^2\theta_{23} = 1.10^{+1.39}_{-0.60}, \quad \tan^2\theta_{13} = 0.006^{+0.001}_{-0.006}, \quad \tan^2\theta_{12} = 0.43^{+0.47}_{-0.22}\,, \tag{1}$$

$$\Delta m_{12}^2 = 6.9^{+0.36}_{-0.36}\cdot 10^{-5}\ \mathrm{eV}^2, \quad \Delta m_{13}^2 = 2.6^{+1.2}_{-1.2}\cdot 10^{-3}\ \mathrm{eV}^2\,. \tag{2}$$

Furthermore, for the lightest neutrino mass we assume $m_1 \leq 0.03\,\mathrm{eV}$, where $m_1 \simeq 0\,\mathrm{eV}$ corresponds to the case of a hierarchical spectrum, while $m_1 \simeq 0.03\,\mathrm{eV}$ approaches the degenerate case.

3 mSUGRA Benchmark Scenarios

For our numerical investigations we focus on the mSUGRA benchmark scenarios proposed in [11] for linear collider studies. These models are consistent with direct SUSY searches, Higgs searches, $b \to s\gamma$, and astrophysical constraints. Recently, the precision measurement of the cosmic microwave background by the WMAP experiment combined with previously available data provided a refined estimate of the density of cold dark matter in the universe. Assuming that most of the dark matter is composed of neutralino LSPs, the smaller neutralino relic density led to a shift of previous benchmark points [12] towards lower values of the universal scalar mass m_0. In Table 1 we specify the mSUGRA parameters of the benchmark scenarios B', C', G', and I'. These are the only models of [11] with left-handed sleptons which are light enough to be pair-produced at e^+e^- colliders with c.m. energies $\sqrt{s} = 500 \div 800\,\text{GeV}$.

Table 1. Parameters of selected mSUGRA post WMAP benchmark scenarios [11]. The sign of μ is chosen to be positive and A_0 is set to zero. Given are also the masses of the heaviest charged slepton and the lightest neutralino, that is the LSP

Scenario	$m_{1/2}$/GeV	m_0/GeV	$\tan\beta$	$\tilde{m}_6$/GeV	$m_{\tilde{\chi}_1^0}$/GeV
B'	250	60	10	192	98
C'	400	85	10	291	163
G'	375	115	20	291	153
I'	350	175	35	310	143

4 SUSY Seesaw Mechanism and Slepton Mass Matrix

If three right-handed neutrino singlet fields ν_R are added to the MSSM particle content, one gets the following additional terms in the superpotential [4]:

$$W_\nu = -\frac{1}{2}\nu_R^{cT} M \nu_R^c + \nu_R^{cT} Y_\nu L \cdot H_2 \,. \tag{3}$$

Here, Y_ν is the matrix of neutrino Yukawa couplings, M is the right-handed neutrino Majorana mass matrix, and L and H_2 denote the left-handed lepton and hypercharge $+1/2$ Higgs doublets, respectively. At energies much below the mass scale of the right-handed neutrinos, W_ν leads to the mass matrix

$$M_\nu = m_D^T M^{-1} m_D = Y_\nu^T M^{-1} Y_\nu (v \sin\beta)^2 \,, \tag{4}$$

for the left-handed neutrinos. Thus, light neutrino masses are naturally obtained if the typical scale of the Majorana mass matrix M is much larger than the scale of the Dirac mass matrix $m_D = Y_\nu \langle H_2^0 \rangle$, where $\langle H_2^0 \rangle = v \sin\beta$ is the appropriate Higgs v.e.v. with $v = 174\,\text{GeV}$ and $\tan\beta = \frac{\langle H_2^0 \rangle}{\langle H_1^0 \rangle}$. Diagonalization of M_ν by the unitary MNS matrix

$$U = \text{diag}(e^{i\phi_1}, e^{i\phi_2}, 1) V(\theta_{12}, \theta_{13}, \theta_{23}, \delta) \;, \tag{5}$$

where $\phi_{1,2}$ and δ are the Majorana and Dirac phases, respectively, and θ_{ij} are the mixing angles, leads to the light neutrino mass eigenvalues m_i:

$$U^T M_\nu U = \text{diag}(m_1, m_2, m_3) \;. \tag{6}$$

The other neutrino mass eigenstates are too heavy to be observed directly. However, they give rise to virtual corrections to the slepton mass matrices that can be responsible for observable lepton-flavor violating effects. In particular, the 6×6 mass matrix of the charged sleptons is given by

$$m_{\tilde{l}}^2 = \begin{pmatrix} m_{\tilde{l}_L}^2 & (m_{\tilde{l}_{LR}}^2)^\dagger \\ m_{\tilde{l}_{LR}}^2 & m_{\tilde{l}_R}^2 \end{pmatrix} \tag{7}$$

with

$$\left(m_{\tilde{l}_L}^2\right)_{ab} = \left(m_L^2\right)_{ab} + \delta_{ab}\left(m_{l_a}^2 + m_Z^2 \cos 2\beta \left(-\frac{1}{2} + \sin^2\theta_W\right)\right) \tag{8}$$

$$\left(m_{\tilde{l}_R}^2\right)_{ab} = \left(m_R^2\right)_{ab} + \delta_{ab}\left(m_{l_a}^2 - m_Z^2 \cos 2\beta \sin^2\theta_W\right) \tag{9}$$

$$\left(m_{\tilde{l}_{LR}}^2\right)_{ab} = A_{ab} v \cos\beta - \delta_{ab} m_{l_a} \mu \tan\beta \;. \tag{10}$$

When $m_{\tilde{l}}^2$ is renormalized from the GUT scale M_X to the electroweak scale one obtains, in mSUGRA,

$$m_L^2 = m_0^2 \mathbf{1} + \left(\delta m_L^2\right)_{\text{MSSM}} + \delta m_L^2 \tag{11}$$

$$m_R^2 = m_0^2 \mathbf{1} + \left(\delta m_R^2\right)_{\text{MSSM}} + \delta m_R^2 \tag{12}$$

$$A = A_0 Y_l + \delta A_{\text{MSSM}} + \delta A \;, \tag{13}$$

where m_0 is the common soft SUSY-breaking scalar mass and A_0 the common trilinear coupling. The terms $(\delta m_{L,R}^2)_{\text{MSSM}}$ and δA_{MSSM} are well-known flavor-diagonal corrections. In addition, the evolution generates off-diagonal terms in $\delta m_{L,R}^2$ and δA^2 which in leading-log approximation are given by [5]

$$\delta m_L^2 = -\frac{1}{8\pi^2}\left(3m_0^2 + A_0^2\right)\left(Y_\nu^\dagger L Y_\nu\right) \tag{14}$$

$$\delta m_R^2 = 0 \tag{15}$$

$$\delta A = -\frac{3A_0}{16\pi^2}\left(Y_l Y_\nu^\dagger L Y_\nu\right) \tag{16}$$

with

$$L_{ij} = \ln\left(\frac{M_X}{M_i}\right)\delta_{ij} \; , \tag{17}$$

and M_i, $i = 1, 2, 3$ being the eigenvalues of the Majorana mass matrix M. In the above, we have chosen a basis in which the charged lepton Yukawa couplings and M are diagonal.

The product of the neutrino Yukawa couplings $Y_\nu^\dagger L Y_\nu$ entering these corrections can be determined as follows [4]. By inverting (6), one obtains

$$Y_\nu = \frac{1}{v \sin\beta} \mathrm{diag}\left(\sqrt{M_1}, \sqrt{M_2}, \sqrt{M_3}\right) R \, \mathrm{diag}\left(\sqrt{m_1}, \sqrt{m_2}, \sqrt{m_3}\right) U^\dagger \; , \tag{18}$$

where R is an unknown complex orthogonal matrix. For real R and degenerate right-handed Majorana masses, R as well as ϕ_1 and ϕ_2 drop out from the product $Y_\nu^\dagger L Y_\nu$. Using then the neutrino data sketched in Sect. 2 as input the result is evolved to the unification scale M_X. In what follows we refer to this illustrative case which suffices for the present discussion. For small neutrino masses, $m_i^2 \ll \Delta m_{ij}^2$, the above procedure yields

$$\left(Y_\nu^\dagger L Y_\nu\right)_{ab} \approx \frac{M_R}{v^2 \sin^2\beta}\left(\sqrt{\Delta m_{12}^2} U_{a2} U_{b2}^* + \sqrt{\Delta m_{23}^2} U_{a3} U_{b3}^*\right) \ln\frac{M_X}{M_R} \; . \tag{19}$$

Upon diagonalization, the flavor off-diagonal corrections in (11)–(13) generate flavor-violating couplings of the slepton mass eigenstates.

5 Charged Lepton Flavor Violation

At low energies, the flavor off-diagonal correction (14) induces the radiative decays $l_i \to l_j\gamma$. From the photon penguin diagrams shown in Fig. 2 with charginos/sneutrinos or neutralinos/charged sleptons in the loop, one derives the decay rates [4, 5]

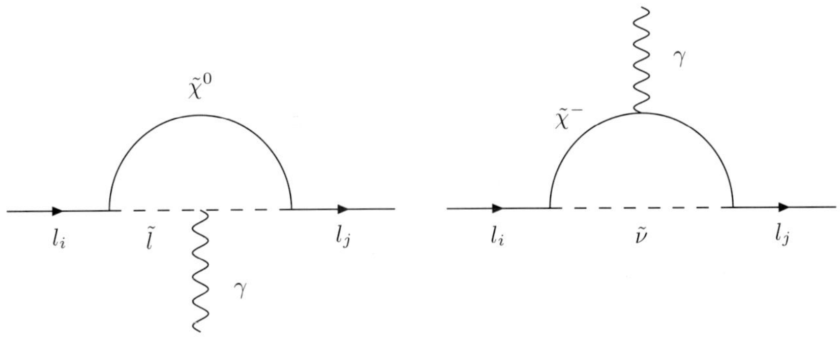

Fig. 2. Diagrams for $l_i^- \to l_j^-\gamma$ in the MSSM

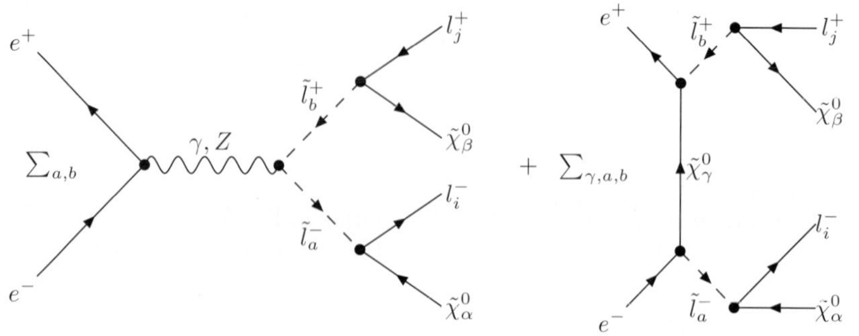

Fig. 3. Diagrams for $e^+e^- \to \tilde{l}_b^+\tilde{l}_a^- \to l_j^+l_i^-\tilde{\chi}_\alpha^0\tilde{\chi}_\beta^0$

$$\Gamma(l_i \to l_j\gamma) \propto \alpha^3 m_{l_i}^5 \frac{|(\delta m_L)_{ij}^2|^2}{\tilde{m}^8} \tan^2\beta\,, \tag{20}$$

where $\tilde{m}$ characterizes the sparticle masses in the loop. Because of the dominance of the penguin contributions, the process $\mu \to 3e$, and also μ-e conversion in nuclei is directly related to $\mu \to e\gamma$, e.g.,

$$\frac{Br(\mu \to 3e)}{Br(\mu \to e\gamma)} \approx \frac{\alpha}{8\pi}\frac{8}{3}\left(\ln\frac{m_\mu^2}{m_e^2} - \frac{11}{4}\right)\,. \tag{21}$$

At high energies, a feasible test of LFV is provided by the process $e^+e^- \to \tilde{l}_b^+\tilde{l}_a^- \to l_j^+l_i^-\tilde{\chi}_\alpha^0\tilde{\chi}_\beta^0$. From the Feynman graphs displayed in Fig. 3, one can see that LFV can occur in production and decay vertices. For sufficiently narrow slepton widths $\Gamma_{\tilde{l}}$ and degenerate masses, the cross-section can be approximated by

$$\sigma_{i\neq j}^{\text{pair}} \propto \frac{|(\delta m_L)_{ij}^2|^2}{\tilde{m}^2\Gamma_{\tilde{l}}^2}\,\sigma(e^+e^- \to \tilde{l}_j^+\,\tilde{l}_i^-)Br(\tilde{l}_j^+ \to l_j^+\,\tilde{\chi}_0)Br(\tilde{l}_i^- \to l_i^-\,\tilde{\chi}_0)\,. \tag{22}$$

where $\sigma(e^+e^- \to \tilde{l}_j^+\,\tilde{l}_i^-)$ can be replaced by the flavor-diagonal cross-section for slepton pair production. The flavor change is described by the factor in front of the r.h.s. of (22) resulting in a single mass insertion.

In the following numerical study we have not assumed slepton degeneracy and have summed the amplitudes for the complete $2 \to 4$ processes coherently over the intermediate slepton mass eigenstates.

6 Numerical Results

Figure 4 illustrates the dependence of $Br(\mu \to e\gamma)$ and $Br(\tau \to \mu\gamma)$ on the Majorana mass M_R. The spread of the predictions reflects the uncertainties

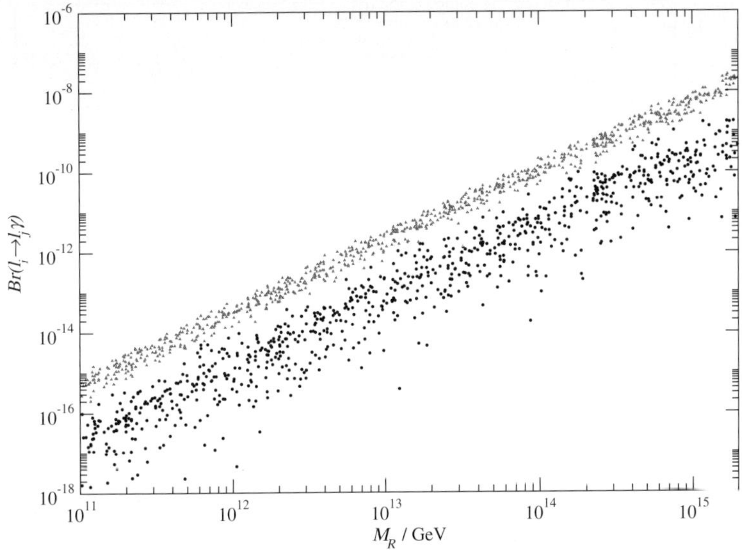

Fig. 4. Branching ratio of $\tau \to \mu\gamma$ (*upper band*) and $\mu \to e\gamma$ (*lower band*) in the mSUGRA scenario B'

in the neutrino data. The update of neutrino and SUSY parameters leads to only a slight decrease of $Br(\mu \to e\gamma)$ as compared to the previous results published in [7]. One sees that a branching ratio in the range between the present bound and the sensitivity limit of the new PSI experiment, that is $10^{-11} \gtrsim Br(\mu \to e\gamma) \gtrsim 10^{-13}$, would point at a value of M_R between $5 \cdot 10^{12}$ GeV and $5 \cdot 10^{14}$ GeV. On the other hand, $Br(\tau \to \mu\gamma)$ is more strongly affected by the smaller value of m_0 in the new benchmark models [11], resulting in a reduction by a factor of about 5 as compared to the results in [7]. Even if $Br(\tau \to \mu\gamma) = 10^{-8}$ is reached, the goal of SUPERKEKB and LHC [13], one would only probe $M_R \gtrsim 10^{15}$ GeV [7]. Nevertheless it is interesting to note that $\tau \to \mu\gamma$ is much less affected by the neutrino uncertainties than $\mu \to e\gamma$.

Analogously, Fig. 5 shows the M_R dependence of the cross-sections for $e^+e^- \to \mu^+e^- + 2\tilde{\chi}_1^0$ and $e^+e^- \to \tau^+\mu^- + 2\tilde{\chi}_1^0$. In this case, the μe channel is enhanced by both the larger Δm_{12}^2 and the smaller m_0 in the new parameter set, resulting in a cross-section one order of magnitude larger than what was found in [8]. For the $\tau\mu$ final state, on the other hand, the effect of the smaller Δm_{23}^2 is compensated by the enhancement due to the smaller value of m_0, so that the net effect is negligible. As can be seen, for a sufficiently large Majorana mass M_R the LFV cross-sections can reach several fb. The τe channel is strongly suppressed by the small mixing angle θ_{13}, and therefore more difficult to observe.

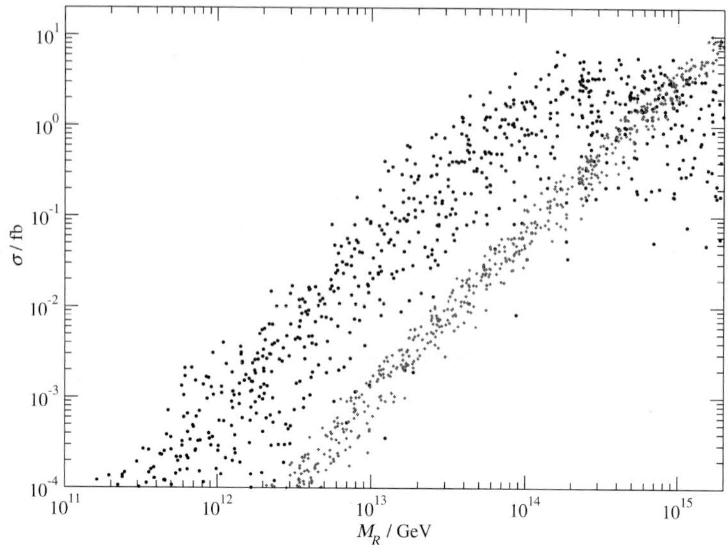

Fig. 5. Cross-sections for $e^+e^- \to \mu^+e^- + 2\tilde{\chi}_1^0$ (*upper band*) and $e^+e^- \to \tau^+\mu^- + 2\tilde{\chi}_1^0$ (*lower band*) at $\sqrt{s} = 500\,\text{GeV}$ for the mSUGRA scenario B'

The Standard Model background mainly comes from W-pair production, W production via t-channel photon exchange, and τ-pair production. A 10 degree beam pipe cut and cuts on the lepton energy and missing energy reduce the SM background cross-sections to less than 30 fb for μe final states and less than 10 fb for $\tau\mu$ final states. If one requires a signal to background ratio $S/\sqrt{S+B} = 3$, and assumes an integrated luminosity of 1000 fb^{-1}, a signal cross-section of 0.1 fb could only afford a background of about 1 fb. Whether or not such a low background can be achieved by applying selectron selection cuts, for example, on the acoplanarity, lepton polar angle and missing transverse momentum has to be studied in a dedicated simulation. For lepton flavor conserving processes one has found that the SM background to slepton pair production can be reduced to about 2-3 fb at $\sqrt{s} = 500\,\text{GeV}$ [14].

The MSSM background is dominated by chargino/slepton production with a total cross-section of 0.2-5 fb and 2-7 fb for μe and $\tau\mu$ final states, respectively, depending on the SUSY scenario and the collider energy. Here, only the direct processes are accounted for. However, the MSSM background in the τe channel can also contribute to the μe channel via the decay $\tau \to \mu\nu_\mu\nu_\tau$. If $\tilde{\tau}_1$ and $\tilde{\chi}_1^+$ are very light, like in scenarios B' and I', this background can be as large as 20 fb. On the other hand, such events typically contain two neutrinos in addition to the two LSPs which are also present in the signal events. Thus, after τ decay one has altogether six invisible particles instead of two, which may allow to eliminate also this particularly dangerous

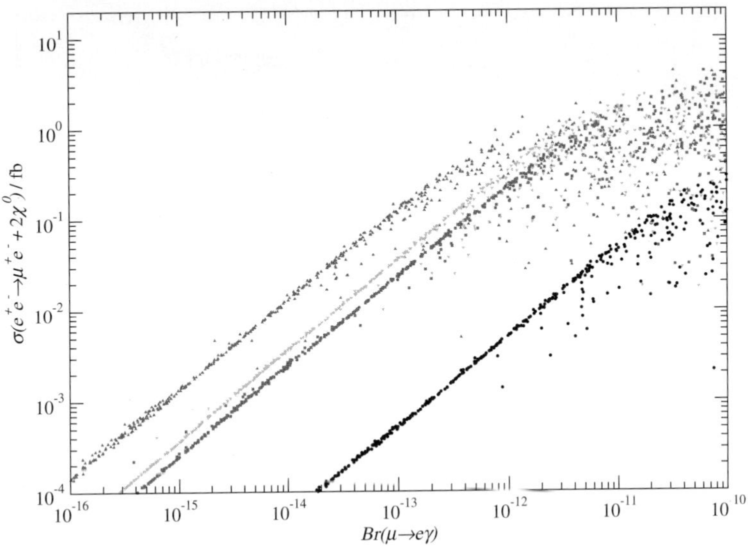

Fig. 6. Correlation between $Br(\mu \to e\gamma)$ and $\sigma(e^+e^- \to \mu^+e^- + 2\tilde{\chi}_1^0)$ at $\sqrt{s} =$ 800 GeV for the mSUGRA scenarios (*from left to right*) C', G', B' and I'

MSSM background by cutting on various distributions. But also here this needs to be studied in a careful simulation.

Particularly interesting and useful are the correlations between LFV in radiative decays and slepton pair production. Such a correlation is illustrated in Fig. 6 for $e^+e^- \to \mu^+e^- + 2\tilde{\chi}_1^0$ and $Br(\mu \to e\gamma)$. One sees that the neutrino uncertainties drop out, while the sensitivity to the mSUGRA parameters remains. Furthermore, while models C', G' and I' are barely affected by the change in the new parameter set as compared to the set used in [8], in model B' $\sigma(e^+e^- \to \mu^+e^- + 2\tilde{\chi}_1^0)$ for a given $Br(\mu \to e\gamma)$ is by a factor 10 larger than in the previous benchmark point B. An observation of $\mu \to e\gamma$ with a branching ratio smaller than 10^{-11} would thus be compatible with a cross-section as large as 1 fb for $e^+e^- \to \sum_{b,a} \tilde{l}_b^+ \tilde{l}_a^- \to \mu^+e^- + 2\tilde{\chi}_1^0$, at least in model C', G' and B'. On the other hand, no signal at the future PSI sensitivity of 10^{-13} would constrain this channel to less than 0.1 fb. The correlation of $Br(\tau \to \mu\gamma)$ and $\sigma(e^+e^- \to \tau^+\mu^- + 2\tilde{\chi}_1^0)$ is shown in Fig. 7. $Br(\tau \to \mu\gamma) < 3 \cdot 10^{-7}$ does not rule out cross-sections in the $\tau\mu$ channel of 1 fb and larger. However, one has to keep in mind that these correlations depend very much on the SUSY scenario.

7 Conclusions

SUSY seesaw models leading to the observed neutrino masses and mixings can be tested by lepton-flavor violating processes involving charged leptons. We

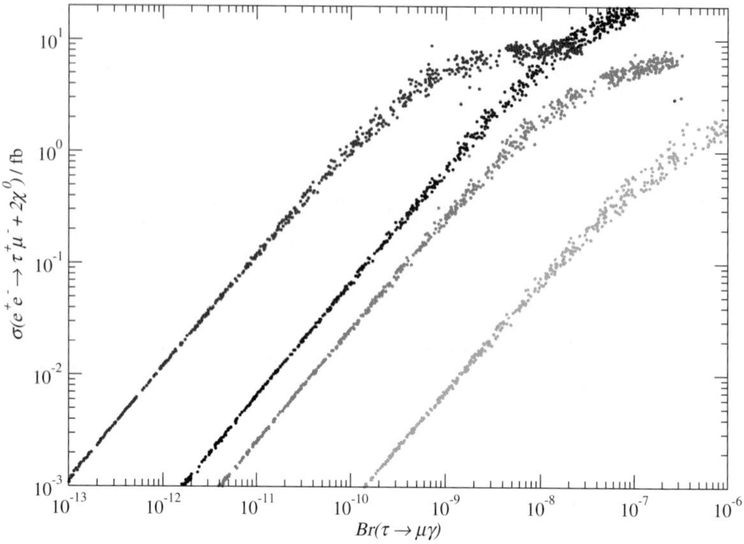

Fig. 7. Correlation between $Br(\tau \to \mu\gamma)$ and $\sigma(e^+e^- \to \tau^+\mu^- + 2\tilde{\chi}_1^0)$ at $\sqrt{s} =$ 800 GeV for the mSUGRA scenarios (*from left to right*) C', B', G' and I'

have presented an updated analysis of the prospects for radiative rare decays $l_i \to l_j\gamma$ and slepton pair production and decay $e^+e^- \to \tilde{l}_b^+\tilde{l}_a^- \to l_j^+l_i^- + \not{E}$. Assuming the most recent global fits to neutrino oscillation experiments [10] we have illustrated the impact of the uncertainties in the neutrino parameters. Furthermore, using post-WMAP mSUGRA scenarios [11] we have investigated the dependence of LFV signals on the mSUGRA parameters. For scenario B' our results can be summarized as follows. A measurement of $Br(\mu \to e\gamma) \approx 10^{-13}$ would probe M_R in the range $5 \cdot 10^{12} \div 5 \cdot 10^{13}$ GeV, while a measurement of $Br(\tau \to \mu\gamma) \approx 10^{-8}$ would allow to determine $M_R \simeq 10^{15}$ GeV within a factor of 2. Furthermore, $Br(\mu \to e\gamma) = 10^{-13} \div 10^{-11}$ implies $\sigma(e^+e^- \to \mu^+e^- + 2\tilde{\chi}_1^0) = 0.02 \div 2$ fb at $\sqrt{s} = 800$ GeV, while $Br(\tau \to \mu\gamma) = 10^{-8} \div 3 \cdot 10^{-7}$ predicts $\sigma(e^+e^- \to \tau^+\mu^- + 2\tilde{\chi}_1^0) = 1 \div 10$ fb, again at $\sqrt{s} = 800$ GeV. Hence, linear collider searches are nicely complementary to searches for rare decays at low energies and at the LHC.

Acknowledgements

This work was supported by the Bundesministerium für Bildung und Forschung (BMBF, Bonn, Germany) under the contract number 05HT4WWA2.

References

1. S. T. Petcov, Sov. J. Nucl. Phys. **25**, 340 (1977) [Yad. Fiz. **25**, 641 (1977 ERRAT,25,698.1977 ERRAT,25,1336.1977)]. S. M. Bilenkii, S. T. Petcov and B. Pontecorvo, Phys. Lett. B **67**, 309 (1977).
2. F. Borzumati and A. Masiero, Phys. Rev. Lett. **57**, 961 (1986).
3. J. Hisano, T. Moroi, K. Tobe and M. Yamaguchi, Phys. Rev. D **53**, 2442 (1996) [arXiv:hep-ph/9510309]. R. Barbieri, L. Hall and A. Strumia, Nucl. Phys. B **445**, 219 (1995); G. K. Leontaris and N. D. Tracas, Phys. Lett. B **431**, 90 (1998); W. Buchmuller, D. Delepine and F. Vissani, Phys. Lett. B **459**, 171 (1999); M. E. Gomez, G. K. Leontaris, S. Lola and J. D. Vergados, Phys. Rev. D **59**, 116009 (1999); S. F. King and M. Oliveira, Phys. Rev. D **60**, 035003 (1999); W. Buchmuller, D. Delepine and L. T. Handoko, Nucl. Phys. B **576**, 445 (2000); J. Ellis, M. E. Gomez, G. K. Leontaris, S. Lola and D. V. Nanopoulos, Eur. Phys. J. C **14**, 319 (2000); J. L. Feng, Y. Nir and Y. Shadmi, Phys. Rev. D **61**, 113005 (2000); J. Sato and K. Tobe, Phys. Rev. D **63**, 116010 (2001); D. F. Carvalho, M. E. Gomez and S. Khalil, JHEP **0107**, 001 (2001). J. Sato, K. Tobe and T. Yanagida, Phys. Lett. B **498**, 189 (2001); S. Davidson and A. Ibarra, JHEP **0109**, 013 (2001). P. Ciafaloni, A. Romanino and A. Strumia, Nucl. Phys. B **458**, 3 (1996); J. Hisano, T. Moroi, K. Tobe and M. Yamaguchi, Phys. Lett. B **391**, 341 (1997); J. Hisano, D. Nomura, Y. Okada, Y. Shimizu and M. Tanaka, Phys. Rev. D **58**, 116010 (1998); J. Hisano, D. Nomura and T. Yanagida, Phys. Lett. B **437**, 351 (1998); S. W. Baek, N. G. Deshpande, X. G. He and P. Ko, Phys. Rev. D **64**, 055006 (2001); G. Barenboim, K. Huitu and M. Raidal, Phys. Rev. D **63**, 055006 (2001); X. J. Bi and Y. B. Dai; S. Lavignac, I. Masina and C. A. Savoy, Phys. Lett. B **520**, 269 (2001). A. Kageyama, S. Kaneko, N. Shimoyama and M. Tanimoto, Phys. Rev. D **65** 096010 (2002) [arXiv:hep-ph/0112359].
4. J. A. Casas and A. Ibarra, Nucl. Phys. B **618**, 171 (2001) [arXiv:hep-ph/0103065].
5. J. Hisano and D. Nomura, Phys. Rev. D **59**, 116005 (1999) [arXiv:hep-ph/9810479].
6. N. Arkani-Hamed, H. C. Cheng, J. L. Feng and L. J. Hall, Phys. Rev. Lett. **77**, 1937 (1996) [arXiv:hep-ph/9603431], N. Arkani-Hamed, J. L. Feng, L. J. Hall and H. C. Cheng, Nucl. Phys. B **505**, 3 (1997) [arXiv:hep-ph/9704205]; H. C. Cheng, arXiv:hep-ph/9712427; M. Hirouchi and M. Tanaka, Phys. Rev. D **58**, 032004 (1998) [arXiv:hep-ph/9712532]; J. L. Feng, Int. J. Mod. Phys. A **13**, 2319 (1998) [arXiv:hep-ph/9803319]; J. Hisano, M. M. Nojiri, Y. Shimizu and M. Tanaka, Phys. Rev. D **60**, 055008 (1999) [arXiv:hep-ph/9808410]; D. Nomura, Phys. Rev. D **64**, 075001 (2001) [arXiv:hep-ph/0004256]; M. Dine, Y. Grossman and S. Thomas, eConf **C010630**, P332 (2001) [Int. J. Mod. Phys. A **18**, 2757 (2003)] [arXiv:hep-ph/0111154]. N. Arkani-Hamed, J. L. Feng, L. J. Hall and H. Cheng, Nucl. Phys. B **505** (1997) 3 [hep-ph/9704205]. J. Hisano, M. M. Nojiri, Y. Shimizu and M. Tanaka, Phys. Rev. D **60** (1999) 055008 [hep-ph/9808410]. M. Guchait, J. Kalinowski and P. Roy, Eur. Phys. J. C **21** (2001) 163 [arXiv:hep-ph/0103161]. Phys. Rev. D **66** (2002) 015003 [arXiv:hep-ph/0201284]. Acta Phys. Polon. B **32** (2001) 3755. Acta Phys. Polon. B **33** (2002) 2613 [arXiv:hep-ph/0207051]. J. J. Cao, T. Han, X. Zhang and G. R. Lu, Phys. Rev. D **59**, 095001 (1999) [arXiv:hep-ph/9808466].

J. Cao, Z. Xiong and J. M. Yang, Eur. Phys. J. C **32**, 245 (2004) [arXiv:hep-ph/0307126]. M. Cannoni, S. Kolb and O. Panella, Phys. Rev. D **68**, 096002 (2003) [arXiv:hep-ph/0306170].

7. F. Deppisch, H. Päs, A. Redelbach, R. Ruckl and Y. Shimizu, Eur. Phys. J. C **28**, 365 (2003) [arXiv:hep-ph/0206122].
8. F. Deppisch, H. Päs, A. Redelbach, R. Rückl and Y. Shimizu, accepted for publication in Phys. Rev. D, arXiv:hep-ph/0310053.
9. Homepage Hitoshi Murayama, http://hitoshi.berkeley.edu/neutrino/.
10. M. Maltoni, T. Schwetz, M. A. Tortola and J. W. F. Valle, Phys. Rev. D **68**, 113010 (2003) [arXiv:hep-ph/0309130].
11. M. Battaglia, A. De Roeck, J. R. Ellis, F. Gianotti, K. A. Olive and L. Pape, arXiv:hep-ph/0306219.
12. M. Battaglia et al., Eur. Phys. J. C **22** (2001) 535 [arXiv:hep-ph/0106204].
13. L. Serin and R. Stroynowski, ATLAS Internal Note (1997); T. Ohshima, talk at the 3rd Workshop on Neutrino Oscillations and their Origin (NOON2001), 2001, ICRR, Univ. of Tokyo, Kashiwa, Japan; D. Denegri, private communication.
14. R. Becker and C. Vander Velde, IIHE-93-08 *Prepared for Working Group on e+ e- Collisions at 500-GeV: The Physics Potential, Munich, Germany, 20 Nov 1992.*

Sterile Neutrino Dark Matter in the Galaxy

Neven Bilić[1,2], Gary B. Tupper[1], and Raoul D. Viollier[1]

[1] Institute of Theoretical Physics and Astrophysics, Department of Physics, University of Cape Town, Private Bag, Rondebosch 7701, South Africa
viollier@science.uct.ac.za

[2] Rudjer Bošković Institute, P.O. Box 180, 10002 Zagreb, Croatia
bilic@thphys.irb.hr

1 Introduction

The accurate WMAP data [1] have recently provided us with compelling evidence that the Universe is approximately critical on the largest scale. About 73% of its total energy resides is in unclustered dark energy (perhaps in the form of a cosmological constant) and approximately 27% is in nonrelativistic dark matter which clusters gravitationally. Included in this dark matter component is between 4 and 5% of the total energy in baryonic matter, mostly in the form of dark gas and dust. About 0.5% of the total energy accounts for baryonic matter contained in stars. The total energy of the microwave background and relativistic or nonrelativistic active neutrinos is less than about 0.1% of the critical energy of the Universe today.

There is also independent evidence for the existence of dark matter on galactic scales, which is extracted from the motion of gas, stars, globular clusters and dwarf galaxies. Indeed, at large distances from the galactic center, the circular velocity tends towards a constant value for nearly all the galaxies. Interpreting this result in a Newtonian context with a spherically symmetric matter distribution, the galactic matter density $\rho(r)$ and the enclosed galactic mass $M(r)$ must scale as $\rho(r) \propto r^{-2}$ and $M(r) \propto r$, respectively, for large distances r from the galactic center. Thus, based on Newton's laws, one concludes that the (mainly baryonic) galactic disks are surrounded by nearly spherically symmetric halos, which dominate the gravitational field at large distances. These halos are presumably made of non-baryonic matter, as dark stars or MACHO's cannot account for all the halo dark mass in the Galaxy, and diffuse baryonic matter would presumably fall into the galactic disks on time scales much shorter than the age of the Universe.

Dark matter distributions are best studied in low-surface-brightness galaxies where dark matter is supposed to be more prominent than in bright galaxies. The dark matter distributions of these galaxies are best fitted with the so-called pseudo-isothermal profile [2] given by

$$\rho_{\mathrm{ISO}}(r) = \frac{\rho_0}{1 + (r/R_c)^2} , \tag{1}$$

which exhibits a flat core of radius R_c and central density ρ_0. An alternative parameterization is the Navarro-Frenk-White profile [3] given by

$$\rho_{\mathrm{NFW}}(r) = \frac{\rho_i}{(1+r/R_s)^2\, r/R_s} \,. \tag{2}$$

However, this halo density exhibits a cusp of the form r^{-1} at the center, and a r^{-3} behaviour at large distances. These two limiting behaviours do not seem to reflect the observations, although the Navarro-Frenk-White profile is supported by dynamical cold dark matter simulations.

There is possible further evidence for dark matter at the centers of galaxies. Indeed, Schödel et al. [4] reported recently a new set of data including the corrected old measurements [5] on the projected positions of the star S2(S0-2) that was observed during the last decade with the ESO telescopes in La Silla (Chile). The combined data suggest that S2(S0-2) is moving on a Keplerian orbit with a period of 15.2 yr around the enigmatic strong radiosource Sgr A* at or near the center of our Galaxy that is widely believed to harbour a black hole with a mass of about $2.6 \times 10^6 M_\odot$ [5, 6]. The salient feature of the new adaptive optics data is that, between April and May 2002, S2(S0-2) apparently sped past the point of closest approach with a velocity $v \sim 6000\,\mathrm{km/s}$ at a distance of about 17 light-hours (123 AU) from Sgr A* [4].

Another star, S0-16, which was observed during the last few years by Ghez et al. [7] with the Keck telescope in Hawaii, made recently a spectacular U-turn, crossing the point of closest approach at an even smaller distance of 8.32 light-hours or (60 AU) from Sgr A* with a velocity $v \sim 9000\,\mathrm{km/s}$. Ghez et al. [7] thus conclude that the gravitational potential around Sgr A* has approximately r^{-1} form, for radii larger than 60 AU, corresponding to 1169 Schwarzschild radii of 26 light-seconds (0.051 AU) for a $2.6 \times 10^6 M_\odot$ black hole. Although the baryonic alternatives are presumably ruled out, this still leaves some room for the interpretation of the supermassive compact dark object at the Galactic center in terms of a finite-size non-baryonic dark matter object rather than a black hole. In fact, the supermassive black hole paradigm may eventually only be proven or ruled out by comparing it with credible alternatives in terms of finite-size non-baryonic objects [8].

If R-parity is conserved, even though this is not a direct consequence of supersymmetry, the dark matter particle could be the lightest supersymmetric particle, a neutral fermion with spin 1/2 and mass of about 100 GeV. Alternatively, it could be the axion, a pseudo-scalar particle with a mass of about 10^{-5} eV, which was introduced to explain the enigmatic CP conservation of the strong interactions, based on the breaking of the global Peccei-Quinn symmetry. Further dark matter particle candidates are the axino [9] or the gravitino [10], which would have masses between 1 keV and about 100 keV in soft supersymmetry breaking scenarios. Finally, the dark matter particle could be any sterile neutrino in the mass range between 1 keV and 1 MeV that is mixed with at least one of the active neutrinos with mixing angles at the level of $\theta \sim 10^{-7}$.

The purpose of this paper is to explore, using the example of sterile neutrino dark matter, the implications of the recent cosmological and astrophysical observations, considering in particular the degenerate fermion ball scenario of the supermassive compact dark objects which was developed during the last decade [8, 11, 12, 13, 14, 15, 16].

2 Stellar-Dynamical Constraints for Fermion Balls

In a self-gravitating ball of degenerate fermionic matter, the gravitational pressure is balanced by the degeneracy pressure of the fermions due to the Pauli exclusion principle. Nonrelativistically, this scenario is described by the Lane-Emden equation with polytropic index $p = 3/2$. Thus the radius R and mass M of a ball of self-gravitating, nearly non-interacting degenerate fermions scale as [12]

$$\begin{aligned} R &= \left[\frac{91.869\,\hbar^6}{m^8 G^3}\left(\frac{2}{g}\right)^2 \frac{1}{M}\right]^{1/3} \\ &= 3610.66\ \mathrm{ld}\left(\frac{15\,\mathrm{keV}}{mc^2}\right)^{8/3}\left(\frac{2}{g}\right)^{2/3}\left(\frac{M_\odot}{M}\right)^{1/3} . \end{aligned} \tag{1}$$

Here 1.19129 ld = 1 mpc = 206.265 AU, and m is the fermion mass. The degeneracy factor $g = 2$ describes either spin 1/2 fermions (without antifermions) or spin 1/2 Majorana fermions ($\equiv$ antifermions). For Dirac fermions and antifermions, or spin 3/2 fermions (without antifermions), we have $g = 4$. Using the canonical value $M = 2.6 \times 10^6 M_\odot$ and $R \leq 60$ AU for the supermassive compact dark object at the Galactic center, we obtain a minimal fermion mass of $m_{\mathrm{min}} = 76.0\,\mathrm{keV}/c^2$ for $g = 2$, or $m_{\mathrm{min}} = 63.9\,\mathrm{keV}/c^2$ for $g = 4$.

The maximal mass for a degenerate fermion ball, calculated in a general relativistic framework based on the Tolman-Oppenheimer-Volkoff equations, is the Oppenheimer-Volkoff (OV) limit [13]

$$M_{\mathrm{OV}} = 0.38322\,\frac{M_{\mathrm{Pl}}^3}{m^2}\left(\frac{2}{g}\right)^{1/2} = 2.7821 \times 10^9 M_\odot \left(\frac{15\,\mathrm{keV}}{mc^2}\right)^2 \left(\frac{2}{g}\right)^{1/2} , \tag{2}$$

where $M_{\mathrm{Pl}} = (\hbar c/G)^{1/2} = 1.2210 \times 10^{19}$ GeV is the Planck mass. Thus, for $m_{\mathrm{min}} = 76.0\,\mathrm{keV}/c^2$ and $g = 2$, or $m_{\mathrm{min}} = 63.9\,\mathrm{keV}/c^2$ and $g = 4$, we obtain

$$M_{\mathrm{OV}}^{\mathrm{max}} = 1.083 \times 10^8 M_\odot . \tag{3}$$

In this scenario all supermassive compact dark objects with mass $M > M_{\mathrm{OV}}^{\mathrm{max}}$ must be black holes, while those with $M \leq M_{\mathrm{OV}}^{\mathrm{max}}$ are fermion balls.

Choosing as the OV-limit the canonical mass of the compact dark object at the center of the Galaxy, $M_{\rm OV}^{\rm min} = 2.6 \times 10^6\ M_\odot$, yields a maximal fermion mass of $m_{\rm max} = 491\,{\rm keV}/c^2$ for $g = 2$, or $m_{\rm max} = 413\,{\rm keV}/c^2$ for $g = 4$. In this ultrarelativistic limit, there is little difference between the black hole and degenerate fermion ball scenarios, as the radius of the fermion ball is 4.45 compared to 3 Schwarzschild radii for the radius of the event horizon of a non-rotating black hole of the same mass. In fact, varying the fermion mass between $m_{\rm min}$ and $m_{\rm max}$, one can smoothly interpolate between a fermion ball of the largest acceptable size and a fermion ball of the smallest possible size, at the limit between fermion balls and black holes.

The masses of the supermassive compact dark objects discovered so far at the centers of both active and inactive galaxies are all in the range [17]

$$10^6 M_\odot \lesssim M \lesssim 3 \times 10^9 M_\odot \ . \tag{4}$$

Thus, as $M_{\rm OV}^{\rm max}$ falls into this range as well, we need both supermassive fermion balls ($M \leq M_{\rm OV}^{\rm max}$) and black holes ($M > M_{\rm OV}^{\rm max}$) to describe the observed phenomenology. At first sight, such a hybrid scenario does not seem to be particularly attractive. However, it is important to note that a similar scenario is actually realized in Nature, with the co-existence of neutron stars which have masses $M \leq M_{\rm OV}^n$, and stellar-mass black holes with mass $M > M_{\rm OV}^n$, as observed in stellar binary systems in the Galaxy [18]. Here the Oppenheimer-Volkoff limit $M_{\rm OV}^n$, which includes the nuclear interaction of the neutrons, is somewhat uncertain due to the unknown equation of state. But the consensus of the experts [18] is that it must be in the range

$$1.4\ M_\odot \leq M_{\rm OV}^n \lesssim 3\ M_\odot \ . \tag{5}$$

None of the observed neutron stars have masses larger than $1.4\ M_\odot$, while there are at least nine candidates for stellar-mass black holes larger than $3\ M_\odot$ [18]. It is thus conceivable that Nature allows for the co-existence of supermassive fermion balls and black holes as well. Of course, we would expect characteristic differences in the properties of supermassive fermion balls and black holes. Similarly, pulsars and stellar-mass black holes are quite different, as pulsars have a strong magnetic field and a hard baryonic surface, while black holes are surrounded by an immaterial event horizon instead. However, one may also argue that the astrophysical differences between supermassive black holes and fermion balls close to the OV-limit are not so easy to detect because both objects are of non-baryonic nature.

3 Cosmological Constraints for Sterile Neutrino Dark Matter

If the supermassive compact dark object at the Galactic center is indeed a degenerate fermion ball of mass $M = 2.6 \times 10^6 M_\odot$ and radius $R \leq 60$ AU, the fermion mass must be in the range

$$\begin{aligned} 76.0\,\mathrm{keV}/c^2 \;\leq\; m \;\leq\; 491\,\mathrm{keV}/c^2 \;\; \text{for} \;\; g = 2 \\ 63.9\,\mathrm{keV}/c^2 \;\leq\; m \;\leq\; 413\,\mathrm{keV}/c^2 \;\; \text{for} \;\; g = 4\,. \end{aligned} \tag{6}$$

It would be most economical if this particle could represent the dark matter particle of the Universe, as well. The conjectured fermion could be a sterile neutrino ν_s which does not participate in the weak interactions. We will now assume that its mass and degeneracy factor is $m_s = 76.0\,\mathrm{keV}/c^2$ and $g_s = 2$, corresponding to the largest fermion ball that is consistent with the stellar-dynamical constraints. In order to make sure that this fermion is actually produced in the early Universe it must be mixed with at least one active neutrino, e.g., the ν_e. Indeed, for an electron neutrino asymmetry,

$$L_{\nu_e} = \frac{n_{\nu_e} - n_{\bar{\nu}_e}}{n_\gamma} \sim 10^{-2} \tag{7}$$

and a mixing angle $\theta_{es} \sim 10^{-7}$ [19], incoherent resonant and non-resonant active neutrino scattering in the early Universe produces sterile neutrino matter amounting to the required fraction $\Omega_m h^2 = \left(0.135^{+0.008}_{-0.009}\right)$ [1], of the critical density of the Universe today. Here n_{ν_e}, $n_{\bar{\nu}_e}$ and n_γ are the electron neutrino, electron antineutrino and photon number densities, respectively. An electron neutrino asymmetry of $L_{\nu_e} \sim 10^{-2}$ is compatible with the observational limits on ^{4}He abundance, radiation density of the cosmic microwave background at decoupling, and formation of the large scale structure [20, 21] which constrain the electron neutrino asymmetry to the range

$$-4.1 \times 10^{-2} \;\leq\; L_{\nu_e} \;\leq 0.79\,. \tag{8}$$

At this stage it is interesting to note that incoherent resonant scattering of active neutrinos produces quasi-degenerate sterile neutrino matter, while incoherent non-resonant active neutrino scattering yields sterile neutrino matter that has approximately a thermal spectrum [19]. Quasi-degenerate sterile neutrino matter may contribute towards the formation of the supermassive compact dark objects at the galactic centers, while thermal sterile neutrino matter is mainly contributing to the dark matter of the galactic halos [22]. In fact it has been recently shown [14], that an extended cloud of degenerate fermionic matter will eventually undergo gravitational collapse and form a degenerate supermassive fermion ball in a few free-fall times, if the collapsed mass is below the OV limit. During the formation, the binding energy of the nascent fermion ball is released in the form of high-energy ejecta at every bounce of the degenerate fermionic matter through a mechanism similar to gravitational cooling that is taking place in the formation of degenerate boson stars [14]. If the mass of the collapsed object is above the OV limit, the collapse inevitably results in a supermassive black hole.

4 Observability of Degenerate Sterile Neutrino Balls

The mixing of the sterile neutrino with at least one of the active neutrinos necessarily causes the main decay mode of the ν_s into three active neutrinos [23] with a lifetime of

$$\tau\,(\nu_s \to 3\nu) = \frac{192\,\pi^3}{G_F^2\, m_s^5\, \sin^2\,\theta_{es}} = \frac{\tau(\mu^- \to e^- + \bar{\nu}_e + \nu_\mu)}{\sin^2\theta_{es}} \left(\frac{m_\mu}{m_s}\right)^5 , \tag{9}$$

which is presumably unobservable as the available neutrino energy is too small. Here $\tau(\mu^- \to e^- + \bar{\nu}_e + \nu_\mu)$ and m_μ are the lifetime and mass of the muon. However, there is a subdominant radiative decay mode of the sterile into an active neutrino and a photon with a branching ratio [24]

$$B(\nu_s \to \nu\gamma) = \frac{\tau(\nu_s \to 3\nu)}{\tau(\nu_s \to \nu\gamma)} = \frac{27\,\alpha}{8\pi} = 0.7840 \times 10^{-2}\,, \tag{10}$$

where $\alpha = e^2/\hbar c$ is the fine structure constant. The lifetime of this potentially observable decay mode is thus

$$\tau\,(\nu_s \to \nu\gamma) = \frac{8\pi}{27\,\alpha}\,\frac{1}{\sin^2\theta_{es}}\left(\frac{m_\mu}{m_s}\right)^5 \tau\,(\mu^- \to e^- + \bar{\nu}_e + \nu_\mu) \tag{11}$$

yielding, for $\theta_{es} = 10^{-7}$ and $m_s = 76.0\,\mathrm{keV}/c^2$, a lifetime of $\tau(\nu_s \to \nu\gamma) = 0.46 \times 10^{19}$ yr.

Although the X-ray luminosity due to the radiative decay of diffuse sterile neutrino dark matter in the Universe is presumably not observable, because it is well below the X-ray background radiation at this energy [19], it is perhaps possible to detect such hard X-rays in the case of sufficiently concentrated dark matter objects. In fact, this could be the smoking gun for both the existence of the sterile neutrino and the fermion balls. For instance, a ball of $M = 2.6 \times 10^6 M_\odot$ consisting of degenerate sterile neutrinos of mass $m_s = 76.0\,\mathrm{keV}/c^2$ [15], degeneracy factor $g_s = 2$, and mixing angle $\theta_{es} = 10^{-7}$ would emit 38 keV photons with a luminosity

$$L_X = \frac{Mc^2}{2\tau(\nu_s \to \nu\gamma)} = 1.6 \times 10^{34}\ \mathrm{erg/s}\,, \tag{12}$$

within a radius of 60 AU (8.32 light hours or 7.6×10^{-3} arcsec) of Sgr A*, assumed to be at a distance of 8 kpc. The current upper limit on X-ray emission from the vicinity of Sgr A* is $\nu L_\nu \sim 3 \times 10^{35}$ erg/s, for an X-ray energy of $E_X \sim 60\,\mathrm{keV}$ [25], where $L_\nu = \mathrm{d}L/\mathrm{d}\nu$ is the spectral luminosity. Thus the X-ray line at 38 keV could presumably only be detected if either the angular or the energy resolution or both, of the present X-ray detectors are increased.

Acknowledgements

This work was supported by the South African National Research Foundation (NRF GUN-2053794), the Research Committee of the University of Cape Town and the Foundation for Fundamental Research (FFR PHY-99-01241).

References

1. WMAP Collaboration, C.L. Bennett et al: Astrophys. J. Suppl. **148**, 1 (2003); D.N. Spergel et al: Astrophys. J. Suppl. **148**, 175 (2003)
2. W.J.G. de Blok, S.S. McGaugh and V.C. Rubin: Astron. J. **122**, 2381 (2001) astro-ph/0107326; ibid **122**, 2396 (2001); astro-ph/0107366
3. J.F. Navarro, C.S. Frenk and S.D.M. White: Astrophys. J. **462**, 563 (1996)
4. R. Schödel et al: Nature **419**, 694 (2002)
5. A. Eckart et al: Mon. Not. R. Astron. Soc. **331**, 917 (2002)
6. A. Ghez et al: Astrophys. J. **509**, 678 (1998)
7. A. Ghez et al: Invited talk at the Galactic Center Conference, 4-8 October, 2002, Kalua-Kona (Hawaii).
8. F. Munyaneza, D. Tsiklauri and R.D. Viollier: Astrophys. J. **509** L105 (1998); ibid **526**, 744 (1999)
9. T. Goto and M. Yamaguchi: Phys. Lett. B **276**, 123 (1992)
10. D.H. Lyth: Phys. Lett. B **488**, 417 (2000)
11. R.D. Viollier, D. Trautmann and G.B. Tupper: Phys. Lett. B **306**, 79 (1993)
12. R.D. Viollier: Prog. Part. Nucl. Phys. **32**, 51 (1994)
13. N. Bilić, F. Munyaneza and R.D. Viollier: Phys. Rev. D **59**, 024003 (1999)
14. N. Bilić, R.J. Lindebaum, G.B. Tupper and R.D. Viollier: Phys. Lett. B **515**, 105 (2001)
15. F. Munyaneza and R.D. Viollier: Astrophys. J. **564**, 274 (2002)
16. N. Bilić, F. Munyaneza, G.B. Tupper and R.D. Viollier: Prog. Part. Nucl. Phys. **48**, 291 (2002)
17. J. Kormendy and L.G. Ho: astro-ph/0003268; L.C. Ho and J. Kormendy: astro-ph/0003267, Published in *Encyclopedia of Astronomy and Astrophysics*, (Institute of Physics Publishing, January 2001)
18. R. Blandford and N. Gehrels: Physics Today, June 1999, p 40
19. K. Abazajian, G.M. Fuller and M. Patel: Phys. Rev. D **64**, 023501 (2001)
20. K. Kang and G.S. Steigman: Nucl. Phys. B **372**, 494 (1992)
21. S. Esposito, G. Mangano, G. Miele and O. Pisanti: JHEP **09**, 038 (2000); M. Orito, T. Kajino, G.J. Matthews and R.N. Boyd: astro-ph/0005446
22. N. Bilić, G.B. Tupper and R.D. Viollier: Dark Matter in the Galaxy. In *Particle Physics in the New Millenium*, ed by J. Trampetić, J. Wess (Springer Lecture Notes in Physics, Berlin Heidelberg New York 2003) pp 24–38; astro-ph/0111366
23. F. Boehm and P. Vogel: *Physics of Massive Neutrinos*, (Cambridge University Press, New York 1987); V. Barger, R.J.N. Phillips and S. Sarkar: Phys. Lett. B **352**, 365 (1995); ibid **356**, 617(E) (1995)
24. P.B. Pal and L. Wolfenstein: Phys. Rev. D **25**, 766 (1982)
25. R. Mahadevan, Nature **394**, 651 (1998); R. Narayan et al: Astrophys. J. **492**, 554 (1998)

Supernovae and Dark Energy

Ariel Goobar

Physics Department, Stockholm University, AlbaNova, 106 91 Stockholm, Sweden
ariel@physto.se

1 Introduction

In the *Standard Model* of cosmology the Universe started with a *Big Bang*. The expansion of an isotropic and homogeneous Universe is described by the Friedmann-Lemâitre-Robertson-Walker model (or FLRW model, for short).

The free parameters of the FLRW model are the energy contributions from radiation, matter and vacuum fluctuations. At the present epoch, the energy density in the form of radiation ρ_{rad} can be neglected in comparison with the matter density ρ_m, and the Friedmann equation for the Hubble parameter (H) becomes:

$$H^2 \equiv \left(\frac{\dot{a}}{a}\right)^2 = \frac{8\pi G}{3}\rho_m + \frac{\Lambda}{3} - \frac{k}{a^2} , \tag{1}$$

where $a(t)$ is the growing scale factor of the Universe and $k = -1,0$ or 1 represent the three possible geometries for the Universe: open, flat or closed.

Thus, the expansion rate of the Universe depends on the matter density, the cosmological constant ($\Lambda - 8\pi G\rho_{vac}$) and the geometry of the Universe. It is also customary to rewrite equation (1) so that it instead contains the fractional energy density contributions at the present epoch ($z = 0$). We thus introduce the definitions:

$$\Omega_{\mathrm{M}} \equiv \frac{8\pi G}{3H_0^2}\rho_m^0, \quad \Omega_\Lambda \equiv \frac{\Lambda}{3H_0^2} \quad \Omega_K \equiv \frac{-k}{a_0^2 H_0^2}$$

There are only two independent contributions to the energy density since in the FLRW model:

$$\Omega_M + \Omega_\Lambda + \Omega_K = 1 \tag{2}$$

The time evolution of the scale factor a and thus the *fate* of the universe as determined by the two independent cosmological parameters. A large cosmological constant, for example, leads to rapid "inflation" of the universe.

The *deceleration* parameter (at $z = 0$), q_0, is defined as:

$$q_0 = \frac{\Omega_M}{2} - \Omega_\Lambda , \tag{3}$$

thus, a *negative* value of q_0 implies that the rate of expansion of the Universe is increasing, i.e. the expansion is accelerating.

In the next section we generalize the discussion as to also include contribution from any arbitrary energy form characterized by the the relation between its pressure and density.

2 Cosmological Parameters from "Standard Candles"

A source of known strength, a *standard candle* can be used to measure relative distances to provide information on the cosmological parameters, see [7]. Ω_M and Ω_X denote the present-day energy density parameters of ordinary matter $\Omega_M(z)$ and "Dark Energy". The Dark Energy is characterized by the equation of state parameter, $w(z)$, where $p_X = w \cdot \rho_X$. For the specific case of the cosmological constant, $w = -1$, i.e. $p_\Lambda = -\rho_\Lambda$.

The apparent magnitude m of a supernova at redshift z is then given by

$$m(z) = \mathcal{M} + 5\log_{10}\left[d'_L(z)\right] , \tag{4}$$

$$\mathcal{M} = 25 + M + 5\log_{10}(c/H_0) , \tag{5}$$

where M is the absolute magnitude of the supernova, and $d'_L \equiv H_0\, d_L$ is the H_0-*independent* luminosity distance, where H_0 is the Hubble parameter[1]. Hence, the intercept $\mathcal{M}$ contains the "nuisance" parameters M and H_0 that apply equally to all magnitude measurements (in this section we do not consider possible evolutionary effects $M = M(z)$). The H_0-independent luminosity distance d'_L is given by

$$d'_L = \begin{cases} (1+z)\frac{1}{\sqrt{-\Omega_K}}\sin(\sqrt{-\Omega_K}\, I), & \Omega_k < 0 \\ (1+z)\, I, & \Omega_K = 0 \\ (1+z)\frac{1}{\sqrt{\Omega_K}}\sinh(\sqrt{\Omega_K}\, I), & \Omega_K > 0 \end{cases} \tag{6}$$

$$\Omega_K = 1 - \Omega_M - \Omega_X , \tag{7}$$

$$I = \int_0^z \frac{dz'}{H'(z')} , \tag{8}$$

$$H'(z) = H(z)/H_0 = \sqrt{(1+z)^3\,\Omega_M + f(z)\,\Omega_X + (1+z)^2\,\Omega_K} , \tag{9}$$

$$f(z) = \exp\left[3\int_0^z dz'\,\frac{1+w(z')}{1+z'}\right] , \tag{10}$$

As the measurements are performed through broad-band filters one has to correct for the fact that different parts of the supernova spectrum are detected depending on the redshift z of the source. For example, at a redshift $z \sim 0.5$

[1] In the expression for $\mathcal{M}$, the units of c and H_0 are km s^{-1} and km s^{-1} Mpc^{-1}, respectively.

the light captured with a red (R) filter at a telescope at Earth originates from the blue (B) part of the spectrum. This so called "K-correction" is preferentially done using blue (B) absolute magnitudes in the resframe and V,R,I filters for the observation of supernovae with increasing redshift, as shown in [11].

3 Current Results

Two collaborations, the SCP [15, 24] and the High-Z team [21, 19, 5, 12, 25], have been searching for high-redshift Type Ia supernovae with the aim to measure cosmological parameters. Both groups find that the data is consistent with the existence of some "Dark Energy" form that is accelerating the rate of expansion of the universe at present, i.e. $q_0 < 0$, as shown in Fig. 1. The Hubble diagram for high-z supernovae found by the SCP along with low-z supernovae from the Calán/Tololo [9] and CfA [20] supernova Surveys indicates that supernovae at $z \sim 0.5$ are 0.2–0.5 magnitudes too faint to be

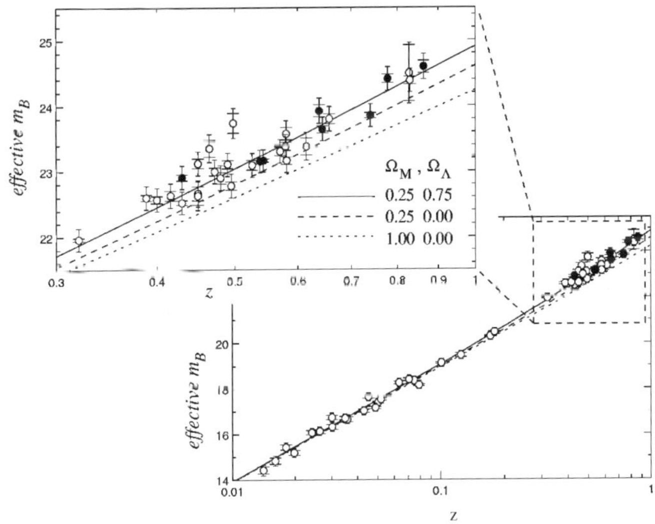

Fig. 1. Hubble diagram for high-redshift Type Ia supernovae from the *Supernova Cosmology Project* [12], and low-redshift Type Ia supernovae from the CfA [20] and Calán/Tololo Supernova Survey [9]. Filled circles represent supernovae measured with the Hubble Space Telescope, i.e. in general with higher accuracy. The curves are the theoretical effective $m_B(z)$ for a range of cosmological models with and without a cosmological constant. It is called "effective m_B" because the measured intensity corresponds to the restframe B-band (blue). Because of cosmological redshift, the photons are observed at longer wavelengths. The best fit to the data (for a flat Universe) corresponds to the FLRW Universe with $(\Omega_M, \Omega_\Lambda) = (0.25, 0.75)$, as shown in [12]

consistent with an open or flat universe with $\Lambda = 0$. The theoretical curves for a universe with no cosmological constant are shown as dotted (open) and dashed (flat) lines. The solid line shows the best fit-cosmology for which the total mass-energy density $\Omega_\mathrm{M} + \Omega_\Lambda = 1$. The best fit value for the mass density in a flat universe is ($\Omega_\Lambda = 1 - \Omega_\mathrm{M}^\mathrm{flat}$):

$$\Omega_\mathrm{M}^\mathrm{flat} = 0.25^{+0.07}_{-0.06} \pm 0.04 \, ,$$

where the first uncertainty is statistical and the second due to known systematics. The details of the estimation of systematic errors such as from extinction, Mamlquist bias and brightness evolution of type Ia supernovae can be found in [15, 12].

The supernova results are in good agreement with what is found from a varity of independent techniques. The CMB anisotropies at scales 1° or smaller as measured by the WMAP, BOOMERANG, MAXIMA and DASI collaborations [22, 3, 4, 17] give a firm constrain on the geometry of the universe indicating that the sum of all energy densities, i.e. $\Omega_M + \Omega_X$ must be unity with only 2% unceratinty. Constraints on the matter density Ω_M from cluster abundances [1, 2] and large-scale structure [14, 10] has left cosmology with a concordance model with $\Omega_M \approx 0.3$ and $\Omega_\Lambda \approx 0.7$, as shown in Fig. 2.

4 How Much Better Can We Do With Type Ia SNe?

Figure 3 shows the degenaracy in the CL-region of the $\Omega_M - \Omega_\Lambda$ parameter space defined by observations at *single* redshifts, ranging from $z = 0.2$ to 1.8, assuming an accuracy of $\Delta m = 0.02$ mag in the measured mean. In Fig. 3, a hypothetical data-set including supernovae at $z = 0.2$–1.8 is used to demonstrate how the major axis of the confidence region could be dramatically shrunk. Clearly, enlarging the redshift range of the followed supernovae has the potential of refining our understanding of the cosmological parameters.

5 The Next Generation of SN Experiments

In Fig. 3 we demonstrated how the accuracy in the magnitude–redshift method increases as supernovae at higher redshifts are added to the sample. In particular, at redshifts above $z \sim 1$ one can study the transition from acceleration to deceleration as the mass density term contribution, enhanced by the the shrinking volume as $(1+z)^3$, overtakes the effect of Ω_Λ, as shown in Fig. 4.

Several projects with the aim to discover thousends of high-z supernovavae are being proposed. One of the most interesting ones is the SNAP satellite [30], a 2-m telescope equipped with an optical and NIR mosaic camera with a field of view of ~0.7 square deg.

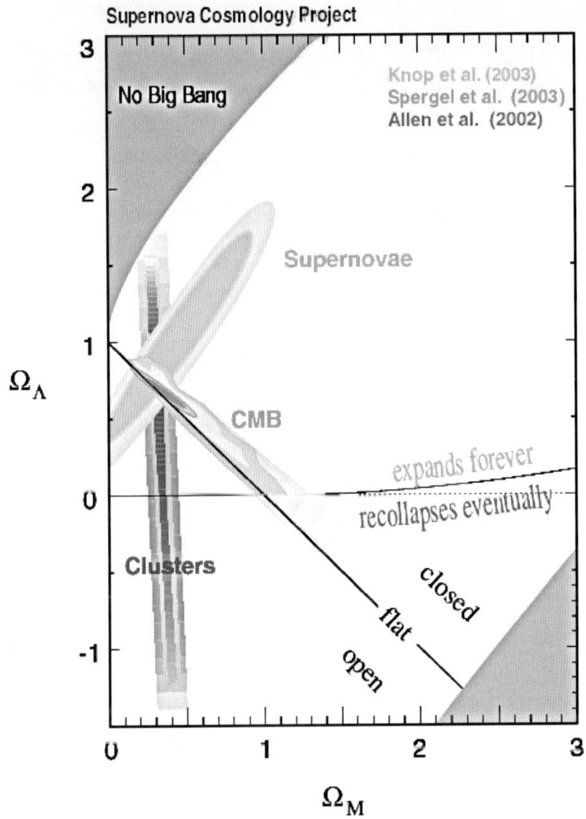

Fig. 2. The "Concordance Model". The combination of SNIa data, galaxy cluster information and CMB anisotropies indicate that we live in a universe with $\Omega_M \approx 0.3$ and $\Omega_\Lambda \approx 0.7$

In addition of having the capability of discovering about 2500 SNe a year up to a redshift $z \sim 2$, the design of the SNAP satellite also includes an integral field spectrograph. This will allow for detailed spectoscopic studies of the supernovae and their host galaxies. Thus, systematic uncertainties on the measured supernova brightnesses are supposed to stay below 0.02 mag in which case one can expect to measure Ω_M and Ω_Λ simultaneously to about 2% and 5% respectively, as shown in Fig. 5.

6 The Quintessence Alternative

The exciting results from the SCP and High-Z teams suggest that the method can be used to further improve our knowledge of cosmological parameters with Type Ia supernovae. While the existence of an energy form with

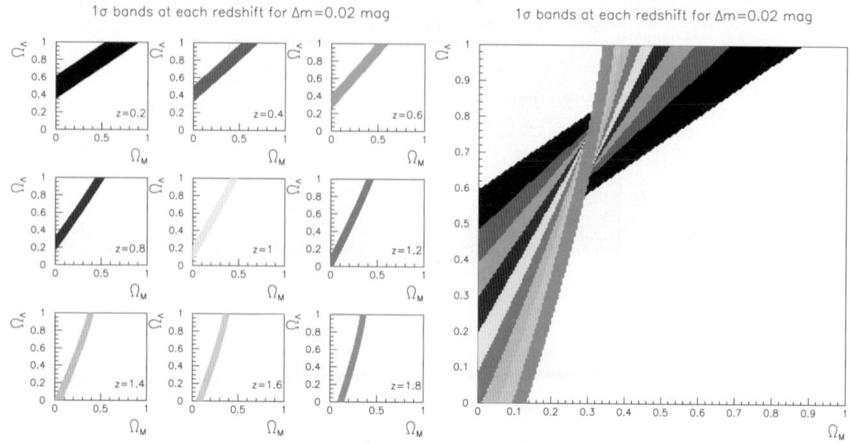

Fig. 3. *Left*: 68% CL-regions in the $\Omega_M - \Omega_\Lambda$ parameter space defined by each redshift bin ($\Delta z = 0.2$) assuming a total uncertainty in the mean brightness of $\Delta m = 0.02$ /bin. *Right*: The bands are superimposed. The resulting CL region is defined by the common area

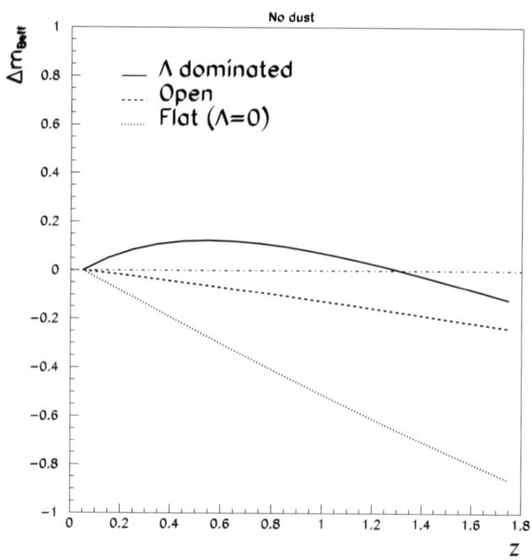

Fig. 4. Differential magnitude for three cosmologies, $\Omega_M, \Omega_\Lambda = (0.3, 0.7)$ (*solid line*), (0.2,0) (*dashed line*) and (1,0) (*dotted line*), compared with an empty universe, $(\Omega_M, \Omega_\Lambda) = (0, 0)$ (*horizontal, dash-dotted line*)

negative pressure is strongly supported by the present data, it is not clear that the Dark Energy really is identical with the cosmological constant. Alternative solutions have been proposed. E.g. Steinhardt [23] suggests that

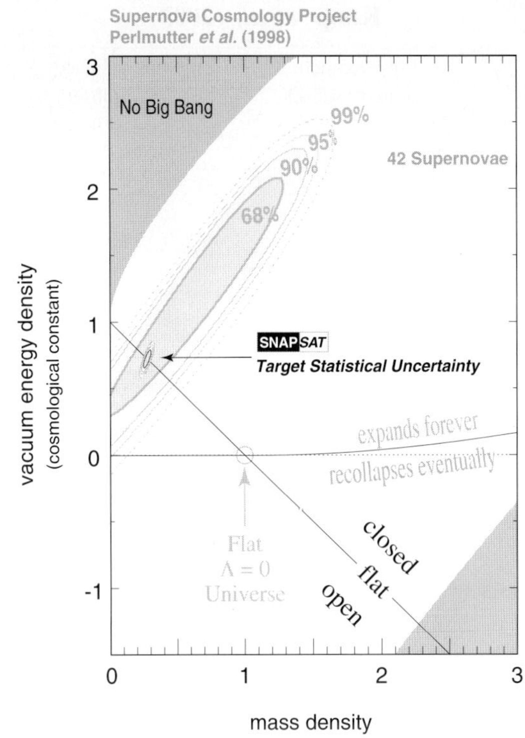

Fig. 5. Target uncertainty for the SNAP satellite experiment (*small ellipses*) compared to the published results in [15]

the effect might be caused by a different type of matter characterized by an equation of state $p = w(z)\rho$, where $w > -1$, as shown in equation 10. The "quintessence" models were proposed to circumvent the two fundamental problems of the cosmological constant: a) a value of $\Omega_\Lambda \sim 0.7$ is about 122 orders of magnitude from the naive theoretical calculation(!) b) It seems somewhat unnatural that we happen to live in a time when $\frac{\Omega_\Lambda}{\Omega_M} \approx 2$ since this ratio depends on the third power of the redshift. For instance, at the epoch of radiation decoupling $\frac{\Omega_\Lambda}{\Omega_M} \sim 10^9$. In "quintessence" models, the "Dark Energy" density tracks the development of the leading energy term making both comparable.

Figure 6 from [12] shows the most recent fits of the Dark Energy state parameter (w) vs Ω_M to the SCP Type Ia SN data assuming that the universe is flat, as indicated by the microwave anisotropy measurements. The cosmological constant ($w = -1$) is comparible with the data, $w = -1.05^{+0.15}_{-0.20}$, yet the uncertainties are still large enough so that a varying dark energy density cannot be excluded.

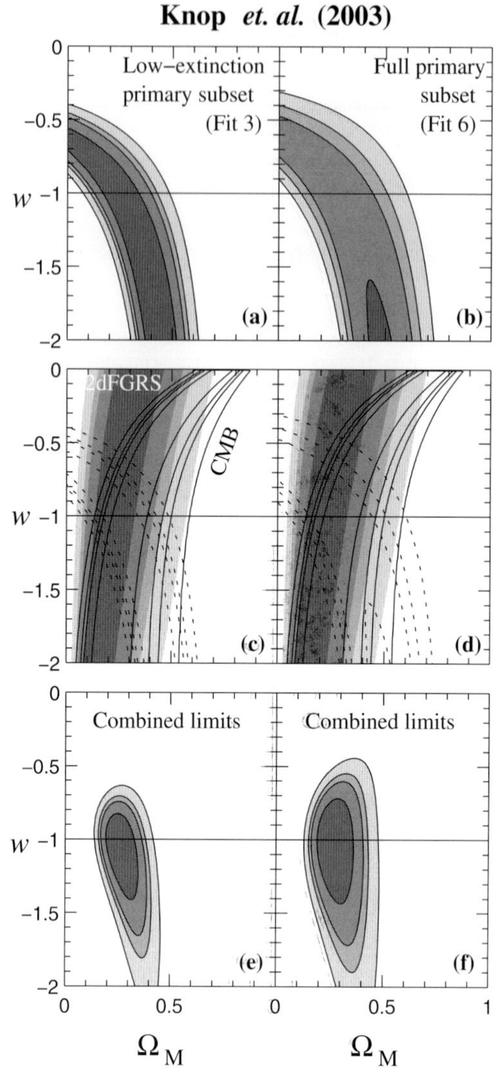

Fig. 6. Joint measurements of Ω_M and constant w assuming a flat universe. The confidence regions plotter are 68%, 90%, 95% and 99%. The left column (panels a, c, and e) shows fits to the low-extinction primary subset of the data in [12]. The right column (panels b, d and f) shows fits to the subset of the data with individual host-galaxy extinction corrections applied to each supernova. The upper panels (and b) show the confidence intervals from the SCP supernovae alone. The middle panels (c and d) overlay this (*dotted lines*) with measurements from 2dFGRS (*filled contours*) [10] and combined CMB measurements (*solid contours*) [22]. The bottom pannels (e and f) combine three confidence regions to provide a combined measurement of Ω_m and w

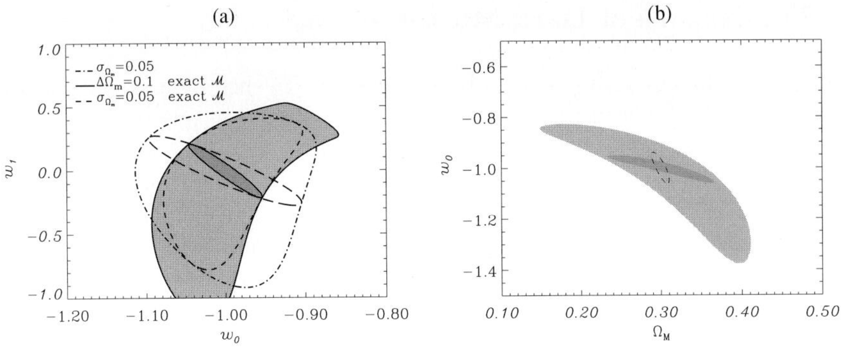

Fig. 7. (**a**) *Left*: 68.3% confidence regions for (w_0, w_1) in the one-year SNAP scenario. The elongated ellipses correspond to the assumption of exact knowledge of Ω_m: the *dash-dot-dot-dotted line* is with exact $\mathcal{M}$ and the *long-dashed line* corresponds to no knowledge of $\mathcal{M}$. The larger, non-elliptic regions assume prior knowledge of Ω_m: the *dash-dotted line* assumes that Ω_m is known with a Gaussian prior for which $\sigma_{\Omega_m - \text{prior}} = 0.05$; the *short-dashed line* assumes the same prior and exact knowledge of $\mathcal{M}$; finally, the *solid line* is with Ω_m confined to the interval $\Omega_m \pm 0.1$ and exact knowledge of $\mathcal{M}$. (**b**) *Right*: 68.3% CL region of $\Omega_\text{M} - w_0$ fit from lensed supernovae in the SNAP 3-yeardata. The dark region shows the smaller confidence region that would result if h would be exactly known from independent measurements. The dashed line shows the expected statistical uncertainty from a 3 year SNAP data sample of Type Ia SNe

The situation becomes even more complicated once we try to measure the time evolution of the equation of state parameter. Assming a linear expansion, $w(z) = w_0 + w_1 \cdot z$, is sufficient for the small redshift range $z < 2$, one additional parameter has to be considered. Figure 7(a) (from [6]) shows the fit of simulated data corresponding to one year of the SNAP satellite. The accuracy on the estimate of the nature of the Dark Energy will depend on independent knowledge, especially, of the Ω_M from e.g. weak lensing measurements. The SNAP satellite, with is large field of view, will also provide extremely accurate measurements of cosmic shear. In addition, dedicated low-z supernova searches will be required in order to bound the intercept of the Hubble diagram, $\mathcal{M}$.

Strongly gravitationally lensed SNe could be detected in large numbers in SNAP, probably on the order of several hundred [8]. Time-delay measurements of lensed SNe are potentially interesting as they provide independent measurements of cosmological parameters, mainly H_0, but also the energy density fractions and the equation of state of dark energy. The results are independent of, and would therefore complement the Type Ia program, as shown in Fig. 7(b).

7 The Nature of Dark Matter

With Type Ia supernovae it may be also possible to shed light on the nature of Dark Matter. Gravitational lensing in the inhomogeneous path that the beam of high-z supernovae follow from the source to us, affects the dispersion of the data points in the Hubble diagram. Thus, with a large sample of high-z supernovae, it is possible to measure the fraction of compact objects in the universe from the residuals of the Hubble diagram. While the compact objects are likely to be of astrophysical nature, e.g. faint stars or black holes, a smooth Dark Matter component would indicate that the missing mass is in the form of particles, such as the lightest stable supersymmetric particles. In [13] we used Monte-Carlo simulations to show that with one year of SNAP data, the fraction of compact objects can be measured with 5% absolute precision.

8 Summary and Conclusions

Observational cosmology is arguably one of the most exciting fields in physics at the moment. Techniques developed during the last years have provided new and unexpected results: the energy density of the universe seems to be dominated by the Einstein's cosmological constant (Λ), or possibly some even more exotic form of *dark energy*. Within the next decade, several measurement techniques are likely to provide conclusive evidence for the nature of the energy form that is currently causing the universe to expand at an accelerated rate, and for ther nature of the *dark matter*, opening a new era of precission observational cosmology.

Acknowledgements

I am very greatful to the organizers for inviting me to such a pleasent conference.

References

1. N. A. Bahcall, and X. Fan, ApJ. 504, 1, 1998.
2. R. G. Carlberg et al., ApJ. 516, 552, 1998.
3. P. de Bernardis et al., Nature 404, 955, 2000.
4. A. Balbi et al., ApJ, 545,1, 2000; erratum: ApJ, 558, 145, 2001.
5. P. Garnavich et al. 1998, ApJ, 493, L53
6. G. Goliath, R. Amanullah, P. Astier, A. Goobar and R. Pain, A&A, 380, 6, 2001.
7. A. Goobar and Perlmutter S., ApJ, 450, 14, 1995.

8. A. Goobar, Mörtsell, E., R.Amanullah, and P. Nugent, 2002, A&A, 393, 25
9. M. Hamuy, M.M. Phillips, J. Maza, N.B. Suntzeff, R.A. Schommer and R. Aviles, 1996, AJ, 112, 2391.
10. E. Hawkins et al., 2003, MNRAS, 346, 78.
11. A. Kim, A. Goobar and S. Perlmutter, PASP, 108, 190, 1996.
12. R. Knop et al., 2003, ApJ, 598, 102.
13. E. Mörtsell, A. Goobar and L. Bergström, 2001, ApJ, 559, 53.
14. J.A. Peacock et al., 2001, Nature, 410, 169.
15. S. Perlmutter et al., 1999, ApJ. 517, 565.
16. S. Perlmutter et al., The SNAP Science Proposal, `http://snap.lbl.gov` .
17. C. Pryke et al., 2002, ApJ, 568, 46.
18. A.G. Riess, W.H. Press and R.P. Kirshner, 1996, ApJ, 473, 88.
19. A.G. Riess et al., 1998, AJ, 116, 1009.
20. A.G. Riess et al. 1999, ApJ. 117,707–724
21. B.P. Schmidt et al. 1998, ApJ, 507, 46.
22. D.N. Spergel et al., 2003, ApJS, 148, 175.
23. P.J. Steinhardt, Proc. of the Nobel Symposium "Particle Physics and the Universe", L. Bergström, P. Carlson and C. Fransson (eds.), T85, 177, 2000.
24. M. Sullivan et al., 2003, MNRAS, 340, 1057.
25. J.L. Tonry et al., 2003, ApJ, 594, 1.

Semiclassical Cosmology with Running Cosmological Constant

Joan Solà

Dep. Estructura i Constituents de la Matèria, Facultat de Física, Universitat de Barcelona, and C.E.R. for Astrophysics, Particle Physics and Cosmology*, Diagonal 647, 08028, Barcelona, Spain
sola@ifae.es

1 Introduction

Astrophysical measurements tracing the rate of expansion of the Universe with high-z Type Ia supernovae indicate that $\Omega_\Lambda^0 \sim 70\%$ of the critical energy density ρ_c^0 of the Universe is cosmological constant (CC) or a dark energy candidate with a similar dynamical impact on the expansion rate of the Universe [1]. Specifically, the CC value experimentally determined from Type Ia supernovae at high-z reads (in natural units) [2]:

$$\Lambda_0 = \Omega_\Lambda^0\,\rho_c^0 \simeq 6\,h_0^2 \times 10^{-47}\,\mathrm{GeV}^4\,. \tag{1}$$

Remarkably, already in 1975 a bound of this order existed in the literature based on structure formation [3]. Notice that the energy scale E_Λ associated to this energy density lies in the millielectronvolt range:

$$E_\Lambda \equiv \Lambda_0^{1/4} = \mathcal{O}\left(10^{-3}\right)\,\mathrm{eV}\,. \tag{2}$$

One would like to have a physical reason for it. In [4] an attempt was made to relate this CC scale to the physics of very light neutrinos in that mass range, which is not excluded at present [5]. As $\Lambda_0 > 0$ our old Universe can be in accelerated expansion. Independent from these supernovae measurements, the CMB anisotropies, including the recent data from the WMAP satellite, lead to $\Omega_0 = 1.02 \pm 0.02$ [6], where $\Omega_0 \equiv \Omega_M^0 + \Omega_\Lambda^0$. Recall that

$$\Omega_M^0 + \Omega_\Lambda^0 + \Omega_K^0 = 1 \tag{3}$$

is the famous sum rule obeyed by the cosmological parameters (see e.g. [1]). It represents a convenient form to rewrite the Friedmann-Lemaître equation,

$$H^2 \equiv \left(\frac{\dot{a}}{a}\right)^2 = \frac{8\pi G}{3}\,(\rho + \Lambda) - \frac{k}{a^2}\,, \tag{4}$$

where the curvature parameter Ω_K^0 in (3) is related to the spatial curvature k in (4) by $\Omega_K^0 = -k/H_0^2\,a_0^2$. When combining the CMB measurement with the

* Associated with Instituto de Ciencias del Espacio-CSIC.

dynamically determined value (from clusters of galaxies) of the matter density (viz. $\Omega_M^0 \simeq 30\%$), leads to an outstanding conclusion: the rest of the present energy budget (a large gap of order 70% of the critical density ρ_c^0) must be encoded in the parameter Ω_Λ^0. In the light of these results, little room is left for our Universe to be spatially curved ($|\Omega_K| \leq 2\%$), and indeed it suggests that our Universe is spatially flat, $k = 0$, as expected from inflation. Hence the CMB measurements and the high-z supernovae data are in concordance with the value of Λ_0 (if one accepts the data on Ω_M^0 from clusters). On the face of it, the situation appears to be quite consistent from the experimental point of view[1].

What about the theoretical situation? In Quantum Field Theory (QFT) we have long expected that the vacuum fluctuations should induce a non-vanishing value for Λ [8, 9], and the question is whether in realistic QFT's we have a prediction for Λ_0 in the ballpark of the measured value (1). Sadly, the answer is no. For, in the context of the Standard Model (SM) of electroweak interactions, this measured CC should be the sum of the original vacuum CC in Einstein's equations, Λ_{vac}, and the induced contribution from the vacuum energy of the quantum fields:

$$\Lambda = \Lambda_{vac} + \Lambda_{ind} \,. \tag{5}$$

What is the expected value for Λ_{ind} in the SM? From the VEV of the Higgs field, $v = \langle\phi\rangle \simeq 246\,\text{GeV}$, and the mass of the Higgs particle, M_H, one can compute the VEV of the Higgs potential, which generates the induced CC [9]. At the tree-level it can be written as follows [10]:

$$\Lambda_{ind} = \langle V_{cl} \rangle = -\frac{1}{8}\, M_H^2\, v^2 \,. \tag{6}$$

From the current LEP 200 numerical bound on the Higgs boson mass, $M_H > 114.1\,\text{GeV}$, one finds $|\Lambda_{ind}| \gtrsim 1.0 \times 10^8\,\text{GeV}^4$. Clearly, $|\Lambda_{ind}|$ is 55 orders of magnitude larger than the observed CC value (1). Moreover, the Higgs potential gets renormalized at higher order in perturbation theory, and therefore it is the value of the effective Higgs potential V_{eff} that matters at the quantum level. The quantum corrections δV by themselves are already much larger than (1). Finally, we also note that in general the induced term may also get contributions from strong interactions, the so-called quark and gluon vacuum condensates. These are also huge as compared to (1), but are much smaller than the electroweak contribution (6).

The incommensurable discrepancy between Λ_{ind} and Λ_0 constitutes the so-called "old" cosmological constant problem (CCP) [9, 11, 12]. It enforces an unnaturally exact fine tuning of the original cosmological term Λ_{vac} in the vacuum action that has to cancel the induced counterpart Λ_{ind} within a precision (in the SM) of one part in 10^{55}. This big conundrum has triggered

[1] See, however, [7] for a more critical attitude.

many theoretical proposals. On the first place there is the longstanding idea of identifying the dark energy component with a dynamical scalar field [13, 14]. More recently this approach took the popular form of a "quintessence" field, χ, slow–rolling down its potential [15], and variations thereof [16]. The main advantage of the quintessence models is that they could explain the possibility of an evolving vacuum energy. This may become important in case such evolution will be someday observed. Furthermore, a plethora of suggestions came along with string theory developments [17] and anthropic scenarios [18].

There are however other (less exotic) possibilities, which should be taken into account. In a series of recent papers [10, 4], the idea has been put forward that already in standard QFT it should not make much sense to think of the CC as a constant, even if taken as a parameter at the classical level, because the Renormalization Group (RG) effects may shift away the prescribed value, in particular if the latter is assumed to be zero. Thus, in the RG approach one takes a point of view very different from the quintessence proposal, as we deal all the time with a "true" cosmological term. It is however a variable one, and therefore a time-evolving, or redshift dependent: $\Lambda = \Lambda(z)$. Although we do not have a QFT of gravity where the running of the gravitational and cosmological constants could ultimately be substantiated, a semiclassical description within the well established formalism of QFT in curved space-time (see e.g. [19, 20]) should be a good starting point. Then, by looking at the CCP from the RG point of view, the CC becomes a scaling parameter whose value should be sensitive to the entire energy history of the Universe – in a manner not essentially different to, say, the electromagnetic coupling constant, $e = e(\mu)$, which runs with the energy scale μ.

The canonical form of renormalization group equation (RGE) for the Λ parameter at high energy is well known – see e.g. [20, 21]. However, at low energy decoupling effects of the massive particles [22] may change significantly the structure of this RGE, with important phenomenological consequences. This idea has been elaborated recently by several authors from various interesting points of view [10, 23, 24]. It is not easy to achieve a RG model where the CC runs smoothly without fine tuning at the present epoch. In [25, 26, 27] a successful attempt in this direction has been made, which is based on the possible existence of physics near the Planck scale. In the following we sketch the main features and implications of this "RG-cosmology".

2 The Model

Consider a free field of spin J, mass M_J and multiplicities $(n_c,\ n_J)$ in an external gravitational field – e.g. $(n_c, n_{1/2}) = (3, 2)$ for quarks, $(1, 2)$ for leptons and $(n_c, n_{0,1}) = (1, 1)$ for scalar and vector fields. At high energy scale μ, the corresponding contribution to the β-function $\mathrm{d}\Lambda/\mathrm{d}\ln\mu$ for the CC is the following [10]:

$$\beta_\Lambda(\mu \gg M_J) \;=\; \frac{(-1)^{2J}}{(4\pi)^2}\,(J+1/2)\,n_c\,n_J\;M_J^4\,. \tag{7}$$

At low energies ($\mu \ll M_J$) this contribution is suppressed due to the decoupling [22]. At an arbitrary scale μ the contribution of a particle with mass M_J should be multiplied by a form factor $F(\mu/M_J)$. At high energies $F(\mu \gg M_J) \simeq 1$ because there must be correspondence between the minimal subtraction scheme and the physical mass-dependent schemes of renormalization at high energies. In the low-energy regime $\mu \ll M_J$ one can expand the function F into powers of μ/M_J:

$$F\Big(\frac{\mu}{M_J}\Big) \;=\; \sum_{n=1}^{\infty} k_n \Big(\frac{\mu}{M_J}\Big)^{2n}\,. \tag{8}$$

Two relevant observations are in order. First, the term $n = 0$ must be absent, because it would lead to the non-decoupling of M_J, with untenable phenomenological implications on the CC value. Indeed, for those terms in the vacuum action where the derivation of the function $F(\mu/M_J)$ is possible [28], the $n = 0$ terms are really absent. Second, in the cosmology context we set the RG scale μ to the value of the expansion parameter H. The latter should define the typical energy range of the cosmological gravitons associated to the FLRW metric at any given cosmological time (see [10]). Hence, general covariance implies that the number of metric derivatives (resulting into powers of H) must be even, and so there are no terms with odd powers in the expansion (8). No other restrictions for the coefficients k_n can be seen. Obviously, in the $H \ll M_J$ regime the most relevant coefficient should be k_1.

At the very low (from the particle physics point of view) energies $\mu = H_0 \sim 1.5 \times 10^{-42}\,$GeV, the relation $(\mu/M_J)^2 \ll 1$ is satisfied for all massive particles: starting from the lightest neutrino, whose presumed mass is $m_\nu \approx 10^{30}\,H_0$, up to the unknown heaviest particle $M_+ \lesssim M_P$. As far as we suppose the usual form of decoupling for the CC, all the contributions in (7) are suppressed by the factor of $(\mu/M_J)^2$. In the rest of this article we develop a cosmological model based on the hypothesis that the overall $n = 1$ coefficient is different from zero, as suggested by naturalness (no fine tuning of the various contributions). Then the total β-function for the CC in the present-day Universe is, in a very good approximation, dominated by a quadratic law of the heaviest masses:

$$\frac{\mathrm{d}\Lambda}{\mathrm{d}\ln\mu} = \beta_\Lambda^t \;\equiv\; \sum_{M_J=m_\nu}^{M_+} \;\beta_\Lambda(M_J) \;\simeq\; \frac{1}{(4\pi)^2}\;\sigma\,M^2\,\mu^2\,. \tag{9}$$

Here M represents the effective high mass scale relevant for the CC value at present, and $\sigma = \pm 1$ indicates the sign of the CC β-function, depending on whether the fermions ($\sigma = -1$) or bosons ($\sigma = +1$) dominate at the highest energies. The quadratic dependence on the masses M_J makes β^t_Λ highly sensitive to the particle spectrum near the Planck scale while the spectrum at lower energies has no impact whatsoever on β^t_Λ.

Having no experimental data about the highest energies, the numerical choice of σM^2 is model-dependent. For example, the fermion and boson contributions in (9) might cancel due to supersymmetry (SUSY) and the total β-function becomes non-zero at lower energies due to SUSY breaking. In this case, the value of M^2 depends on the scale of this breaking, and the sign σ depends on the way SUSY is broken. In particular, the SUSY breaking near the Fermi scale leads to a negligible β^t_Λ, while the SUSY breaking at a GUT scale (particularly at a scale near the Planck mass) provides a significant β^t_Λ. Another option is to suppose some kind of string transition into QFT at the Planck scale. Then the heaviest particles would have the masses comparable to the Planck mass M_P and represent the remnants, e.g., of the massive modes of a superstring. Let us clarify that the mass of each particle may be indeed smaller than M_P, and the equality, or even the effective value $M \gtrsim M_P$, can be achieved due to the multiplicities of these particles. With these considerations in mind, our very first observation is that the natural value of the β-function (9) at the present time is

$$\left|\beta^t_\Lambda\right| = \frac{1}{(4\pi)^2}\, M^2 \cdot \mu^2 = \frac{c}{(4\pi)^2}\, M_P^2 \cdot H_0^2 \sim 10^{-47}\,\mathrm{GeV}^4\,, \tag{10}$$

where c is some coefficient. For $c = \mathcal{O}(1-10)$ the β-function for Λ is very much close to the value of Λ itself at present, (1). Therefore, our RGE (9) suggests that the natural energy scale E_Λ associated to the present value of the CC – see (2) – arises from the geometrical mean of two extreme scales in our Universe: H_0 (the value of μ at present) and M_P. Indeed,

$$E_\Lambda \equiv \Lambda_0^{1/4} \simeq \sqrt{M_P\, H_0} = \mathcal{O}\left(10^{-3}\right)\,\mathrm{eV}\,. \tag{11}$$

On the other hand the Friedmann-Lemaître equation (4) determines another similar scale in our present Universe, even in the absence of Λ. Assuming $k=0$ in (4), the energy scale associated to the matter density at present is $\left(\rho_M^0\right)^{1/4} \sim \sqrt{M_P\, H_0}$, where we used $G = 1/M_P^2$. Remarkably enough, our basic RGE (9) leads to (10), (11) independently of the Friedmann-Lemaître equation, and therefore our framework helps to explain the coincidence between the matter density and the CC at the present time, providing a reason for the characteristic millielectronvolt energy scale (2) common to both: $\Lambda_0^{1/4} \sim \left(\rho_M^0\right)^{1/4} \sim 10^{-3}\,\mathrm{eV}$.

Consider the implications for the Friedmann-Lemaître-Robertson-Walker (FLRW) cosmological models coupled to to our basic RGE (9). The first step is to derive the CC dependence from the redshift parameter z, defined as $1+z = a_0/a$, where a_0 is the present-day scale factor. Using the identification of the RG scale μ with H, we reach the equation

$$\frac{\mathrm{d}\Lambda}{\mathrm{d}z} = \frac{1}{H}\frac{\mathrm{d}H}{\mathrm{d}z}\,\beta_\Lambda^t = \frac{1}{2}\frac{\sigma\,M^2}{(4\pi)^2}\frac{\mathrm{d}H^2}{\mathrm{d}z}\,. \tag{12}$$

In order to construct the cosmological model, we shall use, along with (12), the Friedmann-Lemaître equation (4). Furthermore, the energy conservation law provides the third necessary equation

$$\frac{\mathrm{d}\Lambda}{\mathrm{d}t} + \frac{\mathrm{d}\rho}{\mathrm{d}t} + 3H\,(\rho + p) = 0\,, \tag{13}$$

where p is the matter/radiation pressure. As we shall consider both MD (matter dominated) and RD (radiation dominated) regimes, it is useful to solve the equations (12), (4), (13) using an arbitrary equation of state $p = \alpha\rho$, with $\alpha = 0$ for MD and $\alpha = 1/3$ for RD. The time derivative in (13) can be easily traded for a derivative in z via $\mathrm{d}/\mathrm{d}t = -H\,(1+z)\,\mathrm{d}/\mathrm{d}z$. Hence we arrive at a coupled system of ordinary differential equations in the z variable. The solution for the matter-radiation energy density and CC is completely analytical. For the remaining of the paper we shall present the results only for the spatially flat case ($k = 0$ in (4))[2]. For the MD epoch it reads as follows:

$$\rho(z;\nu) = \rho_M^0\,(1+z)^\zeta\,; \quad \Lambda(z;\nu) = \Lambda_0 + \rho_M^0\,\frac{\nu}{1-\nu}\,\left[(1+z)^\zeta - 1\right]\,. \tag{14}$$

Here we have introduced the following dimensionless coefficients:

$$\nu = \frac{\sigma\,M^2}{12\pi M_P^2}\,, \qquad \zeta = 3\,(1-\nu)\,(\alpha+1)\,. \tag{15}$$

In the limit $\nu \to 0$ we recover the standard result for $\rho(z)$ with constant Λ [1]. The formulas above represent the universal solution at low energies, when all massive particles decouple according to (8). Substituting the above obtained density functions $\rho(z;\nu)$ and $\Lambda = \Lambda(z;\nu)$ in (4) we get the explicit ν-dependent Hubble function $H(z;\nu)$. For flat universes it gives

$$H^2(z;\nu) = H_0^2\left\{1 + \Omega_M^0\,\frac{(1+z)^{3\,(1-\nu)} - 1}{1-\nu}\right\}\,, \tag{16}$$

whereas the standard result in this case is [1]

$$H^2(z) = H_0^2\,\left\{1 + \Omega_M^0\,\left[(1+z)^3 - 1\right]\right\}\,, \tag{17}$$

which is indeed recovered from (16) in the limit $\nu = 0$.

Consider next the nucleosynthesis epoch when the radiation dominates over the matter, and derive the restriction on the single (independent) parameter ν of our model. In the RD regime, the solution for the density (14) can be rewritten in terms of the temperature as

[2] See [27] for more general formulae.

$$\rho_R(T) = \frac{\pi^2}{30}\, g_*\, \left(r^{-\nu} T\right)^4\,, \tag{18}$$

with $r \equiv T/T_0$, $T_0 \simeq 2.75\,K = 2.37 \times 10^{-4}\,\mathrm{eV}$ being the present CMB temperature, $g_* = 2$ for photons and $g_* = 3.36$ if we take the neutrinos into account. It is easy to see that the size of the parameter ν gets restricted, because for $\nu \geq 1$ the density of radiation would be the same or even below the one at the present Universe. On the other hand near the nucleosynthesis time we have $T \gg T_0$, and then one obviously has

$$\Lambda_R(T) \simeq \frac{\nu}{1-\nu}\, \frac{\pi^2}{30}\, g_*\, \left(r^{-\nu} T\right)^4\,. \tag{19}$$

It follows that in order not to be ruled out by the nucleosynthesis, the ratio of the CC and the energy density at that time has to satisfy

$$|\,\Lambda_R\,/\,\rho_R\,| \sim |\,\nu\,/\,(1-\nu)\,| \simeq |\nu| \ll 1\,. \tag{20}$$

A nontrivial range could e.g. be $0 < |\nu| \leq 0.1$. In view of the definition (15), this implies $M \lesssim 2M_P$. Hence, the nucleosynthesis constraint coincides with our general will to remain in the framework of the effective approach. It is remarkable that the two constraints, which come from very different considerations, lead to the very same restriction on the unique free parameter of the model. The canonical choice $M = M_P$, corresponds to

$$|\nu| = \nu_0 \equiv \frac{1}{12\,\pi} \simeq 2.6 \times 10^{-2}\,. \tag{21}$$

3 Numerical Results

After the nucleosynthesis restriction on the ν-parameter, the question is whether there is still some room for useful phenomenological considerations at the present matter epoch. Fortunately, the answer is yes. The evolution of the matter density and of the CC is shown in detail in Fig. 1a,b for the flat case. These graphics illustrate (14). As a result of allowing a non-vanishing β_Λ-function for the CC (equivalently, $\nu \neq 0$) there is a simultaneous, correlated variation of the CC with the matter density. In the phenomenologically most interesting case $|\nu| < 1$ we always have a null density of matter and a finite (positive) CC in the long term future, while for the far past yields $\Lambda = \pm\infty$ depending on the sign of ν. In all these situations the matter density safely tends to $+\infty$. One may worry whether having infinitely large CC and matter density in the past may pose a problem to structure formation. From Fig. 1a,b it is clear that there should not be a problem at all since in our model the CC remains always smaller than the matter density in the far past, and in the radiation epoch, say $z > 1000$, we reach the safe limit (20). Actually the time where $\Lambda(z;\nu)$ and $\rho_M(z;\nu)$ become similar is very recent.

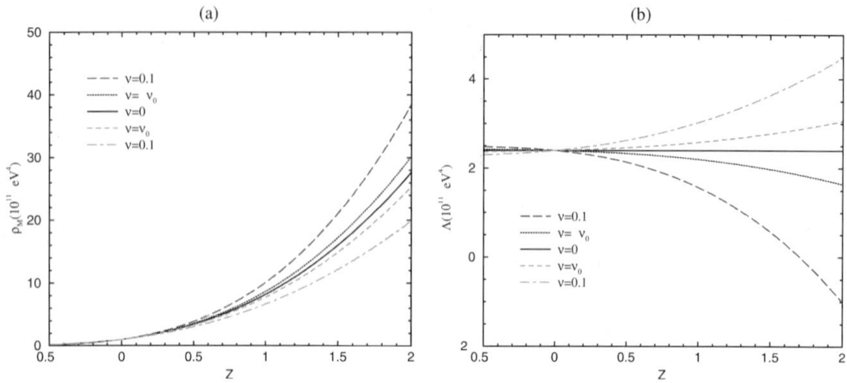

Fig. 1. Future and past evolution of the matter density $\rho_M(z;\nu)$, (**a**), and the cosmological constant $\Lambda(z;\nu)$, (**b**), for a flat Universe ($k = 0$) and for different values of the fundamental parameter ν of our model (ν_0 is defined in (21)). In both cases $\Omega_M^0 = 0.3$ and $\Omega_\Lambda^0 = 0.7$, with $h_0 = 0.65$. The $\nu = 0$ line represents the standard model case, and the remaining curves represent deviations from this case

It should be clear that our approach based on a variable CC departs from all kind of quintessence-like approaches, in which some slow–rolling scalar field χ substitutes for the CC. In these models, the dark energy is tied to the dynamics of the self-conserved χ field; i.e. in contrast to (13) there is no transfer between χ-dark energy and ordinary forms of energy. The phenomenological equation of state is defined by $p_\chi = w_\chi \rho_\chi$. In order to get accelerated expansion in an epoch characterized by $p = 0$ and $\rho \to 0$ in the future, one must require $-w_- \leq w_\chi \leq -1/3$, where usually $w_- \geq -1$ in order to have a canonical kinetic term for χ[3]. The particular case $w_\chi = -1$ corresponds to a quintessence field exactly mimicking the cosmological constant term. Although p_χ and ρ_χ are related to the energy-momentum tensor of χ, the dynamics of this field is unknown because the quintessence models do not have an explanation for the value of the CC. Comparing with the standard model case $\nu = 0$ (cf. Fig. 1a,b), we see that for a negative cosmological index ν the matter density grows faster towards the past ($z \to \infty$) while for a positive value of ν the growing is slower than the usual $(1+z)^3$. Looking towards the future ($z \to -1$), the distinction is not appreciable because for all ν the matter density goes to zero. The opposite result is found for the CC, since then it is for positive ν that $\Lambda(z;\nu)$ grows in the past, whereas in the future it has a different behavior, tending to different (finite) values in the cases $\nu < 0$ and $0 < \nu < 1$, while it becomes $-\infty$ for $\nu \geq 1$ (not shown).

One can use the so-called magnitude-redshift relation [1] to test our model from various simulated distributions of Type Ia supernovae at high redshift, including the one foreseen by SNAP [30]. This analysis was performed in

[3] One cannot completely exclude "phantom matter-energy" ($w_- < -1$) and generalizations thereof [17].

Table 1. Determination of ν with SNAP data and with other two distributions. In all cases we assume a flat Universe. When a prior on Ω_M and its error σ_{Ω_M} is assumed, we use $\Omega_M = 0.3 \pm 0.03$. See [27]

Distribution	Data	σ_{Ω_M}	ν	σ_ν
SNAP (1 year)	50 SNe $0 < z < 0.2$ 1800 SNe $0.2 < z < 1.2$ 50 SNe $1.2 < z < 1.4$ 15 SNe $1.4 < z < 1.7$	None	0.1	± 0.10
SNAP	as above	0.03	0.1	± 0.06
SNAP	3 years	None	0.1	± 0.06
SNAP	3 years	0.03	0.1	± 0.04
Distr.1	50 SNe $0 < z < 0.2$ 2000 SNe $0.2 < z < 1.7$	0.03	0.1	± 0.05
Distr.2	250 SNe $0 < z < 1$ 1750 SNe $1 < z < 2$	0.03	0.1	± 0.02

[27]. Table 1 summarizes the results that we obtained. Specifically we used the magnitude data from the SCP (Supernova Cosmology Project) [2]. The set included 16 low-redshift supernovae from the Calán/Tololo survey and 38 high-redshift supernovae used in the main fit of [2]. In order to determine the cosmological parameters we have performed a χ^2-statistic test, where χ^2 is defined by the difference between the theoretical apparent magnitude and the observed one (see [27] for details). The existing sample of SN Ia data is amply compatible with $\nu \neq 0$, but it does not pin down a narrow interval of values for this parameter. The results of the fit for the parameter ν for the SNAP distribution, with and without prior on the value of Ω_M^0, are shown in Fig. 12 of [27]. The results from alternative distributions are shown in Fig. 13 of that reference and quantified in Table 1. Fits were made under the same assumptions and so the results represent the expected accuracy in the parameters using all the information to come. From SNAP we can determine ν to ± 0.06 for $\nu = 0.1$ for a given prior $\Omega_M = 0.3 \pm 0.03$. Distribution 1 in Table 1 is very similar to the SNAP one but with most of the data homogeneously distributed between $z = 0.2$ and $z = 1.7$. Distribution 2 extends data up to redshift $z = 2$. We see that we can determine ν to within $\pm(20 - 60)\%$ for $\nu = 0.1$, depending on the distribution. Smaller values of ν imply smaller precision. The situation is similar to the determination of the evolution of the equation of state for the quintessence field χ, which can be parametrized in terms of two parameters as follows:

$$\frac{p_\chi}{\rho_\chi} \equiv w_\chi = w_0 + w_1(1-a) = w_0 + w_1 \frac{z}{1+z} \,. \tag{22}$$

Finding a non-vanishing value of w_1 implies a redshift evolution of the equation of state for the χ field [4]. For completeness, consider the modification on the Hubble parameter introduced by quintessence models. In the flat case it is easy to compute

$$H^2(z; w_0, w_1) = H_0^2 \left\{ \Omega_M^0 \,(1+z)^3 \right. \\ \left. + (1-\Omega_M^0)(1+z)^{3(1+w_0+w_1)} \exp\left[-3\, w_1 \frac{z}{1+z}\right] \right\} . \tag{23}$$

This equation reduces to the standard one, (17), for $w_\chi = -1$ ($w_0 = -1, w_1 = 0$), as expected. Comparison between (23) and (16) can be useful to identify the differences between RG models and quintessence models of the dark energy, as these formulae enter directly in the data fits.

If one performs a general (model-independent) fit of the present SN Ia data to quintessence models, leaving free the two parameters w_0 and w_1 in (23), one finds that values $w_0 > -1/3$ (decelerated Universe) are ruled out at a high significance level (for $\Omega_M^0 < 0.4$), thereby supporting the existence of dark energy. Nevertheless, the very same fit is highly insensitive to w_1 [31, 32]. On the other hand SNAP will be able to determine (Ω_M^0, w_0) to within small errors (3%, 5%), and will significantly improve the determination of the time-variation parameter w_1, but only up to 30% at most [33]. The situation with quintessence is therefore comparable to the determination of the ν parameter in our RG-cosmology.

4 Conclusions

I have discussed a semiclassical FLRW type of cosmological model based on a running cosmological constant Λ with the scale $\mu = H$. If the decoupling quantum effects on Λ have the usual form as for the massive fields, then we can get a handle on the variation of Λ at infrared energies without resorting to any low-energy *ad hoc* scalar field (quintessence and the like). The CC is mainly driven, without fine tuning, by the "relic" quantum effects from the physics of the highest available scale (the Planck scale), and its value naturally lies in the acceptable range. It is remarkable that all relevant information about the unknown world of the high energy physics is accumulated into a single parameter ν. Furthermore, we have shown that the next generation of supernovae experiments, like SNAP, should be sensitive to ν within its allowed range. A non-vanishing value of ν produces a cubic dependence of the CC on z at high redshift, which should be well measurable by that

[4] The difficulties of measuring the parameter w_1 are well known, see e.g. [31, 32].

experiment, if it is really there. If these experiments will detect the redshift dependence of the CC similar to that which is predicted in our work, we may suspect that some relevant physics is going on just below the Planck scale. If, on the contrary, they unravel a static CC, this may imply the existence of a desert in the particle spectrum below the Planck scale, which would be no less noticeable. In this respect let us not forget that the popular notion of GUT's (perhaps in the form of string physics) near the Planck scale remains, at the moment, as a pure (though very much interesting!) theoretical speculation, which unfortunately is not supported by a single piece of experimental evidence up to now. Our framework may allow to explore hints of these theories directly from astrophysical/cosmological experiments which are just round the corner. If the results are positive, it would suggest a quantum field theoretical link between the largest scales in cosmology and the shortest distances in high energy physics.

Acknowledgements

Above all I am obliged to Ilya L. Shapiro for sharing his knowledge on this subject on which we have been working together intermittently for many years. Also to E.V. Gorbar, B. Guberina, M. Reuter and H. Stefancic for fruitful discussions. I am grateful to S. Béjar for his help in doing the plots of Fig. 1, and to C. España-Bonet and P. Ruiz-Lapuente for their collaboration in the numerical analysis of [27]. This work has been supported in part by MECYT and FEDER under project FPA2001-3598. Last, but not least, the author wishes to express his gratitude to the organizers of the workshop for the kind invitation to this magnificent, fully Mediterranean, conference and for the generous financial support provided.

References

1. P.J.E. Peebles, *Principles of Physical Cosmology* (Princeton Univ. Press, 1993); T. Padmanabhan, *Structure Formation in the Universe* (Cambridge Univ. Press, 1993); E.W. Kolb, M.S. Turner, *The Early Universe* (Addison-Wesley, 1990); J.A. Peacock, *Cosmological Physics* (Cambridge Univ. Press, 1999).
2. S. Perlmutter et al., *Astrophys. J.* **517** (1999) 565; A.G. Riess et al., *Astronom. J.* **116** (1998) 1009.
3. Ya.B. Zeldovich, I.D. Novikov, *Structure and evolution of the universe* (Moscow, Izdatel'stvo Nauka, 1975).
4. I. Shapiro and J. Solà, *Phys. Lett.* **475** (2000) 236.
5. M. Maltoni, T. Schwetz, M.A. Tortola, J.W.F. Valle, `hep-ph/0305312`
6. P. de Bernardis et al., *Nature* **404** (2000) 955; C.L. Bennett et al., `astro-ph/0302207`; D.N. Spergel et al., `astro-ph/0302209`.
7. A. Blanchard, M. Douspis, M. Rowan-Robinson, S. Sarkar, `astro-ph/0304237`.
8. Ya.B. Zeldovich, *Letters to JETPh.* **6** (1967) 883.

9. S. Weinberg, *Rev. Mod. Phys.,* **61** (1989); *ibid.* Relativistic Astrophysics, ed. J.C. Wheeler and H. Martel, *Am. Inst. Phys. Conf. Proc.* **586** (2001) 893.
10. I. Shapiro and J. Solà, *JHEP* 0202 (2002) 006.
11. V. Sahni, A. Starobinsky, *Int. J. of Mod. Phys.* **9** (2000) 373, `astro-ph/9904398`; S.M. Carroll, *Living Rev. Rel.* **4** (2001) 1 `astro-ph/0004075`; T. Padmanabhan, *Phys. Rept.* **380** (2003) 235, `hep-th/0212290`; M.S. Turner, *Int. J. Mod. Phys.* **A17** (2002) 3446, `astro-ph/0202007`.
12. J. Solà, *Nucl. Phys. Proc. Suppl.* **95** (2001) 29.
13. A.D. Dolgov, in: *The very Early Universe*, Ed. G. Gibbons, S.W. Hawking, S.T. Tiklos (Cambridge U., 1982).
14. R.D. Peccei, J. Solà and C. Wetterich, *Phys. Lett.* **B 195** (1987) 183; L.H. Ford, *Phys. Rev.* **D 35** (1987) 2339; C. Wetterich, *Nucl. Phys.* **B 302** (1988) 668; J. Solà, *Phys. Lett.* **B 228** (1989) 317; J. Solà, Int.J. Mod. Phys. A5 (1990) 4225.
15. R.R. Caldwell, R. Dave, P.J. Steinhardt, *Phys. Rev. Lett.* **80** (1998) 1582; P.J.E. Peebles, B. Ratra, *Astrophys. J. Lett.* **325** L17 (1988).
16. P.J.E. Peebles, B. Ratra, *Rev. Mod. Phys.* **75**(2003) 599.
17. E. Witten, in: Sources and detection of dark matter and dark energy in the Universe, ed. D.B. Cline (Springer, Berlin, 2001), p. 27.; L. Randall and R. Sundrum, *Phys. Rev. Lett.* **83** (1999) 4690.
18. J.D. Barrow, F.J. Tipler, *The Anthropic Cosmological Principle* (Clarendon Press, Oxford, 1986); S. Weinberg, *Phys. Rev.* **D61** (2000) 103505.
19. N.D. Birrell and P.C.W. Davies, *Quantum Fields in Curved Space*, Cambridge Univ. Press (Cambridge, 1982).
20. I.L. Buchbinder, S.D. Odintsov and I.L. Shapiro, *Effective Action in Quantum Gravity*, IOP Publishing (Bristol, 1992).
21. L. Brown, *Quantum Field Theory* (Cambridge Univ. Press, 1992).
22. T. Appelquist and J. Carazzone, *Phys. Rev.* **D11** (1975) 2856.
23. A. Babic, B. Guberina, R. Horvat and H. Stefancic, *Phys. Rev.* **D65** (2002) 085002; B. Guberina, R. Horvat, H. Stefancic *Phys. Rev.* **D67** (2003) 083001.
24. E. Bentivegna, A. Bonanno, M. Reuter, *JCAP* **01** (2004) 001, `astro-ph/0303150`; A. Bonanno, M. Reuter, *Phys. Rev.* **D 65** (2002) 043508.
25. I.L. Shapiro, J. Solà, C. España-Bonet, P. Ruiz-Lapuente, *Phys. Lett.* **B 574** (2003) 149.
26. I.L. Shapiro and J. Solà, *Cosmological constant, renormalization group and Planck scale physics*, *Nucl. Phys. B Proc. Supp.* **127** (2004) 71, `hep-ph/0305279`.
27. C. España-Bonet, P. Ruiz-Lapuente, I.L. Shapiro, J. Solà, *Testing the running of the cosmological constant with Type Ia Supernovae at high-z*, *JCAP* **02** (2004) 006, `hep-ph/0311171`,
28. E.V. Gorbar, I.L. Shapiro, *JHEP* 02 (2003) 021 and *JHEP* 06 (2003) 004.
29. R.R. Caldwell, *Phys. Lett.* **545** (2002) 23; H. Stefancic, `astro-ph/0310904` and `astro-ph/0312484`.
30. See all the information in: http://snap.lbl.gov/.

31. J. A. Frieman, D. Huterer, Eric V. Linder, M. S. Turner, *Phys. Rev.* **D67** (2003) 083505, astro-ph/0208100.
32. T. Padmanabhan, T.R. Choudhury, *Mon. Not. Roy. Astron. Soc.* **344** (2003)823, astro-ph/0212573; T. D. Saini, T. Padmanabhan, S. Bridle, *Mon. Not. Roy. Astron. Soc.* **343** (2003) 533, astro-ph/0301536.
33. E. V. Linder *AIP Conf. Proc.*, **655** (2003) 193.

Limits on New Inverse-Power Law Forces

Dennis E. Krause[1,2] and Ephraim Fischbach[2]

[1] Department of Physics, Wabash College, Crawfordsville, IN 47933
krausedOwabash.edu

[2] Physics Department, Purdue University, West Lafayette, 47907
ephraim@physics.purdue.edu

1 Introduction

In recent years table-top force experiments have proven to be useful tools in the search for physics beyond the Standard Model. Many theoretical models which hope to extend the Standard Model predict the existence of new forces or extra dimensions that might produce deviations from known force laws. It is remarkable that in many of these theories the new potential energies come in either of two forms, Yukawas or inverse-power laws (IPLs). While much attention has been focused on the search for new Yukawa forces, IPL forces are also of significant theoretical interest as will be shown below.

The goal of this paper is to discuss the current limits on IPL forces from experiments probing a wide range of length scales. We will begin by listing examples of theoretical models which give rise to IPL potentials. This will suggest the appropriate phenomenology to be used in analyzing the results of experiments searching for new IPL forces. Then, after reviewing current constraints on new IPL forces, we will discuss future work for obtaining more stringent limits.

2 Examples and Phenomenology of Inverse Power Law Forces

In classical physics the most familiar IPL interactions are the electrostatic and gravitational forces which have potential energies that are proportional to $1/r$. One can show in general that the $1/r$ potentials arise whenever two particles exchange a single massless boson [1]. For example, the exchange of a massless spin-0 or spin-2 boson leads to a purely attractive potential energy of the form

$$V_{S,T}(r) = -\frac{g_1 g_2}{4\pi}\frac{1}{r}\,, \qquad \text{[1-Boson (Scalar, Tensor) Exchange]}\,, \tag{1}$$

where g_i is the coupling of the ith particle to the boson. (Unless stated explicitly, we have set $\hbar = c = 1$.) The electrostatic potential is an example of an interaction arising from the exchange of a massless vector (spin-1 boson), and has the general form

$$V_V(r) = \pm\frac{f_1 f_2}{4\pi}\frac{1}{r}\,, \qquad \text{[1-Boson (Vector) Exchange]}\,, \tag{2}$$

where the interaction may be attractive or repulsive, and f_i is the coupling strength (charge).

While $1/r$ potentials are most familiar, there are many examples of potentials of the more general form

$$V_n(r) \propto \frac{1}{r^n}\,, \tag{3}$$

where n is an integer. Here is a partial list for $n > 1$:
$n = 2$:

$$V_s^{(4a)}(r) = -\frac{g_1^2 g_2^2}{16\pi^2}\left[\frac{m_1^2+m_1m_2+m_2^2}{m_1m_2(m_1+m_2)}\right]\frac{1}{r^2} \qquad \text{2-Scalar Exchange [2]} \tag{4}$$

$$V_{2\gamma}^{(4a)}(r) = +\frac{e_1^2e_2^2}{32\pi^2(m_1+m_2)}\frac{1}{r^2} \qquad \text{2-Photon Exchange [2]} \tag{5}$$

$$V_{\text{gravity}}(r) = -\frac{3G^2(m_1+m_2)}{c^2}\frac{1}{r^2} \qquad \text{General Relativity [3]} \tag{6}$$

$n = 3$:

$$V_{RS}(r) = -\frac{2Gm_1m_2}{3k^2}\frac{1}{r^3} \qquad \text{Randall-Sundrum Model [4]} \tag{7}$$

$$V_{\text{quantum}}(r) = -\frac{41Gm_1m_2}{10\pi}\frac{L_{Pl}^2}{r^3} \qquad \text{Quantum Gravity [3]} \tag{8}$$

$$V_{ps}^{(4)}(r) = -\frac{g_1^2g_2^2}{64\pi^3 m_1m_2}\frac{1}{r^3} \qquad \text{2-Pseudoscalar Exchange (Yukawa coupling) [5, 6]} \tag{9}$$

$$V_s^{(4b)}(r) = -\frac{g_1^2g_2^2}{16\pi^2 m_1m_2}\frac{1}{r^3} \qquad \text{2-Scalar Exchange [2]} \tag{10}$$

$$V_{2\gamma}^{(4b)}(r) = -\frac{7}{96\pi^3}\frac{e_1^2e_2^2}{m_1m_2}\frac{1}{r^3} \qquad \text{2-Photon Exchange [2]} \tag{11}$$

$n = 5$:

$$V_{ps}^{(4)}(r) = +\frac{3g_1^2g_2^2}{128\pi^3m_1^2m_2^2}\frac{1}{r^5} \qquad \text{2-Pseudoscalar Exchange (derivative coupling) [6]} \tag{12}$$

$$V_{\nu\bar{\nu}}(r) = \frac{G_F^2}{4\pi^3}\frac{1}{r^5} \qquad \text{2-Neutrino Exchange [7]} \tag{13}$$

$n = 6, 7$:

$$V_{2\gamma}(r) \sim \begin{cases} -\dfrac{e^2 a^5}{r^6} & \text{(non-retarded)} \\ -\dfrac{e^2 a^6}{r^7} & \text{(retarded)} \end{cases} \quad \text{2-Photon Exchange (neutral atoms) [8]} \tag{14}$$

$n = 11, 13$:

$$V_{3\gamma}(r) \sim \begin{cases} -\dfrac{e^2 a^{10}}{r^{11}} & \text{(non-retarded)} \\ -\dfrac{e^2 a^{12}}{r^{13}} & \text{(retarded)} \end{cases} \quad \text{3-Photon Exchange (neutral atoms) [8]} \tag{15}$$

The possibility that there are large extra compact spatial dimensions implies that the inverse power behavior of Newtonian gravity will also depend on N, the number of these new dimensions. For two point masses separated by a distance r, one finds that the gravitational potential energy is given by [9]:

$$V_{\text{gravity}}(r) = \begin{cases} -\dfrac{Gm_1 m_2}{r}, & r \gg R\,, \\ -\dfrac{G_{4+N} m_1 m_2}{r^{1+N}}, & r \ll R\,, \end{cases} \tag{16}$$

where R is the size of the extra dimensions, G is the usual Newtonian gravity constant, and G_{4+N} is the gravity constant in $4 + N$ spacetime dimensions ($G = G_4$). In such models, it is convenient to express the gravitational constants in terms of the characteristic mass scales: $G = 1/M_{Pl}$, where $M_{Pl} \sim 10^{19}$ GeV, and $G_{4+N} = 1/M_*^{2+N}$, where $M_* \sim 1$ TeV. By using these mass scales and equating the two different forms of $V_{\text{gravity}}(r)$ at $r = R$, one obtains the relations

$$G_{4+N} = GR^N \tag{17}$$

$$M_{Pl}^2 = M_*^{2+N} R^N\,. \tag{18}$$

Therefore, one can write

$$V_{\text{gravity}}(r) = -\frac{G_{4+N} m_1 m_2}{r^{1+N}} = -\frac{Gm_1 m_2}{r}\left(\frac{R}{r}\right)^N, \quad r \ll R\,, \tag{19}$$

where the size of the extra dimensions depends on N:

$$R = \frac{M_{Pl}^2}{M_*^{2+N}} \sim 10^{(32/N)-19}\ \text{m}\,. \tag{20}$$

It is interesting to note that in each of these examples, the IPL potential may be written in the form,

$$V_n(r) \propto \frac{1}{r}\left(\frac{r_0}{r}\right)^{n-1} , \tag{21}$$

where r_0 is some distance scale associated with the interacting system. In most cases, $r_0 \sim 1/m$, where m is the mass of the interacting particles. In Randall-Sundrum model, the IPL correction to Newtonian gravity (7) has the length scale $r_0 \sim 1/k$, where $1/k$ is the scale at which space is warped. The quantum gravity correction to Newtonian gravity involves the Planck length L_{Pl}, so $r_0 \sim L_{Pl}$ For the 2-neutrino-exchange potential given by (13), $r_0 \sim \sqrt{G_F}$, where G_F is the Fermi constant. The dispersion potentials given by (14) and (15), which arise from an induced electric dipole moment, depend on a, the size of the interacting atoms. Finally, the extra dimensional-modified gravity potential given by (19) depends on the size of the extra dimensions.

When searching for IPL forces, one typically wishes to express an IPL as generally as possible so that a given experiment can be used to constrain a broad range of theoretical models. Presently two different phenomenological expressions of the general IPL potential $V_n(r)$ are used, both taking advantage of generic form exhibited by (21). Since most new interactions couple to nucleons, the first phenomenological form of the general IPL potential for two particles separated by a distance r is written as [20]

$$V_n^{(1)}(r) = \Lambda_n \frac{B_1 B_2}{r}\left(\frac{r_0}{r}\right)^{n-1} , \tag{22}$$

where Λ_n is a dimensionless constant and B_i is the baryon number of the ith particle. In addition, it is customary to define the length scale r_0 as

$$r_0 \equiv 10^{-15}\ \mathrm{m} , \tag{23}$$

since in this case, $r_0 \sim 1/m_N$, where m_N is the mass of the nucleon. As seen in the examples listed above, the form given by (22) is naturally suited to characterize IPL forces between nucleons, since in this case typically $r_0 \sim 1/m_N$. This may seem overly restrictive, but for the case of macroscopic bodies in which the force couples only to electrons, one can usually set $B_i \simeq 2Z_i$, where Z_i is the atomic number of the ith particle.

In order to avoid difficulties when dealing with couplings that are not proportional to B or Z, a second phenomenological form is used in which the coupling is proportional to an interacting particle's mass. In this case, one writes [21, 22]

$$V_n^{(2)}(r) = \alpha_n \left(\frac{G m_1 m_2}{r}\right)\left(\frac{r_0}{r}\right)^{n-1} , \tag{24}$$

where α_n is a dimensionless constant, G is the usual Newtonian gravitational constant, and the length scale r_0 is still defined by (23). By equating (22) with (24), one obtains the conversion formula

$$\alpha_n \simeq \Lambda_n \left(\frac{M_{Pl}}{m_H}\right)^2 = 1.7 \times 10^{38} \Lambda_n \ , \tag{25}$$

where $M_{Pl} = \sqrt{1/G} = 2.18 \times 10^{-8}$ kg is the Planck mass, and $m_H = 1.67 \times 10^{-27}$ kg is the mass of ${}_1\mathrm{H}^1$.

The fact that a distance scale r_0 is associated with IPL potentials seems at odds with the fact that power-law potentials are scale-invariant. To understand how both can be true, it proves useful to contrast the IPL potential given by (24) with the analogous Yukawa potential between two particles, which may be written as [1]

$$V_Y(r) = \alpha_Y \left(\frac{Gm_1m_2}{r}\right) \mathrm{e}^{-r/\lambda} \ . \tag{26}$$

Like (24), the Yukawa potential given by (26) has a dimensionless strength constant α_Y and a length scale λ, but this is were the similarities end. In a force experiment, one can explore separately the dependences of α_Y and λ, but this is not true for the IPL constants α_n and r_0 which always appear together as the product $\alpha_n r_0^{n-1}$. There is no experiment which can determine α_n and r_0 separately. Therefore, r_0 does not establish a distance scale for an IPL force in the same way as λ does for a Yukawa force, and so it may be arbitrarily defined as in (23).

On the other hand, the power n in the IPL potential plays a role that is similar in some ways to the Yukawa range λ. Like λ, n can be investigated separately from α_n. In addition, n acts like the range of an IPL potential; larger powers of n give a shorter-range potential, although "range" is used here in a more qualitative sense, since $V_n(r)$ is scale-invariant.

3 Current Limits on Inverse Power Law Forces

One of the simplest approaches to setting limits on new forces is to measure the force between two test bodies and compare the result to what was expected theoretically from known forces. The agreement between the experimental value F_{exp} and theory F_{th} can be used to constrain a new force F_X by

$$|F_X| \le |F_{\mathrm{exp}} - F_{\mathrm{th}}| \ . \tag{27}$$

For IPL forces, we may write $F_X = F_n = \Lambda_n \mathcal{F}_n$, where $\mathcal{F}_n$ is simply the IPL force between the two test bodies for the case $\Lambda_n = 1$. Then the limit on Λ_n can be obtained from (27):

$$|\Lambda_n| \le \frac{|F_{\mathrm{exp}} - F_{\mathrm{th}}|}{|\mathcal{F}_n|} \ . \tag{28}$$

Table 1 gives limits on Λ_n obtained from tests of Newtonian gravity over length scales $>10^{-2}$ m. For test body separations $\sim 10^{-10}$–10^{-5} m, the

Casimir force becomes the dominant force after electrostatic and magnetic effects have been eliminated. The limits in Table 2 were obtained from tests of the Casimir force. For separations $< 10^{-10}$ m, atomic systems have been used to obtain the limits given in Table 3. The best limits for each power n are given in Table 4. Additional limits, including those obtained from less reliable nuclear physics experiments, can be found in the comprehensive review by Bracci et al. [32].

Table 1. Limits on the IPL parameter Λ_n from (22) obtained from gravity experiments. Allowed values of Λ_n are less than these limits

Experiment	Λ_1	Λ_2	Λ_3	Λ_4	Λ_5	Ref.
Eötvös [11]	10^{-45}	10^{-23}	10^{-2}			[20]
Braginsky [12]	10^{-47}	10^{-20}	10^{7}			[20]
Long [13]		10^{-26}	10^{-12}			[20]
Su [14]	1×10^{-47}	1×10^{-26}	1×10^{-12}			[15]
Gundlach [15]	5.8×10^{-48}	4×10^{-30}	6×10^{-16}			[15]
Smith [16]		2.4×10^{-30}	3.4×10^{-16}	4.5×10^{-2}		[16]
Spero [17, 18]		7.7×10^{-30}	7.7×10^{-17}	9.9×10^{-4}	1.4×10^{10}	[22]
Mitrofanov [19]		4.0×10^{-28}	4.7×10^{-16}	7.5×10^{-4}	1.2×10^{9}	[22]

Table 2. Limits on the IPL parameter Λ_n from (22) obtained from Casimir force experiments. The entries marked by * were obtained for a slightly different definition of Λ_n which assumes $B \simeq 2Z$. Allowed values of Λ_n are less than these limits

Experiment	Λ_1	Λ_2	Λ_3	Λ_4	Λ_5	Ref.
Various [23, 24, 25]	1×10^{-40} *	1×10^{-27} *	5×10^{-15} *	3×10^{-3} *	8×10^{7} *	[28]
Lamoreaux [26]		1.1×10^{-26}	1.6×10^{-14}	3.6×10^{-3}		[29]
Mohideen[27]				3		[30]

Table 3. Limits on the IPL parameter Λ_n from (22) obtained from atomic and molecular physics experiments. Allowed values of Λ_n are less than these limits

Experiment	Λ_3	Λ_4	Λ_5	Λ_6	Λ_7	Ref.
Bystritsky [31]	1.5×10^{-2}	7.5				[32]
Zavattini [32]	1×10^{-4}	2×10^{-3}	4×10^{-2}	1		[32]
Barnes [33]	5×10^{-4}	10^{-2}	10^{-1}	1	8	[20]
Bamberger [34]	2×10^{-4}	3×10^{-3}	5×10^{-2}	0.5	3	[20]
Ebersold [35]	7×10^{-4}	2×10^{-2}	0.5	8	100	[20]

Table 4. Best current limits on the IPL parameters Λ_n and α_n from all experiments. Allowed values are less than these limits

n	Experiment Type	Λ_n	α_n	Limit Ref.
1	Gravity	5.8×10^{-48}	9.9×10^{-10}	[15, 16]
2	Gravity	2.4×10^{-30}	4.1×10^{8}	[16]
3	Gravity	7.7×10^{-17}	1.3×10^{22}	[22]
4	Gravity	7.5×10^{-4}	1.3×10^{35}	[22]
5	Atomic	4×10^{-2}	6.8×10^{36}	[32]
6	Atomic	0.5	8.5×10^{37}	[20]
7	Atomic	3	5×10^{38}	[20]

The pattern that emerges from Tables 1–4 is that the best limits on higher powers of n come from shorter distance experiments. This should be expected since IPL forces with larger powers of n grow more rapidly as the separations decrease when they act between point particles. However, the IPL force dependence on separation for macroscopic bodies used in experiments may differ substantially from point particles. For example, Fig. 1 illustrates the IPL forces for $n = 1$–5 between two parallel plates. One sees that for $n \leq 3$, an IPL force has little or no dependence on the plate separation. Hence, for these forces, there is no advantage to performing an experiment using parallel plates at shorter separations since the force is not significantly larger. In addition, for higher powers of n, the macroscopic IPL force exhibits a weaker dependence on separation than for the same force acting between point particles.

One can now compare the existing limits on α_n with the values predicted for large compact extra dimensions. Using (19) and (24), one finds (for $N \geq 1$)

$$\alpha_{n=N+1} = \left(\frac{R}{r_0}\right)^N \sim 10^{32-4N} . \tag{29}$$

The values for $n = 2$–7 are tabulated in Table 5. Comparing Tables 4 and 5, we see that extra dimensional models with $N = 1$ and 2 are excluded by force experiments.

4 Discussion

As we have shown, there are many models which include IPL potentials and so there is significant motivation to use short-ranged force experiments to search for them. Unlike Yukawa forces which have a characteristic length λ, IPL forces are scale invariant so there is no characteristic length which suggests a length scale for an experiment. However, the IPL power n acts like λ in that better limits on α_n are obtained from experiments acting over shorter distance scales. While the best current limits on IPL forces come from

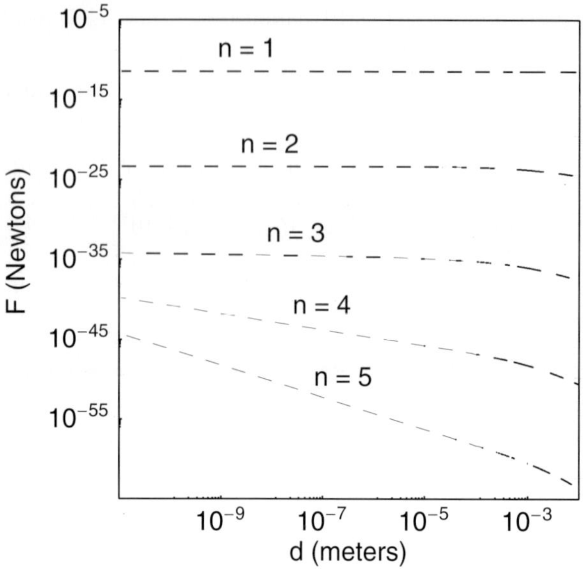

Fig. 1. IPL forces acting between two square copper parallel plates 1 cm × 1 cm × 1 mm separated by a distance d. Here it is assumed $\alpha_n = 1$

Table 5. Comparison of the predicted values of $\alpha_{n=N+1}$ for N large compact extra dimensions and best current limits obtained from Table 4

N	n	$\alpha_{n=N+1}$	Limit on α_n
1	2	10^{28}	$< 4.1 \times 10^{8}$
2	3	10^{24}	$<1.3 \times 10^{22}$
3	4	10^{20}	$<1.3 \times 10^{35}$
4	5	10^{16}	$<6.8 \times 10^{36}$
5	6	10^{12}	$<8.5 \times 10^{37}$
6	7	10^{8}	$<5 \times 10^{38}$

gravity and atomic physics experiments, a new class of Casimir experiments designed specifically to constrain new forces are under development [36, 37] which may be able to significantly improve these limits. In addition, since the limits from atomic physics experiments are now decades old, it is hoped that these can be improved substantially within the next few years.

Acknowledgement

One of the authors, E.F., wishes to acknowledge the support of the U.S. Department of Energy under Contract No. DE-AC02-HER071428.

References

1. E. Fischbach and C. L. Talmadge, *The Search for Non-Newtonian Gravity* (Springer/AIP Press, New York, 1999).
2. J. Sucher and G. Feinberg, in *Long-Range Casimir Forces*, edited by. F. S. Levin and D. A. Micha (Plenum, New York, 1993).
3. N. E. J. Bjerrum-Bohr, J. F. Donoghue, and B. R. Holstein, Phys. Rev. D **67**, 084033 (2003).
4. J. Garriga and T. Tanaka, Phys. Rev. Lett. **84**, 2778 (2000).
5. V. M. Mostepanenko and I. Yu. Sokolov, Sov. J. Nucl. Phys. **46**, 685 (1987).
6. F. Ferrer and J. A. Grifols, Phys. Rev. D **58**, 096006 (1998).
7. G. Feinberg and J. Sucher, Phys. Rev. **166**, 1638 (1968).
8. L. Spruch, Phys. Today **39**, 37 (November 1986).
9. N. Arkani-Hamed, S. Dimopoulus, and G. Dvali, Phys. Today **55**, 35 (February 2002).
10. G. Feinberg, J. Sucher, C.-K. Au, Phys. Rep. **180**, 83 (1989).
11. R. v. Eötvös, D. Pekar, and E. Fekele, Ann. Phys. (Leipzig) **68**, 11 (1922).
12. V. B. Braginsky and V. I. Panov, Zh. Eksp. Teor. Fiz. **61**, 873 (1971) [Sov. Phys. JETP **34**, 464 (1972)].
13. D. R. Long, Nature **260**, 417 (1976).
14. Y. Su, B. R. Heckel, E. G. Adelberger, J. H. Gundlach, M. Harris, G. L. Smith, and H. E. Swanson, Phys. Rev. D **50**, 3614 (1994).
15. J. H. Gundlach, G. L. Smith, E. G. Adelberger, B. R. Heckel, and H. E. Swanson, Phys. Rev. Lett. **78**, 2523 (1997).
16. G. L. Smith, C. D. Hoyle, J. H. Gundlach, E. G. Adelberger, B. R. Heckel, and H. E. Swanson, Phys. Rev. D **61**, 022001 (1999).
17. R. Spero, J. K. Hoskins, R. D. Newman, J. Pellam, and J. Schultz, Phys. Rev. Lett., 1645 (1980).
18. J. K. Hoskins, R. D. Newman, R. Spero, and J. Schultz, Phys. Rev. D **32**, 3084 (1985).
19. V. P. Mitrofanov and O. I. Ponomareva, Sov. Phys. JETP **67**, 1963 (1988). [Translation of Zh. Eksp. Teor. Fiz. **94**, 16 (1988).]
20. G. Feinberg and J. Sucher, Phys. Rev. D **20**, 1717 (1979).
21. R. D. Newman (unpublished).
22. E. Fischbach and D. E. Krause, Phys. Rev. Lett. **83**, 3593 (1999). This updates the earlier analysis by V. M. Mostepanenko and I. Yu. Sokolov, Phys. Lett. A **146**, 373 (1990).
23. B. V. Derjaguin, I. I. Abrikosova, and E. M. Lifschitz, Quart. Rev. **10**, 295 (1968).
24. S. Hunklinger, Proc. Conf. Physics of Adhesion, Karlsruhe, 48 (1969).
25. Y. N. Israelachvili and D. Tabor, Proc. R. Soc. **A331**, 19 (1972).
26. S. K. Lamoreaux, Phys. Rev. Lett. **78**, 5 (1997).
27. U. Mohideen and A. Roy, Phys. Rev. Lett. **81**, 4549 (1998).
28. V. M. Mostepanko and I. Yu. Sokolov, Phys. Lett. A **125**, 405 (1987).
29. M. Bordag, B. Geyer, G. L. Klimchitskaya, and V. M. Mostepanenko, Phys. Rev. D **58**, 075003 (1998). This updates an earlier analysis by G. L. Klimchitskaya, E. R. Bezerra de Mello, V. M. Mostepanenko, Phys. Lett. A **236**, 280 (1997).

30. M. Bordag, B. Geyer, G. L. Klimchitskaya, and V. M. Mostepanenko, Phys. Rev. D **60**, 055004 (1999).
31. V. M. Bystritsky, et al., Sov. Phys. JETP **49**, 232 (1979).
32. L. Bracci, G. Fiorentini, and R. Tripiccione, Nucl. Phys. **B217**, 215 (1983).
33. P. D. Barnes, et al., Nucl. Phys. **A231**, 477 (1974).
34. A. Bamberger, et al., Phys. Lett. **33B**, 233 (1970).
35. P. Ebersold, et al., Nucl. Phys. **A296**, 493 (1978).
36. E. Fischbach, D. E. Krause, R. S. Decca, and D. López, Phys. Lett. A **318**, 165 (2003).
37. R. S. Decca, D. López, H. B. Chan, E. Fischbach, G. L. Klimchitskaya, D. E. Krause, and V. M. Mostepanenko, J. Low. Temp. Phys. **135**, 63 (2004).

Quantum Gravity Phenomenology and Lorentz Violation

Ted Jacobson[1], Stefano Liberati[2] and David Mattingly[3]

[1] Institut d'Astrophysique de Paris, 98bis bd Arago, 75014 Paris France, and Department of Physics, University of Maryland, College Park, MD 20742 USA
jacobson@umd.edu
[2] SISSA, Via Beirut 2-4, 34014 Trieste, Italy and INFN Trieste
liberati@sissa.it
[3] Department of Physics, University of California at Davis
mattingly@physics.ucdavis.edu

In most fields of physics it goes without saying that observation and prediction play a central role, but unfortunately quantum gravity (QG) has so far not fit that mold. Many intriguing and ingenious ideas have been explored, but it seems safe to say that without both observing phenomena that depend on QG, and extracting reliable predictions from candidate theories that can be compared with observations, the goal of a theory capable of incorporating quantum mechanics and general relativity will remain unattainable.

Besides the classical limit, there is one observed phenomenon for which quantum gravity makes a prediction that has received encouraging support: the scale invariant spectrum of primordial cosmological perturbations. The quantized longitudinal linearized gravitational mode, albeit slave to the inflaton and not a dynamically independent degree of freedom, plays an essential role in this story [1].

What other types of phenomena might be characteristic of a quantum gravity theory? Motivated by tentative theories, partial calculations, intimations of symmetry violation, hunches, philosophy, etc, some of the proposed ideas are: loss of quantum coherence or state collapse, QG imprint on initial cosmological perturbations, scalar moduli or other new fields, extra dimensions and low-scale QG, deviations from Newton's law, black holes produced in colliders, violation of global internal symmetries, and violation of spacetime symmetries. It is this last item, more specifically the possibility of Lorentz violation (LV), that is the focus of this article.

From the observational point of view, technological developments are encouraging a new look at the possibility of LV. Increased detector size, spaceborne instruments, technological improvement, and technique refinement are permitting observations to probe higher energies, weaker interactions, lower fluxes, lower temperatures, shorter time resolution, and longer distances. It comes as a welcome surprise that the day of true quantum gravity observations may not be so far off [2]

1 Lorentz Violation?

Lorentz symmetry is linked to a scale-free nature of spacetime: unbounded boosts expose ultra-short distances, and yet nothing changes. However, suggestions for Lorentz violation have come from: the need to cut off UV divergences of quantum field theory and of black hole entropy, tentative calculations in various QG scenarios (e.g. semiclassical spin-network calculations in Loop QG, string theory tensor VEVs, non-commutative geometry, some brane-world backgrounds), and the possibly missing GZK cutoff on ultra high energy (UHE) cosmic rays.

The GZK question has generated a lot of interest, and is to date the only phenomenon that might point to a breakdown of standard physics in quantum gravity, hence we take a moment to discuss it. The idea [3] is that in collisions of ultra high energy protons with cosmic microwave background photons there can be sufficient energy in the center of mass frame to create a pion, leading to the the reaction

$$p + \gamma_{\rm CMB} \to p + \pi \; . \tag{1}$$

In this way, the initial proton energy is degraded with an attenuation length of about 50 Mpc. Since plausible astrophysical sources for UHE particles are located at distances larger than 50 Mpc, one expects a cutoff in the cosmic ray proton energy spectrum at around 5×10^{19} eV for protons coming from beyond a few megaparsecs. If Lorentz symmetry is violated, then the energy threshold for this reaction could be lowered, raised, or removed entirely, or an upper threshold where the reaction cuts off could even be introduced (see e.g. [4] and references therein).

One of the experiments measuring the UHE cosmic ray spectrum, the AGASA experiment, has not seen the cutoff. An analysis [5] from January 2003 concluded that the cutoff was absent at the 2.5 sigma level, while another experiment, Hi-Res, is consistent with the cutoff but at a lower confidence level. The question should be answered in the near future by the AUGER observatory, a combined array of 1600 water Čerenkov detectors and 24 telescopic air flouresence detectors under construction on the Argentine pampas [6]. The new observatory will see an event rate one hundred times higher, with better systematics.

Trans-GZK cosmic rays are not the only window of opportunity we have to detect or constrain Lorentz violation induced by QG effects. In fact, many phenomena accessible to current observations/experiments are sensitive to possible violations of Lorentz invariance. A partial list is

- sidereal variation of LV couplings as the lab moves with respect to a preferred frame or directions
- long baseline dispersion and vacuum birefringence (e.g. of signals from gamma ray bursts, active galactic nuclei, pulsars, galaxies)
- new reaction thresholds (e.g. photon decay, vacuum Čerenkov effect)

- shifted thresholds (e.g. photon annihilation from blazars, GZK reaction)
- maximum velocity (e.g. synchrotron peak from supernova remnants)
- dynamical effects of LV background fields (e.g. gravitational coupling and additional wave modes)

We conclude this section with a brief historical overview including some of the more influential papers but by no means complete. Suggestions of possible LV in particle physics go back at least to the 1960's, when a number of authors wrote on that idea [7][1]. The possibility of LV in a metric theory of gravity was explored beginning at least as early as the 1970's [9]. Such theoretical ideas were pursued in the '70's and '80's notably by Nielsen and several other authors on the particle theory side [10], and by Gasperini [11] on the gravity side. A number of observational limits were obtained during this period [12].

Towards the end of the 80's Kostelecky and Samuel [13] presented evidence for possible spontaneous LV in string theory, and motivated by this explored LV effects in gravitation. The role of Lorentz invariance in the "trans-Planckian puzzle" of black hole redshifts and the Hawking effect was emphasized in the early 90's [14]. This led to study of the Hawking effect for quantum fields with LV dispersion relations commenced by Unruh [15] and followed up by others. Early in the third millenium this line of research led to work on the related question of the possible imprint of trans-Planckian frequencies on the primordial fluctuation spectrum [16].

Meanwhile the consequences of LV for particle physics were being explored using LV dispersion relations e.g. by Gonzalez-Mestres [17], and a systematic extension of the standard model of particle physics incorporating all possible LV in the renormalizable sector was developed by Colladay and Kostelecký [18]. This latter work provided a framework for computing the observable consequences for any experiment and led to much experimental work setting limits on the LV parameters in the lagrangian [19]. Around the same time Coleman and Glashow suggested the possibility that LV was the culprit in the possibly missing GZK cutoff [20], and explored many other consequences of renormalizable, isotropic LV leading to different limiting speeds for different particles [21].

Also at that time it was pointed out by Amelino-Camelia et al. [22] that the sharp high energy signals of gamma ray bursts could reveal LV photon dispersion suppressed by one power of energy over the mass $M \sim 10^{-3} M_{\mathrm{P}}$, tantalizingly close to the Planck mass. Shortly afterwards Gambini and Pullin [23] argued that semiclassical loop quantum gravity suggests just such LV. (Some later work supported this notion, but a recent paper by Kozameh

[1] It is amusing to note that Kirzhnits and Chechin in [7] explore the possibility that an apparent missing cutoff in the UHE cosmic ray spectrum could be explained by something that looks very similar to the recently proposed "doubly special relativity" [8].

and Parisi [3] argues the other way.) In any case the theory is not under enough control at this time to make any definite statements.

A very strong constraint on photon birefringence was obtained by Gleiser and Kozameh [25] using UV light from distant galaxies, and if the recent measurement of polarized gamma rays from a GRB hold up to further scrutiny this constraint will be further strengthened dramatically [26, 27]. Further stimulus came from the suggestion [28] that an LV threshold shift might explain the apparent under-absorption on the cosmic IR background of TeV gamma rays from the blazar Mkn501, however it is now believed by many that this anomaly goes away when a corrected IR background is used [29].

The extension of the effective field theory framework to include LV dimension 5 operators was introduced by Myers and Pospelov [21], and used to strengthen prior constraints. Also this framework was used to deduce a very strong constraint [31] on the possibility of a maximum electron speed less than the speed of light from observations of synchrotron radiation from the Crab Nebula.

2 Theoretical Framework for LV

Various different theoretical approaches to LV have been taken to further pursue the ideas summarized above. Some researchers restrict attention to LV described in the framework of effective field theory (EFT), while others allow for effects not describable in this way, such as those that might be due to stochastic fluctuations of a "space-time foam". Some restrict to rotationally invariant LV, while others consider also rotational symmetry breaking. Both true LV as well as "deformed" Lorentz symmetry (in the context of so-called "doubly special relativity"[8]) have been pursued. Another difference in approaches is whether one allows for distinct LV parameters for different particle types, or proposes a more universal form of LV.

The rest of this article will focus on just one of these approaches, namely LV describable by standard EFT, assuming rotational invariance, and allowing distinct LV parameters for different particles. In exploring the possible phenomenology of new physics, it seems useful to retain enough standard physics so that a) clear predictions can be made, and b) the possibilities are narrow enough to be meaningfully constrained.

This approach is not universally favored. For example a sharp critique appears in [32]. Therefore we think it is important to spell out the motivation for the choices we have made. First, while of course it may be that EFT is not adequate for describing the leading quantum gravity phenomenology effects, it has proven itself very effective and flexible in the past. It produces local energy and momentum conservation laws, and seems to require for its validity just locality and local spacetime translation invariance above some length scale. It describes the standard model and general relativity (which are presumably not fundamental theories), a myriad of condensed matter systems at appropriate length and energy scales, and even string theory (as perhaps most impressively verified in the calculations of black hole entropy and Hawking radiation rates). It is true that, e.g., non-commutative geometry (NCG) seems to lead to EFT with problematic IR/UV mixing, however this more likely indicates a physically unacceptable feature of such NCG rather than a physical limitation of EFT.

The assumption of rotational invariance is motivated by the idea that LV may arise in QG from the presence of a short distance cutoff. This suggests a breaking of boost invariance, with a preferred rest frame, but not necessarily rotational invariance. Since a constraint on pure boost violation is, barring a conspiracy, also a constraint on boost plus rotation violation, it is sensible to simplify with the assumption of rotation invariance at this stage.

Finally why do we choose to complicate matters by allowing for different LV parameters for different particles? First, EFT for first order Planck suppressed LV (see Sect. 2.1) requires this for different polarizations or spin states, so it is unavoidable in that sense. Second, we see no reason a priori to expect these parameters to coincide. The term "equivalence principle" has been used to motivate the equality of the parameters. However, in the presence of LV dispersion relations, particles with different masses travel on different trajectories even if they have the same LV parameters [33, 4]. Moreover, different particles would presumably interact differently with the spacetime microstructure since they interact differently with themselves and with each other. An example of this occurs in the braneworld model discussed in [34], and an extreme version occurs in the proposal of [35] in which only certain particles feel the spacetime foam effects. (Note however that in this proposal the LV parameters fluctuate even for a given kind of particle, so EFT would not be a valid description.)

2.1 Deformed Dispersion Relations

A simple approach to a phenomenological description of LV is via deformed dispersion relations. If rotation invariance and integer powers of momentum are assumed in the expansion of $E^2(\boldsymbol{p})$, the dispersion relation for a given particle type can be written as

$$E^2 = p^2 + m^2 + \Delta(p) , \tag{2}$$

where p is the magnitude of the three-momentum, and

$$\Delta(p) = \tilde{\eta}_1 p^1 + \tilde{\eta}_2 p^2 + \tilde{\eta}_3 p^3 + \tilde{\eta}_4 p^4 + \cdots \tag{3}$$

Let us introduce two mass scales, $M = 10^{19}\,\mathrm{GeV} \approx M_{\mathrm{Planck}}$, the putative scale of quantum gravity, and μ, a particle physics mass scale. To keep mass dimensions explicit we factor out possibly appropriate powers of these scales, defining the dimensionful η's in terms of corresponding dimensionless parameters. It might seem natural that the p^n term with $n \geq 3$ be suppressed by

$1/M^{n-2}$, and indeed this has been assumed in most work. But following this pattern one would expect the $n = 2$ term to be unsuppressed and the $n = 1$ term to be even more important. Since any LV at low energies must be small, such a pattern is untenable. Thus either there is a symmetry or some other mechanism protecting the lower dimension oprators from large LV, or the suppression of the higher dimension operators is greater than $1/M^{n-2}$. This is an important issue to which we return later in this article.

For the moment we simply follow the observational lead and insert at least one inverse power of M in each term, viz.

$$\tilde{\eta}_1 = \eta_1 \frac{\mu^2}{M}, \qquad \tilde{\eta}_2 = \eta_2 \frac{\mu}{M}, \qquad \tilde{\eta}_3 = \eta_3 \frac{1}{M}, \qquad \tilde{\eta}_4 = \eta_4 \frac{1}{M^2}. \tag{4}$$

In characterizing the strength of a constraint we refer to the η_n without the tilde, so we are comparing to what might be expected from Planck-suppressed LV. We allow the LV parameters η_i to depend on the particle type, and indeed it turns out that they *must* sometimes be different but related in certain ways for photon polarization states, and for particle and antiparticle states, if the framework of effective field theory is adopted. In an even more general setting, Lehnert [36] studied theoretical constraints on this type of LV and deduced the necessity of some of these parameter relations.

This general framework allows for superluminal propagation, and spacelike 4-momentum relative to a fixed background metric. It has been argued [37] that this may lead to problems with causality and stability, but we do not share this opinion. In the context of a LV theory, there can be a preferred reference frame. As long as the physics is guaranteed to be causal and the states all have positive energy in the preferred frame, we cannot see any room for such problems to arise.

2.2 Effective Field Theory and LV

The standard model extension (SME) of Colladay and Kostelecký [18] consists of the standard model of particle physics plus all Lorentz violating renormalizable operators (i.e. of mass dimension ≤ 4) that can be written without changing the field content or violating the gauge symmetry. For illustration, the leading order terms in the QED sector are the dimension three terms

$$-b_a \bar{\psi} \gamma_5 \gamma^a \psi - \frac{1}{2} H_{ab} \bar{\psi} \sigma^{ab} \psi \tag{5}$$

and the dimension four terms

$$-\frac{1}{4} k^{abcd} F_{ab} F_{cd} + \frac{i}{2} \bar{\psi} (c_{ab} + d_{ab} \gamma_5) \gamma^a \overset{\leftrightarrow}{D^b} \psi, \tag{6}$$

where the dimension one coefficients b_a, H_{ab} and dimensionless k^{abcd}, c_{ab}, and d_{ab} are constant tensors characterizing the LV. If we assume rotational invariance then these must all be constructed from a given unit timelike vector u^a and the Minkowski metric η_{ab}, hence $b_a \propto u_a$, $H_{ab} = 0$, $k^{abcd} \propto u^{[a}\eta^{b][c}u^{d]}$, c_{ab} and $d_{ab} \propto u_a u_b$. Such LV is thus characterized by just four numbers.

The study of Lorentz violating EFT in the higher mass dimension sector was initiated by Myers and Pospelov [21]. They classified all LV dimension five operators that can be added to the QED Lagrangian and are quadratic in the fields, rotation invariant, gauge invariant, not reducible to lower and/or higher dimension operators using the field equations, and contribute p^3 terms to the dispersion relation. Again, just three parameters arise:

$$\frac{\xi}{M} u^m F_{ma}(u \cdot \partial)(u_n \tilde{F}^{na}) + \frac{1}{M} u^m \bar{\psi} \gamma_m (\zeta_1 + \zeta_2 \gamma_5)(u \cdot \partial)^2 \psi \tag{7}$$

where $\tilde{F}$ denotes the dual of F. All of these terms violate CPT symmetry as well as Lorentz invariance. Thus if one knew CPT were preserved, these LV operators would be forbidden.

In the limit of high energy $E \gg m$, the photon and electron dispersion relations following from QED with the above terms are [21, 26]

$$\omega^2_{R,L} = k^2 \pm \frac{2\xi}{M} k^3 \tag{8}$$

$$E^2_{\pm} = p^2 + m^2 + \frac{2(\zeta_1 \pm \zeta_2)}{M} p^3 \,. \tag{9}$$

The photon subscripts R and L refer to right and left circular polarization, hence these necessarily have opposite LV parameters. The electron subscripts $\pm$ refer to the helicity, which can be shown to be a good quantum number in the presence of these LV terms [26]. Moreover, if we write $\eta_\pm = 2(\zeta_1 \pm \zeta_2)$ for the LV parameters of the two electron helicities, those for positrons are given by [26]

$$\eta_\pm^{\text{positron}} = -\eta_\mp^{\text{electron}} \,. \tag{10}$$

2.3 Un-Naturalness of Small LV at Low Energy

As discussed above in subsection 2.1, if LV operators of dimension $n > 4$ are suppressed, as we have imagined, by $1/M^{n-2}$, LV would feed down to the lower dimension operators and be strong at low energies [21, 38, 39], unless there is a symmetry or some other mechanism that protects operators of dimension four and less from strong LV. What symmetry (other than Lorentz invariance, of course!) could that possibly be?

In the Euclidean context, a discrete subgroup of the Euclidean rotation group suffices to protect the operators of dimension four and less from violation of rotation symmetry. For example [40], consider the "kinetic" term in the EFT for a scalar field with hypercubic symmetry, $M^{\mu\nu}\partial_\mu\phi\partial_\nu\phi$. The only tensor $M^{\mu\nu}$ with hypercubic symmetry is proportional to the Kronecker delta $\delta^{\mu\nu}$, so full rotational invariance is an "accidental" symmetry of the kinetic operator.

If one tries to mimic this construction on a Minkowski lattice admitting a discrete subgroup of the Lorentz group, one faces the problem that each point has an infinite number of neighbors related by the Lorentz boosts. For the action to share the discrete symmetry each point would have to appear in infinitely many terms of the discrete action, presumably rendering the equations of motion meaningless.

Another symmetry that could do the trick is three dimensional rotational symmetry together with a symmetry between different particle types. For example, rotational symmetry would imply that the kinetic term for a scalar field takes the form $(\partial_t\phi)^2 - c^2(\nabla\phi)^2$, for some constant c. Then for multiple scalar fields, a symmetry relating the fields would imply that the constant c is the same for all, hence the kinetic term would be Lorentz invariant with c playing the role of the speed of light. Unfortunately this mechanism does not work in nature, since there is no symmetry relating all the physical fields. Perhaps under some conditions a partial symmetry could be adequate, e.g. grand unified gauge and/or super symmetry.

We are thus in the uncomfortable position of lacking any theoretical realization of the Lorentz symmetry breaking scheme upon which constraints are being imposed. This does not mean that no realization exists, but it is worrisome. If none exists, then our parametrization (4) is misleading, since there should be more powers of $1/M$ suppressing the higher dimension terms, likely rendering any constraints on those terms uninteresting.

3 Constraints

Observable effects of LV arise, among other things, from 1) sidereal variation of LV couplings due to motion of the laboratory relative to the preferred frame, 2) dispersion and birefringence of signals over long travel times, 3) anomalous reaction thresholds. We will often express the constraints in terms of the dimensionless parameters η_n introduced in (4). An order unity value might be considered to be expected in Planck suppressed LV.

The possibility of interesting constraints in spite of Planck suppression arises in different ways for the different types of observations. In the laboratory experiments looking for sidereal variations, the enormous number of atoms allow a resonance frequency to be measured extremely accurately. In the case of dispersion or birefringence, the enormous propagation distances would allow a tiny effect to accumulate. In the anomalous threshold case, the creation of a particle with mass m would be strongly affected by a LV term when the momentum becomes large enough for this term to be comparable to the mass term in the dispersion relation.

Consider first the case $n = 2$. For the $n = 2$ term in (3,4), the absence of a strong threshold effect yields a constraint $\eta_2 \lesssim (m/p)^2(M/\mu)$. If we consider protons and put $\mu = m = m_p \sim 1\,\mathrm{GeV}$, this gives an order unity constraint when $p \sim \sqrt{mM} \sim 10^{19}\,\mathrm{eV}$. Thus the GZK threshold, if confirmed, can give an order unity constraint, but multi-TeV astrophysics yields much weaker constraints. The strongest laboratory constraints on dimension three and four operators come from clock comparison experiments using noble gas masers [41]. The constraints limit a combination of the coefficients for dimension three and four operators for the neutron to be below 10^{-31} GeV (the dimension four coefficients are weighted by the neutron mass, yielding a constraint in units of energy). Astrophysical limits on photon vacuum birefringence give a bound on the coefficients of dimension four operators of 10^{-32} [46].

For $n = 3$ the constraint from the absence of a strong effect on energy thresholds involving only electrons and photons is of order

$$\eta_3 \lesssim (10\,\mathrm{TeV}/p)^3 \, . \tag{11}$$

Thus we can obtain order unity and even much stronger constraints from high energy astrophysics, as discussed shortly.

3.1 Summary of Constraints on LV in QED at $O(E/M)$

Since we do not assume universal LV coefficients, different constraints cannot be combined unless they involve just the same particle types. To achieve the strongest combined constraints it is thus preferable to focus on processes involving a small number of particle types. It also helps if the particles are very common and easy to observe. This selects electron-photon physics, i.e. QED, as a useful arena.

The current constraints on the three LV parameters at order E/M – one in the photon dispersion relation and two in the electron dispersion relation – will now be summarized. These are equivalent to the parameters in the dimension five operators (7) written down by Myers and Pospelov.

First, the constraint $|\eta_+ - \eta_-| < 4$ on the difference between the positive and negative electron helicity parameters was deduced by Myers and Pospelov [21] using a previous spin-polarized torsion pendulum experiment [44] that looked for diurnal changes in resonance frequency. (They also determined a numerically stronger constraint using nuclear spins, however this involves four different LV parameters, one for the photon, one for the up-down quark doublet, and one each for the right handed up and down quark singlets. It also requires a model of nuclear structure.)

In Fig. 1 (from [26]) constraints on the photon (ξ) and electron (η) LV parameters are plotted on a logarithmic scale to allow the vastly differing strengths to be simultaneously displayed. For negative parameters minus the logarithm of the absolute value is plotted, and a region of width 10^{-18} is excised around each axis. The synchrotron and Čerenkov constraints are known to apply only for at least one $\eta_\pm$. The IC and synchrotron Čerenkov lines are

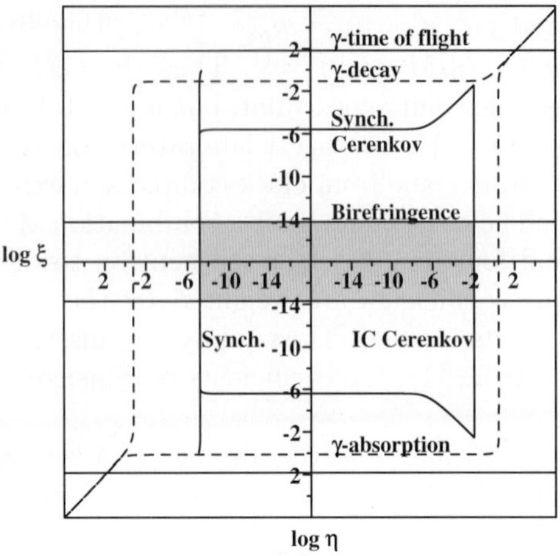

Fig. 1. Constraints on LV in QED at $O(E/M)$ (figure from [26])

truncated where they cross. Prior photon decay and absorption constraints are shown in dashed lines since they do not account for the EFT relations between the LV parameters.

Vacuum Bifrefringence

The birefringence constraint arises from the fact that the LV parameters for left and right circular polarized photons are opposite (9). The phase velocity thus depends on both the wavevector and the helicity. Linear polarization is therefore rotated through an energy dependent angle as a signal propagates, which depolarizes any initially linearly polarized signal. Hence the observation of linearly polarized radiation coming from far away can constrain the magnitude of the LV parameter. This effect has been used to constrain LV in the dimension three (Chern-Simons) [45], four [46] and five [25, 26, 27] terms. The constraint shown in the figure derives from the recent report [47] of a high degree of polarization of MeV photons from GRB021206. The data analysis has been questioned [48] and defended [49], so we shall have to wait and see if it is confirmed. The next best constraint on the dimension five term is $|\xi| \lesssim 2 \times 10^{-4}$, and was deduced by Gleiser and Kozameh [25] using UV light from distant galaxies.

Photon Time of Flight

The γ time of flight constraint arises from an energy dependent dispersion in the arrival time at Earth for photons originating in a distant event [50, 22], which was previously exploited for constraints [51, 52, 53]. The dispersion of the two polarizations is larger since the difference in group velocity is then $2|\xi|p/M$ rather than $\xi(p_2 - p_1)/M$, but the time of flight constraint remains many orders of magnitude weaker than the birefringence one from polarization rotation. In Fig. 1 we use the EFT improvement of the constraint of [52] which yields $|\xi| < 63$.

Vacuum Čerenkov Effect, Inverse Compton Electrons

In the presence of LV the process of vacuum Čerenkov radiation $e \rightarrow e\gamma$ can occur. The inverse Compton (IC) Čerenkov constraint uses the electrons of energy up to 50 TeV inferred via the observation of 50 TeV gamma rays from the Crab nebula which are explained by IC scattering. Since the vacuum Čerenkov rate is orders of magnitude higher than the IC scattering rate, that process must not occur for these electrons [21, 4]. The threshold for vacuum Čerenkov radiation depends in general on both ξ and η, however in part of the parameter plane the threshold occurs with emission of a soft photon, so ξ is irrelevant. This produces the vertical IC Čerenkov line in Fig. 1. One can see from (11) that this yields a constraint on η of order $(10\,\mathrm{TeV}/50\,\mathrm{TeV})^3 \sim 10^{-2}$. It could be that only one electron helicity produces the IC photons and the other loses energy by vacuum Čerenkov radiation. Hence we can infer only that at least one of η_+ and η_- satisfies the bound.

Crab Synchrotron Emission

A complementary constraint was derived in [31] by making use of the very high energy electrons that produce the highest frequency synchrotron radiation in the Crab nebula. For negative values of η the electron has a maximal group velocity less than the speed of light, hence there is a maximal synchrotron frequency that can be produced regardless of the electron energy [31]. Observations of the Crab nebula reveal synchrotron radiation at least out to 100 MeV (requiring electrons of energy 1500 TeV in the Lorentz invariant case), which implies that at least one of the two parameters $\eta_\pm$ must be greater than -7×10^{-8} (this constraint is independent of the value of ξ). We cannot constrain both η parameters in this way since it could be that all the Crab synchrotron radiation is produced by electrons of one helicity. Hence for the rest of this discussion let η stand for whichever of the two η's satisfies the synchrotron constraint.

This must be the same η as satisfies the IC Čerenkov constraint discussed above, since otherwise the energy of these synchrotron electrons would be below 50 TeV rather than the Lorentz invariant value of 1500 TeV. The Crab spectrum is well accounted for with a single population of electrons responsible for both the synchrotron radiation and the IC γ-rays. If there were enough extra electrons to produce the observed synchrotron flux with thirty times less energy per electron, then the electrons of the other helicity which would be producing the IC γ-rays would be too numerous [26]. It is important that the *same* η, i.e. either η_+ or η_-, satisfies both the synchrotron and the IC Čerenkov constraints. Otherwise, both constraints could have been satisfied by having one of these two parameters arbitrarily large and negative, and the other arbitrarily large and positive.

Vacuum Čerenkov Effect, Synchrotron Electrons

The existence of these synchrotron producing electrons can be exploited to improve on the vacuum Čerenkov constraint. For a given η satisfying the synchrotron bound, some definite electron energy $E_{\rm synch}(\eta)$ must be present to produce the observed synchrotron radiation. (This is higher for negative η and lower for positive η than the Lorentz invariant value [31].) Values of $|\xi|$ for which the vacuum Čerenkov threshold is lower than $E_{\rm synch}(\eta)$ for either photon helicity can therefore be excluded [26]. (This is always a hard photon threshold, since the soft photon threshold occurs when the electron group velocity reaches the low energy speed of light, whereas the velocity required to produce any finite synchrotron frequency is smaller than this.) For negative η, the Čerenkov process occurs only when $\xi < \eta$ [4, 54], so the excluded parameters lie in the region $|\xi| > -\eta$.

Photon Decay and Photon Absorption

Previously obtained constraints from photon decay $\gamma \rightarrow e^+e^-$ and absorption $\gamma\gamma \rightarrow e^+e^-$ must be re-analyzed to take into account the different dispersion for the two photon helicities, and the different parameters for the two electron helicities, but there is a further complication: both these processes involve positrons in addition to electrons. Previous constraint derivations have assumed that these have the same dispersion, but that need not be the case [36]. As discussed above, for the $O(E/M)$ corrections this is indeed not so [26]. Taking into account the above factors could not significantly improve the strength of the constraints (which is mainly determined by the energy of the photons). We indicate here only what the helicity dependence of the photon dispersion implies, thus neglecting the important role of differing parameters for electrons, positrons and their helicity states.

The strongest limit on photon decay came from the highest energy photons known to propagate, which at the moment are the 50 TeV photons observed from the Crab nebula [4, 54]. Since their helicity is not measured, only those values of $|\xi|$ for which *both* helicities decay could be ruled out. The photon absorption constraint came from the fact that LV can shift the standard QED threshold for annihilation of multi-TeV γ-rays from nearby blazars such as Mkn 501 with the ambient infrared extragalactic photons [4, 54, 55, 56, 57, 58, 59]. LV depresses the rate of absorption of one photon helicity and increases it for the other. Although the polarization of the γ-rays is not measured, the possibility that one of the polarizations is essentially unabsorbed appears to be ruled out by the observations which show the predicted attenuation [59].

3.2 Constraints at $O(E^2/M^2)$?

As previously mentioned, CPT symmetry alone could exclude the dimension five LV operators that give $O(E/M)$ modifications to particle dispersion relation, and in any case the constraints on those have become nearly definitive. Hence it is of interest to ask about the dimension five and six operators that give $O(E^2/M^2)$ corrections. We close with a brief discussion of the constraints that might be possible on those, i.e. constraints at $O(E^2/M^2)$.

As discussed above, the strength of constraints can be estimated by the requirement $\eta_4 p^4/M^2 \lesssim m^2$, which yields

$$\eta_4 \lesssim \left(\sqrt{\frac{m}{1\,\mathrm{eV}}}\frac{100\,\mathrm{TeV}}{p}\right)^4 . \tag{12}$$

Thus for electrons, an energy around 10^{17} eV is needed and we are probably not going to see any effects directly from such electrons. For protons an energy $\sim 10^{18}$ eV is needed. This is well below the UHE cosmic ray energy cutoff, hence if and when Auger [6]f confirms the identity of UHE cosmic rays as protons at the GZK cutoff, we will obtain a constraint of order $\eta_4 \lesssim 10^{-5}$ from the absence of vacuum Čerenkov radiation for 10^{20} eV protons! Also, from the fact that the GZK threshold is not shifted, we will obtain a constraint of order $\eta_4 \gtrsim -10^{-2}$, assuming equal η_4 values for proton and pion.

Impressive constraints might also be obtained from the absence of neutrino vacuum Čerenkov radiation: putting in 1 eV for the mass in (12) yields an order unity constraint from 100 TeV neutrinos, but only if the Čerenkov *rate* is high enough. The rate will be low, since it proceeds only via the non-local charge structure of the neutrino. Recent calculations [60] have shown that the rate is not high enough at that energy. However, for 10^{20} eV UHE neutrinos, which may be observed by the EUSO and OWL planned satellite observatories, the emission rate will be high enough to derive a strong constraint. The exact value depends on the emission rate, which has not yet been computed. For a *gravitational* Čerenkov reaction, the rate (which is lower but easier to compute than the electromagnetic rate) would be high enough for a neutrino from a distant source provided $\eta_4 \gtrsim 10^{-2}$. Hence in this case one might obtain a constraint of order $\eta_4 \lesssim 10^{-2}$, or stronger in the electromagnetic case.

A time of flight constraint at order $(E/M)^2$ might be possible [61] if gamma ray bursts produce UHE ($\sim 10^{19}$ eV) neutrinos, as some models predict, via limits on time of arrival differences of such UHE neutrinos vs. soft photons (or gravitational) waves. Another possibility is to obtain a vacuum birefringence constraint with higher energy photons [27] (although such a constraint would be less powerful since the parameters for opposite polarizations need not be opposite at order $(E/M)^2$). If future GRB's are found to be polarized at $\sim$100 MeV, that could provide a birefringence constraint $|\xi_{4+} - \xi_{4-}| \lesssim 1$.

4 Conclusion

At present there are only hints, but no compelling evidence for Lorentz violation from quantum gravity. Moreover, even if LV is present, the use of EFT for its low energy parametrization is not necessarily valid. Nevertheless, we believe that the constraints derived from the simple ideas discussed here are still important. They allow tremendous advances in observational reach to be applied in a straightforward manner to limit reasonable possibilities that might arise from fundamental Planck scale physics. Such guidance is especially welcome for the field of quantum gravity, which until the past few years has had little connection with observed phenomena.

References

1. V. F. Mukhanov, H. A. Feldman and R. H. Brandenberger, Phys. Rept. **215**, 203 (1992).
2. G. Amelino-Camelia, Lect. Notes Phys. **541**, 1 (2000) [arXiv:gr-qc/9910089].
3. K. Greisen, Phys. Rev. Lett. **16**, 748 (1966);
   G. T. Zatsepin and V. A. Kuzmin, JETP Lett. **4**, 78 (1966) [Pisma Zh. Eksp. Teor. Fiz. **4**, 114 (1966)].
4. T. Jacobson, S. Liberati and D. Mattingly, Phys. Rev. D **67**, 124011 (2003) [arXiv:hep-ph/0209264].
5. D. De Marco, P. Blasi and A. V. Olinto, Astropart. Phys. **20**, 53 (2003) [arXiv:astro-ph/0301497].
6. http://www.auger.org/
7. See, e.g. P. A. M. Dirac, Nature **168**, 906–907 (1951); J. D. Bjorken, Ann. Phys. **24**, 174 (1963); P. Phillips, Physical Review **146**, 967 (1966); D.I. Blokhintsev, Usp. Fiz. Nauk. **89**, 185 (1966) [Sov. Phys. Usp. **9**, 405 (1966)]; L.B. Rédei, Phys. Rev. 162, 1299-1301 (1967); T. G. Pavlopoulos, Phys. Rev. **159**, 1106 (1967); D.A. Kirzhnits and V.A. Chechin, Yad. Fiz. **15**, 1051 (1972) [Sov. J. Nucl. Phys. **15**, 585 (1972)].
8. G. Amelino-Camelia, Int. J. Mod. Phys. D **11**, 1643 (2002) [arXiv:gr-qc/0210063].
9. C.M. Will and K. Nordvedt, Jr., Astrophys. J. **177**, 757 (1972); K. Nordvedt, Jr. and C.M. Will, Astrophys. J. **177**, 775 (1972); R.W. Hellings and K. Nordvedt, Jr., Phys. Rev. **D7**, 3593 (1973).
10. H. B. Nielsen and M. Ninomiya, Nucl. Phys. B **141**, 153 (1978); S. Chadha and H. B. Nielsen, Nucl. Phys. B **217**, 125 (1983); H. B. Nielsen and I. Picek, Nucl. Phys. B **211**, 269 (1983) [Addendum-ibid. B **242**, 542 (1984)]; J. R. Ellis, M. K. Gaillard, D. V. Nanopoulos and S. Rudaz, Nucl. Phys. B **176**, 61 (1980). A. Zee, Phys. Rev. D **25**, 1864 (1982).
11. See, for example, M. Gasperini, Class. Quantum Grav. **4**, 485 (1987); Gen. Rel. Grav. **30**, 1703 (1998); and references therein.
12. See M. Haugan and C. Will, Physics Today, May 1987; C.M. Will, *Theory and Experiment in Gravitational Physics* (Cambridge Univ. Press, 1993), and references therein.
13. V. A. Kostelecky and S. Samuel, Phys. Rev. D **39**, 683 (1989).

14. T. Jacobson, Phys. Rev. D **44**, 1731 (1991).
15. W. G. Unruh, Phys. Rev. D **51**, 2827-2838 (1995), [arXiv:gr-qc/9409008].
16. See e.g. J. Martin and R. Brandenberger, Phys. Rev. D **68**, 063513 (2003) [arXiv:hep-th/0305161] and references therein.
17. L. Gonzalez-Mestres, *"Lorentz symmetry violation and high-energy cosmic rays,"*, [arXiv:physics/9712005].
18. D. Colladay and V. A. Kostelecky, Phys. Rev. D **58**, 116002 (1998), [arXiv:hep-ph/9809521].
19. V. A. Kostelecky, Proceedings of the *"Second Meeting on CPT and Lorentz Symmetry"*, Bloomington, Usa, 15-18 August 2001". Singapore, World Scientific (2002).
20. S. R. Coleman and S. L. Glashow, *"Evading the GZK cosmic-ray cutoff,"* [arXiv:hep-ph/9808446].
21. S. R. Coleman and S. L. Glashow, Phys. Rev. D **59**, 116008 (1999) [arXiv:hep-ph/9812418].
22. G. Amelino-Camelia, J. R. Ellis, N. E. Mavromatos, D. V. Nanopoulos and S. Sarkar, Nature **393**, 763 (1998) [arXiv:astro-ph/9712103].
23. R. Gambini and J. Pullin, Phys. Rev. D **59**, 124021 (1999) [arXiv:gr-qc/9809038].
24. C. N. Kozameh and M. F. Parisi, *"Lorentz invariance and the semiclassical approximation of loop quantum gravity,"* [arXiv:gr-qc/0310014].
25. R. J. Gleiser and C. N. Kozameh, Phys. Rev. D **64**, 083007 (2001) [arXiv:gr-qc/0102093].
26. T. A. Jacobson, S. Liberati, D. Mattingly and F. W. Stecker, *"New limits on Planck scale Lorentz violation in QED,"* [arXiv:astro-ph/0309681].
27. I. G. Mitrofanov, Nature **426**, 139 (2003).
28. R. J. Protheroe and H. Meyer, Phys. Lett. B **493**, 1 (2000) [arXiv:astro-ph/0005349].
29. A. K. Konopelko, A. Mastichiadis, J. G. Kirk, O. C. de Jager and F. W. Stecker, Astrophys. J. **597**, 851 (2003) [arXiv:astro-ph/0302049].
30. R. C. Myers and M. Pospelov, Phys. Rev. Lett. **90**, 211601 (2003) [arXiv:hep-ph/0301124].
31. T. Jacobson, S. Liberati and D. Mattingly, Nature **424**, 1019 (2003) [arXiv:astro-ph/0212190].
32. G. Amelino-Camelia, *"Improved limit on quantum-spacetime modifications of Lorentz symmetry from observations of gamma-ray blazars,"* [arXiv:gr-qc/0212002]; *"A perspective on quantum gravity phenomenology,"* [arXiv:gr-qc/0402009].
33. E. Fischbach, M. P. Haugan, D. Tadic and H. Y. Cheng, Phys. Rev. D **32**, 154 (1985).
34. C. P. Burgess, J. Cline, E. Filotas, J. Matias and G. D. Moore, JHEP **0203**, 043 (2002) [arXiv:hep-ph/0201082].
35. J. R. Ellis, N. E. Mavromatos, D. V. Nanopoulos and A. S. Sakharov, arXiv:gr-qc/0312044.
36. R. Lehnert, Phys. Rev. D **68**, 085003 (2003) [arXiv:gr-qc/0304013].
37. V. A. Kostelecky and R. Lehnert, Phys. Rev. D **63**, 065008 (2001) [arXiv:hep-th/0012060].
38. A. Perez and D. Sudarsky, *"Comments on challenges for quantum gravity,"* [arXiv:gr-qc/0306113].

39. J. Collins, A. Perez, D. Sudarsky, L. Urrutia and H. Vucetich, *"Lorentz invariance: An additional fine-tuning problem,"* [arXiv:gr-qc/0403053].
40. See e.g. M. Creutz, *Quarks, gluons and lattices* (Cambridge Univ. Press, 1985); G. Moore, *"Informal Lectures on Lattice Gauge Theory,"* http://www.physics.mcgill.ca/~guymoore/latt_lectures.pdf.
41. D. Bear, R. E. Stoner, R. L. Walsworth, V. A. Kostelecky and C. D. Lane, Phys. Rev. Lett. **85**, 5038 (2000) [Erratum-ibid. **89**, 209902 (2002)] [arXiv:physics/0007049].
42. Y. J. Ng, D. S. Lee, M. C. Oh and H. van Dam, Phys. Lett. B **507**, 236 (2001) [arXiv:hep-ph/0010152].
43. R. Aloisio, P. Blasi, A. Galante, P. L. Ghia and A. F. Grillo, Astropart. Phys. **19**, 127 (2003) [arXiv:astro-ph/0205271].
44. B. R. Heckel et al.,Proceedings of the International COnference on Orbis Scientiae, 1999, Coral Gables, Kluwer, 2000.; B. R. Heckel,http://www.npl.washington.edu/eotwash/publications/cpt01.pdf.
45. S. M. Carroll, G. B. Field and R. Jackiw, Phys. Rev. D **41**, 1231 (1990).
46. V. A. Kostelecky and M. Mewes, Phys. Rev. Lett. **87**, 251304 (2001) [arXiv:hep-ph/0111026]; Phys. Rev. D **66**, 056005 (2002) [arXiv:hep-ph/0205211].
47. W. Coburn and S. E. Boggs, Nature **423**, 415 (2003) [arXiv:astro-ph/0305377].
48. R. E. Rutledge and D. B. Fox, *"Re-Analysis of Polarization in the Gamma-ray flux of GRB 021206,"* [arXiv:astro-ph/0310385].
49. S. E. Boggs and W. Coburn, *"Statistical Uncertainty in the Re-Analysis of Polarization in GRB021206,"* [arXiv:astro-ph/0310515].
50. See Pavlopoulos in [7].
51. B. E. Schaefer, Phys. Rev. Lett. **82**, 4964 (1999) [astro-ph/9810479].
52. S. D. Biller et al., Phys. Rev. Lett. **83**, 2108 (1999) [arXiv:gr-qc/9810044].
53. P. Kaaret, *"Pulsar radiation and quantum gravity,"* [arXiv:astro-ph/9903464].
54. T. J. Konopka and S. A. Major, New J. Phys. **4**, 57 (2002) [arXiv:hep-ph/0201184].
55. W. Kluzniak, Astropart. Phys. **11**, 117 (1999).
56. G. Amelino-Camelia and T. Piran, Phys. Rev. D **64**, 036005 (2001) [arXiv:astro-ph/0008107].
57. F. W. Stecker and S. L. Glashow, Astropart. Phys. **16**, 97 (2001) [arXiv:astro-ph/0102226].
58. T. Jacobson, S. Liberati and D. Mattingly, *"Comments on 'Improved limit on quantum-spacetime modifications of Lorentz symmetry from observations of gamma-ray blazars',"* [arXiv:gr-qc/0303001].
59. F. W. Stecker, Astropart. Phys. **20**, 85 (2003) [arXiv:astro-ph/0308214].
60. D. Mattingly and B. McElrath, To be published.
61. G. Amelino-Camelia, Int. J. Mod. Phys. D **12**, 1633 (2003) [arXiv:gr-qc/0305057].

On the Quantum Width of a Black Hole Horizon

Donald Marolf[1]

Physics Department, UCSB, Santa Barbara, CA 93106
marolf@physics.ucsb.edu

1 Introduction

The work presented by the author at Adriatic 2003 comments on arguments [1, 2, 3, 4] which purport to derive fundamental bounds on the entropy of any system from the so-called generalized second law of thermodynamics; i.e. the second law including the Bekenstein-Hawking entropy $S_{BH} = A/4G$ for black holes. A number of interesting loopholes were pointed out stemming from the effect of the thermal atmosphere (see [5, 6]) and an additional "observer dependence" in the concept of entropy (see [7]). However, this work is well described in [5, 6] and [7]. Thus, rather than repeat the story in detail, it seems best to use this proceedings to discuss other thoughts on black hole entropy that are only vaguely related to the material presented at the conference. Nonetheless, a brief summary of that material will be included in the discussion below where it ties in to the current storyline (Sect. 4).

The main focus here will be on the many low energy degrees of freedom near the horizon of a black hole. Our primary interest will be in non-degenerate (i.e., non-extremal) horizons, with the degenerate case being treated only as a limit. Such degrees of freedom give rise to an old puzzle reviewed in [8]. The puzzle is that, when described using quantum field theory in a fixed curved background spacetime, such degrees of freedom give a divergent contribution to the entropy of the black hole's thermal atmosphere[1]. If cut off at the Planck scale, the thermal atmosphere naturally yields [9] an entropy of order the Bekenstein-Hawking entropy $A/4G$ and potentially provides a large correction to the entropy of the black hole. This is closely related (see e.g. [10] and [11]) to the suggestion [12] that the one should consider the entanglement entropy of the quantum fields outside the black hole. In [9], it was in fact suggested that this calculation might in fact account for the entire entropy of the black hole, though this idea appears to suffer from the so-called "species problem" and other issues discussed in [8].

Some time ago, it was argued by Sorkin [13] that one should cut-off the entropy of the thermal atmosphere at an even larger distance from the black hole horizon. Sorkin used a Newtonian model of a 3+1 Schwarzschild black

[1] That is the thermal bath of radiation constantly being emitted and re-absorbed by the black hole due to the Hawking effect.

hole to argue that quantum fluctuations naturally cause the horizon to fluctuate on a scale $L_c \sim (R\lambda_p^2)^{1/3}$ where R is the black hole radius and λ_p is the Planck scale. Since the fluctuations are quantum in nature, we find it natural to describe this as providing a quantum "width" to the horizon. Clearly then, one cannot reliably characterize the region within L_c of the classical horizon location as "outside" the black hole and it is reasonable to suppose a cut-off at this scale on contributions to the entropy of the (external) thermal atmosphere. Quantum fluctuations within L_c of the classical horizon are then perhaps better described as fluctuations of the black hole itself, and may plausibly be assumed to already be included in the Bekenstein-Hawking entropy of the black hole. (Though of course the details of how or whether the full entropy is reflected in a spacetime description remains unclear.)

Since the entropy of the atmosphere scales with A/L_c^2, Sorkin's cut-off would make this contribution a parametrically small correction to the Bekenstein-Hawking entropy of a large black hole. However, one may ask to what extent Sorkin's Newtonian model captures the relevant relativistic physics, and one may also wonder whether a similar result is obtained for black holes with charge or in different numbers of dimensions. In particular, Sorkin's Schwarzschild correction is large enough that, if the same cut-off applied to the extremal case[2], one might expect it to have been seen in stringy studies [14] of sub-leading corrections to the Bekenstein-Hawking result for supersymmetric black holes.

We report here on work to clarify these issues. A general relativistic estimate is provided for the quantum width of the horizon for arbitrary spherical black holes. The result in d spacetime dimensions is

$$L_c^d \sim \frac{\lambda_p^{d-2} R N^{1/2}}{T_H}, \tag{1}$$

where N is the number of effective free fields propagating near the black hole and T_H is the black hole's Hawking temperature. Again we see that A/L_c^{d-1} is parametrically small in comparison with the Bekenstein-Hawking entropy, with the additional feature that the correction also becomes parametrically small[3] as one approaches extremality and $T_H \to 0$. This is then in agreement with the results of [14], which finds no contributions clearly associated with the thermal atmosphere of a certain set of nearly extreme black holes. The full correction NA/L_c^2 can become large only when the number of fields N becomes parametrically large. It is amusing to note that for Schwarzschild black holes in 3+1 dimensions L_c is only just below nuclear length scales for

[2] Viewed as entropy of the thermal atmosphere, one would expect the entropy correction to vanish as $T_H \to 0$, but this is less clear when viewed as an entropy of entanglement across the horizon.

[3] The divergence of L_c is unphysical and signals a breakdown of the near-horizon approximation used below, but nevertheless L_c is much larger for a low T_H black hole than for a black hole of similar size with $T_H \sim 1/R$.

astrophysical black holes and begins to approach atomic length scales for the largest known supermassive black holes.

These ideas and estimates are explained below. We begin in Sect. 2 with a convenient description of the near-horizon degrees of freedom. We then perform the estimate in Sect. 3, which is based on a certain reasonable conjecture with regard to distortions of the black hole horizon by nearby objects. We then discuss a few consequences and compare with the natural expectation based on a "stretched horizon" in Sect. 4. Finally, we include an appendix which provides an estimate which does not rely on the above-mentioned conjecture, but which is clearly overly conservative. Nevertheless, this conservative argument yields qualitatively similar results in $d < 6$ spacetime dimensions.

2 Describing Near-Horizon Degrees of Freedom

In flat spacetime, any mode of any field which occupies a small region of space must have a correspondingly large energy. However, the diverging redshift at a horizon in principle allows for localized degrees of freedom with significantly smaller Killing energies. A particularly well-known modern application of this feature in the context of extremal horizons is the motivation [16] of the famous AdS/CFT conjecture. It will be useful for our discussion to study the near-horizon modes quantitatively. We give a simple description below and refer the reader to, e.g., [12] for a more technical treatment using tortoise (r^*) coordinates.

Let us in particular consider a wavepacket localized within a proper distance L of the horizon, where L is much smaller than the local curvature scale. We focus on the non-extremal case so that, if the wavepacket is also localized on a scale L in the directions along the horizon, the situation is well approximated by a wavepacket near a Rindler horizon in flat space.

Low energy such wavepackets will have essentially no variation on length scales shorter than L, except perhaps at proper distances $s \ll L$ from the horizon where we are free to take the wavepacket to vanish. Thus, they will look much like the larger wavepacket shown in Fig. 1 below. Since the energy density of the wavepacket is redshifted to zero near the horizon, the dominant contribution to the energy of the wavepacket will come from the part a proper distance $s \sim L$ away. Thus, the total energy of the wavepacket is of order $(Lz)^{-1}$, where z is the redshift at $s \sim L$ measured relative to infinity. A simple way to obtain this redshift for arbitrary black holes is to note that the local Unruh temperature at $s \sim L$ is of order $1/L$, but it must also be related through the redshift to the Hawking temperature T_H of the black hole. Thus, $(Lz)^{-1} \sim T_H$, and the wavepacket in question represents an excitation with energy of order T_H.

Note that we obtain such a wavepacket for each L, so that there are an infinite number of modes with energy T_H near any non-degenerate black hole

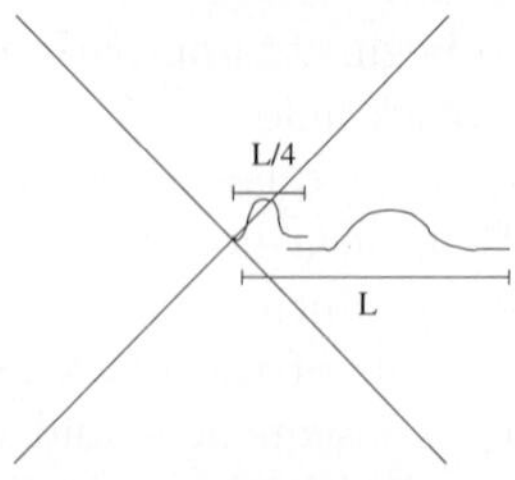

Fig. 1. Two orthogonal low energy modes near a horizon. The larger has proper size $\sim L$ while the smaller is $\sim L/4$. Both have energy $\sim T_H$

horizon. But the thermal atmosphere of the black hole is just this same system at temperature T_H. It is therefore clear that such modes contribute an infinite entropy. In more detail, consider the number of such modes contained between surfaces a distance L and $L + \mathrm{d}L$ from the horizon (for L small compared to the curvature scale). Since we considered wavepackets of size L in each direction, it is useful to take each surface (which has area A) and divide it into A/L^2 cells of size L^2.

In fact, a standard calculation using tortoise (r^*) coordinates shows that a careful choice of wavepacket shape can result in *arbitrarily* low energies, say ϵT_H for $\epsilon \ll 1$, within a proper distance L of the horizon. However, such modes cannot be localized as well on the sphere, so that there are only $\lambda A/L^2$ of them and they contribute only $\mathcal{O}(1)$ changes in coefficients.

Finally, note that in the Rindler approximation a given mode is transformed into the next mode closer to the horizon by scaling it toward the bifurcation surface (see Fig. 1 again). The relevant measure in the radial direction is therefore scale invariant and must be of order $\mathrm{d}L/L$. (This result also follows from the logarithmic relation between the tortoise coordinate (r^*) and proper distance [12].) Thus, the total number of modes between L and $L + \mathrm{d}L$ is of order $\frac{A}{L^3}\mathrm{d}L$. Integrating the total contribution down to some ultraviolet cutoff L_{UV} yields a total contribution of order A/L_{UV}^2. Each mode with energy $\lesssim T_H$ contributes a few bits of entropy, while modes with higher energy do not contribute. This then is the origin of the observation [9] that a Planck scale cutoff yields an entropy of order the Bekenstein-Hawking entropy of the black hole itself. It can also be mapped (see e.g. [10, 11]) to the corresponding result [12] for entanglement entropy.

3 Estimating the Width

The proper interpretation of the thermal atmosphere's entropy is clearly an important issue in black hole thermodynamics. For example, in the most naive interpretation one might expect this to be a next-order correction to the Bekenstein-Hawking entropy S_{BH}. Unfortunately, with a Planck scale cut-off it is of comparable size to the "zero order" contribution S_{BH}. One may also try to cancel this entropy with a renormalization effect (though this creates puzzles when various objects are lowered toward the black hole horizon [8]), or to interpret the entropy of the thermal atmosphere as some sort of "dual" description of the Bekenstein-Hawking entropy itself [9, 12]. Unfortunately, the latter approach suffers from the well-known "species problem" (i.e., it appears to depend on the number of propagating fields near the black hole horizon) and other concerns discussed in [8].

When the issue is phrased in terms of the near horizon modes, Sorkin's potential resolution suggests itself immediately. The presence of an extra particle with fixed energy too close to the horizon will effectively increase the mass of the black hole and will cause the horizon to expand outward and engulf the particle. If this particle is a quantum fluctuation associated with Hawking radiation, then it is natural to describe the effect as providing a non-zero quantum width for the horizon itself.

Consider for simplicity a 3+1 Schwarzschild black hole and let r be its Schwarzschild area-radius coordinate. The large entropy of the thermal atmosphere comes from modes near the horizon with energy T_H. But it is inconsistent to describe a mode with energy T_H supported below $r = R + 2GT_H$ in terms of quantum field theory on a fixed spacetime containing a black hole of radius R: if we add a particle to such a mode, then considering both the particle and the black hole we find a total energy $R/2G + T_H$ concentrated in a region smaller than the corresponding Schwarzschild radius! Thus, if the particle's energy were spherically distributed, one would expect that such a state is better described by a larger black hole (of size $R + 2GT_H$) than by an excitation of the original black hole (of size R). One therefore expects that fluctuations in the occupation numbers of such modes are more properly described as excitations of the horizon, and therefore that their contributions to the entropy are already included in S_{BH}. As a result, if one is considering that part of the thermal atmosphere[4] which acts as a *correction* to S_{BH}, then one should impose a cutoff at some $r_c > R + 2GT_H$. Since $T_H \sim 1/R$ and since for $r \ll R$ the proper distance L from the horizon satisfies $L^2/R \sim r - R$, one indeed arrives at the conclusion that the cutoff must be at least a proper distance of order $L_c \sim \lambda_p$ from the horizon.

On the other hand, as discussed above there are *many* modes near the horizon and, in calculating the entropy of the thermal atmosphere we have considered *all* of these modes to be thermally excited. In particular, even for modes at a given scale L from the horizon, there are R^2/L^2 such modes due to the size of the corresponding sphere around the black hole. Thus, a more careful consideration is in order, and it is to this task that we next turn. Note

[4] This wording has been used because the above observation is not necessarily in disagreement with the idea that at least a part of the thermal atmosphere's entropy is a dual description of S_{BH}. On the other hand, it is not obviously identical. In particular, it does not suffer from the issues discussed in [8].

that it is necessarily *fluctuations* of the occupations numbers that are relevant as the expectation value of the stress-energy tensor remains small even very close to the horizon, with the expected energy density being of order R^{-4}. Thus, a shell of thickness λ_p around the horizon on average contains only an energy of order $(R^2\lambda_p)R^{-4} = \lambda_p R^{-2}$. In contrast, we saw above that an energy of order $T_H \sim R^{-1}$ is required to move the horizon out to this radius. Thus, only departures from the mean can significantly increase the cutoff beyond the naive Planck scale estimate. Note also that energy fluctuations are largest for modes with $E \sim T_H$.

As noted above, the effect of a spherical shell of mass on the black hole horizon is easy to compute. However, a localized disturbance on a scale $L \ll R$ is far from spherically symmetric. One could, of course, average over a sphere's worth of fluctuations to estimate a typical fluctuation in the total mass of the black hole. Such an estimate is discussed in the appendix, but is vastly over-conservative as the averaging will partially cancel positive energy fluctuations on one side of the black hole against negative energy fluctuations on the opposite side of the black hole. Thus, it will dramatically underestimate the *local* fluctuations in the horizon location. In other words, it is clear that the largest effect comes from changes in the *shape* of the horizon as opposed to just the overall *size*.

How then shall we estimate this more localized effect on the horizon? Consider a positive energy fluctuation of length scale L at a corresponding separation from the classical horizon. If this induces a bulge on the horizon which is large enough to capture the fluctuation itself, then it is clear that this must occur on a timescale[5] L. Note that this is also the natural lifetime of the fluctuation. Consider now the center of the fluctuation. On a timescale L the center can receive no information from farther away than L. As a result, it cannot know whether it is indeed part of a homogeneous spherical shell of such fluctuations, or whether it is merely surrounded by an additional layer or so of similar fluctuations[6] (see Fig. 2). Thus, under reasonably common conditions, we should get the right answer (as to whether the horizon bulges outward and engulfs our fluctuation) by supposing that the black hole is in fact surrounded by a spherical shell of such fluctuations and determining whether this shell would add enough mass to the black hole to enlarge the horizon beyond the location of the fluctuations. Note that the shell has thickness L (see Fig. 3); luckily, the calculation is just as easy for thick shells as for thin.

[5] Say, as measured by freely falling observers initially at rest with respect to the black hole. Since we are primarily concerned with the perturbative regime, we may use the metric of the original Black Hole to compute times to leading order.

[6] It may just barely be able to tell whether the neighboring fluctuations have the same sign, but such a clumping will occur a frequency which is not parametrically small, and thus is large enough for our purposes.

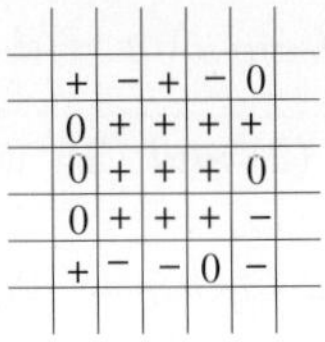

Fig. 2. The central fluctuation (+) happens to be surrounded by other fluctuations (+) of the same sign. Fluctuations of the opposite sign (−) may occur farther away, but the central fluctuation will not feel their influence on a timescale L

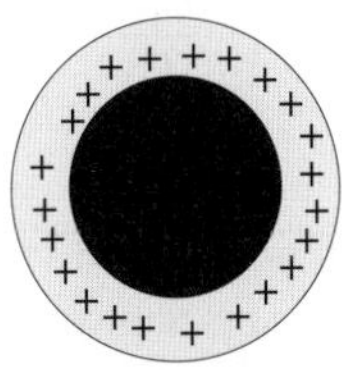

Fig. 3. A shell of fluctuations (+) around the black hole. Since each fluctuation is of size L, a given fluctuation cannot detect the others on a timescale less than L

Let us consider a general spherically symmetric static metric of the form

$$\mathrm{d}s^2 = -g_{tt}(r)\mathrm{d}t^2 + g_{rr}(r)\mathrm{d}r^2 + r^2\mathrm{d}\Omega_{d-2}^2 \,, \tag{2}$$

where as usual $d\Omega_{d-2}^2$ is the metric on the unit $(d-2)$-sphere. We take g_{tt} to have a first-order zero at $r = R$, representing the non-degenerate black hole horizon.

In a sufficiently small region close to the horizon, we may approximate the metric in the r, t directions by the standard Rindler metric:

$$\mathrm{d}s^2 = -\kappa^2\xi^2\mathrm{d}t^2 + \mathrm{d}\xi^2 + R^2\mathrm{d}\Omega_{d-2}^2 + \textit{higher order in } \xi \,, \tag{3}$$

where the constant κ is the surface gravity, and is chosen so that t is properly normalized; e.g., so that $|\partial_t| \to 1$ at infinity if the spacetime is asymptotically flat. The Hawking temperature T_H is then $T_H \sim \kappa$ since we set $\hbar = 1$. Note that for Schwarzschild black holes the description (3) will be valid whenever $r \ll R$, but for nearly extreme black holes we must approach the horizon more closely than the difference in radius between the outer and inner horizons. We will always assume that the fluctuations are close enough to the horizon for (3) to apply[7].

Let us ask how far the horizon moves outward when we add a mass Δm to the black hole but maintain spherical symmetry. The new horizon is where g_{tt} now vanishes. Since it used to vanish at $r = R$, we set

[7] Any cut-off we find will be parametrically small in the Planck length, so only for black holes parametrically close to extremality can this approximation fail.

$$0 = \Delta g_{tt} \approx \partial_m g_{tt} \ \Delta m \ + \partial_r g_{tt} \ (r - R) \ . \tag{4}$$

On the other hand, comparing (2) and (3), we see that $g_{tt} \sim -T_H^2 \xi^2$, so we find

$$\partial_r g_{tt} \sim -2\xi T_H^2 \partial_r \xi = -2\xi T_H^2 \sqrt{g_{rr}} \ . \tag{5}$$

Since g_{tt} vanishes only to first order, we see that $g_{rr} = \frac{c^2}{\xi^2}(1 + O(\xi))$ for some constant c. Thus $r - R \sim \xi^2/2c$. Putting this together with (4) and (5), one finds that adding a mass Δm moves the horizon to a new value of r which, in the original metric, was located at a proper distance

$$\xi = \sqrt{\frac{\partial_m g_{tt}}{T_H^2} \Delta m} \tag{6}$$

from the original horizon.

Unfortunately, it seems that no general theorems concerning the form of $\partial_m g_{tt}$ are available. However, for the Schwarzschild and Reissner-Nordstrom solutions in d spacetime dimensions one finds $\partial_m g_{tt} = (const) \lambda_p^{d-2} r^{-(d-3)}$ where the derivative is taken holding all charges constant[8] and $const$ is a number that depends at most on the dimension; i.e., which is independent of any charges or cosmological constant. We will therefore assume that any new black hole considered follows this pattern.

With this understanding, one finds that the additional mass required to, roughly speaking, "move the horizon outward a proper distance ξ" is given by

$$\Delta m_{sphere} \sim \frac{T_H^2 R^{d-3} \xi^2}{\lambda_p^{d-2}} \ . \tag{7}$$

The subscript *sphere* has been added to remind us that we have considered only spherically symmetric configurations above. For example, the outcome above might be attained by adding a spherical shell of mass Δm_{sphere} to our original black hole.

We now return to considering a single fluctuation of physical size L located at a proper distance L from the classical horizon of our black hole. We have seen that the typical energy of this fluctuation is $\Delta m_{fluct} \sim T_H$, so that a spherical shell of such fluctuations (as shown in Fig. 3) would have a mass $\Delta m_{sphere} \sim T_H (R/L)^{d-2}$. Comparison with (7) then shows that $L > \xi$ if and only if $L > L_c$ where

$$L_c^d \sim \frac{R}{T_H} \lambda_p^{d-2} \ . \tag{8}$$

Putting together our logic above, this L_c represents a rough measure of the quantum width of the horizon and fluctuations within L_c of the classical horizon cannot be cleanly separated from the black hole itself.

[8] This variation models fluctuations in uncharged fields, such as the metric itself. One could of course also consider fluctuations of charged fields.

In general, there will be N propagating degrees of freedom (i.e., helicity states of quantum fields) at the scale L_c. Any combination of bosons and fermions leads to a typical increase in the energy Δm_{fluct} of fluctuations in a given region by a factor of $\sqrt{N}$, so the cut-off length L_c becomes

$$L_c^d \sim \frac{R}{T_H} N^{1/2} \lambda_p^{d-2} \,. \tag{9}$$

The entropy of the thermal atmosphere down to this cut-off is then $S \sim NA/L_c^{d-2}$; i.e., smaller than $S_{BH} \sim A/\lambda_p^{d-2}$ unless N is parametrically large in (R/λ_p). Note that N should include any effective field with a mass $m \lesssim 1/L$, so that in general N depends on L and (9) represents only an implicit solution for the cut-off scale.

4 Discussion

In the above sections we have argued that the horizon of any black hole has an effective "quantum width" given by (9). This supports Sorkin's suggestion [13] that fluctuations of the horizon provide a cut-off on the entropy of the black hole's thermal atmosphere, suppressing this entropy parametrically in comparison with the black hole's Bekenstein-Hawking entropy. Our work provides a fully relativistic treatment and includes both uncharged black holes and those near the extremal limit, in particular giving an explicit dependence of the cut-off on the Hawking temperature separate from the dependence on the size R of the black hole. Our arguments concern general spherical black holes with non-degenerate (i.e., non-extreme) horizons, but an extension to the rotating case and a direct computation for extreme black holes would clearly be of interest. For Schwarzschild black holes in 3+1 dimensions, our estimate of the width L_c is only just below nuclear length scales for astrophysical black holes and begins to approach atomic length scales for the largest known supermassive black holes.

Our estimate of the width is based on the conjecture that whether the horizon bulges outward to engulf a particular fluctuation depends at most on whether the first few layers of surrounding fluctuations have a similar sign (as in Fig. 2), and does not depend on the presence of fluctuations farther away. In this case, we may model the calculation by assuming that there are in fact a sphere's worth of such fluctuations and use simple results for spherically symmetric perturbations of the black hole. As discussed in the Appendix, a skeptic could raise some questions about this conjecture, and a complete proof or counter-example would be much desired. The question should be amenable to study by various perturbative methods, and we look forward to future results in this direction[9].

[9] It would appear that the closest result in the existing literature is provided by [17], which considers two black holes of equal mass as opposed to a single black hole with a small perturbation.

In the absence of such precise results, we may also take comfort from the more conservative estimate given in the appendix. This estimate considers only the average fluctuation in the mass of a spherical shell and so does not rely on any conjectures. On the other hand, it yields the desired result only for $d \leq 5$.

Let us, however, proceed with the estimate (9). One interesting feature is that the entropy of the thermal atmosphere continues to give a large contribution in the presence of a sufficiently large number N of propagating fields. Under such circumstances the black hole will rapidly decay into a fireball of thermal radiation, unless the black hole is close to extremality so that its temperature is low (in which case the thermal atmosphere can be fully excited with only a small fraction of the black hole's energy).

On a related note, we have seen that even for $N = 1$ a large black hole allows of order $(R/L_c)^{d-2}$ modes to be excited with an energy of order T_H. Consider now a particle placed in a mixed state, which occupies an undetermined member of this set of modes. The von-Neumann entropy of the corresponding density matrix is then of order $\ln(R/L_c)$, which is parametrically larger than $E/T_H \sim 1$. The reader might at first wonder what will happen when such a particle falls into the black hole (since the black hole's entropy will increase only by E/T_H by the first law).

Indeed, following the logic of [1], one might expect it to lead to a violation of the generalized second law (GSL). However, as discussed in my talk at the conference (and as described in [5, 6, 7]) other, perhaps unexpected, effects will intervene to save the GSL. Since the particle's energy is T_H, it is clear that external observers will describe the particle as being added to a rather busy thermal state, and not just to the vacuum. Such observers will be most concerned with the *change* in entropy they assign to the process by which one particle falls into the black hole but a thermal state remains outside. Note that since the thermal state is well-occupied, the fundamental indistinguishability of particles may come into play and one cannot guess the answer from a simple model of distinguishable particles. In fact, as discussed in [7], an ensemble of distinguishable particles which allows fluctuations in particle number fails to be well-defined in our present circumstance.

As follows from the corresponding analysis in [7], this change in entropy does not in fact grow arbitrarily large with R/L but instead asymptotes to E/T_H, a value that preserves the second law[10]. It is only observers who fall across the horizon which assign the particle an entropy of order $\ln(R/L)$, but such observers do not see the particle disappear and face no issues with regard to the GSL.

[10] A closely related phenomenon was discovered in [18], which observed that a background of Unruh radiation downgrades the fidelity of quantum teleportation, indicating a loss of access to quantum information by observers who do not cross the horizon.

On the other hand, one can ask interesting questions about how the experiences of such observers could be consistent with a description of the black hole interior in terms of a Hilbert space with a finite number e^S_{BH} of dimensions. We shall not discuss such questions in detail here, as readers will no doubt interpret the outcome according to their pre-existing views of black hole entropy. Some readers will see evidence that S_{BH} counts only "the ways the black hole can interact with the external universe" and not the full set of interior states, while others will see manifestations of "complementarity" between various observers inside the black hole. The latter might be protected by limitations on the transfer of information between such observers analogous to those discussed in [19] for observers comparing Hawking radiation with the original objects falling into a black hole.

Returning now to somewhat firmer ground, perhaps the most interesting observation regarding our result (9) is the discrepancy with naive expectations based on a 'stretched horizon' picture of the black hole. The stretched horizon is typically described as a membrane just outside the black hole's classical horizon having properties that make its classical dynamics indistinguishable from the black hole itself [21]. Various differing proposals have been made for the details to incorporate quantum effects; for example, [19, 20] place the stretched horizon on the symmetry sphere with λ_p^2 more area than the classical horizon while [22] chooses the sphere with string-scale temperature, and [19, 23] choose the Planck temperature sphere. However, at least for Schwarzschild black holes, a membrane of Planck tension and Planck temperature (and thus located at roughly a Planck scale proper distance from the classical horizon) naturally reproduces the thermodynamics (e.g., energy and entropy) of the black hole and so seems to be preferred. On the other hand, since there are no infrared divergences in 2+1 or higher dimensions, typical fluctuations in the location of such a membrane (corresponding to black holes in $d \geq 4$) would also tend to be of intrinsically Planck scale, and thus would be parametrically smaller than our L_c. This suggests that the stretched horizon concept may be in need of refinement; a process that may provide fertile ground for future research.

Acknowledgements

The author would like to thank Raphael Bousso, Ram Brustein, Chris Herzog, Gary Horowitz, Stefan Hollands, Dan Kabat, Luis Lehner, David Lowe, Mark Spradlin, and the participants of the Adriatic 2003 conference for interesting discussions. Special thanks go to Djordje Minic, Simon Ross, and Rafael Sorkin for many discussions during our collaborations, to Gary Horowitz and Eric Poisson for help in locating reference [17], and to Luis Lehner for pointing out a number of typographic errors. This work was supported in part by NSF grants PHY00-98747, PHY99-07949, and PHY03-42749, and by funds from the University of California.

An Overly Conservative Estimate

In Sect. 3, we argued that one could estimate local distortions of the horizon caused by energy fluctuations by considering a spherical shell of such fluctuations surrounding the black hole. The idea was that, due to the short timescales involved, whether or not a given fluctuation was engulfed by a distortion of the horizon should be independent of the existence of neighboring fluctuations. However, the argument clearly falls short of a rigorous proof and must therefore be termed a conjecture. In particular, a skeptic might worry that the presence or absence of neighboring fluctuations over a time in the past of order R could affect the initial conditions in an important way, invalidating the arguments of Sect. 3.

Ultimately, this should be settled by a complete calculation. However, not having such a calculation at hand, it is also of interest to pursue other models of horizon fluctuations not based on the conjecture above. To this end, let us now consider the effect of averaging the fluctuations over a sphere around the black hole. This is clearly overly conservative, as it averages independent positive and negative fluctuations into an overall expansion or contraction of the black hole horizon. However, it provides a similar qualitative behavior to (9) in spacetime dimensions $d \leq 5$ and, in particular, reproduces exactly Sorkin's Newtonian result in $d = 4$.

The idea is a simple extension of our analysis so far. We simply note that there are of order $N(R/L)^{d-2}$ independent fluctuations of size L around the sphere. As a result, a typical fluctuation in the mass contained in this thick shell is of order $\Delta m_{fluct}\sqrt{N(R/L)^{d-2}} = \frac{T_H N^{1/2} R^{(d-2)/2}}{L^{(d-2)/2}}$. Comparing this with (4) yields

$$L_c^{(d+2)/2} \sim \frac{N^{1/2}\lambda_p^{d-2}}{T_H R^{(d-4)/2}} \,. \tag{10}$$

In particular, for a Schwarzschild black hole in $d = 4$ one finds

$$L_c \gtrsim N^{1/6}(R\lambda_p^2)^{1/3} \,, \tag{11}$$

and again this cutoff leads to negligible entropy in the thermal atmosphere. However, since T_H generally scales with $1/R$, we see that (10) fails to suppress the thermal atmosphere's entropy relative to the Bekenstein-Hawking entropy in $d \geq 6$ and that more care will be needed for such cases.

References

1. J. D. Bekenstein, "Black Holes And Entropy," Phys. Rev. D **7**, 2333 (1973); J. D. Bekenstein, "Generalized Second Law Of Thermodynamics In Black Hole Physics," Phys. Rev. D **9**, 3292 (1974).
2. J. D. Bekenstein, "Quantum information and quantum black holes," arXiv:gr-qc/0107049.

3. L. Susskind, "The World as a hologram," J. Math. Phys. **36**, 6377 (1995) [arXiv:hep-th/9409089].
4. G. 't Hooft, "Dimensional Reduction In Quantum Gravity," arXiv:gr-qc/9310026.
5. D. Marolf and R. Sorkin, "Perfect mirrors and the self-accelerating box paradox," Phys. Rev. D **66** (2002) 104004 [arXiv:hep-th/0201255].
6. D. Marolf and R. Sorkin, "On the Status of Highly Entropic Objects", hep-th/0309218.
7. D. Marolf, D. Minic and S. F. Ross, "Notes on spacetime thermodynamics and the observer-dependence of entropy," arXiv:hep-th/0310022.
8. R. M. Wald, "The thermodynamics of black holes," Living Rev. Rel. **4**, 6 (2001) [arXiv:gr-qc/9912119].
9. G. 't Hooft, "On The Quantum Structure Of A Black Hole," Nucl. Phys. B **256**, 727 (1985).
10. S. Mukohyama, "Aspects of black hole entropy," arXiv:gr-qc/9912103.
11. R. Brustein and A. Yarom, "Thermodynamics and area in Minkowski space: Heat capacity of entanglement," arXiv:hep-th/0311029.
12. R. D. Sorkin, "On The Entropy Of The Vacuum Outside A Horizon," Gen. Rel. Grav., proceedings of the GR10 Conference, Padova 1983, ed. B. Bertotti, F. de Felice, A. Pascolini (Consiglio Nazionale della Ricerche, Roma, 1983) Vol. 2; L. Bombelli, R. K. Koul, J. H. Lee and R. D. Sorkin, "A Quantum Source Of Entropy For Black Holes," Phys. Rev. D **34**, 373 (1986); M. Srednicki, "Entropy and area," Phys. Rev. Lett. **71**, 666 (1993) [arXiv:hep-th/9303048].
13. R. D. Sorkin, "How wrinkled is the surface of a black hole?," arXiv:gr-qc/9701056.
14. G. Lopes Cardoso, B. de Wit and T. Mohaupt, "Corrections to macroscopic supersymmetric black-hole entropy," Phys. Lett. B **451**, 309 (1999) [arXiv:hep-th/9812082]; G. Lopes Cardoso, B. de Wit and T. Mohaupt, "Deviations from the area law for supersymmetric black holes," Fortsch. Phys. **48**, 49 (2000) [arXiv:hep-th/9904005]; G. Lopes Cardoso, B. de Wit and T. Mohaupt, "Area law corrections from state counting and supergravity," Class. Quant. Grav. **17**, 1007 (2000) [arXiv:hep-th/9910179].
15. C. Misner, K. Thorne, and J. Wheeler, *Gravitation* (W.H. Freeman and Co., New York, 1970).
16. J. M. Maldacena, "The large N limit of superconformal field theories and supergravity," Adv. Theor. Math. Phys. **2**, 231 (1998) [Int. J. Theor. Phys. **38**, 1113 (1999)] [arXiv:hep-th/9711200]. O. Aharony, S. S. Gubser, J. M. Maldacena, H. Ooguri and Y. Oz, "Large N field theories, string theory and gravity," Phys. Rept. **323**, 183 (2000) [arXiv:hep-th/9905111].
17. G. B. Cook and A. M. Abrahams, "Horizon Structure Of Initial Data Sets For Axisymmetric Two Black Hole Collisions," Phys. Rev. D **46**, 702 (1992).
18. P. M. Alsing, D. McMahon and G. J. Milburn, "Teleportation in a non-inertial frame," arXiv:quant-ph/0311096.; P. M. Alsing and G. J. Milburn, "Teleportation with a uniformly accelerated partner," Phys. Rev. Lett. **91**, 180404 (2003) [arXiv:quant-ph/0302179].
19. L. Susskind, L. Thorlacius, and J. Ugglum, "The stretched horizon and black hole complementarity", Phys. Rev. D **48**, 3743 (1993) [arXiv:hep-th/9306069].
20. L. Susskind and L. Thorlacius, "Gedanken experiments involving black holes," Phys. Rev. D **49**, 966 (1994) [arXiv:hep-th/9308100].

21. K. S. Thorne, R.H. Price, and D. A. Macdonald, *Black Holes: The Membrane Paragdigm*, (Yale University Press, 1986).
22. A. Sen, "Extremal black holes and elementary string states", Mod. Phys. Lett. A **10**, 2081 (1995) [arXiv:hep-th/9204099].
23. N. Ilizuka, D. Kabar, G. Lifschytz, and D. A. Lowe, "Stretched Horizons, Quasiparticles, and Quasinormal modes," Phys. Rev. D **68** 084021 (2003) [arXiv:hep-th/0306209]; N. Ilizuka, D. Kabar, G. Lifschytz, and D. A. Lowe, "Quasiparticle picture of black holes and the entropy area relation," Phys. Rev. D **67** 124001 (2003) [arXiv:hep-th/0212246].

The Internal Structure of Black Holes

Igor D. Novikov[1234]

[1] Theoretical Astrophysics Center, Juliane Maries Vej 30, 2100 Copenhagen, Denmark
[2] Niels Bohr Institute, Blegdamsvej 17, 2100 Copenhagen, Denmark
[3] Astro-Space Center of Lebedev Physical Institute Profsoyuznaya 84/32, Moscow 117997, Russia
[4] NORDITA, Blegdamsvej 17, 2100 Copenhagen, Denmark

1 Introduction

The problem of black holes interior was the subject of a very active investigation last decades. For the systematic discussion of the problems of the internal structure of black holes see [1–8], [50].

A very important point for understanding the problem of black hole's interior is the fact that the path into the gravitational abyss of the interior of a black hole is a progression in time. We recall that inside a spherical black hole, for example, the radial coordinate is timelike. It means that the problem of the black hole interior is an *evolutionary problem*. In this sense it is completely different from a problem of an internal structure of other celestial bodies, stars for example, or planets.

In principle, if we know the conditions on the border of a black hole (on the event horizon), we can integrate the Einstein equations in time and learn the structure of the progressively deeper layers inside the black hole. Conceptually it looks simple, but actually there are principal difficulties which prevent realizing this idea consistently. We will discuss these difficulties.

The serious problem is related to the existence of a singularity inside a black hole. A number of rigorous theorems (see references in [2]) imply that singularities in the structure of spacetime develop inside black holes. Unfortunately these theorems tell us practically nothing about the locations and the nature of the singularities. It is widely believed today that in the singularity inside a realistic black hole the characteristics of the curvature of the spacetime tends to infinity. Close to the singularity, where the curvature of the spacetime approaches the Plank value, the Classical General Relativity is not applicable. We have no a final version of the quantum theory of gravity yet, thus any extension of the discussion of physics in this region would be highly speculative. Fortunately, as we shall see, these singular regions are deep enough in the black hole interior and they are *in the future* with respect to overlying and *preceding* layers of the black hole where curvatures are not so high and which can be described by well-established theory.

2 Spacetime and Physical Fields Inside a Schwarzschild Black Hole

The first attempts to investigate the interior of a Schwarzschild black hole have been made in the late 70's [9,10]. It has been demonstrated that in the absence of external perturbations at late times, those regions of the black hole interior which are located long after the black hole formation are virtually free of perturbations, and therefore it can be described by the Schwarzschild geometry for the region with radius less than the gravitational radius.

The required mathematical analysis was carried out in the paper [9]. The following results were obtained. For scalar perturbations,

$$\Phi \approx D_1 \, t^{-2(l+1)} + D_2 \, t^{-(2l+3)} \ln r \tag{1}$$

where D_1 and D_2 are constants.

For perturbations described by fields with $s \neq 0$ (including metric perturbations), the main term of the r-dependent component has the following form for radiative multipoles $l \geq s$:

$$\Phi_1 \approx D_3 \, t^{-(2l+3)} \, r^{-n} \tag{2}$$

(D_3 and n are constants).

Hence, if r is fixed and $t \to \infty$, radiative modes of perturbations due to external sources are damped out and the spacetime tends to a 'stationary' state described by the Schwarzschild solution. This happens because the gravitational radiation from aspherical initial excitations becomes infinitely diluted as it reaches these regions. But this result is not valid in general case when the angular momentum or the electric charge does not vanish. The reason for that is related to the fact that the topology of the interior of a rotating or/and charged black hole differs drastically from the Schwarzschild one. The key point is that the interior of this black hole possesses a *Cauchy horizon*. This is a surface of infinite blueshift. Infalling gravitational radiation propagates inside the black hole along paths approaching the generators of the Cauchy horizon, and the energy density of this radiation will suffer an infinite blueshift as it approaches the Cauchy horizon.

This infinitely blueshifted radiation together with the radiation scattered on the curvature of spacetime inside a black hole leads to the formation of curvature singularity instead of the regular Cauchy horizon.

We will call this singularity Cauchy horizon singularity. A lot of papers were devoted to investigation of the nature of this singularity. In addition to the papers mentioned above see also [11–26, 51, 52]. Below we consider main processes which are responsible for the formation of the singularity.

3 Infinite Blueshift and Mass Inflation

This section discusses the nonlinear effects which trigger the formation of a singularity at the Cauchy horizon inside a black hole. In the Introduction we emphasized that the problem of the black hole interior is an evolutionary problem, and it depends on the initial conditions at the surface of the black hole for all momenta of time up to infinity. To specify the problem, we will consider an isolated black hole (in asymptotically flat spacetime) which was created as a result of a realistic collapse of a star without assumptions about special symmetries.

The initial data at the event horizon of an isolated black hole, which determine the internal evolution at fairly late periods of time, are known with precision because of the no hair property. Near the event horizon we have a Kerr-Newman geometry perturbed by a dying tail of gravitational waves. The fallout from this tail produces an inward energy flux decaying as an inverse power v^{-2p} of advanced time v, where $p = 2l + 3$ for multipole of order l, see [27–30]. See details in Sect. 4.

Now we should integrate the Einstein equations with the known boundary conditions to obtain the internal structure of the black hole. In general, the evolution with time into the black hole depths looks as follows. The gravitational radiation penetrating the black hole and partly backscattered by the spacetime curvature can be considered, roughly speaking, as two intersecting radial streams of infalling and outgoing gravitational radiation fluxes, the nonlinear interaction of which leads to the formation a non-trivial structure of the black hole interior. However in such a formulation it is a very difficult and still not solved completely mathematical problem. We will consider main achievements in solving it.

What are the processes responsible for formation of the Cauchy horizon singularity? The key factor producing its formation is the infinite concentration of energy density close to the Cauchy horizon as seen by a free falling observer. This infinite energy density is produced by the ingoing radiative "tail".

The second important factor here is a tremendous growth of the black hole internal mass parameter, which was dubbed *mass inflation* [31].

We start by explaining the mechanism responsible for the mass inflation [32–34]. Consider a concentric pair of thin spherical shells in an empty spacetime without a black hole [35]. One shell of mass m_{con} contracts, while the other one of mass m_{exp} expands. We assume that both shells are moving with the speed of light (for example, "are made of photons"). The contracting shell, which initially has a radius greater than the expanding one, does not create any gravitational effects inside it, so that the expanding shell does not feel the existence of the external shell. On the other hand, the contacting shell moves in the gravitational field on the expanding one. The mutual potential of the gravitational energy of the shells acts as a debit (binding energy) on the gravitational mass energy of the external contracting shell. Before the crossing of the shells, the total mass of both of them, measured by an observer outside both shells, is equal to $m_{con} + m_{exp}$ and is constant because the debit of the numerical increase of the negative potential energy is exactly balanced by the increase of the positive energies of photons blueshifted in the gravitational field of the internal sphere.

When shells cross one another, at radius r_0, the debit is transferred from the contracting shell to the expanding one, but the blueshift of the photons in the contraction shell survives. As a result, the masses of both spheres change. The increase of mass m_{con} is called mass inflation.

It is not difficult to extend this result to the shells crossing inside a black hole. For simplicity consider at the beginning a spherical charged black hole.

Ori [18] considered a continuous influx (imitating the "tail" of ingoing gravitational radiation) and the outflux as a thin shell (a very rough imitation of the outgoing gravitational radiation scattered by the spacetime curvature inside a black hole). He specified the mass $m_{in}(v)$ to imitate the Price power-law tail (see Sect. 4) and found that the mass function diverges exponentially near the Cauchy horizon as a result of the ingoing flux with the outgoing crossing of the ingoing shell:

$$m \sim \mathrm{e}^{k_0 v_-} (k_0 v_-)^{-2p}, \qquad v_- \to \infty \,, \tag{3}$$

where v_- is the advanced time in the region lying to the past of the shell, k_0 is constant, the positive constant p depends on perturbations under discussions. Expression (3) describes *mass inflation*. In this model, we have a scalar curvature singularity since the Weyl curvature invariant Ψ_2. (Coulomb component) diverges at the Cauchy horizon. Ori [18] emphasizes that in spite of this singularity, there are coordinates in which the metric is finite at the Cauchy horizon. He also demonstrated that though the tidal force in the reference frame of a freely falling observer grows infinitely its action on the free falling observer is rather modest. According to [18] the rate of growth of the curvature is proportional to

$$\sim \tau^{-2} \left|\ln|\tau|\right|^{-2p} \,, \tag{4}$$

where τ is the observer's proper time, $\tau = 0$ corresponds to the singularity. Tidal forces are proportional to the second time derivatives of the distances between various points of the object. By integrating the corresponding expression twice, one finds that as the singularity is approached ($\tau = 0$), the distortion remains finite.

There is one more effect caused by the outgoing flux. This is the contraction of the Cauchy horizon (which is singular now) with retarded time due to the focusing effect of the outgoing shell-like flux. This contraction continuous until the Cauchy horizon shrinks to $r = 0$, and a stronger singularity occurs. Ori [18] has estimated the rate of approach to this strong singularity $r = 0$.

In the case of realistic rotating black hole both processes the infinite concentration of the energy density and mass inflation near the Cauchy horizon play the key role for formation of the singularity.

4 Decay of Physical Fields Along the Event Horizon of Isolated Black Holes

Behavior of physical fields along the event horizon determines dynamics of these fields inside a black hole and has an impact on the nature of the singularity inside the black hole. We will consider here the behavior of perturbations in the gravitational field. The first guess about the decay of gravitational perturbations outside Schwarzschild black holes was given in [36], a detailed description was given by Price [27,37]. References for subsequent work see in [2] and in the important work [38]. According to [27], any radiative multipole mode l, m of any initially compact linear perturbation dies off outside a black hole at late time as t^{-2l-3}. The mechanism which is responsible for this behavior is the scattering of the field off the curvature of spacetime asymptotically far from the black hole.

In the case of a rotating black hole the problem is more complicated due to the lack of spherical symmetry. This problem was investigated in many works. See analytical analyses, references and criticism in [38], numerical approach in [39]. For individual harmonics there is a power-law decay which is similar to the Schwarzschild case except that at the event horizon the perturbation also oscillates in the Eddington coordinate v along the horizon's null generators proportional to $\sim e^{im\Omega_+ v}$, where Ω_+ is the angular velocity of the black hole rotation, $\Omega_+ = a/2Mr_+$, M and a are mass and specific angular momentum of a black hole correspondingly, $r_+ = M + \sqrt{M^2 - a^2}$. Another important difference from the Schwarzschild case is the following. In the case of the rotating black holes spherical-harmonic modes do not evolve independently. In the linearized theory there is a coupling between spherical harmonics multipole of different l, but with the same m. In the case fully nonlinear perturbations there is the guess that m will not be conserved also. So in the case of arbitrary perturbation the modes with all l which are consistent with the spin weight s of the field will be exited. For the field with spin weight s all modes with $l \geq |s|$ will be exited. Accordingly, the late-time dynamics will be dominated by the mode with $l = |s|$. The falloff rate is then $t^{-(2|s|+3)}$. In the case of the gravitational field it corresponds to $|s| = 2$ and t^{-7}.

5 Nature of the Singularity

As we mentioned in Sect. 3, we can use the initial data at the event horizon which we discussed in the previous Sect. 4 to determine the nature of the singularity inside a black hole. Main processes which are responsible for formation of the singularity were discussed in Sect. 3.

We start from the discussion of the singularity which arises at late time, long after the formation of an isolated black hole.

In general the evolution with time into the black hole deeps looks like the following. There is a weak flux of gravitational radiation into a black hole through the horizon because of small perturbations outside of it. When this radiation approaches the Cauchy horizon it suffers an infinite blueshift. The infinitely blueshift radiation together with the radiation scattered by the curvature of spacetime inside the black hole results in a tremendous growth of the black hole internal mass parameter ("mass inflation", see Sect. 3) and finally leads to formation of the curvature singularity of the spacetime along the Cauchy horizon. The infinite tidal gravitational forces arise here. This result was confirmed by considering different models of the ingoing and outgoing fluxes in the interior of charged and rotating black holes ([6], [4]).

In the case of a rotating black hole the growth of the curvature (and mass function) when we coming to the singularity is modulated by the infinite number of oscillations. This oscillatory behavior of the singularity is related to the dragging of the inertial frame due to rotation of a black hole.

It was shown [18] that the singularity at the Cauchy horizon is quite weak. In particular, the integral of the tidal force in the freely falling reference frame over the proper time remains finite. It means that the infalling object would then experience the finite tidal deformations which (for typical parameters) are even negligible. While an infinite force is extended, it acts only for a very short time. This singularity exists in a black hole at late times from the point of view of an external observer, but the singularity which arises just after the gravitational collapse of a star is much stronger. According to the Tipler's terminology [41] (see also generalization of the classification in [42]) this is a weak singularity. It seems likely that an observer falling into a black hole with the collapsing star encounters a crushing singularity (strong singularity in the Tipler's classification). This is so called Belinsky-Khalatnikov-Lifshitz (BKL) space-like singularity [40]. On the other hand an observer falling into an isolated black hole in a late times generally reaches a weak singularity described above.

The weak Cauchy horizon singularity arises first at very late time (formally infinite time) of the external observer and its null generators propagate deeper into black hole and closer to the event of the gravitational collapse. They are subject of the focusing effect under the action of the gravity of the outgoing scattered radiation (see Sect. 3). Eventually the weak null singularity shrinks to $r = 0$, and strong BKL singularity occurs. This picture was considered in details in the case of a charged spherical black hole but I do not know the strict proof of it in the case of a rotating black hole [4,25]

6 Quantum Effects

As we mentioned in Introduction quantum effects play crucial role in the very vicinity of the singularity. In addition to that the quantum processes probably are important also for the whole structure of a black hole. Indeed, in the previous discussion we emphasized that the internal structure of black holes is a problem of evolution in time starting from boundary conditions on the event horizon for all moments of time up to the infinite future of the external observer.

It is very important to know the boundary conditions up to infinity because we observed that the essential events – mass inflation and singularity formation – happened along the Cauchy horizon which brought information from the infinite future of the external spacetime. However, even an isolated black hole in an asymptotically flat spacetime cannot exist forever. It will evaporate by emitting Hawking quantum radiation. So far we discussed the problem without taking into account this ultimate fate of black holes. Even without going into details it is clear that quantum evaporation of the black holes is crucial for the whole problem.

What can we say about the general picture of the black hole's interior accounting for quantum evaporation? To account for the latter process we have to change the boundary conditions on the event horizon as compared to the boundary conditions discussed above. Now they should include the flux of negative energy across the horizon, which is related to the quantum evaporation. The last stage of quantum evaporation, when the mass of the black hole becomes comparable to the Plank mass $m_{Pl} = (\hbar c/G)^{1/2} \approx 2.2 \times 10^{-5} g$, is unknown. At this stage the spacetime curvature near the horizon reaches l_{Pl}^{-2}, where l_{Pl} is the Plank length:

$$l_{Pl} = \left(\frac{G\hbar}{c^3}\right)^{1/2} \approx 1.6 \times 10^{-33}\,\mathrm{cm}\ . \qquad (5)$$

This means that from the point of view of semiclassical physics a singularity arises here. Probably at this stage the black hole has the characteristics of an extreme black hole, when the external event horizon and internal Cauchy horizon coincide.

As for the processes inside a true singularity in the black hole's interior, they can be treated only in the framework of an unified quantum theory incorporating gravitation, which is unknown. Thus when we discuss any singularity inside a black hole, we should consider the regions with the spacetime curvature bigger than l_{pl}^{-2} as physical singularity from the point of view of semiclassical physics[1]. About different aspects of quantum effects in black holes see also [45,46].

7 Truly Realistic Black Holes

So far we discussed the isolated black holes which were formed as the result of a realistic gravitational collapse without any assumptions about symmetry.

[1] Quantum effects may manifest themselves in the region with the spacetime curvature smaller than l_{pl}^{-2}, see [2,44,45].

Still they are not truly realistic black holes. For the truly realistic black holes we should account matter and radiation falling down through the event horizon at all times up to infinity (or up to the evaporation of a black hole).

The perturbations at the event horizon which arise as a result of the collapse of the non-symmetrical body have a compact support at some initial time. Subsequent perturbations, for example the perturbations which arise from the capture of photons which originate from the relic cosmic background radiation, have non-compact support.

We should account also the difference of the curvature of spacetime in the real Universe from the curvature in the ideal model asymptotically far from the black hole (the scattering of the field in these regions is responsible for the formation of the late-time power-law radiative tails). First steps in the investigation of the truly realistic black holes were done recently.

Burko [25] studied numerically the origin of the singularity in a simple toy model of a spherical charged black hole which was perturbed nonlinearly by a self-gravitating spherical scalar field. This field was specified in such a way that it had a non-compact support. Namely, it grows logarithmically with advanced time along an outgoing characteristic hypersurface. It was demonstrated that in this case the weak null Cauchy horizon singularity was formed. The null generators of the singularity contract with retarded time, and eventually the central spacelike strong singularity forms. Thus in this case the casual structure of the singularity is the same as in the case of the perturbations with a compact support at some initial time. Of course, this example is very far to be a realistic one.

In another work Burko [26] demonstrated numerically that the scalar field can be chosen along an outgoing characteristic hypersurface in such a way that only spacelike strong singularity forms. The scalar field has a non-compact support in this case.

It is an open question whether these results hold also for rotating black holes, and what would be a result in the case of a realistic source of perturbations for realistic black holes.

I want to do the following remark. As I mentioned in Sect. 5, when inside a black hole we come to the singularity close enough, where the spacetime curvature reaches l_{pl}^{-2}, we should consider this region as a singular one from the point of view of semiclassical physics. This means that any details of the classical spacetime structure in the singular quantum region make no sense. This means that if we are interested in the spacetime structure only outside the singular quantum region and want to investigate this structure in some definite region at the singularity we should take into account radiation coming to the event horizon during the restrict period of time t_0 only. All radiation which comes to the border of a black hole later will come to the region under consideration inside the quantum singular region and does not influence on the structure of the spacetime outside it at this place. It is rather easy to estimate this period t_0. It is

$$t_0 \approx 3 \cdot 10^6 \sec \left(\frac{M}{10^9 M_\odot} \right) . \tag{6}$$

8 Can One See What Happens Inside a Black Hole?

Is it possible for a distant observer to receive information about the interior of a black hole? Strictly speaking, this is forbidden by the very definition of a black hole. What we have in mind in asking this question is the following. Suppose there exists a stationary or static black hole. Can we, by using some device, get information about the region lying inside the apparent horizon?

Certainly it is possible if one is allowed to violate the weak energy condition. For example, if one sends into a black hole some amount of "matter" of negative mass, the surface of black hole shrinks, and some of the rays which previously were trapped inside the black hole would be able to leave it. If the decrease of the black hole mass during this process is small, then only a very narrow region lying directly inside the horizon of the former black hole becomes visible.

In order to be able to get information from regions not close the apparent horizon but deep inside an original black hole, one needs to change drastically the parameters of the black hole or even completely destroy it. A formal solution corresponding to such a destruction can be obtained if one considers a spherically symmetric collapse of negative mass into a black hole. The black hole destruction occurs when the negative mass of the collapsing matter becomes equal to the original mass of the black hole. In such a case an external observer can see some region close to the singularity. But even in this case the four-dimensional region of the black hole interior which becomes visible has a four-dimensional spacetime volume of order M^4. It is much smaller than four-volume of the black hole interior, which remains invisible and which is of order M^3T, where T is the time interval between the black hole formation and its destruction (we assume $T \gg M$). The price paid for the possibility of seeing even this small part of the depths of the black hole is its complete destruction.

Does this mean that it is impossible to see what happens inside the apparent horizon without a destructive intervention? We show that such a possibility exists (Frolov and Novikov [47,48]). In particular, in these works we discuss a gedanken experiment which demonstrates that traversable wormholes (if only they exist) can be used to get information from the interior of a black hole practically without changing its gravitational field. Namely, we assume that there exist a traversable wormhole, and its mouths are freely falling into a black hole. If one of the mouths crosses the gravitational radius earlier than the other, then rays passing through the first mouth can escape from the region lying inside the gravitational radius. Such rays would go through the wormhole and enter the outside region through the second mouth, see details in [2,49].

Acknowledgements

This paper was supported in part by the Danish natural Science Research Council through grant No. 9701841 and also in part by Danmarks Grundforskningsfond through its support for establishment of the TAC.

References

1. L.M. Burko, and A. Ori: editors. *Internal Structure of Black Holes and Spacetime Singularity*, (Institute of Physics Publishing, Bristol and Philadelphia, and Israel Physical Society, Jerusalem, 1997)
2. V.P. Frolov, I.D. Novikov: *Black Hole Physics: Basic Concepts and New Developments*, (Kluwer Academic Publisher, Dordrecht 1998)
3. P.R. Brady, S. Droz, and S.M. Morsink: Phys. Rev. D **58** 084034 (1998).
4. A. Ori: Phys. Rev. Letters. **83**, 5423 (1999)
5. A. Ori: Phys. Rev. D, **58**, 084016, (1998)
6. A. Ori: Phys. Rev. D, **61**, 024001, (1999)
7. B.K. Berger: Living Rev. Relativity 2002-1
8. A. Ori: gr-qc/0103012 (2001)
9. A.G. Doroshkevich, and I.D. Novikov: Zh. Eksp. Teor Fiz., **74**, 3 (1978)
10. E. Poisson, and W. Israel: Class. Quantum Grav. **5**, L201, (1998)
11. R. Penrose: *Structure of Space-Time* in C.M. DeWitt and J.A. Wheeler (eds) (Battelle Recontres, Benjamin, NY 1968)
12. J.M. McNamara: Proc. R. Soc. London **A358**, 499, (1978)
13. J.M. McNamara: Proc. R. Soc. London **A364**, 121, (1978)
14. Y. Gürsel, I.D. Novikov, V. Sandberg, and A.A. Starobinsky: Phys. Rev. D, **20**, 1260, (1979)
15. Y. Gürsel, V. Sandberg, I.D. Novikov, and A.A. Starobinsky: Phys. Rev. D, **19**, 413, (1979)
16. R.A. Matzner, N. Zamorano and V.D Sandberg: Phys. Rev. D, **19**, 2821, (1979)
17. S. Chandrasekhar and J.B. Hartle: Proc. R. Soc. London **A384**, 301, (1982)
18. A. Ori: Phys. Rev. Letters. **67**, 789, (1991)
19. A. Ori: Phys. Rev. Letters. **68**, 2117, (1992)
20. P.R. Brady, and J.D. Smith: Phys. Rev. Letters. **75**, 1256, (1995)
21. A. Ori and E.E. Flanagan: Phys. Rev. D. **53**, R1754, (1996)
22. A. Ori: Phys. Rev. D. **55**, 4860, (1997)
23. L.M. Burko and A. Ori: Phys. Rev. D. **57**, R7084, (1998)
24. O. Gurtug and M. Halilsoy: Mustafa Halilsoy, submitted for publication, gr-qc/0203019 (2002).
25. L.M. Burko, Phys. Rev. D **66** 024046 (2002)
26. L.M. Burko, Phys. Rev. Lett. **90** 121101 (2003), Erratum-ibid. **90** 240902 121101 (2003)
27. R.H. Price: Phys. Rev. D. **5**, 2439, (1972)
28. C. Gundlach, R.H. Price, and J. Pullin: Phys. Rev. D. **49**, 883, (1994)
29. C. Gundlach, R.H. Price, and J. Pullin: Phys. Rev. D. **49**, 890, (1994)
30. A. Ori: Gen. Rel. & Grav. **29**, 881, (1997)
31. E. Poisson and W. Israel: Phys. Rev. D. **41**, 1796, (1990)

32. T. Dray, and G. 't Hooft: Commun. Math. Phys. **99**, 613, (1985)
33. C. Barrabès, W. Israel, E. Poisson: Class. Quantum Grav. **7**, L273, (1990)
34. A. Bonanno, S. Droz, W. Israel, and S. Morsink: Proc. R. Soc. London **A450**, 553, (1995)
35. S. Droz, W. Israel, and S. Morsink: Physics World **9**, 34, (1995)
36. A.G. Doroshkevich, Ya. B. Zeldovich and I.D. Novikov: Zh. Eksp. Teor. Fiz. **49**, 170 (1965)
37. R.H. Price: Phys. Rev. D. **5**, 2419 (1972).
38. L. Barack and A. Ori: Phys. Rev. D. **60**, 124005 (1999).
39. L.M. Burko and G. Klama: Phys. Rev. D. **67**, 081502 (2003).
40. V.A. Belinsky, I.M. Khalatnikov and E.M. Lifshitz: Usp. Fiz. Nauk **102**, 463 (1970), [Adv. Phys. **19**, 525 (1970)]
41. F.J. Tipler: Phys. Lett **64A**, 8 (1977)
42. A. Ori: Phys. Rev. D. **61**, 064016, (2000).
43. I.D. Novikov and A.A. Starobinsky: Zh. Eksp. Teor. Fiz **78**, 3 (1980)
44. V.A. Berezin: Preprint NRI p-0183, Moscow (1980)
45. W.G. Anderson, P.R. Brady, W. Israel, and S.M. Morsink: Phys. Lett **70**, 1041 (1993)
46. R. Balbinot and E. Poisson: *Phys. Lett* **70**, 13 (1992)
47. V.P. Frolov and I.D. Novikov: Phys. Rev. D, **42**, 1057 (1990)
48. V.P. Frolov and I.D. Novikov: Phys. Rev. D, **48**, p.1607 (1993)
49. I.V. Artemova and I.D. Novikov: in Proceedings of th 9-th Course of Astrofundamental Physics, The Early Universe and Cosmic Microwave Background: Theory and Observations Palermo- Sicily, (2002) N.Sanchez (eds) in press (2003)
50. I.D. Novikov,: in Proceedings of Texas in Tuscany, XXI Symposium on Relativistic Astrophysics. Ed. R.Bandiera, R.Maiolino, and F.Mannucci; World Scientific, page 77, (2003)
51. Y. Oren and T.Piron: gr-qc/0306078, (2003)
52. M. Dafermos: gr-qc/0307013, (2003)

Microscopic Interpretation of Black Hole Entropy

Maro Cvitan[1], Silvio Pallua[2], and Predrag Prester[3]

[1] mcvitan@phy.hr
[2] pallua@phy.hr
[3] pprester@phy.hr
Theoretical Physics Department, Faculty of Natural Sciences and Mathematics, University of Zagreb, Bijenička c. 32, pp. 331, 10002 Zagreb, Croatia

Summary. It is shown, using conformal symmetry methods, that one can obtain microscopic interpretation of black hole entropy for general class of higher curvature Lagrangians.

1 Introduction

The entropy of black holes can be calculated with the well known Bekenstein-Hawking formula

$$S_{BH} = \frac{A}{4\pi G} , \tag{1}$$

where A represents area of black hole horizon. In fact generalisation of this formula is given in [1] for general interaction of the form

$$L = L(g_{ab}, R_{abcd}, \nabla R_{abcd}, \psi, \nabla\psi, \ldots) . \tag{2}$$

Here ψ refers to matter fields and dots refer to derivatives up to order m. In that case the entropy is given with the relation [1]

$$S = -2\pi \int_{\mathcal{H}\cap C} \hat{\epsilon} E^{abcd} \eta_{ab}\eta_{cd} . \tag{3}$$

Here $\mathcal{H}\cap C$ is a cross section of the horizon, η_{ab} denotes binormal to $\mathcal{H}\cap C$, $\hat{\epsilon}$ is induced volume element on $\mathcal{H}\cap C$ and

$$E^{abcd} = \frac{\partial L}{\partial R_{abcd}} - \nabla_{a_1} \frac{\partial L}{\partial \nabla_{a_1} R_{abcd}} + \ldots (-)^m \nabla_{(a_1 \ldots a_m)} \frac{\partial L}{\partial \nabla_{(a_1 \ldots a_m)} R_{abcd}} . \tag{4}$$

The problem of microscopic description of black hole entropy was approached by different methods like string theory which treated extremal black holes [2] or e.g. loop quantum gravity [3]. An interesting line of approach is based on conformal field theory and Virasoro algebra. One particular formulation was due to Solodukhin who reduced the problem of D-dimesional black holes to effective two-dimensional theory with fixed boundary conditions on the horizon. The effective theory was found to admit Virasoro algebra near horizon. Calculation of its central charge allows then to compute the entropy [4, 5]. An independent formulation is due to Carlip [6, 7, 8, 9] who has shown that under certain simple assumptions on boundary conditions near black hole horizon one can identify a subalgebra of algebra of diffeomorphisms which turns out to be Virasoro algebra. The fixed boundary conditions give rise to central extensions of this algebra. The entropy is then calculated from Cardy formula [10]

$$S_c = 2\pi\sqrt{\left(\frac{c}{6} - 4\Delta_g\right)\left(\Delta - \frac{c}{24}\right)} . \tag{5}$$

Here Δ is the eigenvalue of Virasoro generator L_0 for the state we calculate the entropy and Δ_g is the smallest eigenvalue. In that way the entropy formula (1) for Einstein gravity was reproduced. In this lecture we want to investigate if such microscopic interpretation is possible for more general type of interaction. We shall first treat Gauss-Bonnet gravity using Solodukhin method and then using Carlip method. The latter method will allow us to treat more general cases. These are described with Lagrangian which is allowed to have arbitrary dependence on Riemann tensor but not on its derivatives, more precisely

$$L = L(g_{ab}, R_{abcd}) . \tag{6}$$

In that case the tensor E^{abcd} takes the form

$$E^{abcd} = \frac{\partial L}{\partial R_{abcd}} . \tag{7}$$

We note that interesting new possibilities and open questions arise for interpretation of black hole entropy. For discussion in the Gauss–Bonnet case see e.g. [11, 12, 13, 14, 15, 16].

2 Effective CFT Near the Horizon

Now we turn our attention to particular microscopic derivation of entropy of black hole, which was first done in [4] for the Einstein gravity, and then extended to general D-dimensional Gauss–Bonnet (GB) theories in [5]. General GB action[1] is given by

$$I_{\mathrm{GB}} = -\sum_{m=0}^{[D/2]} \lambda_m \int \mathrm{d}^D x \sqrt{-g}\mathcal{L}_m(g) , \tag{8}$$

where GB densities $\mathcal{L}_m(g)$ are

$$\mathcal{L}_m(g) = \frac{(-1)^m}{2^m} \delta^{\rho_1\sigma_1\cdots\rho_m\sigma_m}_{\mu_1\nu_1\cdots\mu_m\nu_m} R^{\mu_1\nu_1}{}_{\rho_1\sigma_1} \cdots R^{\mu_m\nu_m}{}_{\rho_m\sigma_m} , \tag{9}$$

We take $\lambda_0 = 0$ (cosmological constant), because we shall see that this term is irrelevant for our calculation. Coupling constant λ_1 is related to more familiar D-dimensional Newton gravitational constant G_D through $\lambda_1 = (16\pi G_D)^{-1}$.

We neglect matter and consider S-wave sector of the theory, i.e., we consider only radial fluctuations of the metric. It is easy to show that in this case (8) can

[1] Also known as Lovelock gravity.

be written in the form of an effective two-dimensional "generalised higher-order Liouville theory" given with

$$
\begin{aligned}
I_{\mathrm{GB}} = \Omega_{D-2} \sum_{m=0}^{[D/2]} \lambda_m \frac{(D-2)!}{(D-2m)!} \int d^2x \sqrt{-\gamma}\, r^{D-2m-2} \left[1-(\nabla r)^2\right]^{m-2} \\
\times \Big\{ 2m(m-1)r^2 \left[(\nabla_a \nabla_b r)^2 - (\nabla^2 r)^2\right] \\
+2m(D-2m) r \nabla^2 r \left[1-(\nabla r)^2\right] + m\mathcal{R} r^2 \left[1-(\nabla r)^2\right] \\
-(D-2m)(D-2m-1)\left[1-(\nabla r)^2\right]^2 \Big\} \,. \qquad (10)
\end{aligned}
$$

We now suppose that black hole with horizon *is existing* and we are interested in fluctuations (or, better, quantum states) near it. In the spherical geometry apparent horizon $\mathcal{H}$ (a line in x-plane) can be defined by the condition [17]

$$
(\nabla r)^2\big|_{\mathcal{H}} \equiv \gamma^{ab} \partial_a r \partial_b r \Big|_{\mathcal{H}} = 0 \,. \qquad (11)
$$

Notice that (11) is invariant under (regular) conformal rescalings of the effective two-dimensional metric γ_{ab}. Near the horizon (11) is approximately satisfied. It is easy to see that after partial integration and implementation of horizon condition $(\nabla r)^2 \approx 0$, (10) becomes near the horizon approximately

$$
\begin{aligned}
I_{\mathrm{GB}} = -\Omega_{D-2} \sum_{m=0}^{[D/2]} \lambda_m \frac{(D-2)!}{(D-2m-2)!} \int \mathrm{d}^2x \sqrt{-\gamma}\, r^{D-2m-2} \\
\times \left\{ m(\nabla r)^2 - \frac{m}{(D-2m)(D-2m-1)} \mathcal{R} r^2 + 1 \right\} \,. \qquad (12)
\end{aligned}
$$

If we define

$$
\Phi^2 \equiv 2\Omega_{D-2} \sum_{m=1}^{[D/2]} m\lambda_m \frac{(D-2)!}{(D-2m)!} r^{D-2m} \,, \qquad (13)
$$

and make reparametrizations

$$
\phi \equiv \frac{2\Phi^2}{q\Phi_h} \,, \qquad \tilde{\gamma}_{ab} \equiv \frac{\mathrm{d}\phi}{\mathrm{d}r} \gamma_{ab} \,, \qquad (14)
$$

where q is arbitrary dimensionless parameter, the action (12) becomes

$$
I_{\mathrm{GB}} = \int \mathrm{d}^2x \sqrt{-\tilde{\gamma}} \left[\frac{1}{4} q\Phi_h \phi \tilde{\mathcal{R}} - V(\phi) \right] \qquad (15)
$$

This action can be put in more familiar form if we make additional conformal reparametrization:

$$
\bar{\gamma}_{ab} \equiv \mathrm{e}^{-\frac{2\phi}{q\Phi_h}} \tilde{\gamma}_{ab} \,, \qquad (16)
$$

Now (15) takes the form

$$
I_{\mathrm{GB}} = -\int \mathrm{d}^2x \sqrt{-\bar{\gamma}} \left[\frac{1}{2} \left(\bar{\nabla}\phi \right)^2 - \frac{1}{4} q\Phi_h \phi \bar{\mathcal{R}} + U(\phi) \right] \,, \qquad (17)
$$

which is similar to the Liouville action. The difference is that potential $U(\phi)$ is not purely exponential which means that the obtained effective theory is not exactly conformaly invariant.

Action (17) is of the same form as that obtained from pure Einstein action. In [4] it was shown that if one imposes condition that the metric $\bar{\gamma}_{ab}$ is *nondynamical* then the action (17) describes CFT *near the horizon*[2]. We therefore fix $\bar{\gamma}_{ab}$ near the horizon and take it to be a metric of a static spherically symmetric black hole:

$$d\bar{s}^2_{(2)} \equiv \bar{\gamma}_{ab} dx^a dx^b = -f(w) dt^2 + \frac{dw^2}{f(w)} , \tag{18}$$

where near the horizon $f(w_h) = 0$ we have

$$f(w) = \frac{2}{\beta}(w - w_h) + O\left((w - w_h)^2\right) . \tag{19}$$

We now make coordinate reparametrization $w \to z$

$$z = \int^w \frac{dw}{f(w)} = \frac{\beta}{2} \ln \frac{w - w_h}{f_0} + O(w - w_h) \tag{20}$$

in which 2-dim metric has a simple form

$$d\bar{s}^2_{(2)} = f(z)\left(-dt^2 + dz^2\right) , \tag{21}$$

and the function f behaves near the horizon ($z_h = -\infty$) as

$$f(z) \approx f_0 e^{2z/\beta} , \tag{22}$$

i.e., it *exponentially* vanishes. It is easy to show that equation of motion for ϕ which follows from (17), (21), (22) is

$$\left(-\partial_t^2 + \partial_z^2\right)\phi = \frac{1}{4} q\Phi_h \bar{\mathcal{R}} f + f U'(\phi) \approx O\left(e^{2z/\beta}\right) , \tag{23}$$

and that the "flat" trace of the energy-momentum tensor is

$$-T_{00} + T_{zz} = \frac{1}{4} q\Phi_h \left(-\partial_t^2 + \partial_z^2\right)\phi - f U(\phi) \approx O\left(e^{2z/\beta}\right) , \tag{24}$$

which is exponentially vanishing near the horizon[3]. From (23) and (24) follows that the theory of the scalar field ϕ exponentially approaches CFT near the horizon.

Now, one can construct corresponding Virasoro algebra using standard procedure. Using light-cone coordinates $z_\pm = t \pm z$ right-moving component of energy–momentum tensor near the horizon is approximately

$$T_{++} = (\partial_+\phi)^2 - \frac{1}{2} q\Phi_h \partial_+^2 \phi + \frac{q\Phi_h}{2\beta} \partial_+ \phi . \tag{25}$$

It is important to notice that horizon condition (11) implies that r and ϕ are (approximately) functions only of one light-cone coordinate (we take it to be z_+), which means that only one set of modes (left *or* right) is contributing.

[2] Carlip showed that above condition is indeed consistent boundary condition [7].

[3] Higher derivative terms in (10) make contribution to (24) proportional to $f(\nabla\phi)^2 \approx o(\exp(2z/\beta))$.

Virasoro generators are coefficients in the Fourier expansion of T_{++}:

$$T_n = \frac{\lambda}{2\pi} \int_{-\lambda/2}^{\lambda/2} \mathrm{d}z\, \mathrm{e}^{\mathrm{i}2\pi n z/\lambda} T_{++}\,, \tag{26}$$

where we compactified z-coordinate on a circle of circumference λ. Using canonical commutation relations it is easy to show that Poisson brackets of T_n's are given with

$$i\{T_n, T_m\}_{\mathrm{PB}} = (n-m)T_{n+m} + \frac{\pi}{4} q^2 \Phi_h^2 \left(n^3 + n \left(\frac{\lambda}{2\pi\beta} \right)^2 \right) \delta_{n+m,0}\,. \tag{27}$$

To obtain the algebra in quantum theory (at least in semiclassical approximation) one replaces Poisson brackets with commutators using $[\,,] = i\hbar\{\,,\}_{\mathrm{PB}}$, and divide generators by $\hbar$. From (27) it follows that "shifted" generators

$$L_n = \frac{T_n}{\hbar} + \frac{c}{24} \left(\left(\frac{\lambda}{2\pi\beta} \right)^2 + 1 \right) \delta_{n,0}\,, \tag{28}$$

where

$$c = 3\pi q^2 \frac{\Phi_h^2}{\hbar}\,, \tag{29}$$

satisfy Virasoro algebra

$$[L_n, L_m] = (n-m)L_{n+m} + \frac{c}{12} \left(n^3 - n \right) \delta_{n+m,0} \tag{30}$$

with central charge c given in (29).

Outstanding (and unique, as far as is known) property of the Virasoro algebra is that in its representations a logarithm of the number of states (i.e., entropy) with the eigenvalue of L_0 equal to Δ is asymptoticaly given with Cardy formula (5). If we assume that in our case $\Delta_g = 0$ in semiclassical approximation (more precisely, $\Delta_g \ll c/24$), one can see that number of microstates (purely quantum quantity) is in leading approximation completely determined by (semi)classical values of c and L_0. Now it only remains to determine Δ. In a classical black hole solution we have

$$r = w = w_h + (w - w_h) \approx r_h + f_0 \mathrm{e}^{2z/\beta}\,, \tag{31}$$

so from (14) and (13) follows that near the horizon $\phi \approx \phi_h$. Using this configuration in (26) one obtains $T_0 = 0$, which plugged in (28) gives

$$\Delta = \frac{c}{24} \left(\left(\frac{\lambda}{2\pi\beta} \right)^2 + 1 \right)\,. \tag{32}$$

Finally, using (29) and (32) in Cardy formula (5) one obtains

$$S_{\mathrm{C}} = \frac{c}{12} \frac{\lambda}{\beta} = \frac{\pi}{4} q^2 \frac{\lambda}{\beta} \frac{\Phi_h^2}{\hbar}\,. \tag{33}$$

Let us now compare (33) with classical formula (3), which for GB gravities can be written as [18]

$$S_{\rm GB} = \frac{4\pi}{\hbar} \sum_{m=1}^{[D/2]} m\lambda_m \oint d^{D-2}x \sqrt{\tilde{g}} \mathcal{L}_{m-1}(\tilde{g}_{ij}) \,, \tag{34}$$

Here $\tilde{g}_{ij}$ is induced metric on the horizon, and densities $\mathcal{L}_m$ are given in (9). In the sphericaly symmetric case horizon is a $(D-2)$-dimensional sphere with radius r_h and $R(\tilde{g}_{ij}) = -(D-2)(D-3)/r_h^2$, so (34) becomes

$$S_{\rm GB} = \frac{4\pi}{\hbar} \Omega_{D-2} \sum_{m=1}^{[D/2]} m\lambda_m \frac{(D-2)!}{(D-2m)!} r^{D-2m} = 2\pi \frac{\Phi_h^2}{\hbar} \tag{35}$$

Using this our expression (33) can be written as

$$S_{\rm C} = \frac{q^2}{8} \frac{\lambda}{\beta} S_{\rm GB} \,, \tag{36}$$

so it gives correct result apart from dimensionless coeficient, which can be determined in the same way as in pure Einstein case [7]. First, it is natural to set the compactification period λ equal to period of Euclidean-rotated black hole[4], i.e.,

$$\lambda = 2\pi\beta \,. \tag{37}$$

The relation between eigenvalue Δ of L_0 and c then becomes

$$\Delta = \frac{c}{12} \,. \tag{38}$$

One could be tempted to expect this relation to be valid for larger class of black holes and interactions then those treated so far.

To determine dimensionless parameter q we note that our effective theory given with (17) depends on effective parameters Φ_h and β, and thus one expects that q depends on coupling constants only through dimensionless combinations of them. Thus to determine q one may consider $\lambda_2 = 0$ case and compare expression for central charge (29) with that obtained in [9], which is

$$c = \frac{3A_h}{2\pi\hbar G_D} \,, \tag{39}$$

where $A_h = \Omega_{D-2} r_h^{D-2}$ is the area of horizon. One obtains that

$$q^2 = \frac{4}{\pi} \,. \tag{40}$$

One could also perform boundary analysis of [9] for GB gravity (see Appendix of [5]). This procedure gives $\Delta = \Phi_h^2/\hbar$ which combined with (29) and (38) gives (40).

Using (37) and (40) one finally obtains desired result

$$S_{\rm C} = S_{\rm GB} \,. \tag{41}$$

Let us mention that there were other approaches to calculation of black hole entropy using Virasoro algebra of near-horizon symmetries of effective 2-dim QFT (see [19]).

---

[4] We note that our functions depend only on variable z_+, so the periodicity properties in time t are identical to those in z.

3 Covariant Phase Space Formulation of Gravity

As mentioned before there is another method in which one is not using dimensional reduction. The emphasis will be in assuming appropriate boundary conditions near horizon of black hole. In this approach it will turn out to be useful to use the covariant phase space formulation of gravity [20, 21]. For this reason we shall here review it shortly for any diffeomophism invariant theory with the Lagrangian D-form

$$\mathbf{L}[\Phi] = \epsilon L(\Phi) \ . \tag{42}$$

Here Φ denotes collection of fields, ϵ is the volume D-form. Then one can calculate the variation

$$\delta \mathbf{L}[\phi] = \delta \mathbf{E}[\phi]\delta\phi + d\mathbf{\Theta}[\phi, \delta\phi] \ . \tag{43}$$

The $(D-1)$-form, called symplectic potential for Lagrangians of type (6) was shown in [1] to be

$$\Theta_{pa_1,\dots a_{n-2}} = 2\epsilon_{apa_1\dots a_{n-2}}(E^{abcd}\nabla_d \delta g_{bc} - \nabla_d E^{abcd}\delta g_{bc}) \ . \tag{44}$$

To any vector field ξ we can associate a Noether current $(D-1)$-form

$$\mathbf{J}[\xi] = \mathbf{\Theta}[\phi, \mathcal{L}_\xi \phi] - \xi \cdot \mathbf{L} \ , \tag{45}$$

and the Noether charge $(D-2)$-form

$$\mathbf{J} = \mathrm{d}\mathbf{Q} \ . \tag{46}$$

For all diffeomorphism invariant theories the Hamiltonian is a pure surface term [1]

$$\delta H[\xi] = \int_{\partial C} (\delta \mathbf{Q}[\xi] - \xi \cdot \mathbf{\Theta}[\phi, \delta\phi]) \ . \tag{47}$$

The integrability condition requires that a $(D-1)$-form $\mathbf{B}$ exists with the property

$$\delta \int_{\partial C} \xi \cdot \mathbf{B} = \int_{\partial C} \xi \cdot \mathbf{\Theta} \ , \tag{48}$$

where C is a Cauchy surface. Then (47) can be integrated to give

$$H[\xi] = \int_{\partial C} (\mathbf{Q}[\xi] - \xi \cdot \mathbf{B}) \ . \tag{49}$$

As bulk terms of H vanish, variation of $H[\xi]$ is equal to variations of boundary term $J[\xi]$. As explained in [9, 22], that enables one to obtain the Dirac bracket $\{J[\xi_1], J[\xi_2]\}_D$

$$\{J[\xi_1], J[\xi_2]\}_D = \int_{\partial C} (\xi_2 \cdot \mathbf{\Theta}[\phi, \mathcal{L}_{\xi_1}\phi] - \xi_1 \cdot \mathbf{\Theta}[\phi, \mathcal{L}_{\xi_2}\phi] - \xi_2 \cdot (\xi_1 \cdot \mathbf{L})) \ , \tag{50}$$

and the algebra

$$\{J[\xi_1], J[\xi_2]\}_D = J[\{\xi_1, \xi_2\}] + K[\xi_1, \xi_2] \ , \tag{51}$$

with K as central extension.

Using (44), we get a more explicit form

$$\{J[\xi_1], J[\xi_2]\}_D = \int_{\partial C} \epsilon_{apa_1\cdots a_{n-2}} \left(\xi_2^p E^{abcd}\nabla_d \delta_1 g_{bc} - \xi_1^p \nabla_d E^{abcd}\delta_2 g_{bc} - (1 \leftrightarrow 2)\right) \\ \xi_2 \cdot (\xi_1 \cdot \mathbf{L}) \ . \tag{52}$$

4 Boundary Conditions on Horizon

The main idea of the second approach mentioned in the introduction is to impose existence of Killing horizon and a class of boundary conditions on it proposed by Carlip [9] for Einstein gravity. (For alternative discussions of this method see also [23, 24, 25, 26, 27]). We shall assume the validity of these boundary conditions also for the interactions treated in this paper. The Killing horizon in D-dimensional spacetime M with boundary ∂M has a Killing vector χ^a with the property

$$\chi^2 = g_{ab}\chi^a\chi^b = 0 \quad \text{at} \quad \partial M \,. \tag{53}$$

One defines near horizon spatial vector ϱ_a

$$\nabla_a \chi^2 = -2\kappa\varrho_a \,. \tag{54}$$

We require that variations satisfy

$$\frac{\chi^a\chi^b}{\chi^2}\delta g_{ab} \to 0, \quad \chi^a t^a \delta g_{ab} \to 0 \quad \text{as} \quad \chi^2 \to 0 \,. \tag{55}$$

Here χ^a and ϱ_a are kept fixed, t^a is any unit spacelike vector tangent to ∂M. One considers diffeomorphism generated by vector fields

$$\xi^a = T\chi^a + R\varrho^a \,, \tag{56}$$

Boundary conditions together with the closure of algebra imply

$$R = \frac{\chi^2}{\kappa\varrho^2}\chi^a\nabla_a T \,, \quad \varrho^a\nabla_a T = 0 \,. \tag{57}$$

An additional requirement will be necessary as already explained in [9]. With the help of acceleration of an orbit

$$a^a = \chi^b\nabla_b\chi^a \,, \tag{58}$$

we define

$$\hat\kappa^2 = -\frac{a^2}{\chi^2} \,. \tag{59}$$

We ask that

$$\delta\int_{\partial C}\hat\epsilon\left(\hat\kappa - \frac{\varrho}{|\chi|}\kappa\right) = 0 \,. \tag{60}$$

This condition will (see last section) guarantee existence of generators $H[\xi]$ and for diffeomorphisms (56) will imply

$$\int_{\partial C}\hat\epsilon\dddot{T} = 0 \,, \tag{61}$$

and for one parameter group of diffeomorphisms the orthogonality relations

$$\int_{\partial C}\hat\epsilon T_n T_m = \delta_{n+m,0} \,. \tag{62}$$

In order to calculate central term from (51) we shall use equation (52) where we shall integrate over $(D-2)$-dimensional surface $\mathcal{H}\cap C$ which is the intersection of Killing horizon with the Cauchy surface C. In addition to Killing vector χ^a we introduce other future directed null normal

$$N^a = k^a - \alpha\chi^a - t^a \,, \tag{63}$$

where t^a is tangent to $\mathcal{H} \cap C$, and

$$k^a = -\frac{\left(\chi^a - \varrho^a \frac{|\chi|}{|\varrho|}\right)}{\chi^2} \,. \tag{64}$$

In this way

$$\epsilon_{bca_1 \ldots a_{n-2}} = \epsilon_{a_1 \cdots a_{n-2}} \eta_{bc} \,, \tag{65}$$

and

$$\eta_{ab} = 2\chi_{[b} N_{c]} = \frac{2}{|\chi||\varrho|} \varrho_{[a}\chi_{b]} + t_{[a}\chi_{b]} \,. \tag{66}$$

We proceed now to evaluate the first term of the integrand of (52) in the leading order in χ^2. Using boundary conditions we can derive the following relation

$$\nabla_d \delta g_{ab} = \nabla_d \nabla_a \xi_b + \nabla_d \nabla_b \xi_a = -2\chi_d \chi_a \chi_b \frac{\ddot{T}}{\chi_4} + 2\chi_d \chi_{(a}\varrho_{b)} \left(\frac{\dddot{T}}{\kappa\chi^2\varrho^2} + 2\frac{\kappa\ddot{T}}{\chi^4}\right) \,. \tag{67}$$

After a straightforward calculation and using symmetries of E^{abcd} which are those of Riemann tensor we obtain for the first term in (52)

$$\frac{1}{2} E^{abcd} \eta_{ab} \eta_{cd} \left(2\kappa T_2 \dot{T}_1 - \frac{T_2 \dddot{T}_1}{\kappa}\right) - (1 \leftrightarrow 2) + O(\chi^2) \,. \tag{68}$$

In fact this is the main contribution because we shall show that other terms near horizon are of the order of χ^2. That is obvious for the third term because Lagrangian is expected to be finite on horizon. The second term after using (56) and (66) reads

$$\frac{\chi_{[a}\varrho_{b]}}{\kappa\varrho^2} \left[\frac{1}{\kappa}(\dot{T}_2 \ddot{T}_1 - \dot{T}_1 \ddot{T}_2)\chi_c - (T_1 \ddot{T}_2 - T_2 \ddot{T}_1)\varrho_c\right] \nabla_d E^{abcd} \,. \tag{69}$$

We want to exploit the fact that χ is a Killing vector. For this purpose it would be desirable to connect ∇_d with ∇_χ. We assume that "spatial" derivatives are $O(\chi^2)$ near horizon (see Appendix A of [9]), which implies

$$\nabla_d E^{abcd} = \left(\frac{\chi_d \nabla_\chi}{\chi^2} + \frac{\varrho_d \nabla_\varrho}{\varrho^2}\right) E^{abcd} + O(\chi^2) \,. \tag{70}$$

From (64)

$$\nabla_\varrho = \frac{|\varrho|}{|\chi|} \nabla_\chi - |\varrho||\chi| \nabla_k \,. \tag{71}$$

This last equation because of consistency with (57) implies

$$\nabla_d E^{abcd} = \frac{\chi_d - \varrho_d}{\chi^2} \nabla_\chi E^{abcd} + O(\chi^2) \,. \tag{72}$$

We are able now to exploit the existence of Killing vector

$$\mathcal{L}_\chi E^{abcd} = 0 \,, \tag{73}$$

or

$$\nabla_\chi E^{abcd} - E^{fbcd}\nabla_f \chi^a - E^{afcd}\nabla_f \chi^b - E^{abfd}\nabla_f \chi^c - E^{afcf}\nabla_f \chi^d = 0\,, \tag{74}$$

But due to our boundary conditions up to leading terms in χ^2

$$\nabla^a \chi^b = \frac{\kappa}{\chi^2}(\chi^a \varrho^b - \chi^b \varrho^a)\,, \tag{75}$$

we get

$$\begin{aligned}\nabla_\chi E^{abcd} &= \frac{\chi_{[a}\varrho_{b]}}{\varrho^6}(\alpha\chi_c + \beta\varrho_c)(\chi_d - \varrho_d) \times \\ &\times(\chi_f \varrho^a E^{fbcd} - \chi^a \varrho_f E^{fbcd} - \chi_f \varrho^b E^{facd} + \chi^b \varrho_f E^{facd} \\ &\quad + \chi_f \varrho^c E^{fdab} - \chi^c \varrho_f E^{fdab} + \chi_f \varrho^d E^{fcba} - \chi^d \varrho_f E^{fcba})\,. \end{aligned}\tag{76}$$

Here

$$\alpha = \frac{1}{\kappa}(\dot T_2 \ddot T_1 - \dot T_1 \ddot T_2), \quad \beta = -(T_1 \ddot T_2 - T_2 \ddot T_1)\,, \tag{77}$$

After multiplication we find two classes of terms. One class contains terms like

$$\frac{1}{\chi^2\varrho^2}\chi_e\varrho_f\chi_g\varrho_h E^{efgh} = \frac{1}{\chi^2\varrho^2}\chi_{[a}\varrho_{b]}\chi_{[c}\varrho_{d]}E^{abcd} = \frac{1}{4}\eta_{ab}\eta_{cd}E^{abcd}\,, \tag{78}$$

and such terms are finite but come always in pairs and cancel. All other terms are of the form

$$\frac{1}{\chi^4}\chi_a\chi_b\varrho_c\varrho_d E^{abcd}\,,$$

and due to antisymmetry properties of E^{abcd} they vanish. We conclude that only first term in (52) contributes to $\{J[\xi_1], J[\xi_2]\}_D$. Thus after antisymmetrizing in 1 and 2 we obtain

$$\begin{aligned}\{J[\xi_1], J[\xi_2]\}_D = \frac{1}{2}\int_{\mathcal{H}\cap C} &\hat\epsilon_{a_1\ldots a_{n-2}} E^{abcd}\eta_{ab}\eta_{cd} \times \\ &\times \left[\frac{1}{\kappa}(T_1 \dddot T_2 - T_2 \dddot T_1) - 2\kappa(T_1 \dot T_2 - T_2 \dot T_1)\right]\,. \end{aligned}\tag{79}$$

The Noether charge

$$Q_{c_3\ldots c_n} = -E^{abcd}\epsilon_{abc_3\ldots c_n}\nabla_{[c}\xi_{d]}\,, \tag{80}$$

becomes after similar calculation

$$Q_{c_3\ldots c_n} = -\frac{1}{2}E^{abcd}\eta_{ab}\eta_{cd}\left(2\kappa T - \frac{\ddot T}{\kappa}\right)\hat\epsilon_{c_3\ldots c_n}\,, \tag{81}$$

which gives us

$$J[\{\xi_1, \xi_2\}] = -\frac{1}{2}\int_{\mathcal{H}\cap C}\hat\epsilon_{a_1\ldots a_{n-2}}E^{abcd}\eta_{ab}\eta_{cd} \times \tag{82}$$

$$\times\left[2\kappa(T_1 \dot T_2 - T_2 \dot T_1) - \frac{1}{\kappa}(\dot T_1 \ddot T_2 - \ddot T_1 \dot T_2 + T_1 \dddot T_2 - \dddot T_1 T_2)\right]\,. \tag{83}$$

From (79), (82) and (51) follows central charge

$$K[\xi_1, \xi_2] = -\frac{1}{2}\int_{\mathcal{H}\cap C}\hat\epsilon_{a_1\ldots a_{n-2}}E^{abcd}\eta_{ab}\eta_{cd}\frac{1}{\kappa}(\dot T_1 \ddot T_2 - \ddot T_1 \dot T_2)\,. \tag{84}$$

5 Entropy and Virasoro Algebra

The main idea is that constraint algebra (51) can be connected to the Virasoro algebra of diffeomorphisns of the real line. For that purpose we need to introduce another condition. Denote with v the parameter of orbits of the Killing vector

$$\chi^a \nabla_a v = 1 \, . \tag{85}$$

Let us consider functions T_1, T_2 of v and "Killing angular coordinates" θ_i on horizon such that they satisfy

$$\frac{1}{A}\int_{\mathcal{H}\cap C} \hat{\epsilon} T_1(v,\theta) T_2(v,\theta) = \frac{\kappa'}{2\pi}\int \mathrm{d}v T_1(v,\theta) T_2(v,\theta) \, . \tag{86}$$

Here $A = \int_{\mathcal{H}\cap C} \hat{\epsilon}$ is the area of the horizon and $\frac{2\pi}{\kappa'}$ is the period in variable v of functions $T(v,\theta)$. In particular for rotating black holes

$$\chi^a = t^a + \sum_i \Omega_i \psi_i^a \, , \tag{87}$$

where t^a is time translation Killing vector, ψ_i^a are rotational Killing vectors with corresponding angles ψ_i and angular velocities Ω_i. We shall sometimes, instead of variables t, ψ_i connected with orbits of t^a, ψ_i^a, work with variables (v, θ_i) connected with orbits of $\chi_a, \theta_i^a = \psi_i^a$. Then $v = t$, $\theta_i = \psi_i - \Omega_i v$, and we choose for diffeomorphism defining functions T_n

$$T_n = \frac{1}{\kappa'} \mathrm{e}^{\mathrm{i}n(\kappa' v + \sum l_i \theta_i)} \, , \tag{88}$$

where l_i are integers. These functions are of the form

$$T_n(v,\theta) = \frac{1}{\kappa'} \mathrm{e}^{\mathrm{i}n\kappa' v} f_n(\theta_i) \, . \tag{89}$$

They satisfy

$$\frac{1}{A}\int \hat{\epsilon} T_n T_m = \delta_{n+m,0} \frac{1}{\kappa'^2} \, , \tag{90}$$

and in particular

$$\frac{1}{A}\int \hat{\epsilon} f_n f_m = \delta_{n+m,0} \, . \tag{91}$$

At this point classical Virasoro condition can be checked in the form

$$\{T_m, T_n\} = -i(m-n) T_{m+n} \, . \tag{92}$$

We also see that condition (86) is fulfilled and thus enables us to obtain full Virasoro algebra with nontrivial central term $K[T_m, T_n]$ which can be calculated from (84)

$$iK[T_m, T_n] = \frac{\kappa'}{\kappa} \frac{\hat{A}}{8\pi} m^3 \delta_{m+n,0} \, , \tag{93}$$

where

$$\hat{A} \equiv \frac{1}{8\pi}\int_{\mathcal{H}\cap C} \hat{\epsilon}_{a_1 \ldots a_{n-2}} E^{abcd} \eta_{ab} \eta_{cd} \, . \tag{94}$$

Here we have used the property that metric does not depend on variables θ_i on which diffeormophism defining functions T_n depend. That enabled us to factorize the integral in (84). Finally, we obtain the Virasoro algebra

$$\{J[\xi_1], J[\xi_2]\}_D = (m-n)J[T_{m+n}] + \frac{c}{12} m^3 \delta_{m+n,0} \;, \tag{95}$$

and central charge is equal to

$$\frac{c}{12} = \frac{\hat{A}}{8\pi} \frac{\kappa'}{\kappa} \;. \tag{96}$$

Now we want to calculate the value of the Hamiltonian. This is given with the first term in relation (49) where the second term can be neglected[5]. The first term can be calculated from (81) and

$$T_0 = \frac{1}{\kappa'} \;. \tag{97}$$

Thus

$$J[T_0] = -\int_{\mathcal{H}\cap C} \hat{\epsilon}_{a_1 \dots a_{n-2}} E^{abcd} \eta_{ab} \eta_{cd} \frac{\kappa}{\kappa'} \;, \tag{98}$$

or

$$\Delta \equiv J[T_0] = \frac{\kappa}{\kappa'} \hat{A} \;. \tag{99}$$

We are now able to use Cardy formula (5) and obtain following expression for entropy

$$S = \frac{\hat{A}}{4} \sqrt{2 - \frac{\kappa'^2}{\kappa}} \;. \tag{100}$$

It is remarakable that entropy is proportional to classical classical entropy with a dimensionless constant of proportionality. We assume for the period of functions T_n the period of the Euclidean black hole [9, 5, 28, 29, 30, 31, 32], which implies

$$\frac{c}{12} = \Delta \;, \tag{101}$$

and

$$S = \frac{\hat{A}}{4} = -2\pi \int_{\mathcal{H}\cap C} \hat{\epsilon}_{a_1 \dots a_{n-2}} E^{abcd} \eta_{ab} \eta_{cd} \;. \tag{102}$$

As mentioned in the Introduction this derivation is valid for Lagrangian of general form $L = L(g_{ab}, R_{abcd})$. Let su take the example of Gauss–Bonnet gravity (8) Corresponding tensor E^{abcd} is then

$$E_{ab}{}^{cd} = -\Sigma_{m=0}^{[\frac{D}{2}]} m\lambda_m \frac{(-)^m}{2^m} \delta^{cdc_2d_2 \dots c_m d_m}_{aba_2b_2 \dots a_m b_m} R^{a_2b_2}{}_{c_2d_2} \dots R^{a_m b_m}{}_{c_m d_m} \;. \tag{103}$$

Consequently

$$S = -4\pi \Sigma_m^{[\frac{d}{2}]} m\lambda_m \int \hat{\epsilon} L_{m-1} \;. \tag{104}$$

For the well known case of Einstein gravity

$$S = \frac{A}{4} \;, \tag{105}$$

where A is area of black hole horizon.

[5] As in Einstein case, condition (60) enables us to factorize $\xi \cdot \Theta$ into $E^{abcd}\eta_{ab}\eta_{cd}\delta$(terms that vanish on shell), which together with (48) implies that $\int \xi \cdot B$ vanishes on shell

6 Conclusion

We conclude that idea of conformal symmetry near horizon can be useful to interpret the black hole entropy. This idea can be used in two different ways which are both described in this text and they are consistent with each other when applied to Gauss–Bonnet gravity. The second method, which can be applied in more general cases, consists in assuming appropriate boundary conditions near Killing horizon. One can then identify a subalgebra of diffeomorphism algebra as a Virasoro algebra with nontrivial central charge. From Cardy formula one can then determine the entropy. In this way we obtain the microscopic interpretation of entropy i.e. in terms of the number of states in Hilbert space. This result can be obtained for special cases of Einstein gravity [9], Gauss–Bonnet case [33, 5, 34] and for a more general class of Lagrangians [34]. It is remarkable that in all these cases including the general case treated here one obtains the classical expression for entropy [1]. These results suggest that conformal symmetry and Virasoro algebra could give further insight in exploring quantum mechanical properties of black holes. One is encouraged also to follow this approach due to recent proposals for its physical interpretation from the point of view of induced gravity [35] and an independent geometrical interpretation based on properties of the horizon [36, 37].

Acknowledgements

We would like to acknowledge the financial support under the contract No. 0119261 of Ministry of Science and Technology of Republic of Croatia.

References

1. V. Iyer and R. M. Wald, Phys. Rev. D **50** (1994) 846 [arXiv:gr-qc/9403028]; J. Lee and R. M. Wald, J. Math. Phys. **31**, 725 (1990); T. Jacobson, G. Kang and R. C. Myers, Phys. Rev. D **49** (1994) 6587 [arXiv:gr-qc/9312023].
2. A. Strominger and C. Vafa, Phys. Lett. B **379** (1996) 99 [arXiv:hep-th/9601029]; J. R. David, G. Mandal and S. R. Wadia, Phys. Rept. **369** (2002) 549 [arXiv:hep-th/0203048]; D. Amati and J. G. Russo, Phys. Lett. B **454** (1999) 207 [arXiv:hep-th/9901092].
3. A. Ashtekar, J. Baez, A. Corichi and K. Krasnov, Phys. Rev. Lett. **80**, 904 (1998) [arXiv:gr-qc/9710007]. A. Ashtekar, C. Beetle and S. Fairhurst, Class. Quant. Grav. **16**, L1 (1999) [arXiv:gr-qc/9812065]. A. Ashtekar, J. C. Baez and K. Krasnov, Adv. Theor. Math. Phys. **4**, 1 (2000) [arXiv:gr-qc/0005126]. A. Ashtekar, A. Corichi and K. Krasnov, Adv. Theor. Math. Phys. **3**, 419 (2000) [arXiv:gr-qc/9905089].
4. S. N. Solodukhin, Phys. Lett. B **454**, 213 (1999) [arXiv:hep-th/9812056].
5. M. Cvitan, S. Pallua and P. Prester, Phys. Lett. B **546** (2002) 119 [arXiv:hep-th/0207265].

6. S. Carlip, Phys. Rev. Lett. **88**, 241301 (2002) [arXiv:gr-qc/0203001].
7. S. Carlip, Phys. Lett. B **508**, 168 (2001) [arXiv:gr-qc/0103100].
8. S. Carlip, Phys. Rev. Lett. **82**, 2828 (1999) [arXiv:hep-th/9812013].
9. S. Carlip, Class. Quant. Grav. **16**, 3327 (1999) [arXiv:gr-qc/9906126].
10. J. L. Cardy Nucl. Phys. B **270**, 186 (1986); J. L. Cardy Nucl. Phys. B **275**, 200(1986); H. W. J. Bloete, J. L. Cardy and M. P. Nightingale, Phys. Rev. Lett. **56**, 742 (1986).
11. S. Nojiri, S. D. Odintsov and S. Ogushi, Phys. Rev. D **65** (2002) 023521 [arXiv:hep-th/0108172].
12. M. Cvetic, S. Nojiri and S. D. Odintsov, Nucl. Phys. B **628** (2002) 295 [arXiv:hep-th/0112045].
13. S. Nojiri, S. D. Odintsov and S. Ogushi, Int. J. Mod. Phys. A **17** (2002) 4809 [arXiv:hep-th/0205187].
14. Y. M. Cho and I. P. Neupane, Phys. Rev. D **66**, 024044 (2002) [arXiv:hep-th/0202140].
15. I. P. Neupane, Phys. Rev. D **67**, 061501 (2003) [arXiv:hep-th/0212092].
16. T. Clunan, S. F. Ross and D. J. Smith, arXiv:gr-qc/0402044.
17. J. G. Russo, Phys. Lett. **B359**, 69 (1995).
18. T. Jacobson and R. C. Myers, Phys. Rev. Lett. **70** (1993) 3684 [arXiv:hep-th/9305016];
M. Visser, Phys. Rev. D **48** (1993) 5697 [arXiv:hep-th/9307194].
19. A. Giacomini and N. Pinamonti, JHEP **0302**, 014 (2003); N. Pinamonti and L. Vanzo, Phys. Rev. D **69**, 084012 (2004); A. Giacomini, Ph.D. thesis, hep-th/0403183.
20. C. Crnković, Class. Quant. Grav. **5**, 1557 (1988); E. Witten, Nucl. Phys. B **276**, 291 (1986).
21. B. Julia and S. Silva, arXiv:hep-th/0205072.
22. J. D. Brown and M. Henneaux, Commun. Math. Phys. **104**, 207 (1986).
23. G. Kang, J. I. Koga and M. I. Park, arXiv:hep-th/0402113.
24. O. Dreyer, A. Ghosh and J. Wisniewski, Class. Quant. Grav. **18**, 1929 (2001) [arXiv:hep-th/0101117].
25. M. I. Park and J. Ho, Phys. Rev. Lett. **83**, 5595 (1999) [arXiv:hep-th/9910158]; V. O. Solovev, Phys. Rev. D **61** (2000) 027502 [arXiv:hep-th/9905220].
26. M. I. Park, Nucl. Phys. B **634** (2002) 339 [arXiv:hep-th/0111224].
27. J. I. Koga, Phys. Rev. D **64**, 124012 (2001) [arXiv:gr-qc/0107096].
28. A. Barvinsky, S. Das and G. Kunstatter, Found. Phys. **32**, 1851 (2002) [arXiv:hep-th/0209039].
29. H. A. Kastrup, Phys. Lett. B **385**, 75 (1996) [arXiv:gr-qc/9605038].
30. J. Louko and J. Makela, Phys. Rev. D **54**, 4982 (1996) [arXiv:gr-qc/9605058].
31. J. Makela and P. Repo, Phys. Rev. D **57**, 4899 (1998) [arXiv:gr-qc/9708029].
32. M. Bojowald, H. A. Kastrup, F. Schramm and T. Strobl, Phys. Rev. D **62**, 044026 (2000) [arXiv:gr-qc/9906105].
33. M. Cvitan, S. Pallua and P. Prester, Phys. Lett. B **555** (2003) 248 [arXiv:hep-th/0212029].
34. M. Cvitan, S. Pallua and P. Prester, Phys. Lett. B **571** (2003) 217 [arXiv:hep-th/0306021].
35. V. P. Frolov, D. Fursaev and A. Zelnikov, JHEP **0303** (2003) 038 [arXiv:hep-th/0302207].
36. A. J. M. Medved, D. Martin and M. Visser, arXiv:gr-qc/0403026.
37. A. J. M. Medved, D. Martin and M. Visser, arXiv:gr-qc/0402069.

Dark Matter Experiments at Boulby Mine

Vitaly A. Kudryavtsev

Department of Physics & Astronomy, University of Sheffield,
Sheffield S3 7RH, UK
v.kudryavtsev@sheffield.ac.uk

for the Boulby Dark Matter Collaboration (Sheffield, RAL, Imperial College, Edinburgh, UCLA, Temple, Occidental, Texas A&M, ITEP and Coimbra)

1 Introduction

It is believed that 20% of the Universe may consist of non-baryonic dark matter. Supersymmetric theories provide a good candidate – neutralino or Weakly Interacting Massive Particle (WIMP). Due to very small cross-section of WIMP-nucleus interactions very sensitive and massive detectors are required to detect WIMPs. There are three key requirements for direct dark matter detection technology: (1) low intrinsic radioactive background from detector and surrounding components; (2) good discrimination between electron recoils produced by remaining gamma background and nuclear recoils expected from WIMP interactions; (3) low energy threshold to achieve maximal sensitivity to WIMP-induced nuclear recoils.

The UK Dark Matter Collaboration has been operating dark matter detectors at Boulby mine (North Yorkshire, UK) at a vertical depth of 2800 m w. e. for more than a decade. Three major programmes are currently under way: i) an array of NaI(Tl) crystals (NAIAD) is collecting data, ii) detectors based on liquid xenon, which has a better background discrimination power compared to NaI, have been developed and are either running or being commissioned, iii) a low pressure gas Time Projection Chamber (TPC) with a potential of directional sensitivity has been constructed and is operating at Boulby. Two later projects are carried out in collaboration with international groups from Europe and USA.

2 The NAIAD Experiment

The NAIAD (NaI Advanced Detector) array consists of 7 NaI(Tl) crystals with a total mass of ≈55 kg. Two detectors contain encapsulated crystals, whereas 5 other crystals are unencapsulated. To avoid degradation due to air humidity, the unencapsulated crystals have been sealed in copper boxes filled with dry air. Each crystal is mounted in a 10 mm thick solid PTFE reflector cage and is coupled to two 5 inch diameter photomultiplier tubes (PMTs) through the 4–5 cm light guides at either end. Low background materials are used throughout. PMT signals are digitised using an Acqiris CompactPCI based DAQ system.

Our standard procedure of data analysis involves fitting an exponential to each scintillation pulse. The time constants of these pulses follow a log-normal distribution with a mean time which increases with increasing energy for electron recoils but is practically constant for nuclear recoils. Electron and nuclear recoils give rise to two populations with different mean times. Calibration of the crystals was done using neutron source, which produces nuclear recoils, and gamma-ray source, which produces electron recoils via Compton scattering. Nuclear recoils are also expected from WIMP-nucleus interactions, whereas electron recoils from gammas constitute the main background. Data from about 20 kg×years have been analysed to produce the (preliminary) limit curve shown in Fig. 1.

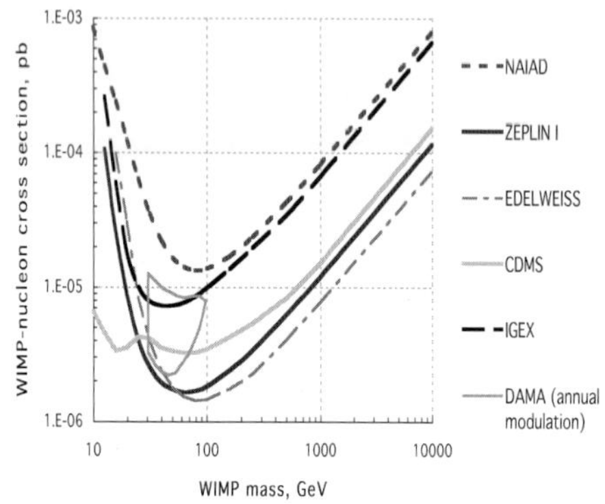

Fig. 1. Preliminary limits for WIMP-nucleon spin-independent cross-section from the NAIAD and ZEPLIN I experiments are shown together with other world-best limits and with a region of parameter space consistent with the DAMA signal

3 Liquid Xenon Experiments

Nuclear recoil discrimination in liquid xenon is feasible by measuring both the scintillation light and the ionisation produced during an interaction, either directly or through secondary recombination. Meanwhile, the chemical inertness and isotopic composition of liquid xenon provide intrinsically low radio-purity and routes, in principle, to further purification using various techniques. The heavy nuclei of xenon also has the advantage of providing a large spin-independent coupling.

The ZEPLIN I detector (ZonEd Proportional scintillation in LIquid Noble gases – shown in Fig. 2) consists of 3.1 kg fiducial mass of liquid Xe incased in a copper vessel and viewed by 3 PMTs through silica windows. The detector itself is enclosed in a 0.93 tonne active scintillator veto, its function being to veto gamma events from the PMTs and the surroundings. ZEPLIN I has better sensitivity than NaIAD due to its improved discrimination at low energies. As with NAIAD, background discrimination is possible due to the difference in time constant between nuclear and electron recoils. The 90% C.L. on the number of nuclear recoils in each 1 keV energy bin is extracted and this is then used to calculate the WIMP-nucleon cross-section as a function of WIMP mass. The preliminary limit on the spin-independent WIMP-nucleon cross-section from 250 kg×days of data is shown in Fig. 1 [1] in comparison with other world-best limits.

Fig. 2. Computer view of the ZEPLIN I set-up showing the target vessel with three PMTs encased in an active veto system

Work is now underway on ZEPLIN II and ZEPLIN III detectors. ZEPLIN II is a two phase detector with a target mass of about 30 kg and a sensitivity to WIMP-nucleon cross-section down to 10^{-7} pb at the minimum of the sensitivity curve. In ZEPLIN II recoils produce both excitation and ionisation. Recombination of electrons and ions produced via ionisation is prevented by the strong electric field. Electrons, drifting in this field towards gas phase, produce a secondary luminescence signal in the gas. For a given primary amplitude an electron recoil produces a much larger secondary signal than a nuclear recoil. This provides ZEPLIN II with greater discrimination power over ZEPLIN I. Computer view of the ZEPLIN II detector is shown in Fig. 3. ZEPLIN III aims to increase background discrimination by increasing electric field through the liquid xenon and, with a fiducial mass of 6 kg, should achieve similar sensitivities as ZEPLIN II.

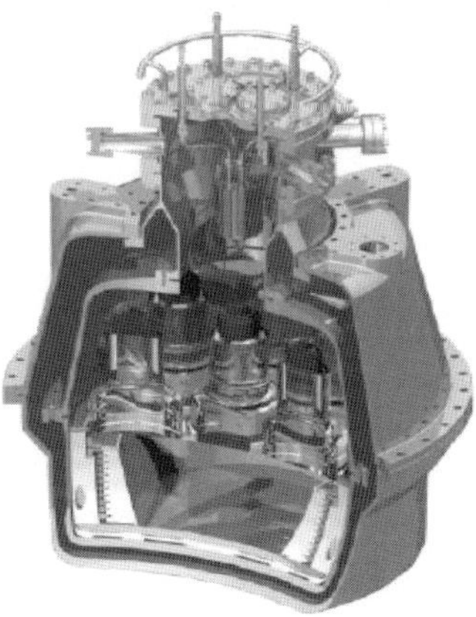

Fig. 3. Computer view of the ZEPLIN II detector

The ZEPLIN programme described above can be viewed as a development sequence aimed at determining the optimum design for a large scale (1–10 tonnes) xenon detector capable of reaching a sensitivity below 10^{-9} pb in 1 year of operation. Such a large mass is required to achieve sufficient signal counts in a year (10–100 events). ZEPLIN MAX is at the R&D stage at present.

4 The DRIFT Experiment

The DRIFT (Directional Recoil Identification From Tracks) detector adopts a different approach to identifying a potential WIMP signal. DRIFT I uses a low pressure CS_2 gas TPC capable of measuring the components of recoil track ranges in addition to their energy. The use of negative ions, notably CS_2, to capture and drift the ionisation electrons reduces diffusion. The detector consists of two $0.5\,m^3$ fiducial volumes defined by 0.5 m long field cages mounted either side of a common anode plane consisting of 512 20 μm stainless steel wires (Fig. 4). Particle tracks are read out with two 1 m long MWPCs, one at either end of the field cages. The difference in track range between electrons, alpha particles and recoils is such that rejection efficiencies as high as 99.9% at 6 keV are possible. After 1 year of operation DRIFT I is expected to reach a sensitivity of $\sim 10^{-6}$ pb. The power of DRIFT comes from its ability to determine the direction of a WIMP induced nuclear recoil. The Earth's motion around the galactic centre means that the Earth experiences a WIMP "wind". As the Earth rotates through this wind the nuclear recoil direction is modulated over a period of a sidereal day, making it a strong signature of a galactic WIMP signal.

Building on DRIFT I the long term objective of the DRIFT programme is to scale-up detectors towards a target mass of 100 kg (DRIFT III) through the development of several intermediate scale detectors (DRIFT II), which can be replicated many times. DRIFT II is proposed to have 30–50 times the sensitivity of DRIFT I through an increase in the volume (several modules) and gas pressure. A higher gas pressure means that the recoil range will be shorter requiring higher spatial

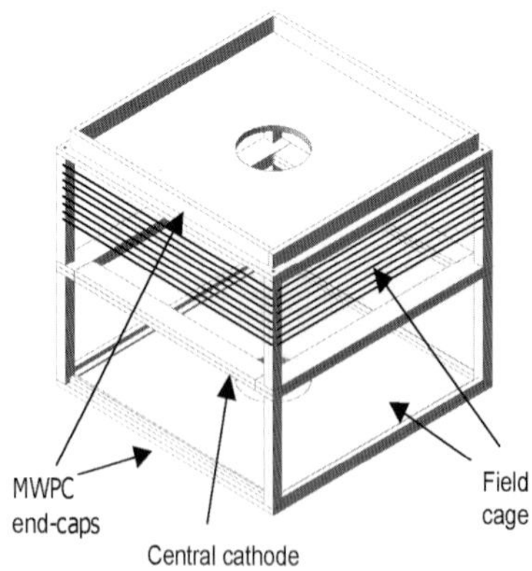

Fig. 4. Schematic of the inner part of the DRIFT I detector

resolution. Alternative read-out schemes are currently being investigated including Gas Electron Multipliers and a MICROMEGAS microstructure detector.

5 Conclusions

Liquid xenon has been demonstrated as an excellent technology for dark matter searches with ZEPLIN I already producing significant sensitivity using pulse shape discrimination. The collaboration is now progressing to two phase operation which shows promise for substantial improvement in sensitivity towards $10^{-7} - 10^{-8}$ pb. There is a route with this technology to reach 10^{-10} pb with a 1 tonne liquid xenon detector. Directional detectors based on low pressure gas provide a unique means of determining the galactic origin of an observed WIMP signal, a significant advantage over conventional dark matter experiments. The DRIFT programme based on this concept is under way to approach this possibility, complementing also the liquid xenon programme through the use of entirely different technology with different target nuclei. More on the Boulby programme can be found in a recent review [2].

References

1. J. C. Barton et al.: Proc. 4th Intern. Workshop on the Identification of Dark Matter (York, UK, 2-6 September 2002), ed. by N. J. C. Spooner and V. A. Kudryavtsev (World Scientific Publishing, 2003), pp. 302–307.
2. N. J. C. Spooner (on behalf of the Boulby Collaboration): Proceedings of the 31st SLAC Summer Institute (Stanford, July 28 – August 8, 2003), http://www-conf.slac.stanford.edu/ssi/2003/lec_notes/spooner.html.

Ultra High Energy Cosmic Rays and the Pierre Auger Observatory

Danilo Zavrtanik[1] and Darko Veberič[1]; for the AUGER Collaboration[2]

[1] Nova Gorica Polytechnic, Vipavska 13, POB 301, 5001 Nova Gorica, Slovenia
[2] Observatorio Pierre Auger, Av. San Martín Norte 304, 5613 Malargüe, Mendoza, Argentina, www.auger.org

1 Ultra-High Energy Cosmic Rays

As confirmed by several experiments using different detection techniques, cosmic rays with energy in excess of 10^{18} eV are a well established fact. In case of protons or heavier particles (nuclei) having ultra-high energies it is believed that the inter- or extra-galactic magnetic fields have only a lesser effect on their trajectories, thus opening a possibility for the "cosmic-ray astronomy". Existence of a small fraction of these Ultra-High Energy Cosmic Rays (UHECR) is nevertheless marked with mystery regarding their origin, composition, means of acceleration, and interactions they are subdued on the path through the space.

1.1 Energy Spectrum

Continuous over more than ten decades (from 1 GeV to and beyond 10^{19} eV) the energy spectrum of cosmic rays exhibits only two prominent features. Centered around the "knee" at $4 \cdot 10^{15}$ eV, the spectrum steepens from power-law with (integral) exponent -2.7 to the one with -3.2. It is believed that cosmic rays with energies reaching as high as 10^{16} eV are produced by diffusive shock acceleration in supernova explosions, and the most energetic, by gaining on interactions with multiple supernova remnants. The maximum attainable energy in the region of the shock can be approximately described by $E \approx ZBL$, where Z is the charge of the cosmic ray particle, B strength of the magnetic field, and L size of the accelerating region. Gathering known astronomic objects in a "Hillas" plot [1] reveals possible candidates for acceleration mechanism (see Fig. 1). Due to inaccessibility of the astronomical data on magnetic field, placement of certain candidate objects on the plot is still only speculative. The second feature around 10^{18} eV is called an "ankle" and may indicate a transition from predominantly galactic to extragalactic source distribution.

During the propagation, the magnetic fields – inter- and extra-galactic alike – are also responsible for deflections of cosmic rays from their original trajectories. Nevertheless, in the case of UHECR, e.g. for those with $E > 10^{18}$ eV, the deflections become small enough to enable correlating arrival directions to the known locations of the astronomical objects (cosmic-ray astronomy). However, this argumentation relays strongly on the assumption of weak magnetic fields of the order of a few nG (10^{-13} T). In contrast, some authors argue [2] that the magnitude of the magnetic fields may be more in the range of μG, producing large deflections even for the UHECR.

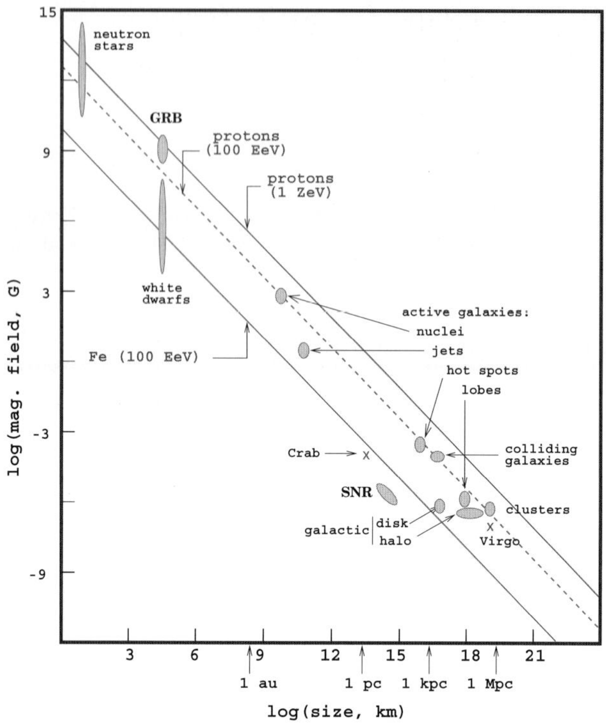

Fig. 1. Candidates for sources of different UHECR must lie above the diagonal lines (for iron with $E = 10^{20}$ eV, for protons with $E = 10^{20}$ eV and $E = 10^{21}$ eV). This Hillas plot is adapted from [1]

While UHECR do not suffer appreciable energy loss during their propagation (even within our galaxy), this ceases to be true for the extremely high energy[1], beyond $E > 10^{19}$ eV. Energy loss due to the interactions with cosmic microwave, infrared, and radio background radiation fields, start to take effects on the cosmic ray spectrum. Soon after discovery of cosmic microwave background (CMB) radiation, a relic of primordial nucleosynthesis, it became the main candidate for inelastic scattering interaction [3]. Space is thus densely filled with microwave photons (several 100 per cm^3) from the black-body radiation at temperature of 2.7 K. In the case where cosmic ray is a proton with energy above a few times 10^{19} eV, a Δ-resonance is excited in the collision with CMB photon, effectively draining proton's energy through the pion or electron-pair production, $p + \gamma \to n + \pi^+$, $p + \pi^0$, or $p + e^+ + e^-$.

Due to the above interactions, a significant drop in the attenuation length of cosmic rays develops for energies exceeding a few times 10^{19} eV. In fact, after a few 10 Mpc ($32.6 \cdot 10^6$ light-years) of propagation through CMB, the energy of a cosmic

[1] $E = 10^{20}$ eV corresponds to the kinetic energy of a tennis ball moving at 100 km/h. However, the momentum of such a cosmic ray is still extremely small and thus can not produce any "macroscopic" effects.

ray drops below 10^{20} eV and becomes almost distance-independent for all initial energies $E > 10^{20}$ eV. Whether or not this feature is observed also in the measured spectrum of cosmic rays depends on the distance of their respective sources. In case of cosmic rays originating predominantly from distances in excess of ~50 Mpc, a GZK cutoff (named after the authors in [3]) should be observed for energies above 10^{19} eV. Furthermore, in case no such cutoff (or at least feature) is observed in the spectrum, sources have to be located in our "local neighborhood", or some new physics, beyond the Standard Model of particle interactions, has to be at work.

We thus have to turn to the experimental measurements of cosmic-ray flux to answer the puzzle of their origin.

Among many past (Volcano Ranch [4], Haverah Park [5], SUGAR [6], Fly's Eye [7]) and present (Yakutsk [8], AGASA [9], HiRes [10]) experiments only AGASA and HiRes have obtained enough statistics in the interesting region of UHECR. In Fig. 2, a flux J (corrected for E^{-3}, the leading power-law dependence) of the two experiments is shown. Next to the obvious discrepancy (of more than a factor of 2) for $E < 3 \cdot 10^{19}$ eV, the data is also conflicting regarding the question of GZK cutoff for $E > 5 \cdot 10^{19}$ eV. The first controversy can be explained with manifestation of different systematic errors due to the diverse experimental methods for cosmic-ray detection. While the AGASA is using an array of more than 100 (plastic) scintillation detectors, covering a ground area of 100 km^2, the HiRes is observing tracks of distant fluorescence light that a cascade of secondary particles leaves in the background of the night sky. Nevertheless, the data seems to be consistent after a systematic shift of the order of 20% or 30% is engaged on either one

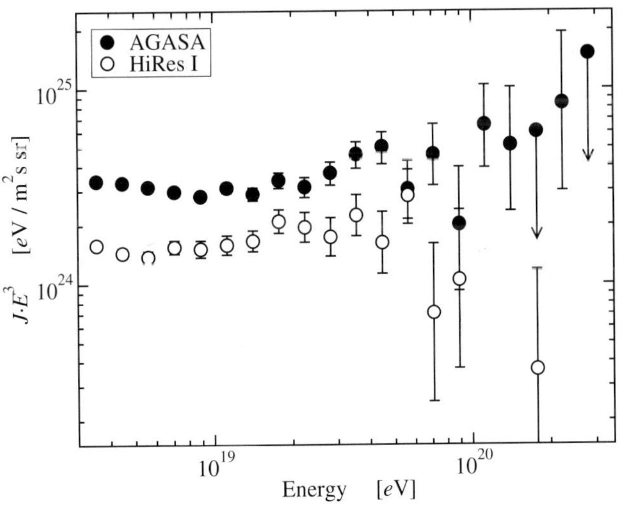

Fig. 2. Energy spectrum of cosmic rays in the ultra-high energy region ($E > 10^{18}$ eV) as measured by two experiments: AGASA (*full circles*), data taken from [9], and HiRes I (*open circles*), data from [10]. AGASA points with arrows are values at 90% confidence level, only. While HiRes seems to be compatible with the GZK cutoff, AGASA certainly is not. Note the large discrepancy of the flux measurements for $E < 3 \cdot 10^{19}$ eV

of them. However, no such explanation can be devised for the respective presence and absence of the GZK feature in the HiRes and AGASA data. Still, the maximum energy observed so far seems to be limited with exposure only. This calls for new experiments that can measure the properties of UHECR with unprecedented statistical precision.

1.2 Sources of UHECR

Almost all of the experiments were analyzing the distribution of cosmic-rays' arrival directions. Next to the fact that cosmic rays are piercing the atmosphere rather isotropically, various enhancements of the distribution have not been reproduced among the experiments. However, some evidence of slight anisotropy, correlated with the galactic structure has been found for $E \sim 10^{18}$ eV, and it seems that the lack of statistically relevant data [11] is the common denominator in the search of super-galactic correlations for higher energies.

2 Status of the Southern Pierre Auger Observatory

According to the conclusions of the previous sections, the cosmic ray community is eagerly awaiting new data on UHECR. Since the rate of post-GZK cosmic rays is so strikingly low, i.e. for $E > 10^{20}$ eV of the order of a few events per km^2 per century, a cosmic-ray observatory of immense proportions is needed in order to properly identify the incident primary particle, to trace back its arrival direction and to pinpoint possible correlation with astronomical objects, to infer the true shape of the spectrum, and all that in a reasonable time span (possibly within a lifetime of a scientist).

In 1992 first ideas for such an observatory have been issued and in 1995 a design work has begun within a collaboration of 16 nations. At the time, more than 300 scientists and technicians are working on the Pierre Auger Project. In order to achieve the full sky coverage, two observatories will have to be built, each placed at intermediate latitudes on both hemispheres. Presently, a southern observatory is under construction near the town of Malargüe, province of Mendoza in Argentina (35°29′S, 69°27′W, ~1400 m a.s.l). In contrast to the other UHECR experiments, the Pierre Auger Observatory (PAO) will use two observational techniques. The first technique, an array of surface detectors, is at the same time the most widely used detection technique[2]. The second technique is based on detection of fluorescence light generated by the charged secondaries in a developing air shower. The PAO fluorescence detector uses a variation of a technique, first used by the Fly's Eye experiment. The PAO is thus by design a unique cosmic-ray detector, strongly relying on the "hybrid" mode of operation [12]. What follows is a description of the two detector parts of the PAO, together with the report on current status and evolution.

[2] The idea is based on the first coincidence measurements by physicists Pierre Auger and Roland Maze in late 1930s.

Fig. 3. One of the many water-Cherenkov detectors deployed to the Pampa Amarilla (semi-)desert. Tanks are filled with more than 10 tons of purified and deionized water. Electronics, radio, GPS, and photomultiplier tubes all run on the independent source of electric power from solar panel, stored in a battery

2.1 Surface Detector

At the time of completion, the surface detector (SD) of each observatory site will be covering an area larger than 3 000 km^2. The SD of the southern observatory consists of 1600 water-Cherenkov units, deployed in a semi-desert like environment in a hexagonal grid with 1.5 km spacing. Each unit (shown in Fig. 3) is filled with more than 10 tons of water, overlooked by three photomultipliers. In order to prevent biological contamination only purified and deionized water can be used. Insuring total operational independence, the electric power is provided by a panel of solar cells and stored for the night-time operation in a battery. Registering muons, electrons and photons from cosmic ray showers reaching ground, signals from the photomultipliers are sampled at 16 bits and 40 MHz and stored in a local memory. Satisfying local trigger conditions, data is sent over the wireless local area network to the concentrators on the perimeter of the array. Signal from the abundant cosmic ray muons is used for the calibration purposes. Timing with relative accuracy of ± 10 ns is supplied by the built-in GPS system. Array trigger conditions will require several units to be hit by the cosmic ray shower. Detection efficiency will thus start around $E = 10^{18}$ eV and reach 100% at 10^{19} eV.

First phase of engineering array operation, consisting of around 40 tanks, has been successfully completed [13]. Gradually, new units are deployed on the site and incorporated in the SD array operation. Number of units and corresponding area covered has in October 2003 grown larger than [13] the size of AGASA ground array, previously the largest cosmic ray experiment in the world. In Fig. 4, present size can be inferred from the schematic plot of deployed and operational units. During all this time, PAO array has been constantly taking data. In Figs. 5 and 6, signal

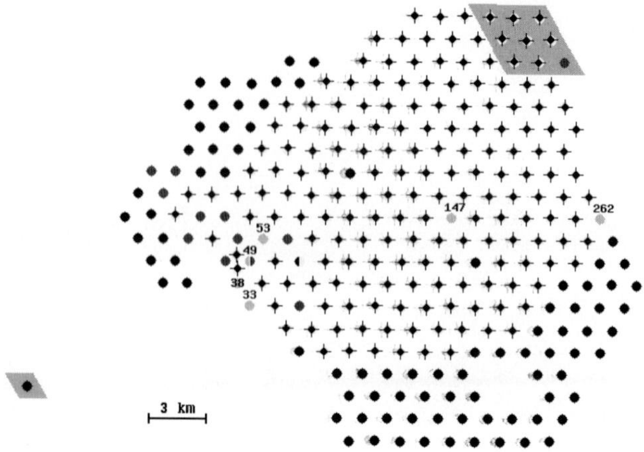

Fig. 4. Status of the surface detector as of January 2004. Currently deployed tanks are shown (315 or ~1/5 of the final number): 219 fully operational (*stars*), and 96 filled with water but with missing electronic parts (*full circles*). There is also a detector for testing and calibration purposes at the operations center in Malargüe (*gray rhomboid, left bottom*)

from the SD and dependence of the particle density on the distance from the core is shown for a specific shower event with $E \sim 4 \cdot 10^{19}$ eV.

2.2 Fluorescence Detector

Secondary particles in the shower cascade excite the air nitrogen molecules. While the nitrogen is subsequently emitting fluorescence light in the range of 300–400 nm, fluorescence photons are effectively indicating the path of the shower through the atmosphere. The parameters of the shower as well as primary particle properties can be inferred [7] from measurements of light intensity at different times of the shower development. Unfortunately, operation is limited to the clear nights with moon fraction not more than 60%, expecting a duty cycle of 10–15%.

The fluorescence detector (FD) will consist of four stations, placed at the perimeter of the SD array. Each station will shelter six wide-angle Schmidt telescopes. Each telescope's mirror with 3.4 m curvature is formed by segments of diamond-cut (no polishing needed) mirrors of 12 m^2 total area. The aperture with 2.2 m radius is covered with UV filter. In the focal plane a camera, segmented in 20×22 photomultipliers, is placed (see Fig. 7). Each pixel is covering $1.5° \times 1.5°$ of sky, making up a 30° field of telescope view. Six telescopes in the station are thus covering $180° \times 30°$ of sky, starting at 2° above the horizon. The signal from the photomultipliers is sampled at 10 MHz and 12 bit with dynamic range of 15 bit. Search for shower patterns in the pixel matrix is built into the hardware. In Fig. 8, an example of the detector's view of the shower track is presented.

Since the FD uses the atmosphere as its calorimeter, and the cameras are only calorimeter's "read-out", sophisticated atmosphere monitoring must also be employed. This includes meteorologic stations, stations for measuring horizontal

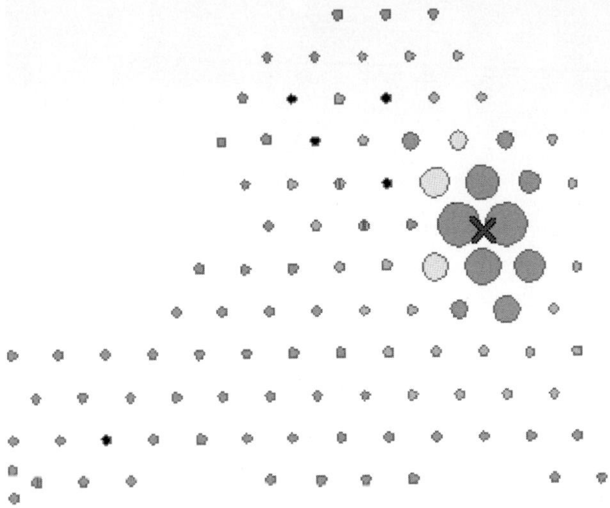

Fig. 5. Signal from the surface detector for a cosmic ray shower with preliminary energy estimation of $4.3 \cdot 10^{19}\,\mathrm{eV} \pm 6\%$ (statistical error only). Radius of the unit's circle is proportional to the measured signal. Cross marks the ground position of the shower core. This was a relatively vertical shower with reconstructed zenith angle of 38.8°

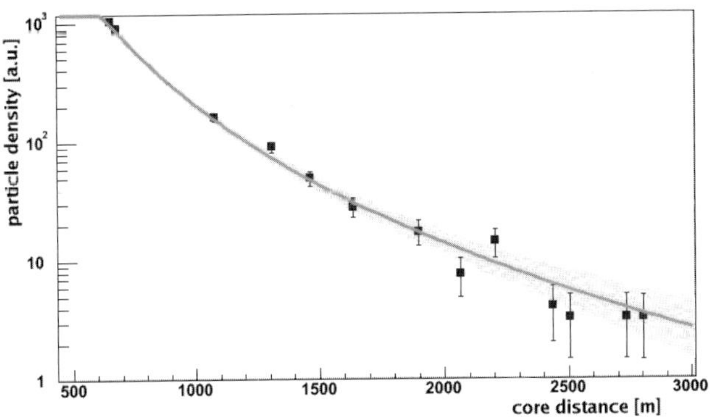

Fig. 6. One of the means to obtain the shower energy is from the shape of lateral distribution of particle density vs. core distance. Signals from 13 Cherenkov surface detectors (see Fig. 5) were used

attenuation length of the light in fluorescence spectral range, and a station firing vertical laser shots that can be seen by all of the FD stations. Furthermore, presently two infrared full-sky cameras are observing cloud coverage and two steerable lidar systems (see Fig. 9) are operational [14]. Note, that the yearly-averaged meteorological conditions, light pollution at night, and flatness of the area (radio propagation) were primary selection criteria in site selection.

Fig. 7. Camera, consisting of 20 × 22 photomultipliers, is facing its diamond-cut mirror

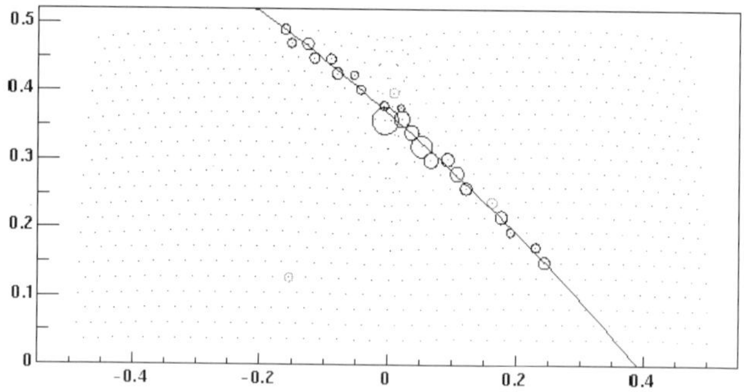

Fig. 8. An example of a shower, seen by two cameras in adjacent telescopes. Size of circles is proportional to the number of photons received. In *solid line* a view of the plane that contains the fluorescence detector and shower is shown

2.3 Future Prospects

The southern PAO is expected to be completed till the end of 2005, when the construction of the northern site is anticipated. At that time, the accuracy of the arrival direction reconstruction for most energetic cosmic rays is expected to be below 0.5°, and the relative error in energy estimation below 10%. Assuming AGASA-like spectrum, with SD alone, 15 000 useful events for $E > 3 \cdot 10^{18}$ eV and 5 000 events for $E > 10^{19}$ eV should be gathered per year. During the operation of the FD, 98% of events are expected to be "hybrid", i.e. detected by both, FD and SD, and 60% of that in "stereo", i.e. at least by two FD stations simultaneously. While useful data is already flowing in, focus of the project's activities is changing to the routine data taking and development of the analysis methods.

Fig. 9. On the roof of the container near the Los Leones fluorescence detector, a steerable telescope of the lidar system can be seen

References

1. A.M. Hillas: Ann. Rev. Astron. Astrophys. **22**, 425 (1984)
2. G.R. Farrar and T. Piran: Phys. Rev. Lett. **84**, 3527 (2000)
3. K. Greisen: Phys. Rev. Lett. **16**, 748 (1966); G. Zatsepin and V. Kuzmin: JETP Lett. **4**, 78 (1966)
4. J. Linsley: Phys. Rev. Lett. **10**, 146 (1963)
5. M.A. Lawrence, R.J.O. Reid, and A.A. Watson: J. Phys. G **17**, 733 (1991)
6. C.J. Bell et al.: J. Phys. A **7**, 990 (1974)
7. R.M. Baltrusaitis et al.: Nucl. Instr. Meth. A **240**, 410 (1985)
8. A.V. Glushkov et al.: Astropart. Phys. **4**, 15 (1995)
9. M. Takeda et al.: Phys. Rev. Lett. **81**, 1163 (1998); Astrophys. J. **522**, 255 (1999)
10. T. Abu-Zayyad et al.: arXiv:astro-ph/0208243 and 0208301
11. M. Nagano and A.A. Watson: Rev. Mod. Phys. **72**, 689 (2000)
12. AUGER Collaboration: *The Pierre Auger Observatory Design Report*, 2nd edition (1997); `http://www.auger.org`
13. J. Abraham et al.: to appear in Nucl. Instr. Meth. A, (2004)
14. A. Filipčič et al.: Astropart. Phys. **18**, 501 (2003)

Self-Accelerated Universe

Boris P. Kosyakov

Russian Federal Nuclear Center–VNIIEF, Sarov 607190, Russia
kosyakov@vniief.ru

1 Introduction

The recent measurements of redshifts for Type Ia supernovae [1] suggest that the Universe expansion is accelerating. To interpret this discovery, one usually write the Friedmann equation

$$H^2 = \left(\frac{\dot{a}}{a}\right)^2 = \frac{2}{A_V^2} + \frac{2A_D}{a^3} + \frac{2A_B}{a^3} + \frac{A_R}{a^4} - k \tag{1}$$

where H is the Hubble expansion parameter, a is the scale factor, A_V, A_D, A_B, and A_R are Friedmann integrals of the motion related to the energy density of vacuum, dark matter, nonrelativistic particles (barions), and radiation, respectively, k is the spatial curvature, with $k = 1, 0, -1$ corresponding to the closed, flat, and open models. Equation (1) is derived from Einstein's equations with a positive cosmological constant

$$R_{\mu\nu} - \frac{1}{2}\,g_{\mu\nu}\,R - \Lambda\,g_{\mu\nu} = -8\pi G\,T_{\mu\nu} \tag{2}$$

using the generic form of the line element for homogeneous and isotropic spacetimes

$$\mathrm{d}s^2 = \mathrm{d}t^2 - a^2 F(r)^2 \mathrm{d}\Omega^2 - a^2 \mathrm{d}r^2\,. \tag{3}$$

Here, t is the proper time, $\mathrm{d}\Omega = \mathrm{d}\theta^2 + \sin^2\theta\,\mathrm{d}\varphi$, $F = \sin r$, r, $\sinh r$ for $k = 1, 0, -1$, respectively, the cosmological constant Λ relates to the vacuum energy density ρ_V as $\Lambda = 8\pi G\rho_V$, and $A_V = (8\pi G\rho_V/3)^{-1/2}$. In the expanding Universe, the scale factor a increases with time. So, there comes a time when the first term in (1) becomes dominant. The asymptotic $a \to \infty$ solution to (1) is

$$a(t) = A_V f(t), \quad f(t) = \cosh(t/A_V),\ \exp(t/A_V),\ \sinh(t/A_V)\ \text{for}\ k = 1, 0, -1\,. \tag{4}$$

This solution describes a cosmological expansion accelerating in time, $\ddot{a} > 0$.

Thus the presence of a positive cosmological constant Λ in (2), which is responsible for the anti-gravitation effect, ensures the accelerating expansion regime of the Universe. At present, this explanation of the large redshift data for distant supernovae is widely accepted. It tacitly assumes that particles (galaxies, clusters, etc.) move along geodesic for the background metric. Recall, however, that galaxies every so often have an internal angular momentum (spin), and a spinning particle is deflected from the geodesic. It is clear that the spacetime curvature is of no concern: a particle with spin can behave in a non-Galilean manner in a flat spacetime.

Let us consider the Frenkel model of a classical spinning particle [2]. Its motion, in the absence of external forces, is governed (see, e.g., [3]) by the equation

$$\mathcal{S}^2\dddot{v}^\mu + M^2 v^\mu = m p^\mu \tag{5}$$

where $\mathcal{S}$ is the spin magnitude, $v^\mu = \dot{z}^\mu$ is the four-velocity, the dot denotes the derivative with respect to the proper time s, p^μ is the four-momentum (which is constant for the free particle), M and m are the mass and rest mass defined as $M^2 = p^2$ and $m = p \cdot v$. For $p^2 > 0$, $p^\mu = \text{const}$, a solution to (5) is

$$z^\mu(s) = \frac{m}{M^2}\,p^\mu s + \frac{\alpha^\mu}{\omega}\cos\omega s + \frac{\beta^\mu}{\omega}\sin\omega s \tag{6}$$

where $\alpha \cdot p = \beta \cdot p = \alpha \cdot \beta = 0$, $\alpha^2 = \beta^2$. This helical world line describes the motion called the Zitterbewegung [4]. For $p^2 < 0$, $p^\mu = \text{const}$, we have

$$z^\mu(s) = -\frac{m}{\mathcal{M}^2}\,p^\mu s + \frac{\alpha^\mu}{\Omega}\cosh\Omega s + \frac{\beta^\mu}{\Omega}\sinh\Omega s \tag{7}$$

where $\mathcal{M}^2 = -p^2$, $\Omega = \mathcal{M}/\mathcal{S}$, and α^μ and β^μ are subjected to the constraint $\alpha^2 = -\beta^2$. This solution describes the motion with increasing velocity.

One may argue that spacelike four-momenta p^μ are highly unnatural for classical particles. While this is a strong objection, it seems reasonable to say that both solutions (6) and (7) support the idea of non-Galilean regimes for free spinning particles.

Another fact deserving of notice is that massive particles can emit gravitational waves. It is conceivable that a massive particle emitting gravitational waves moves in a runaway regime, that is, deviates sharply from a geodesic for the background metric $g_{\mu\nu}$. Runaway solutions offer an alternative explanation for the accelerated expansion of the Universe, without recourse to the cosmological constant hypothesis.

It is interesting to compare a massive particle emitting gravitational waves and a charged particle emitting electromagnetic waves. The nonrelativistic equation of motion for a classical electron, called the Abraham–Lorentz equation (see, e.g., [5]),

$$m\mathbf{a} - \frac{2}{3}e^2\frac{\mathrm{d}\mathbf{a}}{\mathrm{d}t} = \mathbf{f}\,, \tag{8}$$

in the absence of external forces $\mathbf{f} = 0$, becomes

$$\mathbf{a} - \tau_0\frac{\mathrm{d}\mathbf{a}}{\mathrm{d}t} = 0 \tag{9}$$

where

$$\tau_0 = \frac{2e^2}{3m} \approx 10^{-23}\,\mathrm{s}\,. \tag{10}$$

The general solution to (9),

$$\mathbf{a}(t) = \mathbf{A}\exp(t/\tau_0)\,, \tag{11}$$

where $\mathbf{A}$ is the initial acceleration at $t = 0$, describes a runaway motion. For $\mathbf{A} = 0$, we have $\mathbf{a} = 0$, and $\mathbf{v} = \text{const}$. Thus a free electron can behave as both Galilean ($\mathbf{A} = 0$), and non-Galilean ($\mathbf{A} \neq 0$) objects.

Where does the Abraham–Lorentz equation come from? The scheme of its derivation is as follows. We first solve Maxwell's equations

$$\Box A^{\mu}(x) = 4\pi e \int_{-\infty}^{\infty} \mathrm{d}s\, v^{\mu}(s)\, \delta^{4}\big(x - z(s)\big) \tag{12}$$

where the world line of a single charge $z^{\mu}(s)$ is taken to be arbitrary timelike smooth curve. The retarded Liénárd–Wiechert solution

$$A^{\mu}(x) = \frac{\mathrm{e}\, v^{\mu}}{(x-z)\cdot v}\bigg|_{s=s_{\mathrm{ret}}} \tag{13}$$

is regularized and substituted to the equation of motion for a bare charged particle

$$m_0 \mathbf{a} = \mathrm{e}\,(\mathbf{E} + \mathbf{v} \times \mathbf{B}) \ , \tag{14}$$

where m_0 is the bare mass. We then require that the renormalized mass

$$m = \lim_{\epsilon \to 0} \left(m_0(\epsilon) + \frac{e^2}{2\epsilon} \right) \tag{15}$$

be a finite positive quantity. Finally, we arrive at the Abraham–Lorentz equation (8) in the limit of the regularization removal $\epsilon \to 0$.

In order to derive the equation of motion for a particle, which is capable of emitting gravitational waves, one should repeat the essentials of this procedure: to find a retarded solution to Einstein's equations (2) with $\Lambda = 0$ assuming that a given point particle which generates the retarded gravitational field moves along an arbitrary timelike smooth world line, regularize this solution, substitute it to the equation of motion for the bare particle, perform the mass renormalization, and remove the regularization. This will yield the desired equation of motion for the dressed particle, which is apparently different from the equation of a geodesic of the metric $g_{\mu\nu}$. However, this project is highly nontrivial. Even the first stage has still defied implementation: we have no solution to the gravitation equations (2) similar to the Liénárd–Wiechert solution (13) in electrodynamics. So, at the moment we cannot offer a complete explanation for the accelerated expansion of the Universe based on self-accelerated solutions similar to the runaway solution (11) to the Abraham–Lorentz equation. Nevertheless, with results drawn from solvable theories where particles interact with scalar, tensor, Yang–Mills, and linearized gravitational fields, and dimensional considerations, we can construct with some degree of certainty the form of the equation of motion for a particle emiting gravitation waves.

2 Dressed Particles

The relativistic generalization of the Abraham–Lorentz equation (8) is the Lorentz–Dirac equation (see, e.g., [6])

$$ma^{\lambda} - \frac{2}{3}\, e^2 \left(\dot{a}^{\lambda} + v^{\lambda} a^2 \right) = f^{\lambda} \ . \tag{16}$$

It accounts for the dynamics of a synthesized object whose inertia is characterized by the quantity m, defined in (15) where mechanical and electromagnetic field terms contribute. We will call this object the dressed charged particle. The state of the dressed particle is specified by the four-coordinate of the singular field point z^μ and the four-momentum

$$p^\mu = m\,v^\mu - \frac{2}{3}\,e^2 a^\mu \tag{17}$$

assigned to this point [7]. The singular point is governed by the equation

$$\overset{v}{\perp}(\dot{p} - f) = 0 \tag{18}$$

where $\overset{v}{\perp}$ is the projection operator

$$\overset{v}{\perp}_{\mu\nu} = \eta_{\mu\nu} - \frac{v_\mu v_\nu}{v^2}\,, \tag{19}$$

and f^μ is an external four-force applied to the point z^μ. Indeed, substitution of (17) in (18) gives the Lorentz–Dirac equation (16). On the other hand (18) is nothing but Newton's second law in a coordinate-free form.

Teitelboim [7] was able to show that the Lorentz–Dirac equation (16) is equivalent to the local energy-momentum balance,

$$\dot{p}^\mu + \dot{\mathcal{P}}^\mu + \dot{\wp}^\mu = 0\,, \tag{20}$$

where the four-momentum of the dressed particle p^μ is defined in (17), the radiation four-momentum $\mathcal{P}^\mu$ is represented by the Larmor formula,

$$\mathcal{P}^\mu = -\frac{2}{3}\,e^2 \int_{-\infty}^{s} \mathrm{d}\tau\, v^\mu a^2\,, \tag{21}$$

and the four-momentum $\wp^\mu$ relates to the integral of the external Lorentz four-force,

$$\wp^\mu = -\int_{-\infty}^{s} \mathrm{d}\tau\, f^\mu\,. \tag{22}$$

The balance equation (20) reads: the four-momentum $\mathrm{d}\wp^\mu = -f^\mu \mathrm{d}s$ which has been extracted from an external field during the period of time $\mathrm{d}s$ is distributed between the four-momentum of the dressed particle, $\mathrm{d}p^\mu$, and the four-momentum carried away by the radiation $\mathrm{d}\mathcal{P}^\mu$.

Let f^μ be zero, (16) is satisfied by

$$v^\mu(s) = \alpha^\mu \cosh(w_0\tau_0 \mathrm{e}^{s/\tau_0}) + \beta^\mu \sinh(w_0\tau_0 \mathrm{e}^{s/\tau_0}) \tag{23}$$

where α^μ and β^μ are constant four-vectors that meet the conditions $\alpha\cdot\beta = 0$, $\alpha^2 = -\beta^2 = 1$, w_0 is an initial acceleration magnitude, and τ_0 is given by (10). The solution (23) describes a runaway motion, which becomes a uniform Galilean motion for $w_0 = 0$. We see from (11) and (23) that the class of Galilean world lines is distinct from the class of runaway world lines. A dressed particle may either show itself as a Galilean object or execute perpetually a self-accelerated motion, none of the Galilean objects is able to become self-accelerated and vice versa. The non-Galilean behavior is an innate feature of some species of dressed particles.

It is often asserted that the solution (23) is "unphysical". The fact that because a free particle continually accelerates and continually radiates seems contrary to energy conservation. We note that a mechanical object with the four-momentum

$p^\mu = mv^\mu$ is understood in such claims. By contrast, proceeding from the concept of a dressed particle with the four-momentum defined in (17), we have the balance equation (20). There is no contradiction with energy conservation in this framework: the energy variation of the dressed particle dp^0 is equal to the energy carried away by the radiation $-\dot{\mathcal{P}}^0 ds$. A subtlety is that the dressed particle energy

$$p^0 = m\gamma\left(1 - \tau_0\,\gamma^3\,\mathbf{a}\cdot\mathbf{v}\right) \tag{24}$$

is not positive definite. The indefiniteness of expression (24) means that increase of the velocity magnitude $|\mathbf{v}|$ need not be accompanied by increase of p^0.

For $m = 0$, which is another reasonable option of the mass renormalization (15), the first term of (16) disappears. Take $f^\mu = 0$, then (16) becomes

$$(\overset{v}{\perp}\,\dot{a})^\mu = 0,$$

which is the equation that describes a relativistic uniformly accelerated motion [6]. The world line of a free dressed particle with $m = 0$ is a hyperbola

$$v^\mu(s) = \alpha^\mu \cosh w_0 s + \beta^\mu \sinh w_0 s, \quad \alpha\cdot\beta = 0, \quad \alpha^2 = -\beta^2 = 1\,. \tag{25}$$

The curvature $k = w_0 = \text{const}$ of such a world line may be arbitrary. Thus the non-Galilean regime for massless dressed particle is a hyperbolic motion.

It follows from (17) that

$$M^2 = p^2 = m^2\left(1 + \tau_0^2\,a^2\right)\,. \tag{26}$$

For $\tau_0^2\,a^2 < -1$, the dressed particle turns to a tachyonic state, that is, the state with $p^2 < 0$, during the period of time

$$\Delta s = -\frac{\tau_0}{2}\,\log\left|\tau_0^2 a^2(0)\right|\,.$$

This observation gives insight into why the runaway motion of dressed charged particles was never observed. The period of time over which a self-accelerated electron possesses a timelike four-momentum is tiny. From (10) we find $\Delta s \sim \tau_0 \sim 10^{-23}$ s for electrons, and still shorter for other charged elementary particles. All primordial self-accelerated particles with such τ_0 have long been in tachyonic states. However, we have not slightest notion of how tachyons can be experimentally recorded. It seems plausible that primordial self-accelerated particles, which transmuted into tachyons, represent part of dark matter.

If a cosmological object is considered as a dressed particle emitting gravitational waves, the characteristic period τ_0 may be found to be comparable with the inverse current Hubble scale H^{-1}. The self-acceleration of such an object can indeed be observed at the present time. For clarity, the experimental value of this scale is $H^{-1} = (46 \pm 4)\cdot 10^{16}$ s.

We now turn to a dressed colored particle in the cold QCD phase [8]. The equation of motion for a dressed quark with the color charge Q in an external SU($\mathcal{N}$) Yang–Mills field $F^{\mu\nu}$ is

$$m\left[a^\mu + \lambda\left(\dot{a}^\mu + v^\mu a^2\right)\right] = \mathrm{tr}(QF^{\mu\nu})\,v_\nu \tag{27}$$

where m is the renormalized mass, and

$$\lambda = \frac{8}{3mg^2}\left(1 - \frac{1}{N}\right), \tag{28}$$

g is the Yang–Mills coupling constant. For $F^{\mu\nu} = 0$, the general solution to (27)

$$v^\mu(s) = \alpha^\mu \cosh(w_0 \lambda\, \mathrm{e}^{-s/\lambda}) + \beta^\mu \sinh(w_0 \lambda\, \mathrm{e}^{-s/\lambda}) \tag{29}$$

describes a self-decelerated motion. By contrast, in the hot QCD phase, dressed quarks may execute a runaway motion like (23) or (25).

At first sight, the self-deceleration is an innocent phenomenon, because the motion becomes almost indistinguishable from Galilean in the short run. However, the presence of self-decelerations actually jeopardizes the consistency of the theory: the acceleration increases exponentially as $s \to -\infty$, and the intensity of the energy gain grows along with it in this limit. The Yang–Mills field energy at any finite instant is therefore divergent.

Let us consider a dressed particle interacting with a massless scalar field [9]. The equation of motion for a dressed particle in an external scalar field ϕ is

$$\frac{\mathrm{d}}{\mathrm{d}s}(m + g\phi)\, v^\mu - \frac{1}{3}\, g^2 (\dot{a}^\mu + a^2 v^\mu) = g\, \partial^\mu \phi \tag{30}$$

where m is the renormalized mass, and g is the coupling constant.

Likewise, the equation of motion for a dressed particle interacting with an external tensor field $\phi_{\alpha\beta}$ in the linearized gravity [10, 9] is

$$\frac{\mathrm{d}}{\mathrm{d}s}\left[v^\mu \left(1 - \frac{1}{2} v^\alpha v^\beta \phi_{\alpha\beta} - \frac{1}{2}\phi^\alpha_{\ \alpha}\right) + v_\alpha \phi^{\alpha\mu}\right] + \frac{11}{3}\, Gm(\dot{a}^\mu + a^2 v^\mu)$$
$$= \frac{1}{2} v^\alpha v^\beta \partial^\mu \phi_{\alpha\beta} - \frac{1}{4}\partial^\mu \phi^\alpha_{\ \alpha} \tag{31}$$

where G is the gravitational constant, and m is the renormalized mass.

If we define the Abraham factor

$$\Gamma^\mu = \dot{a}^\mu + v^\mu a^2, \tag{32}$$

the results regarding the equations of motion for dressed particles can be summarized in the following table

Table 1. Abraham term in different theories

Scalar field	Vector field	Yang–Mills (cold phase)	Tensor field $\phi_{\mu\nu} = \lambda\, \frac{v_\mu v_\nu}{\rho}$	Tensor field $\phi_{\mu\nu} = \lambda\, \frac{\eta_{\mu\nu}}{\rho}$	Linearized gravity
$\frac{1}{3} g^2 \Gamma^\mu$	$\frac{2}{3} e^2 \Gamma^\mu$	$-\frac{2}{3}\lvert Q^2\rvert \Gamma^\mu$	$-\frac{5}{3}\lambda^2 \Gamma^\mu$	$-\frac{1}{3}\lambda^2 \Gamma^\mu$	$-\frac{11}{3} Gm^2 \Gamma^\mu$

3 Discussion

For dimensional reasons, we may expect that the equation of motion for a dressed particle emitting gravitational waves should include a covariant generalization of the Abraham term. Meanwhile the sign of the Abraham term in the linearized gravity is negative which implies that the non-Galilean regime of massive particle emitting gravitational waves is self-decelerating, not self-accelerating. Moreover, one can estimate the characteristic time τ_0 in (31) taking m to be a typical cluster mass, $\tau_0 = Gm \sim 10^8$ s ~ 3 light years. It is very far from $\tau_0 \sim H^{-1} \sim 2 \cdot 10^{10}$ light years. These results may appear discouraging in respect to the alternative explanation of the accelerated motion of distant cosmological object. Recall, however that the nonlinearity peculiar to Einstein's equation may change the situation drastically, as the Yang–Mills theory suggests [8]. Two phases are frequent among nonlinear systems: hadron and quark-gluon phases in QCD, laminar and turbulent phases in hydrodynamics are examples. It is conceivable that gravity may also reveal two phases which are classified according to whether the emission of gravitational waves is attended with energy gains or energy losses. In the former phase, the motion of dressed particles is self-decelerated, and in the latter phase, it is self-accelerated. Contributions from interactions with other classical fields (electromagnetic, dilaton, gluon) may also have a dramatic effect on the sign and the magnitude of the aggregate Abraham term. In addition, when on the subject of galaxies and clusters, we should take into account their spins. The description of a radiating charged particle with spin is a challenging problem; it was discarded here to make the key idea as simple as possible (for a review see [11]).

Finally, a remarkable fact is that the characteristic time $\tau_0 = Gm$ with m being the total visible mass of the Universe is of order of the inverse current Hubble scale H^{-1}. Does this mean that the Universe as a whole executes a self-accelerated motion?

I would like to thank I. D. Novikov for helpful comments. This work was supported in part by ISTC under the Project # 840.

References

1. A. G. Riess et al., "Observational evidence from supernovae for an accelerating universe and a cosmological constant," Astron. J. **116**, 1009, 1998; astro-ph/9805201; S. Perlmutter et al., "Measurement of Omega and Lambda from 42 high-redshift supernovae," Astrophys. J. **517**, 565, 1999; astro-ph/9812133.
2. J. Frenkel, "Die Elektrodynamik des Rotierenden Electrons," Z. Phys. **37**, 243, 1926.
3. B. P. Kosyakov, "On inert properties of particles in classical theory," hep-th/0208035.
4. E. Schrödinger, "Über die kräftfreie Bewegung in der relativistischen Quantenmechanik," Sitzungber. Preuss. Acad. Wiss. **24**, 418, 1930.
5. J. D. Jackson, *Classical Electrodynamics* (New York: Wiley, 1962, 1975, and 1998).
6. F. Rohrlich, *Classical Charged Particles* (Reading: Addison-Wesley, 1965, and 1990).
7. C. Teitelboim, "Splitting of Maxwell tensor: Radiation reaction without advanced fields," Phys. Rev. D **1**, 1572, 1970.
8. B. P. Kosyakov, "Exact solutions in the Yang-Mills-Wong theory," Phys. Rev. D **57**, 5032, 1998; hep-th/9902039.

9. A. O. Barut and D. Villarroel, "Radiation reaction and mass renormalization in scalar and tensor fields and linearized gravitation," J. Phys. A **8**, 156, 1975.
10. P. Havas and J. N. Goldberg, "Lorentz-invariant equations of motion of point masses in the general theory of relativity," Phys. Rev. **128**, 398, 1962.
11. E. G. P. Rowe and G. T. Rowe, "The classical equation of motion for a spinning point particle with charge and magnetic moment," Phys. Rep. **149**, 287, 1987.

Charge and Isospin Fluctuations in High Energy pp-Collisions

Mladen Martinis[1] and Vesna Mikuta-Martinis[2]

[1] martinis@rudjer.irb.hr
[2] vmikuta@rudjer.irb.hr
Rudjer Bošković Institute, Zagreb, Croatia

Charge and isospin event-by-event fluctuations in high-energy pp-collisions are predicted within the Unitary Eikonal Model, in particular the fluctuation patterns of the ratios of charged-to-charged and neutral-to-charged pions. These fluctuations are found to be sensitive to the presence of unstable resonances, such as ρ and ω mesons. We predict that the charge-fluctuation observable D_{UEM} should be restricted to the interval $8/3 \leq D_{UEM} \leq 4$ depending on the ρ/π production ratio. Also, the isospin fluctuations of the DCC-type of the ratio of neutral-to-charged pions are suppressed if pions are produced together with ρ mesons.

1 Introduction

In a single central ultrarelativistic collisions at RHIC and LHC more then 2400 hadrons are created [1], presenting remarkable opportunity to study event-by-event fluctuations of various hadronic observables. Such single event analysis with large statistics may reveal new physical phenomena usually hidden when averages over a large statistical sample of events are made [2]. Recently, the study of event-by-event fluctuations of charged particles in high-energy pp and heavy-ion collisions has gained a considerable attention [3, 4, 5, 6, 8]. The idea was to find an adequate measure that can differentiate a quark-gluon plasma (QGP) from a hadron gas (HG). So far, no consideration has been given to the fluctuations generated by the phase transition (PT) itself [8].

The number of particles produced in relativistic pp and heavy-ion collisions can differ dramatically from collision to collision due to the variation of impact parameter (centrality dependence), energy deposition (leading particle effect), and other dynamical effects [3]. The fluctuations can also be influenced by novel phenomena such as the formation of disoriented chiral condensates (DCCs) [4, 15, 18, 19, 20, 21, 22, 23] in consequence of the transient restoration of chiral symmetry. It is generally accepted that much larger fluctuations of the neutral-to-charged pion ratio than expected from Poisson-statistics could be a sign of the DCC formation. However, such fluctuations are possible even without invoking the DCC formation if, for example, pions are produced semiclassically and constrained by global conservation of isospin [12, 13, 14, 15, 16]. In this paper, we present results of an event-by-event analysis of charged-charged and neutral-charged pion fluctuations as a function of the ρ/π production ratio in pp-collisions. Our study of these fluctuations is performed within the Unitary Eikonal Model (UEM) [24, 27].

2 Coherent Production of π and ρ-Mesons

At high energies most of the pions are produced in the nearly baryon-free central region of the phase space. The energy available for their production is

$$E_{had} = \frac{1}{2}\sqrt{s} - E_{leading} \tag{1}$$

which at fixed total c.m. energy $\sqrt{s}$ varies from event-to-event. Within the UEM the N-pion contribution to the s-channel unitarity in the central region can be written as an integral over the relative impact parameter $\boldsymbol{b}$ between two incident leading particles:

$$\sigma_N(s) = \frac{1}{4s^2}\int \mathrm{d}^2 b \prod_{i=1}^{N} \mathrm{d}q_i \mid T_N(s, \boldsymbol{b}; q_1 \ldots q_N) \mid^2 , \tag{2}$$

where $\mathrm{d}q = \mathrm{d}^2 q_T \mathrm{d}y/2(2\pi)^3$. If the isospin of the incoming (outgoing) leading particle-system is II_3 $(I'I'_3)$, then the N-pion production amplitude becomes [24]

$$iT_N(s, \boldsymbol{b}; q_1 \ldots q_N) = 2s\langle I'I'_3; q_1 \ldots q_N \mid \hat{S}(s, \boldsymbol{b}) \mid II_3\rangle , \tag{3}$$

where $\hat{S}(s, \boldsymbol{b})$ denotes the $\hat{S}$-matrix in the isospace of the leading particles.

The coherent emission of pions or clusters of pions, such as ρ and ω mesons, in b-space of leading particles is described by the factorized form of the scattering amplitude, T_N. In that case the $\hat{S}(s, \boldsymbol{b})$-matrix has the following generic form:

$$\hat{S}(s, \boldsymbol{b}) = \int \mathrm{d}^2\boldsymbol{n} \mid \boldsymbol{n}\rangle \hat{D}(s, \boldsymbol{b})\langle \boldsymbol{n} \mid , \tag{4}$$

where $\mid \boldsymbol{n}\,\rangle$ represents the isospin-state vector of the two-leading-particle system. The quantity $\hat{D}(s, \boldsymbol{b})$ is the unitary coherent-state displacement operator defined in our case as

$$D(s, \boldsymbol{b}) = \exp[a^\dagger(s, \boldsymbol{b}) - a(s, \boldsymbol{b})] \tag{5}$$

with

$$a^\dagger(s, \boldsymbol{b}) = \sum_{c=\pi,\rho} \int \mathrm{d}q J_c(s, \boldsymbol{b}; q)\boldsymbol{n}\boldsymbol{a_c}^\dagger(q) \; . \tag{6}$$

where, J_c denotes a classical source function of the cluster c. The cluster decays into pions outside the region of strong interactions (i.e. the final-state interaction between pions is neglected).

The isospin (I', I'_3) of the outgoing leading particle system varies from event-to-event. If the probabilities ω_{I',I'_3} of producing (I', I'_3) states are known, we can sum over all $(I'I'_3)$ to obtain a probability distribution of producing N_+, N_- and N_0 pions from a given initial isospin state:

$$P_{II_3}(N_+N_-N_0 \mid N)C_{II_3}(N) = \\ \sum_{I'I'_3} \omega_{I',I'_3} \int \mathrm{d}^2 b \mathrm{d}q_1 \mathrm{d}q_2 \ldots \mathrm{d}q_N \mid \langle I'I'_3, N_+N_-N_0 \mid \hat{S}(s, \boldsymbol{b}) \mid II_3\rangle \mid^2) \tag{7}$$

where

$$\begin{aligned} N &= N_+ + N_- + N_0 \\ N_+ &= n_{\pi^+} + n_{\rho^+} + n_{\rho^0} \\ N_- &= n_{\pi^-} + n_{\rho^-} + n_{\rho^0} \\ N_0 &= n_{\pi^0} + n_{\rho^+} + n_{\rho^-} \end{aligned} \tag{8}$$

and $C_{II_3}(N)$ is the corresponding normalization factor. This is now our basic equation for calculating various pion-multiplicity distributions, pion-multiplicities, and pion-correlations between definite charge combinations. In the following, we consider fluctuations of the π_+/π_- and π_0/N ratios in pp-collisions, $(I = I_3 = 1)$.

3 Charge and Isospin Fluctuations

A suitable measure of the charge fluctuations was suggested in [3]. It is related to the fluctuation of the ratio $R_{ch} = N_+/N_-$ and the observable to be studied is

$$D \equiv \langle N_{\rm ch}\rangle \langle \delta R_{\rm ch}^2 \rangle = 4\frac{\langle \delta Q^2 \rangle}{\langle N_{\rm ch}\rangle} \, . \tag{9}$$

where $N_{\rm ch} = N_+ + N_-$, $Q = N_+ - N_-$ and $\langle \delta Q^2 \rangle = \langle Q^2 \rangle - \langle Q \rangle^2$.

Our prediction of the D quantity within the UEM, when both π and ρ mesons are produced is

$$D_{UEM} = \frac{8}{2 + \frac{\overline{n}_\pi}{\overline{n}_\pi + \overline{n}_\rho}} \, , \tag{10}$$

where $\overline{n}_\pi = \int dq \mid J_\pi(q) \mid^2$ denotes the average number of directly produced pions, and similarly $\overline{n}_\rho$ denotes the average number of ρ mesons which decay into two short-range correlated pions. The total number of emitted pions is

$$\overline{N} = \overline{n}_\pi + 2\overline{n}_\rho \, . \tag{11}$$

It was argued [3] that the value of D may be used to distinguish the hadron gas ($D_{\pi-{\rm gas}} \approx 4$) from the quark-gluon plasma ($D_{\rm QGP} \simeq 3/4$). It is expected that $D_{\pi-{\rm gas}} \simeq 3$ if appropriate corrections for resonance production are taken into account [28].

The UEM predicts $D_{UEM} = 4$ if $\overline{n}_\rho = 0$. In that case the pion production is restricted only by the global conservation of isospin. However, if $\overline{n}_\pi = 0$ the UEM predicts $D_{UEM} = 8/3$. This means that D is restricted to the interva $8/3 \leq D_{UEM} \leq 4$.

The preliminary results from CERES , NA49 and STAR collaboration [29, 30, 31], however, indicate that the measured value of D is close to that predicted for hadron gas and differs noticeably from that expected for QGP. This finding is somewhat disturbing since no effect of resonance production is visible in the fluctuations.

The formation of DCC in pp-collisions is expected to lead to different types of isospin fluctuations [18, 19, 20, 21, 22]. Since pions formed in the DCC are essentially classical and form a quantum superposition of coherent states with different orientation in isospin space,large event-by-event fluctuations in the ratio $R_0 = N_0/N$ are expected. The probability distribution of R_0 inside the DCC domain is [15, 18, 19, 21]

$$P_{DCC}(R_0) = \frac{1}{2\sqrt{R_0}} \tag{12}$$

There are a variety of proposed mechanisms other than DCC formation which can also lead to the distribution $P_{DCC}(R_0)$ [23, 24, 25, 26]. The distribution $P_{DCC}(R_0)$ is different from the generic binomial-distribution expected in normal events which assumes equal probability for production of π_+, π_- and π_0 pions.

Following the approach of our earlier papers [24, 27], we have calculated the probability distribution function, $P^{\pi+\rho}(N_0 \mid N)$ for producing N_0 neutral pions. For large N and N_0 such that R_0 is fixed, we find that only $NP^{\pi}(N_0 \mid N)$ is of the form $\sqrt{N/N_0}$ which resembles the DCC-type fluctuations and is typical for coherent pion production.

4 Conclusion

Our general conclusion is that within the UEM the large charge and isospin fluctuations depend strongly on the value of the ρ/π production ratio which fluctuate from event to event. Recent estimate of the ρ/π production ratio at accelerator energies, is $\overline{n}_\rho = 0.10\overline{n}_\pi$ [3].

References

1. F. Sikler, Nucl. Phys. **A661**, (1999) 45c.
2. H. Heiselberg, G.A. Baym, B. Blättel, L.L. Frankfurt and M. Strikman, Phys. Rev. Lett. **67** (1991) 2946; G. Baym, B. Blättel, L.L. Frankfurt, H. Heiselberg and M. Strikman, Phys. Rev C **52** (1995) 1604; G. Baym, G. Friedman and I. Sarcevic, Phys. Lett. B **219**(1989) 205; H. Heiselberg, Phys. Rept. **351** (2001) 161.
3. S. Jeon and V. Koch, Phys. Rev. Lett. **83** (1999) 5435; **85** (2000) 2076.
4. M. Asakawa, U. Heinz, and B. Mueller, Phys. Rev. Lett. **85** (2000) 2072.
5. H. Heiselberg, and A. D. Jackson, Phys. Rev. **C 63** (2001) 064904.
6. C. Gale, V. Topor Pop, and Q. H. Zhang, McGill preprint (2001).
7. For a review, see V. Koch, talk given at Quark Matter 2001, 15th Int. Conf. on Ultra-Relativistic Nucleus-Nucleus Collision, Stony Brook, NY, January 2001.
8. R. Hwa, C. B. Yang, Phys. Lett. **B534** (2002) 69.
9. M. Gazdzicki and S. Mrowczynski, Z. Phys **C54**, (1992) 127.
10. K. Rajagopal and F. Wilczek, Nucl. Phys. B **399** (1993) 395; S. Gavin, Nucl. Phys.A **590** (1995) 163c.
11. M. Stephanov, K. Rajagopal and E. Shuryak, Phys. Rev. Lett. **81** (1998) 4816; Phys. Rev. D **60** (1999) 114028.
12. D. Horn and R. Silver, Ann. Phys. **66** (1971) 509.
13. J.C. Botke, D.J. Scalapino and R.L. Sugar, Phys. Rev. D **9** (1974) 813.
14. P. Pirilä, Acta Phys. Pol. B **8** (1977) 305.

15. I.V. Andreev, Sov. Physica JETP Lett. **33** (1981) 367.
16. M. Martinis and V. Mikuta, Phys. Rev. D **12** (1975) 909.
17. L.J. Reinders, Acta Phys. Pol. B **7** (1976) 256.
18. A.A. Anselm, Phys. Lett. B **217** (1989) 169; A.A. Anselm and M. Ryskin, Phys. Lett. B **266** (1991) 482; A.A. Anselm and M. Bander, JETP Lett. **59** (1994) 503; A.A. Anselm and M. Ryskin, Z.Phys. A **358** (1997) 353.
19. J.D. Bjorken, Int. J. Mod. Phys. A **7** (1992) 4819; Acta Phys. Pol. B **23** (1992) 561; J.D. Bjorken, K.L. Kowalski and C.C. Taylor, SLAC-PUB-6109 (April 1993); hep-ph/9309235.
20. J.P. Blaizot and A. Krzywicki, Phys. Rev. D **46** (1992) 246; A. Krzywicki, Phys. Rev. D **48** (1993) 5190; J.P. Blaizot and A. Krzywicki, Phys. Rev. D **50** (1994) 442; J.P. Blaizot and A. Krzywicki, Acta Phys. Polon. B **27** (1996) 1687.
21. K.L. Kowalski and C.C. Taylor, hep-ph/9211282; K. Rajagopal and F. Wilczek:, Nucl. Phys. B **399** (1993) 395; Nucl. Phys. B **404** (1993) 577.
22. K. Rajagopal, hep-ph/9703258 (published in Hirschegg Workshop 1997, 136).
23. MiniMax Collab., T.C. Brooks et al., Phys. Rev. D **61** (2000) 032003.
24. M. Martinis, V. Mikuta-Martinis, A. Švarc and J. Črnugelj, Fizika B **3** (1994) 197; M. Martinis, V. Mikuta-Martinis, A. Švarc and J. Črnugelj, Phys. Rev. D **51** (1995) 2482; M. Martinis, V. Mikuta-Martinis and J. Črnugelj, Phys. Rev. C **52** (1995) 1073.
25. V. Karmanov and A. Kudrjavtsev, ITEP-88, (1983).
26. R.D. Amado et al., Phys. Rev. Lett. **72** (1994) 970.
27. M. Martinis, V. Mikuta-Martinis and J. Črnugelj, Acta Phys. Slov. **49** (1999) 875.
28. M. Doering and V. Koch, nucl-th/0204009.
29. CERES coll., J. Stachel, contribution to 2002 Epiphany Conf., Cracow, January 2002 (to be published in Acta Phys. Pol. B).
30. C. Blume, NA49 coll., Nucl. Phys. **A698**,(2002) 104c.
31. J.G. Reid et al., STAR coll., Nucl. Phys. **A698**,(2002) 611c.

Superluminal Pions in the Linear Sigma Model

Hrvoje Nikolić

Theoretical Physics Division, Rudjer Bošković Institute, P.O.B. 180,
10002 Zagreb, Croatia
hrvoje@thphys.irb.hr

1 Introduction – Superluminal Velocities in General

It is widely believed that a wave cannot propagate faster than $c \equiv 1$. This is a consequence of Lorentz invariance. For example, if the wave is massless, then Lorentz invariance implies the wave equation

$$\partial_t^2 \phi - \nabla^2 \phi = 0 . \tag{1}$$

A plane-wave solution

$$\phi = \mathrm{e}^{\mathrm{i}(\omega t - \mathbf{q}\mathbf{x})} \tag{2}$$

leads to the dispersion relation

$$\omega^2 = \mathbf{q}^2 . \tag{3}$$

The corresponding group velocity is

$$v_g = \frac{\mathrm{d}\omega}{\mathrm{d}q} = 1 . \tag{4}$$

However, if the massless wave propagates through a medium (with which it interacts), then the medium defines a preferred Lorentz frame, that is the frame with respect to which the medium is at rest. Then, Lorentz invariance is broken. In such a case, there is no longer a reason why the wave should propagate with the velocity equal to 1. The best known example is light in a transparent medium, in which it propagates with a velocity slower than 1.

However, owing to the absence of Lorentz invariance, there are also cases in which a massless field propagates with a velocity *faster* than 1. In a class of such examples, the modification of the dispersion relation is induced by the quantum (loop) corrections to the propagator:

$$\frac{1}{\omega^2 - \mathbf{q}^2} \longrightarrow \frac{1}{\omega^2 - \mathbf{q}^2 - \Sigma(\omega, \mathbf{q})} . \tag{5}$$

Owing to the absence of Lorentz invariance, the self-energy $\Sigma(\omega, \mathbf{q})$ is *not* a function of $\omega^2 - \mathbf{q}^2$. The dispersion relation is

$$\omega^2 - \mathbf{q}^2 - \Sigma(\omega, \mathbf{q}) = 0 , \tag{6}$$

which defines the group velocity that may exceed 1. For example, photons in QED in a gravitational background may propagate superluminally [1]. Another example is known as the Scharnhorst effect [2], in which photons propagate superluminally in QED in the vacuum between Casimir plates.

In QED, owing to the smallness of the fine structure constant α, these corrections to the propagation velocity are typically small:

$$v = 1 + \mathcal{O}(\alpha^2)\,. \tag{7}$$

Therefore, it is difficult to experimentally confirm the theoretical predictions in [1] and [2].

It is often argued that superluminal velocities violate causality. If a superluminal signal is moving forwards in time in one Lorentz frame, then there exists another Lorentz frame in which this signal is moving *backwards* in time. The motion backwards in time is a potential source of causal paradoxes. However, to construct a causal paradox (i.e. a closed causal loop), we need at least two *different* superluminal velocities within one Lorentz frame. This is because a causal paradox requires motions *both* forwards and backwards in time. On the other hand, in the examples above, for a given reference frame and a given medium, the propagation velocity is *unique*. Therefore, there are no causal paradoxes! More details on the issue of causality in systems with superluminal velocities can be found in [3].

2 Superluminal Pions

In the chiral limit, pions are massless. Normally, such pions propagate with the velocity of light. However, similarly to photons, it is possible that they become superluminal under certain conditions. If this occurs, then, owing to strong interactions, we expect a correction to $v = 1$ much larger than that for QED. Here we review our results originally presented in [4], where we study the velocity of massless pions in the framework of the linear σ-model, which is an effective model for the low-energy phase of QCD. We find superluminal velocities at finite temperature and chemical potential, as well as in the vacuum between Casimir plates.

For soft pions, the self-energy at finite temperature can be expanded as

$$\Sigma(q,T) = \Sigma(0,T) + \frac{1}{2!}q^\mu q^\nu \frac{\partial}{\partial q^\mu}\frac{\partial}{\partial q^\nu}\left(\Sigma(q,T) - \Sigma(q,0)\right)\big|_{q=0} + \dots\,. \tag{8}$$

Therefore, the inverse propagator takes the form

$$q_0^2 - \boldsymbol{q}^2 - \Sigma = (a+b)q_0^2 - a\boldsymbol{q}^2 + \dots\,. \tag{9}$$

The pion velocity is then

$$v^2 = \left(1 + \frac{b}{a}\right)^{-1}\,. \tag{10}$$

At one-loop order there are two q-dependent diagrams that contribute to Σ [5]. One of them contains a fermion loop, while the other contains a boson loop having one pion propagator and one σ-particle propagator. Therefore, we write

$$a = 1 + a_B + a_F\,, \quad b = b_B + b_F\,, \tag{11}$$

where [4]

$$a_B = \frac{16g'^2}{m_\sigma^4}\int\frac{d^3p}{(2\pi)^3}\left[\frac{n_B(\omega_\pi)}{4\omega_\pi} + \frac{n_B(\omega_\sigma)}{4\omega_\sigma} - \frac{1}{3}\frac{\omega_\pi^2}{m_\sigma^2}\left(\frac{n_B(\omega_\pi)}{\omega_\pi} - \frac{n_B(\omega_\sigma)}{\omega_\sigma}\right)\right]\,, \tag{12}$$

$$b_B = \frac{16g'^2}{m_\sigma^4} \int \frac{d^3p}{(2\pi)^3} \left[\frac{\omega_\pi n_B(\omega_\pi)}{m_\sigma^2} - \frac{\omega_\sigma n_B(\omega_\sigma)}{m_\sigma^2} + \frac{1}{3}\frac{\omega_\pi^2}{m_\sigma^2} \left(\frac{n_B(\omega_\pi)}{\omega_\pi} - \frac{n_B(\omega_\sigma)}{\omega_\sigma} \right) \right] \tag{13}$$

$$a_F = N_c g^2 \int \frac{d^3p}{(2\pi)^3} \frac{n_F(\omega_F)}{p^2 \omega_F} , \tag{14}$$

$$b_F = -N_c g^2 \int \frac{d^3p}{(2\pi)^3} \frac{m_F^2}{p^2} \frac{n_F(\omega_F)}{\omega_F^3} . \tag{15}$$

Here n_F and b_F are the Fermi-Dirac and the Bose-Einstein distribution, respectively, at finite temperature and finite chemical potential of fermions. The masses m_σ and m_F depend on the chiral condensate σ as explained in [4, 5], while g and g' are the coupling constants of the linear sigma model [5]. We see that the sign of b_F is negative, which is the technical reason that superluminal velocities occur for certain values of temperature and chemical potential.

In finite-temperature field theory, the euclidean time is compactified with the period $\beta = 1/T$. Therefore, the results above can be easily modified to study the case $T = \mu = 0$, but with the z-coordinate compactified. The compactification length is $\beta = L$. Essentially, time and the z-coordinate exchange their roles. The inverse propagator is then

$$a\left(q_0^2 - q_x^2 - q_y^2\right) - (a+b)q_z^2 . \tag{16}$$

The pion velocity in the compact direction is

$$v_{||}^2 = 1 + \frac{b'}{a'} . \tag{17}$$

For antiperiodic fermions,

$$a' = a, \quad b' = b , \tag{18}$$

while for periodic fermions we find

$$a' = 1 + a_B - \bar{a}_F, \quad b' = b_B - \bar{b}_F , \tag{19}$$

$$\bar{a}_F = 2N_c g^2 \int \frac{d^3p}{(2\pi)^3} \frac{n_B(\omega_F)}{p^2 \omega_F} , \tag{20}$$

$$\bar{b}_F = -2N_c g^2 \int \frac{d^3p}{(2\pi)^3} \frac{m_F^2}{p^2} \frac{n_B(\omega_F)}{\omega_F^3} . \tag{21}$$

Note that the Bose-Einstein distribution n_B appears in the last two expressions for the contribution of the fermion loop. This is because we take the periodic bondary condition for fermion fields [4], while at finite temperature one must take the antiperiodic boundary condition for fermions.

By numerical evaluation of the integrals given above, we find superluminal velocities for certain values of temperature, chemical potential and compactification length. The numerical results are presented in detail in [4], which we omit here because of the space limitation.

3 Conclusion

Our results suggest that pions might propagate superluminally under certain conditions (finite temperature and chemical potential, or compactified space). This is *not* inconsistent with causality. However, it would be premature to claim with certainty that pions may propagate superluminally because the present calculation is based on several approximations. First, pions are treated in the chiral limit ($m_\pi = 0$). Second, the linear σ model (instead of full QCD) is used. Third, the one-loop approximation is used. Besides, the results refer to soft pions ($q \to 0$) only. It would be interesting to see whether pions would remain superluminal if some of the approximations were improved.

Acknowledgement

This work was supported by the Ministry of Science and Technology of the Republic of Croatia under Contract No. 0098002.

References

1. I. T. Drummond, S. J. Hathrell: Phys. Rev. D **22**, 343 (1980)
2. K. Scharnhorst: Phys. Lett. B **236**, 354 (1990)
3. S. Liberati, S. Sonego, M. Visser: Ann. Phys. **298**, 167 (2002)
4. N. Bilić, H. Nikolić: Phys. Rev. D **68**, 085008 (2003)
5. N. Bilić, H. Nikolić: Eur. Phys. J. C **6**, 515 (1999)

Part II

Strings, Branes, Noncommutative Field Theories and Grand Unification

Comments on Noncommutative Field Theories

Luis Álvarez-Gaumé[1] and Miguel A. Vázquez-Mozo[2]

[1] Theory Division, CERN
1211 Geneva 23, Switzerland
Luis.Alvarez-Gaume@cern.ch
[2] Física Teórica, Universidad de Salamanca,
Plaza de la Merced s/n, 37008 Salamanca, Spain
Miguel.Vazquez-Mozo@cern.ch

1 Introduction

Since its formulation by Alain Connes, noncommutative geometry (NCG) has become a very active and interesting branch of Mathematics [1]. In Physics, NCG has had an early impact in a number of subjects including condensed matter physics [2] and high energy physics [3]. In String Theory, the use of NCG was pioneered by its application by Witten to string field theory [4]. More recently, compactifications of string and M-theory on noncommutative tori were studied in [5]. Although quantum field theories in noncommutative spaces had been the subject of attention [6], a renewed interest in the subject came after the realization by Seiberg and Witten [7] that a certain class of field theories on noncommutative Minkowski space can be obtained as particular low-energy limits in the presence of a constant NS-NS B-field.

Unlike the standard low-energy limit of string theory, the Seiberg-Witten limit leads to a nonlocal effective theory, where the interaction vertices are constructed in terms of the nonlocal Moyal product (see [8] for comprehensive reviews). In physical terms, this nonlocality is due to the extended nature of the low energy excitations, which in fact are rigid rodes whose size depends on the momentum of the state [9]. It is therefore interesting, from the field theoretic point of view, to understand how our ordinary view of field theory changes by the introduction of this particular type of nonlocality. Many standard notions and results require revision, like renormalizability, unitarity, discrete and space-time symmetries, etc. Nonetheless, since these theories are obtained from String Theory, one would expect them to be better behaved than other kinds of nonlocal theories.

One of the more remarkable results in the subject was obtained by Minwalla, van Raamsdonk and Seiberg [10]. These authors realized that quantum theories on noncommutative spaces are afflicted from an endemic mixing of ultraviolet (UV) and infrared (IR) divergences. Even in massive theories the existence of UV divergences induce IR problems, and this leads to a breakdown of the Wilsonian approach to field theory. Contrary to some expectations [2], noncommutativity does not provide a full regularization of UV divergences, but only of a subsector of the Feynman graphs. Hence the issue of renormalizability of NCQFT become rather subtle [12].

In ordinary Quantum Field Theory there are a number of properties that can be derived from general principles collectively called Wightman axioms [13]. Among them we can cite the CPT theorem, the connection between spin-statistics and the cluster decomposition. The extension of some of these properties to NCQFTs is not straightforward [14, 15, 16] and therefore it would be interesting to study whether this kind of nonlocal field theories admit an axiomatic formulation in order to gain a better insight about the extension to NCQFTs of properties like the CPT and spin-statistics theorems [16].

In this lecture we would like to make a number of remarks on noncommutative field theories, in particular those obtained from String Theory through the Seiberg-Witten limit. We will pay special attention to the analysis of the phenomenological viability of this kind of field theories. For that we will focus on noncommutative QED (NCQED). The Standard Model contains Maxwell's theory at low energies and thus the "usual" photon should be recovered in any noncommutative generalization of QED, independently of how the Standard Model is embedded into its noncommutative extension. Among the properties of the QED photon, we will look at its masslessness, and the fact that the speed of light is constant, i.e. independent of the magnitude and direction of the photon momentum [17]. As we will see, it is remarkably difficult to obtain that ordinary electromagnetism is embedded as the low-energy limit of a noncommutative $U(1)$-theory. In particular, due to UV/IR mixing, it is rather common to obtain that one of the photon polarizations remains massless while the other becomes either massive or tachyonic. In ordinary gauge theories vector bosons get masses through the Higgs mechanism. Here the nonlocality of the interaction terms may lead to a massive photon polarization.

In order to give sense to NCQED we define it in terms of a softly broken $\mathcal{N}=4$ noncommutative U(1) gauge theory. This provides a construction that makes sense in the UV and IR, and where we can have control on the UV/IR mixing. Here we find that unless some conditions are satisfied by the soft breaking terms, one of the components of the photon becomes tachyonic. Even when this disaster is avoided, one generically gets a completely unacceptable value for the photon mass, unless one is willing to engage in massive fine-tuning. We will follow the presentation in our paper [16], where a more complete list of references is provided.

Before we proceed we would like to clearly state our point of view. As mentioned above, we will focus here on the type of noncommutative theories that are obtained from string theory via the Seiberg-Witten limit. There are of course other approaches to the problem, and we would like to briefly make a comparison. If one follows the quantization procedure proposed in [18, 19] the results should be the same, because both approaches agree in the case of space-space noncommutativity. Regarding the approach of [20], they extend the Seiberg-Witten map to arbitrary groups, and their actions are obtained order by order in an expansion in powers of θ. Hence if we truncate at a given order, we find the standard commutative Lagrangian, and a collection of corrections corresponding the higher dimension operators. This theory is technically nonrenormalizable and one should not find UV/IR mixing, which occur only after one has summed to all order in θ, in which case we would expect to obtain the same results because the Feynman rules are the same. Other approaches has been studied in [21].

We follow here the "orthodox" string approach, namely we use the Feynman rules that follow from String Theory after we take the Seiberg-Witten limit, in particular we restrict our considerations always to space-space noncommutativity. Since the vertices and Feynman integrands are only modified by sine and cosine functions, the naive degree of divergence of the theory will not change, and one should expect some sort of renormalizability to hold once the UV/IR problems are tamed. It is possible to extend the Seiberg-Witten limit to have time-space noncommutativity, but this does not lead to a field theory but a theory of noncommutative open string [22].

In the next section we give a short overview of some well-known facts about NCQFTs, in particular the UV/IR mixing characteristic of these theories. In Sect. 3 an extension of axiomatic formulation to NCQFTs is briefly discussed as well as the validity of the CPT theorem in this type of theories. Section 4 reviews the IR problems of NCQED and in Sect. 5 we study the construction of such a theory from its softly broken $\mathcal{N} = 4$ supersymmetric extension and the possibility of eliminating tachyonic states in the spectrum. This section concludes with a discussion of the phenomenological prospects of NCQED.

2 Seiberg-Witten Limit, Dipoles and UV/IR Mixing

In [7] it was shown how noncommutative field theories are obtained as a particular low-energy limit of open string theory on D-brane backgrounds in the presence of constant NS-NS B-field. In this case, the endpoints of the open strings behave as electric charges in the presence of an external magnetic field $B_{\mu\nu}$ resulting in a polarization of the open strings. Labeling by $i = 1, \ldots, p$ the D-brane directions and assuming $B_{0i} = 0$, the difference between the zero modes of the string endpoints is given by [9]

$$\Delta X^i = X^i(\tau, 0) - X^i(\tau, \pi) = (2\pi\alpha')^2 g^{ij} B_{jk} p^k \ , \tag{1}$$

where $g_{\mu\nu}$ is the closed string or σ-model metric and p^μ is the momentum of the string. In the ordinary low-energy limit, where $\alpha' \to 0$ while $g_{\mu\nu}$ and $B_{\mu\nu}$ remain fixed, the distance $|\Delta X|$ goes to zero and the effective dynamics is described by a theory of particles, i.e. by a commutative quantum field theory.

There are, however, other possibilities of decoupling the massive modes without collapsing at the same time the length of the open strings. Seiberg and Witten proposed to consider a low-energy limit $\alpha' \to 0$ where both B_{ij} and the open string metric

$$G_{ij} = -(2\pi\alpha')^2 (Bg^{-1}B)_{ij} \tag{2}$$

are kept fixed. Introducing the notation $\theta^{ij} = (B^{-1})^{ij}$, the separation between the string endpoints can be expressed as:

$$\Delta X^i = \theta^{ij} G_{jk} p^k \ , \tag{3}$$

fixed in the low energy limit. The resulting low-energy effective theory is a noncommutative field theory with noncommutative parameter θ^{ij}. In physical terms the Seiberg-Witten limit corresponds to making the string tension go to infinity, while and balancing it with the Lorentz force on the string-ends caused by the external magnetic field. This limit makes the string rigid and with a finite length that depends on its momentum.

The previous analysis was confined to situations in which the B_{0i} components are set to zero. The result is a noncommutative field theory with only space-space noncommutativity. From a purely field-theoretical point of view it is possible to consider also noncommutative theories where the time coordinate does not commute with the spatial ones, i.e. $\theta^{i0} \neq 0$. In this case, however, non-locality is accompanied by a breakdown of unitarity reflected in the fact that the optical theorem is not satisfied [23, 24]. In addition there is no well-defined Hamiltonian formalism (see, however, the alternative approaches of [18, 19]). From the string theory point of view, taking the Seiberg-Witten limit with $B_{0i} \neq 0$ results in a lack of decoupling of closed string modes. In the resulting low energy noncommutative field theory the violation of the optical theorem can be formally solved by including the undecoupled string modes that, however, have negative norm [24]. In the following we will restrict our attention to the case of space-space noncommutativity, $\theta^{0i} = 0$.

In the Seiberg-Witten limit we obtain therefore field theories on a quantum plane, the coordinates x^μ do not commute but rather satisfy:

$$[x^\mu, x^\nu] = \mathrm{i}\theta^{\mu\nu} , \tag{4}$$

with

$$\theta^{\mu\nu} = \begin{pmatrix} 0 & 0 & 0 & 0 \\ 0 & 0 & \theta & 0 \\ 0 & -\theta & 0 & 0 \\ 0 & 0 & 0 & 0 \end{pmatrix} . \tag{5}$$

The action for these field theories looks the same as for commutative theories except that functions are multiplied in terms of the Moyal product:

$$f(x) \star g(x) = f(x) \mathrm{e}^{\frac{i}{2}\theta^{\mu\nu}\overleftarrow{\partial}_\mu \overrightarrow{\partial}_\nu} g(x) . \tag{6}$$

Using the Fourier transform of (6) we can write down the Feynman rules for a scalar field theory containing a $\varphi_\star^n$-vertex. The result is:

$$\begin{aligned} &\int \mathrm{d}^d x \varphi(x)_\star^n \qquad (7) \\ &= \int \frac{\mathrm{d}^d k_1}{(2\pi)^d} \cdots \frac{\mathrm{d}^d k_n}{(2\pi)^d} (2\pi)^d \delta\left(\sum_{j=1}^n k_j\right) \tilde{\varphi}(k_1) \ldots \tilde{\varphi}(k_n) \mathrm{e}^{-\frac{i}{2}\sum_{i<j} k_i \cdot \tilde{k}_j} , \end{aligned}$$

with $\tilde{k}^\mu \equiv \theta^{\mu\nu} k_\nu$ and $\tilde{\varphi}(k)$ the Fourier transform of $\varphi(x)$.

The phases in (7) are at the origin of the UV/IR mixing [10]. Because of these phases depend on the incoming momenta in the vertex, planar and nonplanar Feynman diagrams will have different contributions. As a matter of example one can consider the mass renormalization in $\lambda\varphi_\star^4$-theory. Whereas in the commutative theory there is a single diagram contributing to one loop, the new Feynman rules produce two contributions (see Fig. 1). One of them is the quadratically divergent planar diagram which is identical to the commutative one except for a different combinatorial factor. Together with this there is the "non-planar" contribution (see [10] for details) which has the form:

$$\Pi(p)_{\text{nonplanar}} = \frac{\lambda}{6} \int \frac{\mathrm{d}^4 k}{(2\pi)^4} \frac{\mathrm{e}^{ik.\tilde{p}}}{k^2 + m^2} . \tag{8}$$

As long as the external momentum p, or rather $\tilde{p}$, is nonvanishing, the integral converges due to the rapidly oscillating phase at large loop momentum. Exponentiating

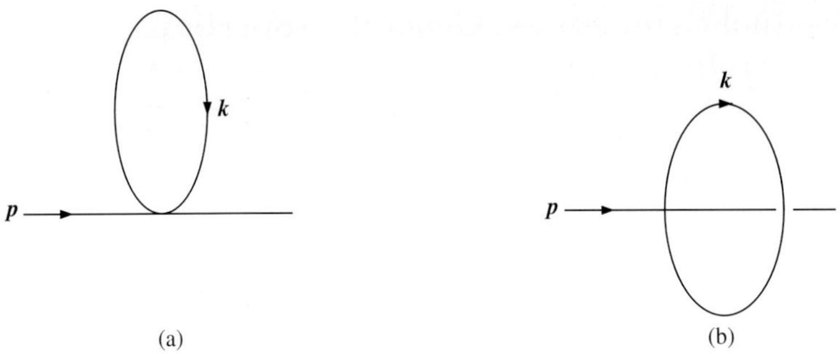

Fig. 1. Planar (a) and nonplanar (b) contributions to the mass renormalization in $\lambda\varphi^4_\star$

the denominator in (8) using a Schwinger parameter and introducing a Schwinger cutoff Λ we find that the nonplanar diagram has an effective momentum-dependent cut-off given by

$$\frac{1}{\Lambda^2_{\text{eff}}} = \frac{1}{\Lambda^2} + \tilde{p}^2 \,, \tag{9}$$

This clearly shows how UV divergences of (8) are transformed into IR ones. After the UV cutoff Λ is sent to infinity, the quadratic divergence will reappear in the IR limit $\tilde{p} \to 0$. At fixed cutoff, on the other hand, the UV/IR mixing reflects in that the two limits $\Lambda \to \infty$ and $\theta \to 0$ do not commute. It is thus clear that in general we will have problems defining low-energy Wilsonian effective actions since UV and IR scales do not decouple.

The phenomenon of UV/IR mixing has some resemblance with Quantum Gravity or String Theory, and is probably one of the reasons why NCQFTs have received so much attention. When we consider General Relativity, an object with a given energy E has two lengths associated with it: one is the Compton wavelength as in ordinary field theory $\hbar/E$. At the same time it also has its Schwarzschild radius $G_N E$. Obviously as the energy grows there is a point where the Schwarzschild radius becomes bigger than the Compton wavelength, and certainly at this point our standard notions of quantum field theory no longer apply.

If we consider the origin of the UV/IR mixing, the analogy is very appealing. In the loops of the NCQFTs we will have particles of very high energy and thus very short Compton wavelength. Since the fundamental objects in the theory are dipoles, the states running in loops have also an associated length of order θp, growing with the energy. If we define the theory with a sharp momentum space cutoff Λ, the longest dipole has size $\Lambda\theta$. This rod-like structure of the noncommutative quanta breaks Lorentz invariance, as it is already clear from the commutation relations (4) with the noncommutativity parameter given by (5). However if we consider momenta $0 \leq p \lesssim 1/(\Lambda\theta)$ Lorentz invariance and the commutative theory should be recovered, since these momenta correspond to length scales bigger than the size of the largest dipole and therefore one cannot probe the "dipolar" structure of the excitations of the noncommutative theory.

3 Residual Symmetries, General Properties and Discrete Symmetries

Looking at the commutation relations (4) we see that NCQFTs are invariant under translations and therefore states can still be labeled by their four-momentum eigenvalues. On the other hand[1], given the form of θ in (5), the Lorentz group is broken from O(1,3) to O(1,1) $\times$ SO(2).

For generic theories with these reduced symmetry one should not expect a connection between spin and statistics, and it should not be hard to construct theories with exotic statistics. It is thus an interesting question whether basic theorems like CPT which hold in local relativistic field theories will continue to hold in this context. Since we consider these theories as obtained from the Seiberg-Witten limit, it is reasonable to ask this question within String Theory. The CPT theorem in string theory has been investigated by several groups [25]. If the parent string theory satisfies the CPT theorem in perturbation theory, and since the constant background B-field is CPT-even, it is reasonable to expect that the noncommutative quantum field theory obtained in the Seiberg-Witten limit should also preserve CPT.

At the level of the noncommutative field theory it is also expected to have CPT-invariance for theories with $\theta^{0i} = 0$. As mentioned above, these theories preserve perturbative unitarity. In ordinary quantum field theory there is an intimate connection between unitarity and CPT-invariance. Indeed, if the condition of asymptotic completeness holds, it can be shown [26, 27] that the S-matrix can be written in terms of the CPT operator of the complete theory Θ and the corresponding one for the asymptotic theory Θ_0:

$$S = \Theta^{-1}\Theta_0 \,. \tag{10}$$

The unitarity of the S-matrix follows then from the antiunitarity of Θ and Θ_0. A theory with CPT invariance is therefore likely to be unitary.

There are several proofs of the CPT theorem for ordinary QFTs. We find however that the deeper and more elegant one is due to Jost [28, 13]. Few ingredients are required. One only needs the theory to satisfy the Wightman axioms, essentially Poincare invariance, uniqueness of the vacuum, positivity of the energy and microscopic causality. With these conditions it is shown that the Wightman functions[2] admit an analytic continuation that is invariant under the complexified Lorentz group. The standard Lorentz group has four sheets, one connected to the identity, and the other three obtained by applying to it P, T and PT. However, the complexified Lorentz group contains only two sheets, and the one obtained by applying the full space-time inversion (PT) is connected with the identity. Expressing invariance under this transformation essentially amounts to the proof of the CPT theorem.

[1] In the general case, $\theta^{\mu\nu}$ is determined by two vectors, the electric and magnetic components θ^{0i} and $\epsilon^{ijk}\theta^{jk}$. If they are not collinear, the Lorentz group is completely broken.

[2] Wightman functions are vacuum expectation values of products of fields without time-ordering, namely

$$W_n(x_1,\ldots,x_n) = \langle\Omega|\Phi(x_1)\ldots\Phi(x_n)|\Omega\rangle$$

where $|\Omega\rangle$ is the true vacuum of the theory.

In spite of the reduced space-time symmetry, this proof can be extended to NCQFTs, at least those with space-space noncommutativity[3]. The key ingredient lies in the fact that for NCQFT microcausality is defined only with respect to the O(1,1) factor of the space-time symmetry group. Given the fact that this group has a structure very similar to the full O(1,3) Lorentz group, Jost's proof is carried out to the noncommutative case without any problem (see [16] for the details).

Hence, although in general one should expect serious problems with nonlocal theories, for the type of nonlocality produced by the Seiberg-Witten limit we nevertheless recover the standard form of the CPT theorem. The spin-statistics connection is more subtle. If the type of representations of the reduced Lorentz group in NCQFTs are obtained as reductions from standard representations, it is likely that the same construction goes through. However, if we start with other representations, not inherited from higher dimensions, exotic statistics may easily occur.

4 The Infrared Problems of Noncommutative QED

As we discussed in Sect. 2, much of the interesting physics of NCQFTs comes from UV/IR mixing. This lack of decoupling of the different scales in the theory might pose a serious problem to phenomenology, since noncommutative effects can show up at low energies interfering with standard model predictions.

Since, apart from gravity, electromagnetism is the only long range interaction at low energies, it seems that QED would be the perfect test bench for the phenomenology of NCQFT. The noncommutative version of QED can be easily constructed by deforming ordinary QED with the introduction of Moyal products. Here we will consider the simplest case of pure NCQED with action

$$S_{\mathrm{NCQED}} = \frac{1}{4g^2}\int \mathrm{d}^4x\, F_{\mu\nu}F^{\mu\nu} \tag{11}$$

where g is the coupling constant and

$$F_{\mu\nu} = \partial_\mu A_\nu - \partial_\nu A_\mu - i(A_\mu \star A_\nu - A_\nu \star A_\mu) \tag{12}$$

Because of the quadratic θ-dependent term in the definition of $F_{\mu\nu}$, the noncommutative photon self-interacts unlike the case of ordinary QED. The corresponding Feynman rules are very similar to those of nonabelian Yang-Mills where the group structure constants are replaced by trigonometric functions of the incoming momenta (see Fig. 2).

At tree level, noncommutative corrections to ordinary QED can be made small and compatible with experimental bounds as long as the noncommutative energy scale $1/\sqrt{\theta}$ is chosen large enough. This is due to the fact that noncommutativity only appears in the form of global θ-dependent phases which have a smooth commutative limit. The situation is radically different once one-loop effects are taken into account, due to UV/IR mixing. The kind of problems encountered show up

[3] A different proof that applies also the the time-space noncommutativity can be found in [14].

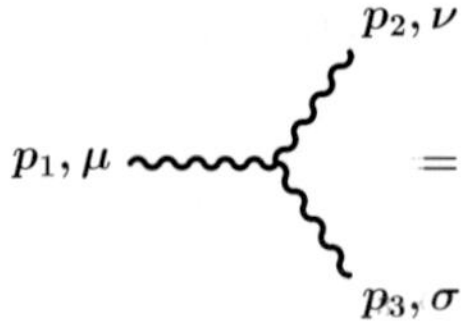

$$= 2g \sin\left(\frac{1}{2}\tilde{p}_1 \cdot p_3\right) [\eta^{\mu\sigma}(p_1 - p_3)^\nu + \text{perm.}]$$

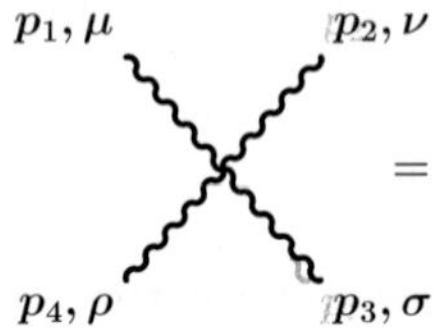

$$= -4ig^2 \Big[\sin\left(\frac{1}{2}\tilde{p}_1 \cdot p_2\right) \sin\left(\frac{1}{2}\tilde{p}_4 \cdot p_3\right) \times (\eta^{\mu\rho}\eta^{\nu\sigma} - \eta^{\mu\sigma}\eta^{\nu\rho}) + \text{perm.}\Big]$$

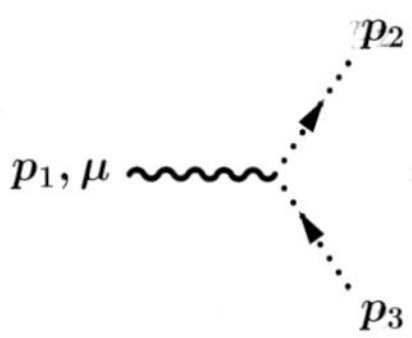

$$= 2g\, p_2^\mu \sin\left(\frac{1}{2}\tilde{p}_1 \cdot p_2\right)$$

Fig. 2. Interaction vertices for NCQED. The *wavy line* represents the photon and the line of points the Fadeev-Popov ghost

in the calculation of the one-loop corrected dispersion relation for the photon in NCQED which results to be [29]

$$\omega(\boldsymbol{p})^2 = \boldsymbol{p}^2 - \frac{2g^2}{\pi^2} \frac{1}{p \circ p}\,, \tag{13}$$

where $p \circ p \equiv \theta^2(p_1^2 + p_2^2)$. The divergence at low transverse momentum in (13) is the result of a UV quadratic divergence that reappears in the IR as a result of UV/IR mixing. It seems therefore that the disastrous situation implied by the previous dispersion relation could be overcome by completing noncommutative QED with another theory softer in the UV.

This was attempted by embedding NCQED into $\mathcal{N} = 1$ noncommutative supersymmetric QED and breaking supersymmetry softly at a scale M by adding a mass term to the gaugino. The resulting dispersion relation, however, eliminates the IR divergence although it leaves behind a finite tachyonic mass for the photon [30]

$$\omega(\boldsymbol{p})^2 \approx \boldsymbol{p}^2 - \frac{g^2 M^2}{2\pi^2}\,. \tag{14}$$

In spite of having removed the divergence, the situation has not improved at all, since the tachyonic mass of the photon only depends on the soft-breaking mass of the gaugino and not on the noncommutative parameter θ. Therefore it cannot be made small by taking the noncommutative energy scale large. On the contrary if, for phenomenological reasons, we set $M \sim 1\,\text{TeV}$ we find a tachyonic photon with a mass much outside the current experimental bounds.

As we already mentioned, the source of all the trouble with NCQED is the mixing of UV and IR scales, which induces IR singularities in nonplanar amplitudes. One possible way to tame this problem is by defining the noncommutative theory with an sharp UV cutoff $\Lambda > 1/\sqrt{\theta}$. In this case, the noncommutative scale gets, in a sense, "corrected" due to one loop effects and noncommutative effects start being relevant already at scales of order $1/(\theta\Lambda)$. Moreover, because of the regularization of the UV singularities provided by the cutoff, the commutative theory is recovered at scales $E \ll 1/(\theta\Lambda)$, including its Lorentz invariance.

Therefore, it seems that this might provide a way to define NCQED at low energies avoiding the problems of the emergence of tachyons. Unfortunately [16], there are additional difficulties associated with this regularization scheme. In particular, apart from the lattice, a sharp cutoff Λ of the type required (either a cutoff in momenta or a Schwinger cutoff) leads to violations of gauge invariance. This can only be avoided by considering "mixed" cutoffs which combines a sharp cutoff with dimensional regularization. In the case of NCQED this scheme works fine for the one-loop polarization tensor where gauge invariance is preserved and the result for ordinary QED recovered at low momentum [16]. Nevertheless, its extension to other amplitudes or higher loops is more problematic.

5 Some Phenomenological Considerations on NCQED

An alternative, that we will pursue here, is to ameliorate the IR problems of NCQED by looking for high energy completion of the theory which would be free of UV divergences. In particular, let us consider $\mathcal{N}=4$ U(1) noncommutative super-Yang-Mills, which is believed to be finite [31] as its commutative counterpart. In this case, instead of a single U(1) gauge vector field we have one $\mathcal{N}=1$ vector multiplet together with three scalar multiplets in the adjoint representation. NCQED can be then recovered at low energies by breaking supersymmetry softly by adding masses M_f to the gauginos and M_s to the scalars [32]. This provides a construction that makes sense in the UV and IR, and where we can have control on the UV/IR mixing.

With this setup, we can proceed to compute the one-loop polarization tensor for the photon $\Pi_{\mu\nu}(p)$. We will work in Euclidean signature and rotate back to Minkowski at the end of the calculation. On symmetry grounds it has the form

$$\Pi_{\mu\nu}(p) = \Pi_1(p)\left(p^2\delta_{\mu\nu} - p_\mu p_\nu\right) + \Pi_2(p)\frac{\tilde{p}_\mu \tilde{p}_\nu}{\tilde{p}^2} \tag{15}$$

where $\tilde{p}^\mu = \theta^{\mu\nu} p_\nu$. It is important to notice that, due to the antisymmetry of $\theta^{\mu\nu}$, the extra piece on the right-hand side in (15) is transverse and the Ward identity $p^\mu \Pi_{\mu\nu}(p) = 0$ is satisfied.

Now we can proceed to compute the functions $\Pi_1(p)$ and $\Pi_2(p)$ at one loop for the theory with soft-breaking mass terms. Using the background field method [9] and working in dimensional regularization in the $\overline{\text{MS}}$ scheme the results are

$$\begin{aligned}\Pi_1(p) = \frac{1}{4\pi^2}\int_0^1 \mathrm{d}x \Bigg\{ &\left[4-(1-2x)^2\right]\left[\frac{1}{2}\log\left(\frac{\Delta_v}{4\pi\mu^2}\right)+K_0\left(\sqrt{\Delta_v}|\tilde{p}|\right)\right] \\ &-\left[1-(1-2x)^2\right]\sum_f\left[\frac{1}{2}\log\left(\frac{\Delta_f}{4\pi\mu^2}\right)+K_0\left(\sqrt{\Delta_f}|\tilde{p}|\right)\right] \\ &-\frac{1}{2}(1-2x)^2\sum_s\left[\frac{1}{2}\log\left(\frac{\Delta_s}{4\pi\mu^2}\right)+K_0\left(\sqrt{\Delta_s}|\tilde{p}|\right)\right]\Bigg\} , \end{aligned} \tag{16}$$

and

$$\begin{aligned}\Pi_2(p) = -\frac{1}{\pi^2}\int_0^1 \mathrm{d}x \Bigg[&\Delta_v K_2\left(\sqrt{\Delta_v}|\tilde{p}|\right) - \sum_f \Delta_f K_2\left(\sqrt{\Delta_f}|\tilde{p}|\right) \\ &+\frac{1}{2}\sum_s \Delta_s K_2\left(\sqrt{\Delta_s}|\tilde{p}|\right)\Bigg] . \end{aligned} \tag{17}$$

Here μ is the dimensional regularization energy scale and the subindices v, f and s indicate respectively the contributions from the vector-ghost system, fermions and scalars. In addition we have defined

$$\begin{aligned}\Delta_v &= x(1-x)p^2 , \\ \Delta_f &= M_f^2 + x(1-x)p^2 , \\ \Delta_s &= M_s^2 + x(1-x)p^2 , \end{aligned} \tag{18}$$

with M_f, M_s the soft-breaking masses.

With (16) and (17) we can easily compute the dispersion relation by looking at the poles of the full propagator $G_{\mu\nu}(p)$, once the one loop 1PI parts are resumed:

$$G_{\mu\nu}(p) = \frac{ig^2}{p^2}\left[\mathbf{1} + \frac{-g^2}{p^2}\mathbf{\Pi}(p) + \left(\frac{-g^2}{p^2}\right)^2\mathbf{\Pi}(p)^2 + \ldots\right]_{\mu\nu} \tag{19}$$

where $\mathbf{1}$ is the 4×4 identity matrix and $\mathbf{\Pi}(p)$ is a matrix notation for the polarization tensor in (15). After a straightforward calculation we find

$$\begin{aligned}G_{\mu\nu}(p) = &\frac{ig^2 p_\mu p_\nu}{p^2} + \frac{ig^2}{p^2\left[1+g^2\Pi_1(p)\right]}\left(\delta_{\mu\nu} - \frac{p_\mu p_\nu}{p^2}\right) \\ &+\left\{\frac{ig^2}{p^2\left[1+g^2\Pi_1(p)\right]+g^2\Pi_2(p)} - \frac{ig^2}{p^2\left[1+g^2\Pi_1(p)\right]}\right\}\frac{\tilde{p}_\mu\tilde{p}_\nu}{\tilde{p}^2} . \end{aligned} \tag{20}$$

Unlike the case of ordinary QED in (20) we have two sources of poles in the full photon propagator. On the one hand we find the usual solution

$$p^2\left[1+g^2\Pi_1(p)\right] = 0 , \tag{21}$$

which gives rise to the usual massless dispersion relation for the photon, $p^2 = 0$. Together with this we also find a second pole associated with photon polarizations along the vector $\tilde{p}^\mu$:

$$p^2\left[1+g^2\Pi_1(p)\right] + g^2\Pi_2(p) = 0 . \tag{22}$$

In order to extract the dispersion relation we perform the rotation back to Minkowski signature by replacing $p^2 \to -p^2$ and $\tilde{p}^2 \to p \circ p$. Using the low momentum expansion of (16) and (17) we find the dispersion relation for the polarizations along $\tilde{p}^\mu$ for low momentum

$$\omega(\boldsymbol{p})^2 \approx \boldsymbol{p}^2 - \frac{g^2}{2\pi^2}\left(\sum_f M_f^2 - \frac{1}{2}\sum_s M_s^2\right) . \tag{23}$$

Unlike the case in which NCQED is completed in the UV by $\mathcal{N} = 1$ U(1) noncommutative super-Yang-Mills [30], here we can avoid a tachyonic photon by appropriately tuning the soft breaking masses in (23), i.e. by demanding

$$\sum_f M_f^2 - \frac{1}{2}\sum_s M_s^2 \leq 0 . \tag{24}$$

Unfortunately, a tuning of this quantity to zero does not result in a massless photon polarization, as we would like to recover at low energies. When the inequality (24) is saturated the leading term in the expansion of $\Pi_2(p)$ around $\boldsymbol{p} = 0$ is negative, and we find a dispersion relation with negative energy squared for low momentum photons. Therefore one is forced to a finite value of the quantity on the left hand side of (24), i.e. to a massive photon polarization. Using the current bounds for the photon mass [34], one has to engage in a massive fine tuning of the soft breaking masses

$$\frac{1}{2}\sum_s M_s^2 - \sum_f M_f^2 \lesssim 10^{-32}\,\mathrm{eV}^2 . \tag{25}$$

This result is not affected by the addition of matter in the fundamental representation of U(1). In the calculation of the one-loop polarization tensor fundamental fields in the loop only contribute to planar diagrams. Since the function $\Pi_2(p)$ in (15) is solely determined by non-planar diagrams the only effect of the fundamental fields is in modifying the running of the coupling constant through the function $\Pi_1(p)$.

Even if the problem of a tachyonic photon can be avoided by this un-natural fine tuning of the mass scales, the dispersion relation of photons with polarizations along $\tilde{p}^\mu$ will be different from the standard relation $\omega(\boldsymbol{p}) = |\boldsymbol{p}|$ of photons with polarizations orthogonal to $\tilde{p}^\mu$. This implies that in the construction of NCQED we are studying here there is a phenomenon of birefringence associated with the fact that the dispersion relations (and therefore the speed of propagation) of photons with different polarizations are different (cf. [35]).

After our analysis we have to conclude that the phenomenological perspectives of NCQED look rather poor. In our attempt to eliminate the tachyonic polarization of the photon we have been lead to massive photon polarizations and birefringence, at the prize also of a huge fine tuning of the masses of the soft breaking masses.

To summarize, here we have studied the problem of making sense out of NCQED at low energies, as derived from string theory in the Seiberg-Witten limit. To ameliorate the hard IR problems that afflict this theory we have completed it in the UV by $\mathcal{N} = 4$ noncommutative U(1) super-Yang-Mills, softly broken by mass terms for the gauginos and scalars. Our conclusions regarding the phenomenological viability of such a theory are, however, rather negative. We found that tachyons can be avoided only by allowing a massive polarization for the photon. This requires also a tremendous fine tuning of the soft-breaking masses. It seems, therefore, that any attempt to extract phenomenology from this theory should be postponed to find a formulation of the theory that can describe at least the rough features of the world we live in.

Acknowledgements

We thank the organizers of the 9th Adriatic Meeting for the opportunity of presenting this work. We are also thankful to J. L. F. Barbón, M. Chaichian, J. M. Gracia-Bondía, K.E. Kunze, D. Lüst, R. Stora and J. Wess for useful discussions. M.A.V.-M. acknowledges support from Spanish Science Ministry Grant FPA2002-02037.

References

1. A. Connes, *Noncommutative Geometry*, Academic Press 1994.
   J. Madore, *An Introduction to Noncommutative Differential Geometry and its Applications*, Cambridge 1995.
   G. Landi, *An Introduction to Noncommutative Spaces and their Geometries*, Springer 1997 [arXiv:hep-th/9701078]
   J. M. Gracia-Bondía, J. C. Varilly and H. Figueroa, *Elements of Noncommutative Geometry*, Birkhäuser 2001.
2. J. Bellisard, A. van Elst and H. Schulz-Baldes, *The Noncommutative Geometry of the Quantum Hall Effect*, [arXiv:cond-mat/9411052].
3. A. Connes and J. Lott, Nucl. Phys. Proc. Suppl. **18B** (1991) 29.
   C. P. Martin, J. M. Gracia-Bondia and J. C. Varilly, Phys. Rept. **294** (1998) 363 [arXiv:hep-th/9605001].
4. E. Witten, Nucl. Phys. **B268** (1986) 253.
5. A. Connes, M. R. Douglas and A. Schwarz, J. High Energy Phys. **02** (1998) 003 [arXiv:hep-th/9711162].
   M. R. Douglas and C. M. Hull, J. High Energy Phys. **9802** (1998) 008 [arXiv:hep-th/9711165].
6. A. González-Arroyo and C. P. Korthals Altes, Phys. Lett. **B131** (1983) 396.
   T. Filk, Phys. Lett. **B376** (1996) 53.
7. N. Seiberg and E. Witten, J. High Energy Phys. **09** (1999) 032 [arXiv:hep-th/9908142].
8. M. R. Douglas and N. A. Nekrasov, Rev. Mod. Phys. **73** (2001) 977 [arXiv:hep-th/0106048].
   R. J. Szabo, Phys. Rept. **378** (2003) 207 [arXiv:hep-th/0109162].
9. M. M. Sheikh-Jabbari, Phys. Lett. **B455** (1999) 129 [arXiv:hep-th/9901080].
10. S. Minwalla, M. van Raamsdonk and N. Seiberg, J. High Energy Phys. **02** (2000) 020 [arXiv:hep-th/9912072].
    M. van Raamsdonk and N. Seiberg, J. High Energy Phys. **03** (2000) 035 [arXiv:hep-th/0002186].

11. H. S. Snyder, Phys. Rev. **71** (1947) 38.
12. M. M. Sheikh-Jabbari, J. High Energy Phys. **06** (1999) 015 [arXiv:hep-th/9903107].
T. Krajewski and R. Wulkenhaar, Int. J. Mod. Phys. **A15** (2000) 1011 [arXiv:hep-th/9903187].
I. Chepelev and R. Roiban, J. High Energy Phys. **05** (2000) 037 [arXiv:hep-th/9911098].
A. Armoni, Nucl. Phys. **B593** (2001) 229 [arXiv:hep-th/0005208].
E. A. Nicholson, *Perturbative Wilsonian formalism for noncommutative gauge theories in the matrix representation,* PhD. Thesis, arXiv:hep-th/0305044.
13. R. Jost, *The General Theory of Quantized Fields*, AMS 1965.
R. F. Streater and A. S. Wightman, *PCT, Spin and Statistics and All That*, Benjamin 1964.
R. Haag, *Local Quantum Physics: Fields, Particles, Algebras*, Springer 1992.
14. M. Chaichian, K. Nishijima and A. Tureanu, Phys. Lett. **B568** (2003) 146 [arXiv:hep-th/0209008].
15. M. M. Sheikh-Jabbari, Phys. Rev. Lett. **84** (2000) 5265 [arXiv:hep-th/0001167].
S. M. Carroll, J. A. Harvey, V. A. Kostelecky, C. D. Lane and T. Okamoto, Phys. Rev. Lett. **87** (2001) 141601 [arXiv:hep-th/0105082].
P. Aschieri, B. Jurčo, P. Schupp and J. Wess, Nucl. Phys. **B651** (2003) 45 [arXiv:hep-th/0205214].
16. L. Álvarez-Gaumé and M. A. Vázquez-Mozo, Nucl. Phys. **B668** (2003) 293 [arXiv:hep-th/0305093].
17. S. M. Carroll, G. B. Field and R. Jackiw, Phys. Rev. **D41** (1990) 1231.
S. R. Coleman and S. L. Glashow, Phys. Lett. **B405** (1997) 249 [arXiv:hep-ph/9703240]; Phys. Rev. **D59** (1999) 116008 [arXiv:hep-ph/9812418].
18. S. Doplicher, K. Fredenhagen and J. E. Roberts, Commun. Math. Phys. **172** (1995) 187 [arXiv:hep-th/0303037].
D. Bahns, S. Doplicher, K. Fredenhagen and G. Piacitelli, Phys. Lett. **B533** (2002) 178 [arXiv:hep-th/0201222].
D. Bahns, S. Doplicher, K. Fredenhagen and G. Piacitelli, Commun. Math. Phys. **237** (2003) 221 [arXiv:hep-th/0301100].
19. Y. Liao and K. Sibold, Eur. Phys. J. **C25** (2002) 469 [arXiv:hep-th/0205269]; ibid. **C25** (2002) 479 [arXiv:hep-th/0206011].
20. J. Madore, S. Schraml, P. Schupp and J. Wess, Eur. Phys. J. **C16** (2000) 161 [arXiv:hep-th/0001203].
B. Jurčo, L. Möller, S. Schraml, P. Schupp and J. Wess, Eur. Phys. J. **C21** (2001) 383 [arXiv:hep-th/0104153].
21. A. Bichl, J. Grimstrup, H. Grosse, L. Popp, M. Schweda and R. Wulkenhaar, J. High Energy Phys. **0106** (2001) 013 [arXiv:hep-th/0104097].
A. A. Bichl, J. M. Grimstrup, H. Grosse, E. Kraus, L. Popp, M. Schweda and R. Wulkenhaar, Eur. Phys. J. **C24** (2002) 165 [arXiv:hep-th/0108045].
H. Bozkaya, P. Fischer, H. Grosse, M. Pitschmann, V. Putz, M. Schweda and R. Wulkenhaar, Eur. Phys. J. **C29** (2003) 133 [arXiv:hep-th/0209253].
22. N. Seiberg, L. Susskind and N. Toumbas, J. High Energy Phys. **06** (2000) 021 [arXiv:hep-th/0005040].
R. Gopakumar, J. M. Maldacena, S. Minwalla and A. Strominger, J. High Energy Phys. **06** (2000) 036 [arXiv:hep-th/0005048].

J. L. F. Barbón and E. Rabinovici, Phys. Lett. B **486** (2000) 202 [arXiv:hep-th/0005073].
R. G. Cai and N. Ohta, Prog. Theor. Phys. **104** (2000) 1073 [arXiv:hep-th/0007106].
23. J. Gomis and T. Mehen, Nucl. Phys. **B591** (2000) 265 [arXiv:hep-th/0005129].
O. Aharony, J. Gomis and T. Mehen, J. High Energy Phys. **09** (2000) 023 [arXiv:hep-th/0006236].
24. L. Álvarez-Gaumé, J. L. F. Barbón and R. Zwicky, J. High Energy Phys. **05** (2001) 057 [arXiv:hep-th/0103069].
25. H. Sonoda, Nucl. Phys. **B326** (1989) 135.
V. A. Kostelecky and R. Potting, Nucl. Phys. **B359** (1991) 545.
A. Pasquinucci and K. Roland, Nucl. Phys. **B473** (1996) 31 [arXiv:hep-th/9602026].
26. C. N. Yang and D. Feldman, Phys. Rev. **79** (1950) 972.
27. R. Haag, Phys. Rev. **112** (1958) 669.
D. Ruelle, Helv. Phys. Acta **35** (1962) 147.
28. R. Jost, Helv. Phys. Acta **30** (1957) 409.
29. F. Ruiz Ruiz, Phys. Lett. **B502** (2001) 274 [arXiv:hep-th/0012171].
K. Landsteiner, E. López and M. H. G. Tytgat, J. High Energy Phys. **06** (2001) 055 [arXiv:hep-th/0104133].
30. C. E. Carlson, C. D. Carone and R. F. Lebed, Phys. Lett. **B549** (2002) 337 [arXiv:hep-ph/0209077].
31. I. Jack and D. R. Jones, Phys. Lett. **B514** (2001) 401 [arXiv:hep-th/0105221].
32. J. G. Taylor, Phys. Lett. **B121** (1983) 386.
J. J. van der Bij and Y. P. Yao, Phys. Lett. **B125** (1983) 171.
33. V. V. Khoze and G. Travaglini, J. High Energy Phys. **01** (2001) 026 [arXiv:hep-th/0011218].
34. K. Hagiwara et al. [Particle Data Group Collaboration], Phys. Rev. **D66** (2002) 010001.
35. Z. Guralnik, R. Jackiw, S. Y. Pi and A. P. Polychronakos, Phys. Lett. **B517** (2001) 450 [arXiv:hep-th/0106044].

Seiberg-Witten Maps and Anomalies in Noncommutative Yang-Mills Theories

Friedemann Brandt

Max-Planck-Institute for Mathematics in the Sciences,
Inselstraße 22-26, 04103 Leipzig, Germany
fbrandt@mis.mpg.de

1 Introduction

We shall discuss two aspects of noncommutative Yang-Mills theories of the type introduced in [9] (see Sect. 2 for a brief review). The first aspect concerns the construction of these theories which is based on so-called Seiberg-Witten mappings (SW maps, for short). These mappings express "noncommutative" fields and gauge transformations in terms of the standard ("commutative") fields and gauge transformations. The mappings have been named after Seiberg and Witten because they were established first in [16] for the particular case of $U(N)$-theories. However, it should be kept in mind that in the present context they are not limited to $U(N)$-theories but extended to other gauge groups. This raises the questions whether and why SW maps exist for general gauge groups, how they can be constructed efficiently and to which extend they are unique resp. ambiguous. These questions are the topic of Sect. 3 which reviews work in collaboration with G. Barnich and M. Grigoriev [3, 4, 5].

Section 4 reports on work in collaboration with C.P. Martín and F. Ruiz Ruiz [6]. It addresses the question whether the gauge symmetries of noncommutative Yang-Mills theories can be anomalous when one applies the standard perturbative approach to (effective) quantum field theories. It is not to be discussed here whether or not such an approach makes sense; currently there is hardly an alternative perspective on these theories in the general case (i.e., for a general gauge group) since the theories are constructed only by means of SW maps and no formulation in terms of "noncommutative" variables is known. Hence, at present we have to content ourselves with a formulation of the "effective type" that is not renormalizable by power counting, i.e., a Lagrangian containing field monomials of arbitrarily high mass dimension. As a consequence, there is no simple argument which can rule out from the outset the possible occurrence of gauge anomalies with mass dimensions larger than 4. This complicates the anomaly discussion as compared to renormalizable Yang-Mills theories whenever the gauge group contains at least one abelian factor since in that case there is an infinite number of candidate gauge anomalies in addition to the well-known chiral gauge anomalies.

2 Brief Review of Noncommutative Yang-Mills Theories

The noncommutative Yang-Mills theories under consideration involve a $\star$-product given by the Weyl-Moyal product,

$$f_1 \star f_2 = f_1 \exp\left(\overleftarrow{\partial_\mu}\,\frac{\mathrm{i}}{2}\,\tau\,\theta^{\mu\nu}\,\overrightarrow{\partial_\nu}\right) f_2\,, \quad \theta^{\mu\nu} = -\theta^{\nu\mu} = \text{constant}\,.$$

τ is a constant deformation parameter that has been introduced for the sake of convenience. The "noncommutative" generalization of the Yang-Mills action reads

$$\hat{I}[\hat{A}] = -\frac{1}{4}\int \mathrm{d}^n x \,\mathrm{Tr}\,(\hat{F}_{\mu\nu}\star\hat{F}^{\mu\nu}), \quad \hat{F}_{\mu\nu} = \partial_\mu\hat{A}_\nu - \partial_\nu\hat{A}_\mu + \hat{A}_\mu\star\hat{A}_\nu - \hat{A}_\nu\star\hat{A}_\mu \tag{1}$$

where $\hat{A}_\mu$ is constructed from "commutative" gauge potentials A_μ by means of a SW map. A_μ lives in the Lie algebra of the gauge group and has the standard Yang-Mills gauge transformations,

$$\delta_\lambda A_\mu = \partial_\mu\lambda + [A_\mu, \lambda] \equiv D_\mu\lambda\,, \tag{2}$$

where λ denotes Lie algebra valued gauge parameters. SW maps, by definition, express the noncommutative gauge potentials $\hat{A}_\mu$ and gauge parameters $\hat{\lambda}$ in terms of A_μ and λ such that (2) induces the noncommutative version of Yang-Mills gauge transformations given by

$$\hat{\delta}_{\hat{\lambda}}\hat{A}_\mu = \partial_\mu\hat{\lambda} + \hat{A}_\mu\star\hat{\lambda} - \hat{\lambda}\star\hat{A}_\mu \equiv \hat{D}_\mu\hat{\lambda}\,. \tag{3}$$

Furthermore we require that $\hat{A}_\mu$ and $\hat{\lambda}$ coincide with A_μ and λ at $\tau = 0$,

$$\hat{A}_\mu = \hat{A}_\mu(A,\tau) = A_\mu + O(\tau), \quad \hat{\lambda} = \hat{\lambda}(\lambda, A, \tau) = \lambda + O(\tau)\,.$$

Hence, SW maps are required to fulfill

$$\delta_\lambda\hat{A}_\mu(A,\tau) = (\hat{\delta}_{\hat{\lambda}}\hat{A}_\mu)(A,\tau)\,.$$

For the inclusion of fermions see, e.g., [4, 9].

3 Analysis of SW Maps

Noncommutative Yang-Mills theories can be regarded as consistent deformations of corresponding commutative Yang-Mills theories. This allows one to apply BRST-cohomological tools to analyse SW maps along the lines of [7]. In the following, we first review briefly the BRST-cohomological approach to consistent deformations and then the results on SW maps.

3.1 Consistent Deformations

Consider an action $I^{(0)}[\varphi]$ with gauge invariance $\delta_\lambda^{(0)}$, i.e. $\delta_\lambda^{(0)} I^{(0)}[\varphi] = 0$. Consistent deformations of $I^{(0)}[\varphi]$ and $\delta_\lambda^{(0)}$ are power series' $I[\varphi,\tau]$ and δ_λ in a deformation parameter τ, such that the deformed action is invariant under the (possibly) deformed gauge transformation,

$$I[\varphi,\tau] = I^{(0)}[\varphi] + \sum_{k\geq 1}\tau^k I^{(k)}[\varphi]\,, \quad \delta_\lambda = \delta_\lambda^{(0)} + \sum_{k\geq 1}\tau^k\delta_\lambda^{(k)}\,, \quad \delta_\lambda I[\varphi,\tau] = 0\,.$$

Two such deformations are called equivalent ($\sim$) if they are related by mere field redefinitions $\hat{\varphi}(\varphi,\tau)$, $\hat{\lambda}(\lambda,\varphi,\tau)$:

$$\hat{I}[\hat{\varphi}(\varphi,\tau),\tau] = I[\varphi,\tau], \quad (\hat{\delta}_{\hat{\lambda}}\hat{\varphi})(\varphi,\lambda,\tau) \approx \delta_\lambda\hat{\varphi}(\varphi,\tau)\,,$$

where $\approx$ is "equality on-shell" (equality for all solutions to the field equations).

Accordingly, a deformation is called trivial if the deformed action and gauge transformations are equivalent to the original action and gauge transformations, i.e., if $I \sim I^{(0)}$ and $\delta \sim \delta^{(0)}$.

We may distinguish two types of nontrivial deformations:

Type I: $I \not\sim I^{(0)}$, $\delta_{\hat{\lambda}} \sim \delta_\lambda^{(0)}$, i.e., the deformation of the action is nontrivial whereas the deformation of the gauge transformations is trivial.

Type II: $I \not\sim I^{(0)}$, $\delta_{\hat{\lambda}} \not\sim \delta_\lambda^{(0)}$, i.e., the deformations of both the action and the gauge transformations are nontrivial.

Notice that in this terminology noncommutative Yang-Mills theories as described in Sect. 2 are type I deformations of Yang-Mills theories because SW maps are field redefinitions that bring the noncommutative gauge transformations back to the standard (commutative) form, i.e., the deformation of the gauge transformations is trivial.

3.2 BRST-Cohomological Approach to Consistent Deformations

The BRST-cohomological approach to consistent deformations [8] is most conveniently formulated in the so-called field-antifield formalism [9]. The "fields" ϕ^a of that formalism are the fields φ^i occurring in the action $I[\varphi]$, ghost fields C^α corresponding to the nontrivial gauge symmetries of the action, as well as ghost fields of higher order ("ghosts for ghosts") if the gauge transformations are reducible. Each field is accompanied by an antifield ϕ_a^* according to definite rules which are not reviewed here. In particular this allows one to define the so-called antibracket (,) of functions or functionals of the fields and antifields according to

$$(F,G) = \int \mathrm{d}^n x\, F\Big(\frac{\overleftarrow{\delta}}{\delta\phi^a(x)}\,\frac{\overrightarrow{\delta}}{\delta\phi_a^*(x)} - \frac{\overleftarrow{\delta}}{\delta\phi_a^*(x)}\,\frac{\overrightarrow{\delta}}{\delta\phi^a(x)}\Big)G\,.$$

A central object of the formalism is the master action S. Its importance originates from the fact that it contains both the action $I[\varphi]$ and all information about its gauge symmetries, such as the gauge transformations, their commutator algebra, reducibility relations etc. In particular the gauge transformations occur in S via terms $\varphi_i^*\delta_C\varphi^i$ where $\delta_C\varphi^i$ is a gauge transformation of φ^i with ghost fields C in place of gauge parameters λ. The information about the gauge symmetry is encoded in the master equation $(S,S)=0$,

$$S[\phi,\phi^*] = I[\varphi] + \int \mathrm{d}^n x\, \varphi_i^*\delta_C\varphi^i + \ldots \text{ such that } \underbrace{(S,S)=0.}_{\text{master equation}}$$

In particular S defines the BRST differential s via the antibracket with S. The master equation $(S,S)=0$ implies that s squares to zero ($s^2=0$),

$$s = (S,\ \cdot\ ) \quad (\Rightarrow\ s^2 = 0)\ .$$

These properties of S make it so useful in the context of consistent deformations. Indeed, the fact that S contains both the action and the gauge transformations allows one to analyse consistent deformations in terms of the single object S that has to satisfy the master equation,

$$S = S^{(0)} + \sum_{k\geq 1} \tau^k S^{(k)}, \quad (S,S) = 0\ .$$

The first relation to BRST-cohomology can be established by differentiation of the master equation with respect to the deformation parameter:

$$(S,S) = 0 \overset{\frac{\partial}{\partial\tau}}{\Rightarrow} \left(S, \frac{\partial S}{\partial \tau}\right) = 0 \ \Leftrightarrow\ s\, \frac{\partial S}{\partial \tau} = 0\ .$$

This shows that $\partial S/\partial\tau$ is a cocycle of s. The second relation to BRST-cohomology derives from the fact that field redefinitions (of φ and/or the gauge parameters) translate into anticanonical transformations $\hat{\phi}(\phi,\phi^*,\tau)$, $\hat{\phi}^*(\phi,\phi^*,\tau)$ (these are transformations generated via the antibracket by some functional Ξ). This implies:

$$\frac{\mathrm{d}\hat{\phi}}{\mathrm{d}\tau} = (\Xi,\hat{\phi}),\ \frac{\mathrm{d}\hat{\phi}^*}{\mathrm{d}\tau} = (\Xi,\hat{\phi}^*)\ \Rightarrow\ \frac{\mathrm{d}S(\hat{\phi},\hat{\phi}^*,\tau)}{\mathrm{d}\tau} = \frac{\partial S}{\partial\tau} - (S,\Xi) = \frac{\partial S}{\partial \tau} - s\,\Xi\ .$$

As a consequence, master actions of equivalent deformations are related as follows:

$$S \sim S'\ \Rightarrow\ \frac{\partial S}{\partial\tau} - \frac{\partial S'}{\partial\tau} = s\,\Xi\ .$$

This shows that consistent deformations are determined by the BRST-cohomology $H(s)$ in ghost number 0 since $\partial S/\partial\tau$ (i) has to be a BRST-cocycle, (ii) is defined only up to a BRST-coboundary, and (iii) has ghost number 0 (S has ghost number 0 according to the standard ghost number assignments).

3.3 BRST-Cohomological Analysis of SW Maps

To describe SW maps in the field-antifield formalism we denote the "noncommutative fields" by $\hat{\phi}$ and the "commutative" fields by ϕ. Actually we enlarge the setup here as compared to Sect. 2: all the fields $\hat{\phi}$ and ϕ take values in the enveloping algebra of the Lie algebra of the gauge group, resp. some representation $\{T_A\}$ thereof. The superfluous fields ϕ (those that do not belong to the Lie algebra of the gauge group) are set to zero at the end of the construction, see [4] for details. Dropping again the fermions, we have

$$\{\hat{\phi}^a\} = \{\hat{A}^A_\mu, \hat{C}^A\}, \quad \{\phi^a\} = \{A^A_\mu, C^A\}\ .$$

The "noncommutative" master action reads

$$S[\hat{\phi},\hat{\phi}^*,\tau] = \int \mathrm{d}^n x \Big[-\frac{1}{4}\,\mathrm{Tr}\,(\hat{F}_{\mu\nu}\star \hat{F}^{\mu\nu}) + \hat{A}^{*\mu}_A \star (\hat{D}_\mu \hat{C})^A + \hat{C}^*_A \star (\hat{C}\star\hat{C})^A \Big]\ .$$

The existence of a SW map means that the gauge transformations can be brought to the standard Yang-Mills form, which particularly does not depend on τ. In terms of the master action this means that there is an anticanonical transformation $\hat{\phi}(\phi, \phi^*, \tau)$, $\hat{\phi}^*(\phi, \phi^*, \tau)$ which casts $S[\hat{\phi}, \hat{\phi}^*, \tau]$ in the form of an effective type Yang-Mills action $I_{\text{eff}}[A, \tau]$ plus a piece that involves the antifields and encodes gauge transformations of Yang-Mills type (for the enveloping algebra),

$$S[\hat{\phi}(\phi, \phi^*, \tau), \hat{\phi}^*(\phi, \phi^*, \tau), \tau] = \underbrace{I_{\text{eff}}[A, \tau]}_{\text{no antifields}} + \int \mathrm{d}^n x \Big[\underbrace{A^{*\mu} D_\mu C + C^* CC}_{\text{no dependence on } \tau} \Big] ,$$

where indices have been dropped ($A^{*\mu} D_\mu C$ means $A^{*\mu}_A (D_\mu C)^A$ etc). Differentiating with respect to τ und using the properties of anticanonical transformations (see above), we obtain

$$\frac{\partial S}{\partial \tau} - s\,\Xi = \frac{\partial I_{\text{eff}}[A, \tau]}{\partial \tau} , \quad \frac{\mathrm{d}\hat{\phi}}{\mathrm{d}\tau} = (\Xi, \hat{\phi}), \quad \frac{\mathrm{d}\hat{\phi}^*}{\mathrm{d}\tau} = (\Xi, \hat{\phi}^*) .$$

Hence, in order to find and analyse SW maps one may analyse whether $\partial S/\partial \tau$ can be written as a BRST-variation $s\Xi$ up to terms that do not involve antifields. Notice that Ξ gives the SW map. For $\partial S/\partial \tau$ one obtains

$$\begin{aligned} \frac{\partial S}{\partial \tau} = \frac{\mathrm{i}\theta^{\alpha\beta}}{2} \int \mathrm{d}^n x \Big[&\mathrm{Tr}\,(-\hat{F}^{\mu\nu} \star \partial_\alpha \hat{A}_\mu \star \partial_\beta \hat{A}_\nu) \\ &+ \hat{A}^{*\mu} \star \{\partial_\alpha \hat{A}_\mu \stackrel{\star}{,} \partial_\beta \hat{C}\} + \hat{C}^* \star \partial_\alpha \hat{C} \star \partial_\beta \hat{C} \Big] , \end{aligned}$$

where $\{\ \stackrel{\star}{,}\ \}$ denotes the $\star$-anticommutator,

$$\{X \stackrel{\star}{,} Y\} = X \star Y + Y \star X .$$

This expression for $\partial S/\partial \tau$ is indeed BRST-exact up to terms that do not contain antifields. One can infer this by means of so-called contracting homotopies for derivatives of the ghost fields used already in [10, 11]. We shall not review the construction of these homotopies here since this is a somewhat technical matter. Rather, we shall only present the result. It is actually ambiguous as we shall discuss below. In particular it depends on the specific contracting homotopy one uses (there are various options). A particularly nice version of the result is

$$\begin{aligned} \Xi &= \frac{\mathrm{i}}{4}\,\theta^{\alpha\beta} \int \mathrm{d}^n x\,(-\hat{A}^{*\mu}\{\hat{F}_{\alpha\mu} + \partial_\alpha \hat{A}_\mu \stackrel{\star}{,} \hat{A}_\beta\} + \hat{C}^*\{\hat{A}_\alpha \stackrel{\star}{,} \partial_\beta \hat{C}\}) , \\ \frac{\mathrm{d}\hat{A}_\mu}{\mathrm{d}\tau} &= (\Xi, \hat{A}_\mu) = \frac{\mathrm{i}}{4}\,\theta^{\alpha\beta}\{\hat{F}_{\alpha\mu} + \partial_\alpha \hat{A}_\mu \stackrel{\star}{,} \hat{A}_\beta\} , \\ \frac{\mathrm{d}\hat{C}}{\mathrm{d}\tau} &= (\Xi, \hat{C}) = -\frac{\mathrm{i}}{4}\,\theta^{\alpha\beta}\{\hat{A}_\alpha \stackrel{\star}{,} \partial_\beta \hat{C}\} \\ \frac{\mathrm{d}I_{\text{eff}}[\hat{A}(A,\tau),\tau]}{\mathrm{d}\tau} &= \mathrm{i}\theta^{\alpha\beta} \int \mathrm{d}^n x\,\mathrm{Tr}\left(\frac{1}{8}\hat{F}_{\alpha\beta} \star \hat{F}_{\mu\nu} \star \hat{F}^{\mu\nu} - \frac{1}{2}\hat{F}_{\alpha\mu} \star \hat{F}_{\beta\nu} \star \hat{F}^{\mu\nu}\right) . \end{aligned}$$

The expressions for $\mathrm{d}\hat{A}_\mu/\mathrm{d}\tau$ and $\mathrm{d}\hat{C}/\mathrm{d}\tau$ are differential equations for SW maps of the same form as derived in [16] for $U(N)$-theories.

The ambiguities of the result can be described in terms of Ξ as shifts $\Xi + \Delta\Xi$ of Ξ which satisfy

$$0 = s\,(\Delta\Xi) + \text{terms without antifields}\ ,$$

where the terms without antifields yield the shift $\mathrm{d}(\Delta I_{\mathrm{eff}})/\mathrm{d}\tau$ corresponding to $\Delta\Xi$. This is again an equation that can be analysed by cohomological means which are not reviewed here, and we only present the result: the general SW map $\hat{A}_\mu(A,\tau)$, $\hat{\lambda}(\lambda, A, \tau)$ for the gauge fields and gauge parameters can be written as

$$\hat{A}_\mu(A,\tau) = \left[\Lambda^{-1} \star \hat{A}_\mu^{sp} \star \Lambda + \Lambda^{-1} \star \partial_\mu \Lambda\right]_{A_\mu \to A'_\mu(A,\tau)}$$

$$\hat{\lambda}(\lambda, A,\tau) = \left[\Lambda^{-1} \star \hat{\lambda}^{sp} \star \Lambda + \Lambda^{-1} \star \delta_\lambda \Lambda\right]_{A_\mu \to A'_\mu(A,\tau)}$$

where

$\Lambda(A,\tau) = \exp_\star(f^B(A,\tau)T_B)$ with arbitrary $f^B(A,\tau)$,

$\hat{A}_\mu^{sp}(A,\tau)$, $\hat{\lambda}^{sp}(\lambda, A,\tau)$ is a particular SW map ,

${A'}_\mu^{B}(A,\tau) = [A_\mu + W_\mu(A,\tau)]^C R_C^B(\tau)$ where:

$\delta_\lambda W_\mu(A,\tau) = [W_\mu(A,\tau), \lambda]$ (i.e., W_μ is gauge covariant) ,

$T_B \to R_B^C(\tau)\, T_C$ is an (outer) Lie algebra automorphism .

Recall that $\{T_A\}$ is (a representation of) the enveloping algebra of the Lie algebra of the gauge group. Hence, the Lie algebra automorphisms $T_B \to R_B^C(\tau) T_C$ that enter here refer to the Lie algebra of $\{T_A\}$ rather than to the Lie algebra of the gauge group. Without loss of generality one may restrict these automorphisms to outer automorphisms since inner ones are already covered by the Λ-terms. Note that the latter are (field dependent) noncommutative gauge transformations of a special SW map $\hat{A}_\mu^{sp}$.

Hence, SW maps are determined only up to (compositions of) noncommutative gauge transformations of $\hat{A}_\mu$, gauge covariant shifts of enveloping algebra valued gauge fields A_μ, and outer automorphisms of the Lie algebra of the enveloping algebra.

4 Gauge Anomalies

A 1-loop computation, performed with dimensional regularization, yields the following expression for gauge anomalies in four-dimensional noncommutative Yang-Mills theories with chiral fermions [6]:

$$\mathcal{A}[\hat{C}, \hat{A}, \tau] = \int \mathrm{Tr}\left[\hat{C} \star \mathrm{d}\left(\hat{A} \star \mathrm{d}\hat{A} + \frac{1}{2}\hat{A} \star \hat{A} \star \hat{A}\right)\right]\ , \tag{4}$$

where we used differential form notation ($\mathrm{d} = \mathrm{d}x^\mu \partial_\mu$, $\hat{A} = \mathrm{d}x^\mu \hat{A}_\mu$). This expression is reminiscent of anomalies in ordinary (commutative) Yang-Mills theories since it arises from the latter by replacing commutative fields C and A_μ with their noncommutative counterparts and ordinary products with $\star$-products. However, the presence of $\star$-products poses an apparent puzzle: $\mathcal{A}=0$ does not only impose the usual anomaly cancellation conditions $\mathrm{Tr}(T_{(a}T_bT_{c)})=0$ but additional conditions at higher orders in θ, such as $\mathrm{Tr}(T_{[a}T_bT_{c]})=0$. On the other hand all candidate gauge anomalies of noncommutative Yang-Mills theories of the type considered here are known because these theories can be considered Yang-Mills theories of the effective type whose anomalies were exhaustively classified (see [12] for a review). These known results state in particular that the chiral (Bardeen) anomalies exhaust all candidate gauge anomalies when the gauge group is semisimple. According to this result (4) is cohomologically equivalent to a standard chiral anomaly, i.e., all (infinitely many!) θ-dependent terms in (4) are BRST-exact when the gauge group is semisimple.

The situation is more involved when the gauge group contains an abelian factor. In this case there are additional, and in fact infinitely many, candidate anomalies, and it is not obvious from the outset whether or not some of them occur in (4). The question is thus: is (4) always cohomologically equivalent to a standard chiral anomaly, even when the gauge group contains abelian factors? The answer is affirmative, as was shown in [6]. Again, we shall only briefly sketch how this result was obtained and drop all details.

The idea is to differentiate (4) with respect to τ and to show that the resultant expression is BRST-exact. The reason for dealing with $\mathrm{d}\mathcal{A}/\mathrm{d}\tau$ rather than with $\mathcal{A}$ itself is that, as it turns out, $\mathrm{d}\mathcal{A}/\mathrm{d}\tau$ is the BRST-variation of an expression that can be compactly written as an integrated $\star$-polynomial of the noncommutative variables $\hat{A}_\mu$:

$$
\begin{aligned}
\frac{\mathrm{d}\mathcal{A}}{\mathrm{d}\tau} &= s\mathcal{B}_\star\,,\\
\mathcal{B}_\star &= \frac{\mathrm{i}\theta^{\alpha\beta}}{2}\int \mathrm{Tr}\,\Big(\hat{A}_\alpha\star\partial_\beta\mathrm{d}\hat{A}\star\mathrm{d}\hat{A}-\frac{1}{2}\mathrm{d}\hat{A}_\alpha\star\hat{A}_\beta\star\mathrm{d}\hat{A}\star\hat{A}\\
&\quad+\frac{3}{2}\mathrm{d}\hat{A}\star\mathrm{d}\hat{A}_\alpha\star\hat{A}\star\hat{A}_\beta-\frac{1}{2}\mathrm{d}\hat{A}_\alpha\star\hat{A}_\beta\star\hat{A}\star\mathrm{d}\hat{A}\\
&\quad+\partial_\alpha\hat{A}_\beta\star\mathrm{d}\hat{A}\star\hat{A}\star\hat{A}+\text{terms with 5 or 6 }\hat{A}\text{'s}\Big)\,.
\end{aligned}
$$

We remark that $\mathcal{B}_\star$ is not unique (it is determined only up to BRST-cocycles with ghost number 0) and can be written in various ways. Hence, the expression given above is just one particular choice. The desired result for $\mathcal{A}$ is now obtained using $\mathcal{A}(\tau)=\mathcal{A}(0)+\int_0^\tau \mathrm{d}\tau'\mathrm{d}\mathcal{A}/\mathrm{d}\tau'$. This gives

$$
\mathcal{A}=\int \mathrm{Tr}\left[Cd\left(AdA+\frac{1}{2}A^3\right)\right]+s\,\mathcal{B}[A,\tau],\quad \mathcal{B}[A,\tau]=\int_0^\tau \mathrm{d}\tau'\;\mathcal{B}_\star[\hat{A}(A,\tau'),\tau']\,. \tag{5}
$$

Notice that $\mathcal{B}$, in contrast to $\mathcal{B}_\star$, can not be naturally written as an integrated $\star$-polynomial of the noncommutative variables $\hat{A}_\mu$ because of the dependence of $\hat{A}(A,\tau')$ on τ'. (5) shows that $\mathcal{A}$ is indeed given by the standard chiral gauge anomaly $\int \mathrm{Tr}[Cd(AdA+\frac{1}{2}A^3)]$ up to a BRST-exact piece $s\mathcal{B}$. Hence, at least at the 1-loop level, noncommutative Yang-Mills theories do not possess additional gauge anomalies or anomaly cancellation conditions as compared to the corresponding commutative theories, even when the gauge group contains abelian factors (the above results apply to all gauge groups). Notice that $-\mathcal{B}$ is the counterterm that cancels the θ-dependent terms in $\mathcal{A}$.

References

1. B. Jurco, S. Schraml, P. Schupp and J. Wess, Eur. Phys. J. C **17**, 521 (2000) [arXiv:hep-th/0006246].
2. N. Seiberg and E. Witten, JHEP **09**, 032 (1999) [arXiv:hep-th/9908142].
3. G. Barnich, F. Brandt and M. Grigoriev, Fortsch. Phys. **50**, 825 (2002) [arXiv:hep-th/0201139].
4. G. Barnich, F. Brandt and M. Grigoriev, JHEP **08**, 023 (2002) [arXiv:hep-th/0206003].
5. G. Barnich, F. Brandt and M. Grigoriev, Nucl. Phys. B **677**, 503 (2004) [arXiv:hep-th/0308092].
6. F. Brandt, C. P. Martín and F. Ruiz Ruiz, JHEP **07**, 068 (2003) [arXiv:hep-th/0307292].
7. G. Barnich, M. Grigoriev, and M. Henneaux, JHEP **10**, 004 (2001) [arXiv:hep-th/0106188].
8. G. Barnich and M. Henneaux, Phys. Lett. B **311**, 123 (1993) [arXiv:hep-th/9304057].
9. I. A. Batalin and G. A. Vilkovisky, Phys. Lett. B **102**, 27 (1981).
10. F. Brandt, N. Dragon and M. Kreuzer, Nucl. Phys. B **332**, 224 (1990).
11. M. Dubois-Violette, M. Henneaux, M. Talon and C. M. Viallet, Phys. Lett. B **289**, 361 (1992) [arXiv:hep-th/9206106].
12. G. Barnich, F. Brandt and M. Henneaux, Phys. Rept. **338**, 439 (2000) [arXiv:hep-th/0002245].

Renormalisation Group Approach to Noncommutative Quantum Field Theory

Harald Grosse[1] and Raimar Wulkenhaar[2]

[1] Institut für Theoretische Physik der Universität Wien
Boltzmanngasse 5, 1090 Wien, Austria
harald.grosse@univie.ac.at

[2] Max-Planck-Institut für Mathematik in den Naturwissenschaften
Inselstraße 22-26, 04103 Leipzig, Germany
raimar.wulkenhaar@mis.mpg.de

1 Introduction

Quantum field theory on Euclidean or Minkowski space is extremely successful. For suitably chosen action functionals one achieves a remarkable agreement of up to 10^{-11} between theoretical predictions and experimental data. However, combining the fundamental principles of both general relativity and quantum mechanics one concludes that space(-time) cannot be a differentiable manifold [1]. To the best of our knowledge, such a possibility was first discussed in [2].

Since geometric concepts are indispensable in physics, we need a replacement for the space-time manifold which still has a geometric interpretation. Quantum physics tells us that whenever there are measurement limits we have to describe the situation by non-commuting operators on a Hilbert space. Fortunately for physics, mathematicians have developed a generalisation of geometry, baptised noncommutative geometry [3], which is perfectly designed for our purpose. However, in physics we need more than just a better geometry: We need renormalisable quantum field theories modelled on such a noncommutative geometry.

Remarkably, it turned out to be very difficult to renormalise quantum field theories even on the simplest noncommutative spaces [4]. It would be a wrong conclusion, however, that this problem singles out the standard commutative geometry as the only one compatible with quantum field theory. The problem tells us that we are still at the very beginning of *understanding* quantum field theory. Thus, apart from curing the contradiction between gravity and quantum physics, in doing quantum field theory on noncommutative geometries we learn a lot about quantum field theory itself.

2 Field Theory on Noncommutative $\mathbb{R}^D$ in Momentum Space

The simplest noncommutative generalisation of Euclidean space is the so-called noncommutative $\mathbb{R}^D$. Although this space arises naturally in a certain limit of string theory [16], we should not expect that it is a good model for nature. In particular, the noncommutative $\mathbb{R}^D$ does not allow for gravity. For us the main purpose of this space is to develop an understanding of quantum field theory which has a broader range of applicability.

The noncommutative $\mathbb{R}^D$, $D = 2, 4, 6, \dots$, is defined as the algebra $\mathbb{R}^D_\theta$ which as a vector space is given by the space $\mathcal{S}(\mathbb{R}^D)$ of (complex-valued) Schwartz class functions of rapid decay, equipped with the multiplication rule

$$(a \star b)(x) = \int \frac{\mathrm{d}^D k}{(2\pi)^D} \int \mathrm{d}^D y \, a(x + \tfrac{1}{2}\theta{\cdot}k)\, b(x+y)\, \mathrm{e}^{\mathrm{i}k\cdot y} \,, \qquad (1)$$
$$(\theta{\cdot}k)^\mu = \theta^{\mu\nu} k_\nu \,, \quad k{\cdot}y = k_\mu y^\mu \,, \quad \theta^{\mu\nu} = -\theta^{\nu\mu} \,.$$

The entries $\theta^{\mu\nu}$ in (1) have the dimension of an area. The physical interpretation is $\|\theta\| \approx \lambda_P^2$. Much information about the noncommutative $\mathbb{R}^D$ can be found in [6].

A field theory is defined by an action functional. We obtain action functionals on $\mathbb{R}^D_\theta$ by replacing in standard action functionals the ordinary product of functions by the $\star$-product. For example, the noncommutative ϕ^4-action is given by

$$S[\phi] := \int \mathrm{d}^D x \left(\frac{1}{2} \partial_\mu \phi \star \partial^\mu \phi + \frac{1}{2} m^2 \phi \star \phi + \frac{\lambda}{4!} \phi \star \phi \star \phi \star \phi \right) . \qquad (2)$$

The action (2) is then inserted into the partition function which we solve perturbatively by Feynman graphs. Due to $\int \mathrm{d}^D x \; (a \star b)(x) = \int \mathrm{d}^D x \; a(x) b(x)$, the propagator in momentum space is unchanged. For later purpose it is, however, convenient to write it as a double line, 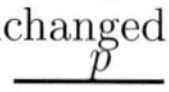 $= (p^2 + m^2)^{-1}$. The novelty are phase factors in the vertices, which we also write in double line notation,

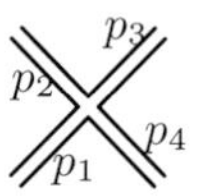

$$= \frac{\lambda}{4!} \, \mathrm{e}^{-\frac{\mathrm{i}}{2} \sum_{i<j} p_i^\mu p_j^\nu \theta_{\mu\nu}} \,. \qquad (3)$$

The double line notation reflects the fact that the vertex (3) is invariant only under cyclic permutations of the legs (using momentum conservation). The resulting Feynman graphs are *ribbon graphs* which depend crucially on how the valences of the vertices are connected. For *planar graphs* the total phase factor of the integrand is independent of internal momenta, whereas *non-planar graphs* have a total phase factor which involves internal momenta. For example, the one-loop contribution to the two-point function splits as follows into a planar part

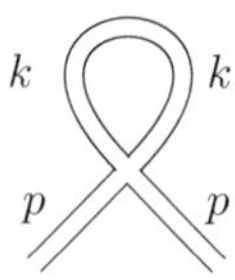

$$= \frac{\lambda}{6} \int \frac{\mathrm{d}^4 k}{(2\pi)^4} \frac{1}{k^2 + m^2} \qquad (4)$$

and a non-planar part

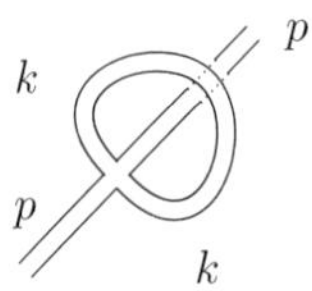

$$= \frac{\lambda}{12} \int \frac{\mathrm{d}^4 k}{(2\pi)^4} \frac{\mathrm{e}^{\mathrm{i}p^\mu k^\nu \theta_{\mu\nu}}}{k^2 + m^2} = \frac{\lambda}{48\pi^2} \sqrt{\frac{m^2}{\tilde{p}^2}} \, K_1\Big(\sqrt{m^2 \tilde{p}^2}\Big) \,, \qquad (5)$$

where $\tilde{p}_\mu := \theta_{\mu\nu} p^\nu$.

Planar graphs are treated as usual. Here, the contribution (4) can entirely be removed by a suitable normalisation condition for the physical mass. The contribution from the non-planar graph (5) is – at first sight – finite, which is a relict of the original motivation that noncommutativity would serve as regulator. The finiteness is important, because the momentum dependence (5) does not appear in the original action (2), which means that a divergence of the form (5) cannot be absorbed by multiplicative renormalisation.

However, the expansion of the modified Bessel function K_1 shows that the contribution (5) behaves $\sim\tilde{p}^{-2}$ for small momenta. If we insert the graph (5) declared as finite as a subgraph into a bigger graph, one easily builds examples (with an arbitrary number of external legs) which lead to non-integrable integrals at small inner momenta. This is the so-called UV/IR-mixing problem [4].

The heuristic argumentation can be made exact: Chepelev and Roiban have proven a power-counting theorem [7, 8] which relates the power-counting degree of divergence to the topology of the ribbon graph. The rough summary of the power-counting theorem is that noncommutative field theories with quadratic divergences become meaningless beyond a certain loop order. The situation is better for field theories with logarithmic UV/IR-divergences, e.g. supersymmetric models. These can be formulated to any loop order. However, the logarithmic IR-divergences at exceptional external momenta are still present so that the correlation functions are unbounded: For every $\delta > 0$ one finds non-exceptional momenta such that $\big|\langle\phi(p_1)\dots\phi(p_n)\rangle\big| > \frac{1}{\delta}$. In the remainder of this article we present an approach which solves these problems.

3 Renormalisation Group Approach to Noncommutative Scalar Models

We have seen that quantum field theories on noncommutative $\mathbb{R}^D$ are not renormalisable by standard Feynman graph evaluations. One may speculate that the origin of this problem is the too naïve way one performs the continuum limit. A way to treat the limit more carefully is the use of flow equations. The idea goes back to Wilson [9]. It was then used by Polchinski [10] to give a very efficient renormalisability proof of commutative ϕ^4-theory. Applying Polchinski's method to the noncommutative ϕ^4-model, we can hope to be able to prove renormalisability to all orders, too. There is, however, a serious problem of the momentum space proof. We have to guarantee that planar graphs only appear in the distinguished interaction coefficients for which we fix the boundary condition at the renormalisation scale Λ_R. Non-planar graphs have phase factors which involve inner momenta. Polchinski's method consists in taking norms of the interaction coefficients, and these norms ignore possible phase factors. Thus, we would find that boundary conditions for non-planar graphs at Λ_R are required. Since there is an infinite number of different non-planar structures, the model is not renormalisable in this way. A more careful examination of the phase factors is also not possible because the cut-off integrals prevent the Gaußian integration required for the parametric integral representation [7, 8].

Fortunately, there is a matrix representation of the noncommutative $\mathbb{R}^D$ where the $\star$-product becomes a simple product of infinite matrices. The price for this simplification is that the propagator becomes complicated, but the difficulties can be overcome.

3.1 Matrix Representation

For simplicity we restrict ourselves to the noncommutative $\mathbb{R}^2$. There exists a matrix base $\{f_{mn}(x)\}_{m,n\in\mathbb{N}}$ of the noncommutative $\mathbb{R}^2$ which satisfies

$$(f_{mn} \star f_{kl})(x) = \delta_{nk} f_{ml}(x) , \qquad \int d^2x \, f_{mn}(x) = 2\pi\theta_1 , \tag{6}$$

where $\theta_1 := \theta_{12} = -\theta_{21}$. In terms of radial coordinates $x_1 = \rho\cos\varphi$, $x_2 = \rho\sin\varphi$ one has

$$f_{mn}(\rho,\varphi) = 2(-1)^m \mathrm{e}^{\mathrm{i}(n-m)\varphi} \sqrt{\tfrac{m!}{n!}} \Big(\sqrt{\tfrac{2\rho^2}{\theta_1}}\Big)^{n-m} L_m^{n-m}\Big(\tfrac{2\rho^2}{\theta_1}\Big) \, \mathrm{e}^{-\frac{\rho^2}{\theta_1}} , \tag{7}$$

where $L_n^\alpha(z)$ are the Laguerre polynomials. See also [6]. The matrix representation was also used to obtain exactly solvable noncommutative quantum field theories [11, 12].

Now we can write down the noncommutative ϕ^4-action in the matrix base by expanding the field as $\phi(x) = \sum_{m,n\in\mathbb{N}} \phi_{mn} f_{mn}(x)$. It turns out, however, that in order to prove renormalisability we have to consider a more general action than (2) at the initial scale Λ_0. This action is obtained by adding a harmonic oscillator potential to the standard noncommutative ϕ^4-action:

$$\begin{aligned} S[\phi] := \int d^2x \, \Big(& \frac{1}{2}\partial_\mu\phi \star \partial^\mu\phi + 2\Omega^2(\tilde{x}^\mu\phi)\star(\tilde{x}_\mu\phi) + \frac{\mu_0^2}{2}\phi\star\phi \\ & + \frac{\lambda}{4!}\phi\star\phi\star\phi\star\phi\Big)(x) \\ = 2\pi\theta_1 & \sum_{m,n,k,l} \Big(\frac{1}{2}G_{mn;kl}\phi_{mn}\phi_{kl} + \frac{\lambda}{4!}\phi_{mn}\phi_{nk}\phi_{kl}\phi_{lm}\Big) , \end{aligned} \tag{8}$$

where $\tilde{x}^\mu := \theta^{\mu\nu} x_\nu$ and

$$G_{mn;kl} := \int \frac{d^2x}{2\pi\theta_1} \Big(\partial_\mu f_{mn} \star \partial^\mu f_{kl} + 4\Omega^2(\tilde{x}^\mu f_{mn})\star(\tilde{x}_\mu f_{kl}) + \mu_0^2 f_{mn}\star f_{kl}\Big) . \tag{9}$$

We view Ω as a regulator and refer to the action (8) as describing a regularised ϕ^4-model. The action (8) could also be obtained by restricting a complex ϕ^4-model with magnetic field [11, 12] to the real part. One finds

$$\begin{aligned} G_{mn;kl} = & \Big(\mu_0^2 + \frac{2}{\theta_1}(1+\Omega^2)(n+m+1)\Big)\delta_{nk}\delta_{ml} \\ & - \frac{2}{\theta_1}(1-\Omega^2)\big(\sqrt{(n+1)(m+1)}\,\delta_{n+1,k}\delta_{m+1,l} - \sqrt{nm}\,\delta_{n-1,k}\delta_{m-1,l}\big) . \end{aligned} \tag{10}$$

The kinetic matrix $G_{mn;kl}$ has the important property that $G_{mn;kl} = 0$ unless $m+k = n+l$. The same relation is induced for the propagator $\Delta_{nm;lk}$ defined by $\sum_{k,l=0}^\infty G_{mn;kl}\Delta_{lk;sr} = \sum_{k,l=0}^\infty \Delta_{nm;lk}G_{kl;rs} = \delta_{mr}\delta_{ns}$. In order to evaluate the propagator we first diagonalise the kinetic matrix $G_{mn;kl}$:

$$G_{m,m+\alpha;l+\alpha,l} = \sum_{y\in\mathbb{N}} U^{(\alpha)}_{my}\big(\mu_0^2 + \tfrac{4\Omega}{\theta}(2y+\alpha+1)\big)U^{(\alpha)}_{yl} , \tag{11}$$

$$U^{(\alpha)}_{ny} = \sqrt{\binom{\alpha+n}{n}\binom{\alpha+y}{y}\Big(\frac{1-\Omega}{1+\Omega}\Big)^{2n+2y+\alpha+1}\Big(\frac{4\Omega}{1-\Omega^2}\Big)^{\alpha+1}} \times M_n\Big(y;1+\alpha,\frac{(1-\Omega)^2}{(1+\Omega)^2}\Big) , \tag{12}$$

where $M_n(y;\beta,c) = {}_2F_1\Big({-n,-y \atop \beta}\Big|1-c\Big)$ are the (orthogonal) Meixner polynomials [13]. A lengthy calculation gives

$$\begin{aligned}\Delta_{mn;kl} = {} & \frac{\theta_1}{2(1+\Omega^2)}\delta_{m+k,n+l} \sum_{v=\frac{|m-l|}{2}}^{\frac{\min(m+l,k+n)}{2}} B\Big(\tfrac{1}{2}+\tfrac{\mu_0^2\theta_1}{8\Omega}+\tfrac{1}{2}(m+k)-v, 1+2v\Big) \\ & \times \sqrt{\binom{n}{v+\frac{n-k}{2}}\binom{k}{v+\frac{k-n}{2}}\binom{m}{v+\frac{m-l}{2}}\binom{l}{v+\frac{l-m}{2}}}\Big(\frac{1-\Omega}{1+\Omega}\Big)^{2v} \\ & \times {}_2F_1\left({1+2v\,,\ \frac{1}{2}+\frac{\mu_0^2\theta_1}{8\Omega}-\frac{1}{2}(m+k)+v \atop \frac{3}{2}+\frac{\mu_0^2}{2\sqrt{1-\omega}\,\mu^2}+\frac{1}{2}(m+k)+v}\middle|\frac{(1-\Omega)^2}{(1+\Omega)^2}\right) .\end{aligned} \tag{13}$$

Here, $B(a,b)$ is the Beta-function and $F\big({a,\,b \atop c};z\big)$ the hypergeometric function.

3.2 The Polchinski Equation for Matrix Models

We summarise here our derivation [14] of the Polchinski equation for the noncommutative ϕ^4-theory in the matrix base. According to Polchinski's derivation of the exact renormalisation group equation [10] we consider the following cut-off partition function:

$$\begin{aligned} Z[J,\Lambda] &= \int \Big(\prod_{a,b} d\phi_{ab}\Big) \exp\big(-S[\phi,J,\Lambda]\big) , \\ S[\phi,J,\Lambda] &= (2\pi\theta_1)\Bigg(\sum_{m,n,k,l}\frac{1}{2}\phi_{mn}G^K_{mn;kl}(\Lambda)\,\phi_{kl} + \sum_{m,n,k,l}\phi_{mn}F_{mn;kl}[\Lambda]J_{kl} \\ &\qquad + \sum_{m,n,k,l}\frac{1}{2}J_{mn}E_{mn;kl}[\Lambda]J_{kl} + L[\phi,\Lambda] + C[\Lambda]\Bigg) , \\ G^K_{mn;kl}(\Lambda) &= \prod_{i\in\{m,n,k,l\}} K\big[\tfrac{i}{\Lambda^2\theta_1}\big]^{-1} G_{mn;kl} . \end{aligned} \tag{14}$$

with $L[0,\Lambda]=0$. The cut-off function $K(x)$ is a smooth decreasing function with $K(x)=1$ for $0\le x\le 1$ and $K(x)=0$ for $x\ge 2$. Accordingly, we define

$$\Delta^K_{nm;lk}(\Lambda) = \prod_{i\in\{m,n,k,l\}} K\big[\tfrac{i}{\Lambda^2\theta_1}\big]\Delta_{nm;lk} \;. \tag{15}$$

The function $C[\Lambda]$ is the vacuum energy and the matrices E and F, which are not necessary in the commutative case, must be introduced because the propagator Δ is non-local. It is in general not possible to separate the support of the sources J from the support of the Λ-variation of K. We would obtain the original problem (without cut-off) for the choice

$$\begin{aligned} L[\phi,\infty] &= \sum_{m,n,k,l} \frac{\lambda}{4!}\phi_{mn}\phi_{nk}\phi_{kl}\phi_{lm} \;, \\ C[\infty] &= 0\;, \qquad E_{mn;kl}[\infty] = 0\;, \qquad F_{mn;kl}[\infty] = \delta_{ml}\delta_{nk}\;. \end{aligned} \tag{16}$$

However, we shall expect divergences in the partition function which require a renormalisation, i.e. additional (divergent) counterterms in $L[\phi,\infty]$.

Following Polchinski [10] we first ask ourselves how to choose L, C, E, F in order to make $Z[J,\Lambda]$ *independent* of Λ. After straightforward calculation one finds the answer

$$\begin{aligned} &\Lambda\frac{\partial L[\phi,\Lambda]}{\partial\Lambda} \\ &= \sum_{m,n,k,l}\frac{1}{2}\Lambda\frac{\partial\Delta^K_{nm;lk}(\Lambda)}{\partial\Lambda}\Big(\frac{\partial L[\phi,\Lambda]}{\partial\phi_{mn}}\frac{\partial L[\phi,\Lambda]}{\partial\phi_{kl}} - \frac{1}{2\pi\theta_1}\Big[\frac{\partial^2 L[\phi,\Lambda]}{\partial\phi_{mn}\,\partial\phi_{kl}}\Big]_\phi\Big)\;, \end{aligned} \tag{17}$$

where $\big[f[\phi]\big]_\phi := f[\phi]-f[0]$. The corresponding differential equations for C, E, F are easy to integrate [14]. Now, instead of computing Green's functions from $Z[J,\infty]$ we can equally well start from $Z[J,\Lambda_R]$, where it leads to Feynman graphs with vertices given by the Taylor expansion coefficients $A^{(V)}_{m_1n_1;\dots;m_Nn_N}$ in

$$L[\phi,\Lambda] = \lambda\sum_{V=1}^{\infty}\big(2\pi\theta_1\lambda\big)^{V-1}\sum_{N=2}^{\infty}\frac{1}{N!}\sum_{m_i,n_i} A^{(V)}_{m_1n_1;\dots;m_Nn_N}[\Lambda]\phi_{m_1n_1}\cdots\phi_{m_Nn_N}\;. \tag{18}$$

These vertices are connected with each other by internal lines $\Delta^K_{nm;lk}(\Lambda)$ and to sources J_{kl} by external lines $\Delta^K_{nm;lk}(\Lambda_0)$. Since the summation variables are cut-off in the propagator (15), loop summations are finite, provided that the interaction coefficients $A^{(V)}_{m_1n_1;\dots;m_Nn_N}[\Lambda]$ are bounded. Thus, renormalisability amounts to prove that for certain initial conditions (parametrised by finitely many parameters!) the evolution of L according to (17) does not produce any divergences.

Inserting the expansion (18) into (17) and restricting to the part with N external legs we get the graphical expression

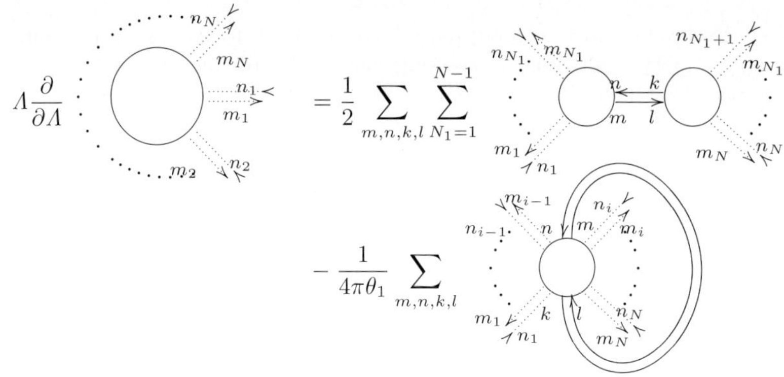

(19)

Combinatorical factors are not shown and symmetrisation in all indices $m_i n_i$ has to be performed. On the rhs of (19) the two valences mn and kl of subgraphs are connected to the ends of a *ribbon* which symbolises the differentiated propagator $\overset{n\ \ k}{\underset{m\ \ l}{\rightleftarrows}} = \Lambda\frac{\partial}{\partial\Lambda}\Delta^K_{nm;lk}$. We see that for the simple fact that the fields ϕ_{mn} carry two indices, the effective action is expanded into ribbon graphs.

In the expansion of L there will occur very complicated ribbon graphs with crossings of lines which cannot be drawn any more in a plane. A general ribbon graph can, however, be drawn on a *Riemann surface* of some *genus* g. In fact, a ribbon graph *defines* the Riemann surfaces topologically through the *Euler characteristic* χ. We have to regard here the external lines of the ribbon graph as amputated (or closed), which means to directly connect the single lines m_i with n_i for each external leg $m_i n_i$. A few examples may help to understand this procedure:

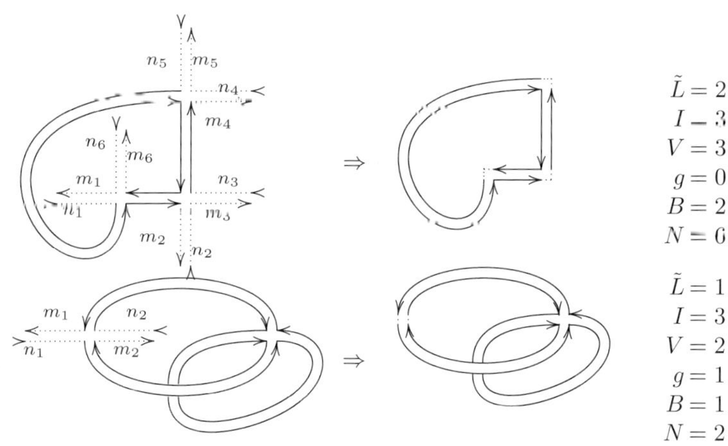

(20)

The genus is computed from the number $\tilde{L}$ of single-line loops, the number I of internal (double) lines and the number V of vertices of the graph according to Euler's formula $\chi = 2 - 2g = \tilde{L} - I + V$. The number B of boundary components of a ribbon graph is the number of those loops which carry at least one external leg. There can be several possibilities to draw the graph and its Riemann surface, but $\tilde{L}, I, V, B$ and thus g remain unchanged. Indeed, the Polchinski equation (17) interpreted as in (19) tells us which external legs of the vertices are connected. It is completely irrelevant how the ribbons are drawn between these legs. In particular, there is no distinction between overcrossings and undercrossings.

We expect that non-planar ribbon graphs with $g > 0$ and/or $B > 1$ behave differently under the renormalisation flow than planar graphs having $B = 1$ and $g = 0$. This suggests to introduce a further grading in g, B in the interactions coefficients $A^{(V,B,g)}_{m_1n_1;\dots;m_Nn_N}$. Technically, our strategy is to apply the summations in (19) either to the propagator or the subgraph only and to maximise the other object over the summation indices. For that purpose one has to introduce further characterisations of a ribbon graph which disappear at the end, see [14].

3.3 ϕ^4-Theory on Noncommutative $\mathbb{R}^2$

First one estimates the A-functions by integrating (17) perturbatively between an initial scale Λ_0 to be sent to ∞ later on and the renormalisation scale Λ_R:

Lemma 1. *The homogeneous parts $A^{(V,B,g)}_{m_1n_1;\dots;m_Nn_N}$ of the coefficients of the effective action describing a regularised ϕ^4-theory on $\mathbb{R}^2_\theta$ in the matrix base are for $2 \leq N \leq 2V+2$ and $\sum_{i=1}^N (m_i - n_i) = 0$ bounded by*

$$\begin{aligned} &\big|A^{(V,B,g)}_{m_1n_1;\dots;m_Nn_N}[\Lambda,\Lambda_0,\Omega,\rho_0]\big| \\ &\qquad \leq \big(\Lambda^2\theta_1\big)^{2-V-B-2g}\Big(\frac{1}{\Omega}\Big)^{3V-\frac{N}{2}+B+2g-2} P^{2V-\frac{N}{2}}\Big[\ln\frac{\Lambda_0}{\Lambda_R}\Big]\,. \end{aligned} \tag{21}$$

We have $A^{(V,B,g)}_{m_1n_1;\dots;m_Nn_N} \equiv 0$ for $N > 2V+2$ or $\sum_{i=1}^N (m_i - n_i) \neq 0$. By $P^q[x]$ we denote a polynomial in x of degree q.

The proof of (21) for general matrix models by induction goes over 20 pages! The formula specific for the ϕ^4-model on $\mathbb{R}^2_\theta$ follows from the asymptotic behaviour of the cut-off propagator (15), (13) and a certain index summation, see [14, 15].

We see from (21) that the only divergent function is

$$\begin{aligned} A^{(1,1,0)}_{m_1n_1;m_2n_2} &= A^{(1,1,0)}_{00;00}\delta_{m_1n_2}\delta_{m_2n_1} \\ &\quad + \Big(A^{(1,1,0)}_{m_1n_1;m_2n_2}[\Lambda,\Lambda_0,\rho^0] - A^{(1,1,0)}_{00;00}\delta_{m_1n_2}\delta_{m_2n_1}\Big)\,, \end{aligned} \tag{22}$$

which is split into the distinguished divergent function

$$\rho[\Lambda,\Lambda_0,\Omega,\rho^0] := A^{(1,1,0)}_{00;00}[\Lambda,\Lambda_0,\Omega,\rho^0] \tag{23}$$

for which we impose the boundary condition $\rho_R := \rho[\Lambda_R,\Lambda_0,\Omega,\rho^0] = 0$ and a convergent part with boundary condition at Λ_0.

The limit $\Omega \to 0$ in (21) is singular. In fact the estimation for $\Omega = 0$ with an optimal choice of the ρ-coefficients (different than (23)!) would grow with $\big(\Lambda\sqrt{\theta_1}\big)^{V-\frac{N}{2}-B-2g+2}$. Since the exponent of Λ can be arbitrarily large, there would be an infinite number of divergent interaction coefficients, which means that the ϕ^4-model is not renormalisable when keeping $\Omega = 0$.

In order to pass to the limit $\Lambda_0 \to \infty$ one has to control the total Λ_0-dependence of the functions $A^{(V,B,g)}_{m_1n_1;\dots;m_Nn_N}[\Lambda,\Lambda_0,\Omega[\Lambda_0],\rho^0[\Lambda_R,\Lambda_0,\rho_R]]$. This leads again to a differential equation in Λ, see [15]. It is then not difficult to see that the regularised ϕ^4-model with $\Omega > 0$ is renormalisable. It turns out that one can even prove more [15]: On can endow the parameter Ω for the oscillator frequency with an Λ_0-dependence so that in the limit $\Lambda_0 \to \infty$ one obtains a standard ϕ^4-model without the oscillator term:

Theorem 1. *The ϕ^4-model on $\mathbb{R}^2_\theta$ is (order by order in the coupling constant) renormalisable in the matrix base by adjusting the bare mass $\Lambda_0^2\rho[\Lambda_0]$ to give $A^{(1,1,0)}_{00;00}[\Lambda_R] = 0$ and by performing the limit $\Lambda_0 \to \infty$ along the path of regulated models characterised by $\Omega[\Lambda_0] = \big(1+\ln\frac{\Lambda_0}{\Lambda_R}\big)^{-1}$. The limit $A^{(V,B,g)}_{m_1n_1;\dots;m_Nn_N}[\Lambda_R,\infty] := \lim_{\Lambda_0\to\infty} A^{(V,B,g)}_{m_1n_1;\dots;m_Nn_N}[\Lambda_R,\Lambda_0,\Omega[\Lambda_0],\rho^0[\Lambda_0]]$ of the expansion coefficients of the effective action $L[\phi,\Lambda_R,\Lambda_0,\Omega[\Lambda_0],\rho^0[\Lambda_0]]$ exists and satisfies*

$$\begin{aligned}
&\Big|\lambda\big(2\pi\theta_1\lambda\big)^{V-1} A^{(V,B,g)}_{m_1n_1;\dots;m_Nn_N}[\Lambda_R,\infty] \\
&\qquad - \big(2\pi\theta_1\lambda\big)^{V-1} A^{(V,V^e,B,g,\iota)}_{m_1n_1;\dots;m_Nn_N}\Big[\Lambda_R,\Lambda_0,\frac{1}{(1+\ln\frac{\Lambda_0}{\Lambda_R})},\rho^0[\Lambda_0]\Big]\Big| \\
&\leq \frac{\Lambda_R^4}{\Lambda_0^2}\Big(\frac{\lambda}{\Lambda_R^2}\Big)^V \left(\frac{(1+\ln\frac{\Lambda_0}{\Lambda_R})}{\Lambda_R^2\theta_1}\right)^{B+2g-1} P^{5V-N-1}\Big[\ln\frac{\Lambda_0}{\Lambda_R}\Big]\,. \qquad (24)
\end{aligned}$$

In this way we have proven that the real ϕ^4-model on $\mathbb{R}^2_\theta$ is perturbatively renormalisable when formulated in the matrix base.

It was important to observe that the cut-off action at Λ_0 is (due to the cut-off) not translation-invariant. We are therefore free to break the translational symmetry of the action at Λ_0 even more by adding a harmonic oscillator potential for the fields ϕ. There exists a Λ_0-dependence of the oscillator frequency Ω with $\lim_{\Lambda_0\to\infty}\Omega = 0$ such that the effective action at Λ_R is convergent (and thus bounded) order by order in the coupling constant in the limit $\Lambda_0 \to \infty$. This means that the partition function of the original (translation-invariant) ϕ^4-model without cut-off and with suitable divergent bare mass can equally well be solved by Feynman graphs with propagators cut-off at Λ_R and vertices given by the bounded expansion coefficients of the effective action at Λ_R. Hence, this model is renormalisable, and in contrast to the naïve Feynman graph approach in momentum space [8] there is no problem with exceptional configurations. This makes clear that the adaptation of Polchinski's renormalisation programme is the preferred method for noncommutative field theories.

3.4 ϕ^4-Theory on Noncommutative $\mathbb{R}^4$

The renormalisation of ϕ^4-theory on $\mathbb{R}^4_\theta$ in the matrix base is performed similarly [16]. We choose a coordinate system in which $\theta_1 = \theta_{12} = -\theta_{21}$ and $\theta_2 = \theta_{34} = -\theta_{43}$ are the only non-vanishing components of θ. Moreover, we assume $\theta_1 = \theta_2$ for simplicity. Then we expand the scalar field according to $\phi(x) = \sum_{m_1,n_1,m_2,n_2\in\mathbb{N}} \phi_{\substack{m_1\ n_1\\ m_2\ n_2}} f_{m_1n_1}(x_1,x_2) f_{m_2n_2}(x_3,x_4)$. The action (8) with integration over $\mathbb{R}^4$ leads then to a kinetic term generalising (10) and a propagator generalising (13). Using estimates on the asymptotic behaviour of that propagator one proves the four-dimensional generalisation of Lemma 1 on the power-counting degree of the N-point functions. For $\Omega > 0$ one finds that all non-planar graphs ($B > 1$ and/or $g > 0$) and all graphs with $N \geq 6$ external legs are convergent.

The remaining infinitely many planar two- and four-point functions have to be split into a divergent ρ-part and a convergent complement. Using some sort of locality for the propagator (13), which is a consequence of its derivation from Meixner polynomials, one proves that

$$
\begin{aligned}
A^{\text{planar}}_{\substack{m_1 n_1 ; k_1 l_1 \\ m_2 n_2 ; k_2 l_2}} & - \delta_{m_1 l_1}\delta_{n_1 k_1}\delta_{m_2 l_2}\delta_{n_2 k_2}\Big(A^{\text{planar}}_{\substack{0\,0 ; 0\,0 \\ 0\,0 ; 0\,0}} \\
& + m_1\Big(A^{\text{planar}}_{\substack{1\,0 ; 0\,1 \\ 0\,0 ; 0\,0}} - A^{\text{planar}}_{\substack{0\,0 ; 0\,0 \\ 0\,0 ; 0\,0}}\Big) + n_1\Big(A^{\text{planar}}_{\substack{0\,1 ; 1\,0 \\ 0\,0 ; 0\,0}} - A^{\text{planar}}_{\substack{0\,0 ; 0\,0 \\ 0\,0 ; 0\,0}}\Big) \\
& + m_2\Big(A^{\text{planar}}_{\substack{0\,0 ; 0\,0 \\ 1\,0 ; 0\,1}} - A^{\text{planar}}_{\substack{0\,0 ; 0\,0 \\ 0\,0 ; 0\,0}}\Big) + n_2\Big(A^{\text{planar}}_{\substack{0\,0 ; 0\,0 \\ 0\,1 ; 1\,0}} - A^{\text{planar}}_{\substack{0\,0 ; 0\,0 \\ 0\,0 ; 0\,0}}\Big)\Big) \\
& - \big(\sqrt{(m_1+1)(n_1+1)}\delta_{m_1+1,l_1}\delta_{n_1+1,k_1}\delta_{m_2 l_2}\delta_{n_2 k_2} \\
& \quad + \sqrt{m_1 n_1}\delta_{m_1-1,l_1}\delta_{n_1-1,k_1}\delta_{m_2 l_2}\delta_{n_2 k_2}\big) A^{\text{planar}}_{\substack{1\,1 ; 0\,0 \\ 0\,0 ; 0\,0}} \\
& - \big(\sqrt{(m_2+1)(n_2+1)}\delta_{m_2+1,l_2}\delta_{n_2+1,k_2}\delta_{m_1 l_1}\delta_{n_1 k_1} \\
& \quad + \sqrt{m_2 n_2}\delta_{m_2-1,2_1}\delta_{n_2-1,k_2}\delta_{m_1 l_1}\delta_{n_1 k_1}\big) A^{\text{planar}}_{\substack{0\,0 ; 0\,0 \\ 1\,1 ; 0\,0}} ,
\end{aligned}
\tag{25}
$$

$$
A^{\text{planar}}_{\substack{m_1 n_1 ; \dots ; m_4 n_4 \\ m_1' n_1' ; \dots ; m_4' n_4'}} - \big(\tfrac{1}{6}\delta_{\substack{n_1 m_2 \\ n_1' m_2'}}\delta_{\substack{n_2 m_3 \\ n_2' m_4'}}\delta_{\substack{n_3 m_4 \\ n_3' m_4'}}\delta_{\substack{n_4 m_1 \\ n_4' m_1'}} + 5 \text{ perm's}\big) A^{\text{planar}}_{\substack{0\,0 ; \dots ; 0\,0 \\ 0\,0 ; \dots ; 0\,0}} , \tag{26}
$$

are convergent functions, thus identifying

$$
\begin{aligned}
\rho_1 &:= A^{\text{planar}}_{\substack{0\,0 ; 0\,0 \\ 0\,0 ; 0\,0}} , \\
\rho_2 &:= A^{\text{planar}}_{\substack{1\,0 ; 0\,1 \\ 0\,0 ; 0\,0}} - A^{\text{planar}}_{\substack{0\,0 ; 0\,0 \\ 0\,0 ; 0\,0}} = A^{\text{planar}}_{\substack{0\,0 ; 0\,0 \\ 1\,0 ; 0\,1}} - A^{\text{planar}}_{\substack{0\,0 ; 0\,0 \\ 0\,0 ; 0\,0}} , \\
\rho_3 &:= A^{\text{planar}}_{\substack{1\,1 ; 0\,0 \\ 0\,0 ; 0\,0}} = A^{\text{planar}}_{\substack{0\,0 ; 0\,0 \\ 1\,1 ; 0\,0}} \\
\rho_4 &:= A^{\text{planar}}_{\substack{0\,0 ; 0\,0 ; 0\,0 ; 0\,0 \\ 0\,0 ; 0\,0 ; 0\,0 ; 0\,0}}
\end{aligned}
\tag{27}
$$

as the distinguished divergent ρ-functions for which we impose boundary conditions at Λ_R. Details will be given in [16].

The function ρ_3 has no commutative analogue. Due to (25) it corresponds to a normalisation condition for the frequency parameter Ω in (10). This means that in contrast to the two-dimensional case we cannot remove the oscillator potential with the limit $\Lambda_0 \to \infty$. In other words, the oscillator potential in (8) is a necessary companionship to the $\star$-product interaction. This observation is in agreement with the UV/IR-entanglement first observed in [4]. Whereas the UV/IR-problem prevents the renormalisation of ϕ^4-theory on $\mathbb{R}^4_\theta$ in momentum space [8], we have found a self-consistent solution of the problem by providing the unique (due to properties of the Meixner polynomials) renormalisable extension of the action.

Acknowledgement

Harald Grosse thanks Josep Trampetić and all the other organisers for the invitation and kind hospitality at the conference.

References

1. S. Doplicher, K. Fredenhagen and J. E. Roberts, "The Quantum structure of space-time at the Planck scale and quantum fields," Commun. Math. Phys. **172** (1995) 187 [arXiv:hep-th/0303037].
2. E. Schrödinger, "Über die Unanwendbarkeit der Geometrie im Kleinen," Naturwiss. **31** (1934) 34.
3. A. Connes, "Noncommutative geometry," Academic Press (1994).
4. S. Minwalla, M. Van Raamsdonk and N. Seiberg, "Noncommutative perturbative dynamics," JHEP **0002** (2000) 020 [arXiv:hep-th/9912072].
5. N. Seiberg and E. Witten, "String theory and noncommutative geometry," JHEP **9909** (1999) 032 [arXiv:hep-th/9908142].
6. V. Gayral, J. M. Gracia-Bondía, B. Iochum, T. Schücker and J. C. Várilly, "Moyal planes are spectral triples," arXiv:hep-th/0307241.
7. I. Chepelev and R. Roiban, "Renormalization of quantum field theories on noncommutative $\mathbb{R}^d$. I: Scalars," JHEP **0005** (2000) 037 [arXiv:hep-th/9911098].
8. I. Chepelev and R. Roiban, "Convergence theorem for non-commutative Feynman graphs and renormalization," JHEP **0103** (2001) 001 [arXiv:hep-th/0008090].
9. K. G. Wilson and J. B. Kogut, "The Renormalization Group And The Epsilon Expansion," Phys. Rept. **12** (1974) 75.
10. J. Polchinski, "Renormalization And Effective Lagrangians," Nucl. Phys. B **231** (1984) 269.
11. E. Langmann, R. J. Szabo and K. Zarembo, "Exact solution of noncommutative field theory in background magnetic fields," Phys. Lett. B **569** (2003) 95 [arXiv:hep-th/0303082].
12. E. Langmann, R. J. Szabo and K. Zarembo, "Exact solution of quantum field theory on noncommutative phase spaces," arXiv:hep-th/0308043.
13. R. Koekoek and R. F. Swarttouw, "The Askey-scheme of hypergeometric orthogonal polynomials and its q-analogue," arXiv:math.CA/9602214.
14. H. Grosse and R. Wulkenhaar, "Power-counting theorem for non-local matrix models and renormalisation," arXiv:hep-th/0305066.
15. H. Grosse and R. Wulkenhaar, "Renormalisation of ϕ^4 theory on noncommutative $\mathbb{R}^2$ in the matrix base," JHEP **0312** (2003) 019 [arXiv:hep-th/0307017].
16. H. Grosse, R. Wulkenhaar, "Renormalisation of ϕ^4 theory on noncommutative $\mathbb{R}^4$ in the matrix base," in preparation.

Noncommutative Gauge Theories via Seiberg-Witten Map

Branislav Jurčo

Sektion Physik, Universität München,Theresienstr. 37, 80333 München, Germany
jurco@lmu.de

1 Introduction

Noncommutative coordinates were for the first time proposed by W. Heisenberg in 1930. He expressed, in his letter to Peierls [1], the hope that uncertainty relations for the coordinates might provide a natural cut-off for divergent integrals of QFT. First analysis of a quantum theory based on noncommutative coordinates was published by H. S. Snyder [2]. Although Pauli [3] considered his work to be mathematically ingenious he rejected it for physical reasons. Most lately noncommutative coordinates appeared in string theory [4]. It was argued that the D-brane world-volume becomes noncommutative in the presence of a nonzero background B-field.

In the first part of this contribution I will try to describe a systematic approach [5] to the construction of noncommutative gauge theories (NCGT) based on the Seiberg-Witten (SW) map [8]. This construction applies to any gauge group in particular to gauge groups of the standard model or GUTs [7], [8]. Starting with an ordinary commutative gauge theory we obtain a noncommutative gauge theory with the same degrees of freedom (fields) in the same multiplets of the gauge group as the original theory. The theory is expanded in the noncommutativity parameter and the noncommutativity presents itself in additional interaction terms in the action (finite number at each order of the noncommutativity parameter). This leads to a rich phenomenology and has consequences for the renormalizability. This construction works for any noncommutativity of coordinates which can be formulated as a deformation quantization (star-product) of usual commutative coordinates.

In the case of an abelian gauge group there is a nice relation to Kontsevich's formality [9] which allows an explicit construction of the SW map in all orders of the noncommutativity parameter [10] and captures some global features of abelian NCGT [11]. This will be discussed in the second part of the present contribution.

2 Noncommutative Spaces and $\star$-Products

Noncommutative coordinates are introduced by non-trivial commutation relations

$$[\widehat{x}^\mu, \widehat{x}^\nu] = \mathrm{i}\theta^{\mu\nu}(\widehat{x}), \qquad \mu, \nu = 1, \dots, n \, . \tag{1}$$

The choice of $\theta^{\mu\nu}(\widehat{x})$ is restricted by antisymmetry of the commutator and the Jacobi identity. Non-trivial examples which will be discussed here are

1. The canonical case:

$$[\widehat{x}^\mu, \widehat{x}^\nu] = \mathrm{i}\theta^{\mu\nu} . \tag{2}$$

$\theta^{\mu\nu}$ is $\widehat{x}$-independent. The canonical commutation relations in the quantum mechanics are of this type.

2. The Lie algebra case:

$$[\widehat{x}^\mu, \widehat{x}^\nu] = \mathrm{i}\theta^{\mu\nu}_\rho \widehat{x}^\rho . \tag{3}$$

Both examples can be understood as deformation quantizations of algebras of commutative coordinates equipped with appropriate Poisson brackets. In the case of canonical commutation relations this is done by the well-known Moyal-Weyl $\star$-product

$$\phi \star \psi(x) = \mathrm{e}^{\frac{i}{2}\frac{\partial}{\partial x^\mu}\theta^{\mu\nu}\frac{\partial}{\partial y^\nu}} \phi(x)\psi(y)|_{y=x} , \tag{4}$$

which corresponds to the symmetric ordering prescription. We will denote the algebra of functions equipped with the Moyal-Weyl $\star$-product as $\mathcal{A}$. The ordinary integral has the following crucial property

$$\int \mathrm{d}^n x (\phi \star \psi)(x) = \int \mathrm{d}^n x (\psi \star \phi)(x) = \int \mathrm{d}^n x \psi(x)\phi(x) . \tag{5}$$

Similarly in the Lie algebra case the symmetric ordering leads to a $\star$-product of the form

$$\phi \star \psi(x) = \mathrm{e}^{\frac{i}{2}x^\mu g_\mu (i\frac{\partial}{\partial y}, \frac{\partial}{\partial z})} \phi(y)\psi(z)|_{y=x=z} , \tag{6}$$

where g_μ is read of from BCH formula

$$\mathrm{e}^{\mathrm{i}k_\mu \widehat{x}^\mu} \mathrm{e}^{\mathrm{i}p_\nu \widehat{x}^\nu} = \mathrm{e}^{\mathrm{i}\left(k_\mu + p_\mu + \frac{1}{2} g_\mu(k,p)\right)\widehat{x}^\mu} . \tag{7}$$

This gives an algebra which is equivalent to the corresponding universal enveloping algebra $\mathcal{U}$.

3 Enveloping Algebra Valued NC Gauge Fields

Let us now fix the gauge group G with generators T_i and its representation V and introduce the noncommutative matter field $\widehat{\psi}(x)$. This is an element of $\mathcal{A} \otimes V$. We keep the "hat" to distinguish it from the corresponding ordinary commutative matter field $\psi(x)$. It transforms under the infinitesimal noncommutative gauge transformation $\delta_{\widehat{\Lambda}}$ as

$$\delta_{\widehat{\Lambda}} \widehat{\psi} = \mathrm{i}\widehat{\Lambda} \star \widehat{\psi} . \tag{8}$$

The noncommutative gauge parameter $\widehat{\Lambda}$ is a function with values in the universal enveloping algebra $\mathcal{U}$, i.e. $\widehat{\Lambda} \in \mathcal{A} \otimes \mathcal{U}$ and $\star$ in the above formula means the Moyal-Weyl $\star$-product tensored with the action of $\mathcal{U}$ on V.

The noncommutative vector potential $\widehat{A}_\mu$ is an element in $\mathcal{A} \otimes \mathcal{U}$ and it transforms under the noncommutative gauge transformation as

$$\delta_{\widehat{\Lambda}} \widehat{A}_\mu = \partial_\mu \widehat{\Lambda} + \mathrm{i}[\widehat{\Lambda}, \widehat{A}_\mu]_\star \,. \tag{9}$$

Here $[.,.]_\star$ has the obvious meaning, it is simply the commutator in $\mathcal{A} \otimes \mathcal{U}$.

The covariant noncommutative field strength $\widehat{F}_{\mu\nu} \in \mathcal{A} \otimes \mathcal{U}$ is defined in analogy with the commutative case

$$\widehat{F}_{\mu\nu} = \partial_\mu \widehat{A}_\mu - \partial_\nu \widehat{A}_\nu - \mathrm{i}[\widehat{A}_\mu, \widehat{A}_\nu]_\star \,. \tag{10}$$

As expected $\widehat{F}_{\mu\nu}$ transform covariantly

$$\delta_{\widehat{\Lambda}} \widehat{F}_{\mu\nu} = \mathrm{i}[\widehat{\Lambda}, \widehat{F}_{\mu\nu}]_\star \,. \tag{11}$$

Finally the covariant derivative D_μ of the matter field $\widehat{\psi} \in \mathcal{A} \otimes V$ is given as

$$\widehat{D}_\mu \widehat{\psi} = \partial_\mu \widehat{\psi} - \mathrm{i}\widehat{A}_\mu \star \widehat{\psi} \,, \tag{12}$$

where again the $\star$ means the Moyal-Weyl $\star$-product tensored with the action of $\mathcal{U}$ on V. So far we have increased the number of gauge fields drastically. Since a generic element of $\mathcal{A} \otimes \mathcal{U}$ is of the form

$$\alpha(x).1 + \alpha^i(x) T_i + \alpha^{ij}(x) : T_i T_j : + \ldots \tag{13}$$

our e.g. noncommutative vector potential contains infinitely many independent fields, coefficients of its expansion in the infinite linear basis of $\mathcal{U}$ (given e.g. by the symmetric ordering prescription). We will reduce the number of fields in the next section using the Seiberg-Witten map. That way we will obtain the same degrees of freedom as in the ordinary commutative gauge theory.

4 SW Map

So far we have defined noncommutative gauge fields $\widehat{\psi}$ and $\widehat{A}_\mu$ only by their transformation properties. However there is an explicit way how to construct them starting from ordinary commutative ones ψ and A_μ. Let us stress again that this construction works for every gauge group G and every of its representations V. What we are looking for are functions (these are the noncommutative fields and gauge parameter as introduced earlier)

$$\widehat{\psi}[\psi, A, \theta], \quad \widehat{A}_\mu[A, \theta], \quad \widehat{\Lambda}[\Lambda, A, \theta] \tag{14}$$

of ordinary commutative fields ψ, $A_\mu = A_\mu^i T_i$ and gauge parameter $\Lambda = \Lambda^i T_i$. Prescriptions (14) should have the property that in the commutative limit $\theta^{\mu\nu} \to 0$ we recover the corresponding commutative quantities and that they intertwine between commutative and noncommutative gauge transformation, (8) and (9) should follow from

$$\delta_\Lambda \psi = \mathrm{i}\Lambda\psi \tag{15}$$

and

$$\delta_\Lambda A_\mu = \partial_\mu \Lambda + \mathrm{i}[\Lambda, A_\mu] \,. \tag{16}$$

The argument of Seiberg and Witten [8] shows that such a map exists in the case of constant θ. We will discuss a different approach later. Here we only give (not unique) formulas for the SW map up to the first order in θ

$$\widehat{\psi} = \psi + \frac{1}{2}\theta^{\mu\nu} A_\mu \partial_\nu \psi + \frac{i}{8}\theta^{\mu\nu}[A_\mu, A_\nu]\psi + \dots , \tag{17}$$

$$\widehat{A}_\mu = A_\mu + \frac{1}{4}\theta^{\nu\rho}\{A_\rho, \partial_\nu A_\mu\} + \frac{1}{4}\theta^{\nu\rho}\{F_{\nu\mu}, A_\rho\} + \dots \tag{18}$$

and

$$\widehat{\Lambda} = \lambda + \frac{1}{4}\theta^{\mu\nu}\{\partial_\mu \Lambda, A_\nu\} + \dots . \tag{19}$$

As usually curly brackets denote the anticommutator in the gauge Lie algebra.

5 Consistency (Cocycle) Condition

There is an obvious consistency condition which is fulfilled by the noncommutative gauge parameters $\widehat{\Lambda}[\Lambda, A, \theta]$. This is simply the statement that the commutator of two noncommutative gauge transformations is again a noncommutative gauge transformation. Explicitly

$$[\widehat{\Lambda}_1, \widehat{\Lambda}_2]_\star = \widehat{\Lambda}_{12} + \mathrm{i}\delta_{\Lambda_2}\widehat{\Lambda}_1 - \mathrm{i}\delta_{\Lambda_1}\widehat{\Lambda}_2 . \tag{20}$$

Here we introduced the notation $\widehat{\Lambda}_1 = \widehat{\Lambda}[\Lambda_1, A, \theta]$, $\widehat{\Lambda}_2 = \widehat{\Lambda}[\Lambda_2, A, \theta]$ and $\widehat{\Lambda}_{12} = \widehat{\Lambda}[[\Lambda_1, \Lambda_2], A, \theta]$. The extra term $\mathrm{i}\delta_{\Lambda_2}\widehat{\Lambda}_1 - \mathrm{i}\delta_{\Lambda_1}\widehat{\Lambda}_2$ on the right hand side is due to the fact that noncommutative gauge parameters depend explicitly on the commutative gauge potential.

The most general solution to the consistency condition (20) up to the first order in θ is the following one:

$$\widehat{\Lambda} = \lambda + \frac{1}{4}\theta^{\mu\nu}\left(c\partial_\mu \Lambda A_\nu + (1-c)A_\nu \partial_\mu \Lambda\right) + \dots . \tag{21}$$

Here c is an arbitrary function on the space-time and the only freedom is in the field redefinitions of A and Λ. If we want hermiticity to be preserved we should take $\mathrm{Re}\, c = \frac{1}{2}$. Using consistency condition (20) we can solve for $\widehat{\Lambda}$ and hence for $\widehat{\psi}$ and $\widehat{A}$ using (8) and (9) order by order in θ.

6 Action

It is now tempting to write down the action for a NCYM theory in the form

$$S = \int \mathrm{d}^4x \frac{-1}{2g^2}\mathrm{Tr}\widehat{F}_{\mu\nu} \star \widehat{F}^{\mu\nu} + \overline{\widehat{\psi}} \star i\widehat{\not{D}} \star \widehat{\psi} \tag{22}$$

and use our formulas for the SW map given in Sect. 4 and find the first order θ-expanded action. Before we do that the following comment is in order. Our noncommutative field strength $\widehat{F}$ is now enveloping algebra valued and traces of $\widehat{F}_{\mu\nu} \star \widehat{F}^{\mu\nu}$

in different representations are not proportional anymore. So in general we would have to consider $\mathrm{Tr} = \sum_R c_R \mathrm{Tr}_R$, a weighted sum of traces over all irreducible representations such that it has correct commutative limit [12]. We need some additional physical criteria to decide what combination of coefficients c_R is the right one. These may include, e.g., renormalization, CPT invariance, anomaly freedom, or any kind of symmetry one might want to impose on the action. Without specifying our choice of Tr the first order θ-expanded kinetic term is given as

$$\begin{aligned} -\frac{1}{4}\mathrm{Tr}\widehat{F}_{\mu\nu} \star \widehat{F}^{\mu\nu} &\sim -\frac{1}{4}\mathrm{Tr}F_{\mu\nu}F^{\mu\nu} \\ +\frac{1}{8}\theta^{\rho\sigma}\mathrm{Tr}F_{\rho\sigma}F_{\mu\nu}F^{\mu\nu} &- \frac{1}{2}\theta^{\rho\sigma}\mathrm{Tr}F_{\mu\rho}F_{\nu\sigma}F^{\mu\nu} + \dots . \end{aligned} \tag{23}$$

Fermionic part of the action gives

$$\begin{aligned} \overline{\widehat{\psi}} \star (\gamma^\mu \widehat{D}_\mu - m) \star \widehat{\psi} &\sim \overline{\psi}(\gamma^\mu D_\mu - m)\psi \\ -\frac{1}{4}\theta^{\mu\nu}\overline{\psi}F_{\mu\nu}(\gamma^\rho D_\rho - m)\psi &+ \frac{1}{2}\theta^{\mu\nu}\overline{\psi}\gamma^\rho F_{\rho\mu}D_\nu\psi + \dots . \end{aligned} \tag{24}$$

We have used $\sim$ to indicate the equality under the integral.

Noncommutative Standatd Model [7], GUTs and C, P, T are discussed in [8]. See also contributions of P. Aschieri, and P. Schupp and J. Trampetić. The noncommutative Standard Model and GUTs are anomaly free [13]. Renormalizability of NCGT described here is discussed in [14].

7 Formality and SW Map in the Abelian Case

In the abelian case we can construct an explicit map that relates the ordinary gauge potential A_μ to the generalized noncommutative gauge potential $\hat{A}$. This can be done in some sense for any Poisson manifold M equipped with an arbitrary Poisson structure θ using Kontsevich's formality maps U_n [9]. $\{.,.\}$ denotes the corresponding Poisson bracket. We start with a semiclassical version, which is described by a formal version of the Moser lemma [15]. For this we introduce a one-parameter family of Poisson structures

$$\theta_t = \sum (-t)^n \theta (F\theta)^n = \theta(1 + tF\theta)^{-1} . \tag{25}$$

Here the multiplication is the ordinary matrix one, e.g., $(F\theta)_\mu^\nu = F_{\mu\sigma}\theta^{\sigma\nu}$. We use θ' for θ_1 and correspondingly $\{.,.\}'$ for $\{.,.\}_1$. Using this and the abelian gauge potential A_μ we construct a vector field $A_{\theta_t} = \theta_t^{\mu\nu} A_\nu \partial_\mu$ and a formal diffeomorphism of our manifold M

$$\rho_A = \mathrm{e}^{A_{\theta_t} + \partial_t} \mathrm{e}^{-\partial_t}|_{t=o} . \tag{26}$$

We will also need the following function

$$\tilde{\Lambda} = \sum_{n=o}^{\infty} \frac{(A_{\theta_t} + \partial_t)^n}{(n+1)!}(\Lambda)|_{t=o} . \tag{27}$$

It is easy to check that diffeomorphism ρ_A has the following nice property under the infinitesimal gauge transformation $A \mapsto A + \mathrm{d}\Lambda$

$$\rho_{A+dA}(f) = \rho_A(f) + \{\rho_A(f), \tilde{\Lambda}\} \tag{28}$$

for any function on M. It also has another important property: it intertwines between the two Poisson brackets $\{.,.\}$ and $\{.,.\}'$

$$\rho_A\{f,g\}' = \{\rho_A(f), \rho_A(g)\}\ . \tag{29}$$

We now apply the diffeomorphism ρ_A to our local coordinate x^μ and write the resulting function (in the case of constant, nondegenerate θ) in the form

$$\rho_A(x^\mu) = x^\mu + \theta^{\mu\nu}\tilde{A}_\nu[A,\theta]\ , \tag{30}$$

we obtain for $\tilde{A}_\mu[A,\theta]$ the semiclassical version of the noncommutative gauge transformation, i.e. in (9) $\hat{}$s will be replaced by $\tilde{}$s and the $\star$-commutator will be replaced by the Poisson bracket $\{.,.\}$. So we can interpret our ρ_A intertwining between θ and θ' as a semiclassical version of the SW map.

Thanks to Kontsevich's formality this can be quantized to obtain the full SW as a formal differential operator intertwining between the corresponding star-products $\star$ and $\star'$. We will need only few of the properties of formality maps U_n. These are graded antisymmetric multilinear maps which when applied to a collection of n polyvector fields $\alpha_1,\ldots,\alpha_n$ of degrees $k_1,\ldots,k_n$ produce polydifferential operators $U_n(\alpha_1,\ldots,\alpha_n)$ of degree $m = 2-2n+\sum k_i$. This means that $U_n(\alpha_1,\ldots,\alpha_n)$ can be applied to m functions $f_1,\ldots,f_m$ and the result is in general a non-zero function $U_n(\alpha_1,\ldots,\alpha_n)(f_1,\ldots,f_m)$.

Properties of the maps U_n allow us to generalize the Moyal-Weyl $\star$-product to any Poisson structure θ. We have the Kontsevich $\star$

$$f\star g = \sum_{n=0}^{\infty}\frac{(i\hbar)^n}{n!}U_n(\theta,\ldots,\theta)(f,g)\ . \tag{31}$$

Similarly we can lift any vector field ξ to obtain a differential operator $\bar{\xi}$

$$\bar{\xi} = \sum_{n=0}^{\infty}\frac{(i\hbar)^n}{n!}U_{n+1}(\xi,\theta\ldots,\theta) \tag{32}$$

and any function f to obtain a new function $\bar{f}$

$$\bar{f} = \sum_{n=0}^{\infty}\frac{(i\hbar)^n}{n!}U_{n+1}(f,\theta\ldots,\theta)\ . \tag{33}$$

If ξ preserves the Poisson bracket θ then $\bar{\xi}$ is a derivation of the corresponding $\star$-product. In particular if $\xi = \{f,.\}$ is the hamiltonian vector field generated by function f then $\bar{\xi} = \frac{1}{i\hbar}[\bar{f},.]_\star$. Equipped with formality maps U_n we can now quantize all Poisson structures θ_t

$$f\star_t g = \sum_{n=0}^{\infty}\frac{(i\hbar)^n}{n!}U_n(\theta_t,\ldots,\theta_t)(f,g) \tag{34}$$

to obtain a one-parameter family of star-products $\star_t$. Also we can lift vector fields A_{θ_t} to differential operators $\bar{A}_{\theta_t}$. Finally we can use these to write down the full SW map in complete analogy with the semiclassical one

$$\mathcal{D}_A = e^{\bar{A}_{\theta_t} + \partial_t} e^{-\partial_t}|_{t=o} \ , \tag{35}$$

$$\widehat{\Lambda} = \sum_{n=o}^{\infty} \frac{(\bar{A}_{\theta_t} + \partial_t)^n}{(n+1)!} (\bar{\Lambda})|_{t=o} \ . \tag{36}$$

Now the Formality guarantees that $\mathcal{D}_A$ has the following nice property under the infinitesimal gauge transformation $A \mapsto A + \mathrm{d}\Lambda$

$$\mathcal{D}_{A+d\Lambda}(f) = \mathcal{D}_A(f) - \mathrm{i}[\mathcal{D}_A(f), \widehat{\Lambda}]_\star \tag{37}$$

for any function on M. Also it intertwines between the $\star$ and $\star'$

$$\mathcal{D}_A(f \star' g) = \mathcal{D}_A(f) \star \mathcal{D}_A(g) \ . \tag{38}$$

In the case of constant and nondegenerate θ

$$\mathcal{D}_A(x^\mu) - x^\mu + \theta^{\mu\nu} \widehat{A}_\nu[A, \theta] \tag{39}$$

gives the SW map in the usual sense.

Note that SW map $\mathcal{D}_A$ was defined so far only locally, it depends on the choice of the gauge potential A. On the other hand the new $\star'$ is defined globally as it depends only on the field strength. So the SW map must have also some global meaning. We touch this briefly in the next section.

Wilson lines and the inverse SW map within this formalism were discussed in [16] for arbitrary θ.

8 Noncommuatative Line Bundles

Usually in the noncommutative geometry vector bundles are defined, in the spirit of the Serre-Swan theorem, as projective modules. Formalism developed in the previous section can be used to define noncommutative line bundles in a more geometric language using noncommutative transition functions. This will give a global meaning to SW map. Globally SW map is Morita equivalence of $\star$ and $\star'$. To formulate this in a more accurate way we have to take in all formulas in the previous section $\hbar A$ instead of A. If we do so then $\mathcal{D}_A \sim id + O(\hbar)$ and it is an equivalence of star-products in the usual sense. With this conventions θ' is only a formal Poisson structure as it depends explicitly on $\hbar$ and $\star$ and $\star'$ have the same classical limit θ, i.e., they are deformation quantizations of the same Poisson structure

The infinitesimal cocycle condition (20) leads in the abelian case to a finite one [11]

$$\widehat{G}[G_1, A - iG_2 dG_2^{-1}] \star \widehat{G}[G_2, A] = \widehat{G}[G_1.G_2, A] \ . \tag{40}$$

where G_i are finite commutative (abelian) gauge parameters. The infinitesimal covariance property of SW map (37) is replaced by the finite one

$$\mathcal{D}_{A-iGdG^{-1}}(f) \star \widehat{G}[G, A] = \widehat{G}[G, A] \star \mathcal{D}_A \ . \tag{41}$$

Let us now consider an ordinary commutative line bundle described locally with respect to some good covering U_i of M by transition functions G_{ij} and equipped with a connection locally given by a collection of one-forms A_i. We have

$$G_{ij}G_{jk} = G_{ik} \tag{42}$$

on $U_i \cap U_j \cap U_k$ and

$$G_{ij} = G_{ji}^{-1} \tag{43}$$

and

$$A_i = A_j - iG_{ij}dG_{jk}^{-1} \tag{44}$$

on $U_i \cap U_j$. We will use the following notation

$$\widehat{G}_{ij} = \widehat{G}[G_{ij}, A_j], \qquad \mathcal{D}_{A_i} = \mathcal{D}_i \ .$$

With this notation we rewrite the cocycle (40) and the covariance (41) conditions as

$$\widehat{G}_{ij} \star \widehat{G}_{jk} = \widehat{G}_{ij} \tag{45}$$

and

$$\mathcal{D}_i(f) \star \widehat{G}_{ij} = \widehat{G}_{ij} \star \mathcal{D}_j(f) \ . \tag{46}$$

Obviously $\widehat{G}_{ij}$ can be interpreted as noncommutative transition functions and $\mathcal{D}$ as "id + noncommutative connection".

For the rest of this contribution we will denote algebras of functions on M equipped with the star-products $\star$ and $\star'$, $\mathcal{A}$ and $\mathcal{A}'$ respectively. In contrary to the commutative case we can introduce the space of sections of our noncommutative line bundles in two different ways. We use $\mathcal{E}$ for the space of sections defined as collections of local functions ψ_i related on double intersections by

$$\psi_i = \widehat{G}_{ij} \star \psi_j \ , \tag{47}$$

whereas $\bar{\mathcal{E}}$ is defined using opposite multiplication

$$\bar{\psi}_i = \bar{\psi}_j \star \widehat{G}_{ji} \ . \tag{48}$$

From its definition it immediately follows that $\mathcal{E}$ is a right $\mathcal{A}$ module. Using (29) and (38) it also follows that it is an left $\mathcal{A}'$ module, the left $\mathcal{A}'$-action is given by $f : \psi_i \mapsto \mathcal{D}_i(f) \star \psi_i$. Left and right actions commute and we have the bimodule ${}_{\mathcal{A}'}\mathcal{E}_{\mathcal{A}}$. We have actually more [11]: ${}_{\mathcal{A}'}\mathcal{E}_{\mathcal{A}}$ is projective as a left $\mathcal{A}'$ and right $\mathcal{A}$ module and it is of finite type if the cover U_i is finite. The same of course holds for the bimodule ${}_{\mathcal{A}}\bar{\mathcal{E}}_{\mathcal{A}'}$ with the roles of $\mathcal{A}$ and $\mathcal{A}'$ interchanged. Finally, it can be shown that ${}_{\mathcal{A}'}\mathcal{E}_{\mathcal{A}}$ and ${}_{\mathcal{A}'}\mathcal{E}_{\mathcal{A}}$ are Morita equivalence bimodules, hence $\star$ and $\star'$ are Morita equivalent star-products. This observation can be used to classify Morita equivalent star-products. All this and more can the interested reader find in [11] or from a different perspective in [17]. Further generalization to the case of Poisson structures twisted by a three-form, which leads to the noncommutative gerbes is described in [18].

References

1. Letter of Heisenberg to Peierls (1930), In: Wolfgang Pauli, Scientific Correspondence, vol. II, ed by Karl von Meyenn (Springer-Verlag 1985) p 15
2. H. S. Snyder: Phys. Rev. **71**, 38 (1947)
3. Letter of Pauli to Bohr (1947), In: Wolfgang Pauli, Scientific Correspondence, vol. II, ed by Karl von Meyenn (Springer-Verlag 1985) p 414
4. C.-S. Chu, P.-M. Ho: Nucl. Phys. **B550**, 151 (1999)
   V. Schomerus: JHEP **9906**, 030 (1999)
5. B. Jurčo, S. Schraml, P. Schupp, J. Wess: Eur. Phys. J. **C17**, 521 (2000)
   B. Jurčo, L. Möller, S. Schraml, P. Schupp, J. Wess: Eur. Phys. J. **C21**, 383 (2001)
6. N. Seiberg, E. Witten: JHEP **9909**, 032 (1999)
7. X. Calmet, B. Jurčo, P. Schupp, J. Wess, M. Wohlgenannt: Eur. Phys. J. **C23**, 363 (2002)
8. P. Aschieri, B. Jurčo, P. Schupp, J. Wess: Nucl. Phys. **B651**, 45 (2003)
9. M. Kontsevich: q-alg/9709040
10. B. Jurčo, P. Schupp, J. Wess: Nucl. Phys. **B604**, 148 (2001)
11. B. Jurčo, P. Schupp, J. Wess: Lett. Math. Phys. **61**, 171 (2002)
12. W. Behr, N.G. Deshpande, G. Duplancic, P. Schupp, J. Trampetic, J. Wess: Eur. Phys. J. **C29**, 441 (2003);
    G. Duplancic, P. Schupp and J. Trampetic: Eur. Phys. J. C **32**, 141 (2003);
    P. Schupp, J. Trampetic, J. Wess and G. Raffelt: hep-ph/0212292, to be published in Eur. Phys. J. C;
    P. Minkowski, P. Schupp and J. Trampetic: hep-th/0302175;
    J. Trampetić: Acta Phys. Pol. **B33**, 4317 (2002).
13. C.P. Martin: Nucl. Phys. **B652**, 72 (2003)
    F. Brandt, C.P. Martin, F. Ruiz Ruiz: JHEP **0307**, 068 (2003)
14. M. Buric, V. Radovanovic: JHEP **0402**, 04 (2004)
    M. Buric, V. Radovanovic: JHEP **0210**, 074 (2002)
    R. Wulkenhaar: JHEP **0203**, 024 (2002)
15. J. K. Moser: Trans. Amer. Math. Soc. **120**, 286 (1965)
16. W. Behr, A. Sykora: hep-th/0312138
    W. Behr, A. Sykora: hep-th/0309145
17. H. Bursztyn: math. QA/0105001
    H. Bursztyn, S. Waldmann: Lett. Math. Phys. **53**, 349 (2000)
    H. Bursztyn, S. Waldmann: math. QA/0106178
    S. Waldmann: math. QA/0107112
18. P. Aschieri, I. Bakovic, B. Jurčo, P.Schupp: hep-th/0206101

The Noncommutative Standard Model and Forbidden Decays

Peter Schupp[1] and Josip Trampetić[2]

[1] International University Bremen, Campus Ring 8, 28759 Bremen, Germany
p.schupp@iu-bremen.de
[2] Theoretical Physics Division, Rudjer Bošković Institute, 10002 Zagreb, Croatia
josipt@rex.irb.hr

1 Introduction

In this contribution we discuss the Noncommutative Standard Model and the associated Standard Model-forbidden decays that can possibly serve as an experimental signature of space-time noncommutativity.

The idea of quantized space-time and noncommutative field theory has a long history that can be traced back to Heisenberg [1] and Snyder [2]. A noncommutative structure of spacetime can be introduced by promoting the usual spacetime coordinates x to noncommutative (NC) coordinates $\hat{x}$ with

$$[\hat{x}^\mu, \hat{x}^\nu] = \mathrm{i}\theta^{\mu\nu} , \tag{1}$$

were $\theta^{\mu\nu}$ is a real antisymmetric matrix. A noncommutativity scale Λ_{NC} is fixed by choosing dimensionless matrix elements $c^{\mu\nu} = \Lambda_{NC}^2 \,\theta^{\mu\nu}$ of order one. The original motivation to study such a scenario was the hope that the introduction of a fundamental scale could deal with the infinities of quantum field theory in a natural way [3]. The mathematical theory that replaces ordinary differential geometry in the description of quantized spacetime is noncommutative geometry [4]. A realization of the electroweak sector of the Standard Model in the framework of noncommutative geometry can be found [5], where the Higgs field plays the role of a gauge boson in the non-commutative (discrete) direction. This model is noncommutative in an extra internal direction but not in spacetime itself. It is therefore not the focus of the present work, although it can in principle be combined with it.

Noncommutativity of spacetime is very natural in string theory and can be understood as an effect of the interplay of closed and open strings. The commutation relation (1) enters in string theory through the Moyal-Weyl star product

$$f \star g = \sum_{n=0}^{\infty} \frac{\theta^{\mu_1\nu_1} \ldots \theta^{\mu_n\nu_n}}{(-2i)^n n!}\, \partial_{\mu_1} \ldots \partial_{\mu_n} f \cdot \partial_{\nu_1} \ldots \partial_{\nu_n} g \,. \tag{2}$$

For coordinate functions: $x^\mu \star x^\nu - x^\nu \star x^\mu = \mathrm{i}\theta^{\mu\nu}$. The tensor $\theta^{\mu\nu}$ is determined by a NS $B^{\mu\nu}$-field and the open string metric $G^{\mu\nu}$ [6], which both depend on a given closed string background. The effective physics on D-branes is most naturally captured by noncommutative gauge theory, but it can also be described by ordinary gauge theory. Both descriptions are related by the Seiberg-Witten (SW) map [8], which expresses noncommutative gauge fields in terms of fields with ordinary "commutative" gauge transformation properties.

The star product formalism in conjunction with the Seiberg-Witten map of fields naturally leads to a perturbative approach to field theory on noncommutative spaces. It is particulary well-suited to study Standard Model-forbidden processes induced by spacetime noncommutativity. This formalism can also be used to study non-perturbative noncommutative effects. In particular cases an algebraic approach may be more convenient for actual computations but the structure of the star product results can still be a useful guideline.

A method for implementing non-Abelian $SU(N)$ Yang-Mills theories on noncommutative spacetime has been proposed in [8, 9, 10, 11]. In [12] this method has been applied to the full Standard Model of particle physics [13] resulting in a minimal non-commutative extension of the Standard Model with structure group $SU(3)_C \times SU(2)_L \times U(1)_Y$ and with the same fields and the same number of coupling parameters as in the original Standard Model. It is the only known approach that allows to build models of the electroweak sector directly based on the structure group $SU(2)_L \times U(1)_Y$ in a noncommutative background. Previously only $U(N)$ gauge theories were under control, and it was thus only possible to consider extensions of the Standard Model. Furthermore there were problems with the allowed charges and with the gauge invariance of the Yukawa terms in the action.

In an alternative approach to the construction of a noncommutative generalization of the Standard Model the usual problems of noncommutative model buildings, i.e., charge quantization and the restriction of the noncommutative gauge group are circumvented by enlarging the gauge group to $U(3) \times U(2) \times U(1)$ [15]. The hypercharges and the electric charges are quantized to the correct values of the usual quarks and leptons, however, there are some open issues with the NC gauge invariance of the Yukawa terms. In principle the two approaches can be combined.

2 The Noncommutative Standard Model

2.1 Noncommutative Yang-Mills

Consider an ordinary Yang-Mills action with gauge group G, where G is a compact simple Lie group, and a fermion multiplet Ψ

$$S = \int \mathrm{d}^4x \, \frac{-1}{2g^2} \, \mathrm{Tr}(F_{\mu\nu} F^{\mu\nu}) + \overline{\Psi} i \not{D} \Psi \tag{3}$$

This action is gauge invariant under

$$\delta\Psi = \mathrm{i}\rho_\Psi(\Lambda)\Psi \tag{4}$$

where ρ_Ψ is the representation of G determined by the multiplet Ψ. The noncommutative generalization of (3) is given by

$$\widehat{S} = \int \mathrm{d}^4x \, \frac{-1}{2g^2} \, \mathrm{Tr}(\widehat{F}_{\mu\nu} \star \widehat{F}^{\mu\nu}) + \overline{\widehat{\Psi}} \star i\widehat{\not{D}}\widehat{\Psi} \tag{5}$$

where the noncommutative field strength $\widehat{F}$ is defined by

$$\widehat{F}_{\mu\nu} = \partial_\mu \widehat{A}_\nu - \partial_\nu \widehat{A}_\mu - \mathrm{i}[\widehat{A}_\mu, \widehat{A}_\nu]_\star \,. \tag{6}$$

The covariant derivative is given by

$$\widehat{D}_\mu \widehat{\Psi} = \partial_\mu \widehat{\Psi} - \mathrm{i}\rho_\Psi(\widehat{A}_\mu) \star \widehat{\Psi} \, . \tag{7}$$

The action (1) is invariant under the noncommutative gauge transformations

$$\hat{\delta}\widehat{\Psi} = \mathrm{i}\rho_\Psi(\widehat{\Lambda}) \star \widehat{\Psi}, \qquad \hat{\delta}\widehat{A}_\mu = \partial_\mu \widehat{\Lambda} + \mathrm{i}[\widehat{\Lambda}, \widehat{A}_\mu]_\star, \qquad \hat{\delta}\widehat{F}_{\mu\nu} = \mathrm{i}[\widehat{\Lambda}, \widehat{F}_{\mu\nu}]_\star \, . \tag{8}$$

If the gauge fields are assumed to be Lie-algebra valued, it appears that only $U(N)$ in the fundamental representation is consistent with noncommutative gauge transformations: Only in this case the commutator

$$[\widehat{\Lambda}, \widehat{\Lambda}']_\star = \frac{1}{2}\{\Lambda_a(x) \stackrel{\star}{,} \Lambda'_b(x)\}[T^a, T^b] + \frac{1}{2}[\Lambda_a(x) \stackrel{\star}{,} \Lambda'_b(x)]\{T^a, T^b\} \tag{9}$$

of two Lie algebra-valued non-commutative gauge parameters $\widehat{\Lambda} = \Lambda_a(x)T^a$ and $\widehat{\Lambda}' = \Lambda'_a(x)T^a$ again closes in the Lie algebra [8, 9]. The fact that a $U(1)$ factor cannot easily be decoupled from NC $U(N)$, can also be seen by noting the interactions of $SU(N)$ gluons and $U(1)$ (hyper) photons in NC Yang-Mills theory [14]. For a sensible phenomenology of particle physics on noncommutative spacetime we need to be able to use other gauge groups. Furthermore, in the special case of $U(1)$ a similar argument show that charges are quantized to values $\pm e$ and zero. These restrictions can be avoided if we allow gauge fields and gauge transformation parameters that are valued in the enveloping algebra of the gauge group.

$$\widehat{\Lambda} = \Lambda^0_a(x)T^a + \Lambda^1_{ab}(x)T^aT^b + \Lambda^2_{abc}(x)T^aT^bT^c + \ldots \tag{10}$$

A priori we now face the problem that we have an infinite number of parameters $\Lambda^0_a(x)$, $\Lambda^1_{ab}(x)$, $\Lambda^2_{abc}(x), \ldots$, but these are not independent. They can in fact all be expressed in terms of the right number of classical parameters and fields via the Seiberg-Witten maps. The non-commutative fields $\widehat{A}$, $\widehat{\Psi}$ and non-commutative gauge parameter $\widehat{\Lambda}$ can be expressed as "towers" built upon the corresponding ordinary fields A, Ψ and ordinary gauge parameter Λ. The Seiberg-Witten maps [16] express non-commutative fields and parameters as local functions of the ordinary fields and parameters,

$$\widehat{A}_\xi[A] = A_\xi + \frac{1}{4}\theta^{\mu\nu}\{A_\nu, \partial_\mu A_\xi\} + \frac{1}{4}\theta^{\mu\nu}\{F_{\mu\xi}, A_\nu\} + \mathcal{O}(\theta^2) \tag{11}$$

$$\widehat{\Psi}[\Psi, A] = \Psi + \frac{1}{2}\theta^{\mu\nu}\rho_\Psi(A_\nu)\partial_\mu\Psi + \frac{i}{8}\theta^{\mu\nu}[\rho_\Psi(A_\mu), \rho_\Psi(A_\nu)]\Psi + \mathcal{O}(\theta^2) \tag{12}$$

$$\widehat{\Lambda}[\Lambda, A] = \Lambda + \frac{1}{4}\theta^{\mu\nu}\{A_\nu, \partial_\mu \Lambda\} + \mathcal{O}(\theta^2) \tag{13}$$

where $F_{\mu\nu} = \partial_\mu A_\nu - \partial_\nu A_\mu - \mathrm{i}[A_\mu, A_\nu]$ is the ordinary field strength. The Seiberg-Witten maps have the remarkable property that ordinary gauge transformations $\delta A_\mu = \partial_\mu \Lambda + \mathrm{i}[\Lambda, A_\mu]$ and $\delta\Psi = \mathrm{i}\Lambda \cdot \Psi$ induce non-commutative gauge transformations (8) of the fields $\widehat{A}$, $\widehat{\Psi}$ with gauge parameter $\widehat{\Lambda}$.

2.2 Standard Model Fields

The Standard Model gauge group is $G_{SM} = SU(3)_C \times SU(2)_L \times U(1)_Y$. The gauge potential A_μ and gauge parameter Λ are valued in Lie(G_{SM}):

$$A_\nu = g'\mathcal{A}_\nu(x)Y + g\sum_{a=1}^{3} B_{\nu a}(x)T_L^a + g_S\sum_{b=1}^{8} G_{\nu b}(x)T_S^b \tag{14}$$

$$\Lambda = g'\alpha(x)Y + g\sum_{a=1}^{3} \alpha_a^L(x)T_L^a + g_S\sum_{b=1}^{8} \alpha_b^S(x)T_S^b\,, \tag{15}$$

where Y, T_L^a, T_S^b are the generators of $u(1)_Y$, $su(2)_L$ and $su(3)_C$ respectively. In addition to the gauge bosons we have three families of left- and right-handed fermions and a Higgs doublet

$$\Psi_L^{(i)} = \begin{pmatrix} L_L^{(i)} \\ Q_L^{(i)} \end{pmatrix}, \qquad \Psi_R^{(i)} = \begin{pmatrix} e_R^{(i)} \\ u_R^{(i)} \\ d_R^{(i)} \end{pmatrix}, \qquad \Phi = \begin{pmatrix} \phi^+ \\ \phi^0 \end{pmatrix} \tag{16}$$

where i = 1, 2, 3 is the generation index and ϕ^+, ϕ^0 are complex scalar fields. We shall now apply the appropriate SW maps to the fields A_μ, $\Psi^{(i)}$, Φ, expand to first order in θ and write the corresponding NC Yang-Mills action [12].

2.3 Noncommutative Yukawa Terms

Special care must be taken in the definition of the trace in the gauge kinetic terms and in the construction of covariant Yukawa terms. The classical Higgs field $\Phi(x)$ commutes with the generators of the $U(1)$ and $SU(3)$ gauge transformations. It also commutes with the corresponding gauge parameters. The latter is no longer true in the noncommutative setting: The coefficients $\alpha(x)$ and $\alpha_b^S(x)$ of the $U(1)$ and $SU(3)$ generators in the gauge parameter are functions and therefore do not $\star$-commute with the Higgs field. This makes it hard to write down covariant Yukawa terms. The solution to the problem is the hybrid SW map [17]

$$\begin{aligned}\widehat{\Phi}[\Phi, A, A'] = {} & \Phi + \frac{1}{2}\theta^{\mu\nu} A_\nu\Big(\partial_\mu\Phi - \frac{i}{2}(A_\mu\Phi - \Phi A'_\mu)\Big) \\ & + \frac{1}{2}\theta^{\mu\nu}\Big(\partial_\mu\Phi - \frac{i}{2}(A_\mu\Phi - \Phi A'_\mu)\Big)A'_\nu + \mathcal{O}(\theta^2)\,. \end{aligned} \tag{17}$$

By choosing appropriate representations it allows us to assign separate left and right charges to the noncommutative Higgs field $\widehat{\Phi}$ that add up to its usual charge [12]. Here are two examples:

$$\begin{array}{cccccccccc} & \overline{\widehat{L}}_L & \star & \rho_L(\widehat{\Phi}) & \star & \widehat{e}_R & \qquad \overline{\widehat{Q}}_L & \star & \rho_Q(\widehat{\Phi}) & \star & \widehat{d}_R \\ Y = & 1/2 & & \underbrace{-1/2+1}_{1/2} & & -1 & \qquad -1/6 & & \underbrace{1/6+1/3}_{1/2} & & -1/3 \end{array} \tag{18}$$

We see here two instances of a general rule: The gauge fields in the SW maps and in the covariant derivatives inherit their representation (charge for Y, trivial or fundamental representation for T_L^a, T_S^b) from the fermion fields $\Psi^{(i)}$ to their left and to their right.

In GUTs it is more natural to first combine the left-handed and right-handed fermion fields and then contract the resulting expression with Higgs fields to obtain a gauge invariant Yukawa term. Consequently in NC GUTs we need to use the hybrid SW map for the left-handed fermion fields and then sandwich them between the NC Higgs on the left and the right-handed fermion fields on the right [18].

2.4 The Minimal NCSM

The trace in the kinetic terms for the gauge bosons is not unique, it depends on the choice of representation. This would not matter if the gauge fields were Lie algebra valued, but in the noncommutative case they live in the enveloping algebra. The simplest choice is a sum of three traces over the $U(1)$, $SU(2)$, $SU(3)$ sectors with $Y = \frac{1}{2}\begin{pmatrix} 1 & 0 \\ 0 & -1 \end{pmatrix}$ in the definition of $\mathbf{Tr_1}$ and the fundamental representation for $\mathbf{Tr_2}$ and $\mathbf{Tr_3}$. This leads to the following gauge kinetic terms

$$\begin{aligned} S_{\text{gauge}} &= -\frac{1}{4}\int \mathrm{d}^4x\, f_{\mu\nu}f^{\mu\nu} - \frac{1}{2}\,\mathrm{Tr}\int \mathrm{d}^4x\, F^L_{\mu\nu}F^{L\mu\nu} \\ &\quad -\frac{1}{2}\,\mathrm{Tr}\int \mathrm{d}^4x\, F^S_{\mu\nu}F^{S\mu\nu} + \frac{1}{4}g_S\,\theta^{\mu\nu}\,\mathrm{Tr}\int \mathrm{d}^4x\, F^S_{\mu\nu}F^S_{\rho\sigma}F^{S\rho\sigma} \\ &\quad -g_S\,\theta^{\mu\nu}\,\mathrm{Tr}\int \mathrm{d}^4x\, F^S_{\mu\rho}F^S_{\nu\sigma}F^{S\rho\sigma} + \mathcal{O}(\theta^2)\,. \end{aligned} \tag{19}$$

Note, that there are no new triple f or triple F^L-terms.

The full action of the Minimal Noncommutative Standard Model is [12]:

$$\begin{aligned} S_{NCSM} &= \int \mathrm{d}^4x \sum_{i=1}^{3} \overline{\widehat{\Psi}}{}^{(i)}_L \star i\widehat{\not{D}}\widehat{\Psi}^{(i)}_L + \int \mathrm{d}^4x \sum_{i=1}^{3} \overline{\widehat{\Psi}}{}^{(i)}_R \star i\widehat{\not{D}}\widehat{\Psi}^{(i)}_R \\ &\quad - \int \mathrm{d}^4x \frac{1}{2g'}\,\mathbf{Tr_1}\widehat{F}_{\mu\nu} \star \widehat{F}^{\mu\nu} - \int \mathrm{d}^4x \frac{1}{2g}\,\mathbf{Tr_2}\widehat{F}_{\mu\nu} \star \widehat{F}^{\mu\nu} \\ &\quad - \int \mathrm{d}^4x \frac{1}{2g_S}\,\mathbf{Tr_3}\widehat{F}_{\mu\nu} \star \widehat{F}^{\mu\nu} + \int \mathrm{d}^4x \Big(\rho_0(\widehat{D}_\mu\widehat{\Phi})^\dagger \star \rho_0(\widehat{D}^\mu\widehat{\Phi}) \\ &\quad -\mu^2\rho_0(\widehat{\Phi})^\dagger \star \rho_0(\widehat{\Phi}) - \lambda\rho_0(\widehat{\Phi})^\dagger \star \rho_0(\widehat{\Phi}) \star \rho_0(\widehat{\Phi})^\dagger \star \rho_0(\widehat{\Phi}) \Big) \\ &\quad - \int \mathrm{d}^4x \Big(\sum_{i,j=1}^{3} W^{ij} \Big((\overline{\widehat{L}}{}^{(i)}_L \star \rho_L(\widehat{\Phi})) \star \widehat{e}^{(j)}_R + \overline{\widehat{e}}{}^{(i)}_R \star (\rho_L(\widehat{\Phi})^\dagger \star \widehat{L}^{(j)}_L) \Big) \\ &\quad + \sum_{i,j=1}^{3} G^{ij}_u \Big((\overline{\widehat{Q}}{}^{(i)}_L \star \rho_{\bar{Q}}(\widehat{\bar{\Phi}})) \star \widehat{u}^{(j)}_R + \overline{\widehat{u}}{}^{(i)}_R \star (\rho_{\bar{Q}}(\widehat{\bar{\Phi}})^\dagger \star \widehat{Q}^{(j)}_L) \Big) \\ &\quad + \sum_{i,j=1}^{3} G^{ij}_d \Big((\overline{\widehat{Q}}{}^{(i)}_L \star \rho_Q(\widehat{\Phi})) \star \widehat{d}^{(j)}_R + \overline{\widehat{d}}{}^{(i)}_R \star (\rho_Q(\widehat{\Phi})^\dagger \star \widehat{Q}^{(j)}_L) \Big) \Big) \end{aligned} \tag{20}$$

where W^{ij}, G^{ij}_u, G^{ij}_d are Yukawa couplings and $\bar{\Phi} = i\tau_2\Phi^*$.

2.5 Non-Minimal Versions of the NCSM

We can use the freedom in the choice of traces in kinetic terms for the gauge fields to construct non-minimal versions of the NCSM. The general form of the gauge kinetic terms is [12, 18]

$$S_{\text{gauge}} = -\frac{1}{2}\int \mathrm{d}^4x \sum_\rho \kappa_\rho \mathrm{Tr}\Big(\rho(\widehat{F}_{\mu\nu}) \star \rho(\widehat{F}^{\mu\nu})\Big)\,, \tag{21}$$

where the sum is over all unitary irreducible inequivalent representations ρ of the gauge group G. The freedom in the kinetic terms is parametrized by real coefficients κ_ρ that are subject to the constraints

$$\frac{1}{g_I^2} = \sum_\rho \kappa_\rho \mathrm{Tr}\Big(\rho(T_I^a)\rho(T_I^a)\Big) , \tag{22}$$

where g_I and T_I^a are the usual "commutative" coupling constants and generators of $U(1)_Y$, $SU(2)_L$, $SU(3)_C$, respectively. Both formulas can also be written more compactly as

$$S_{\text{gauge}} = -\frac{1}{2}\int d^4x \,\mathbf{Tr}\frac{1}{\mathbf{G}^2}\widehat{F}_{\mu\nu}\star\widehat{F}^{\mu\nu} , \qquad \frac{1}{g_I^2} = \mathbf{Tr}\frac{1}{\mathbf{G}^2}T_I^a T_I^a , \tag{23}$$

where the trace **Tr** is again over all representations and **G** is an operator that commutes with all generators T_I^a and encodes the coupling constants. The possibility of new parameters in gauge theories on noncommutative spacetime is a consequence of the fact that the gauge fields are in general valued in the enveloping algebra of the gauge group.

The expansion in θ is at the same time an expansion in the momenta. The θ-expanded action can thus be interpreted as a low energy effective action. In such an effective low energy description it is natural to expect that all representations that appear in the commutative theory (matter multiplets and adjoint representation) are important. We should therefore consider the non-minimal version of the NCSM with non-zero coefficients κ_ρ at least for these representations. The new parameters in the non-minimal NCSM can be restricted by considering GUTs on noncommutative spacetime [18].

2.6 Properties of the NCSM

The key properties of the Noncommutative Standard Model (NCSM) are:

- The known elementary particles can be accomodated with their correct charges as in the original "commutative" Standard Model. There is no need to introduce new fields.
- The noncommutative Higgs field in the minimal NCSM has distinct left and right hyper (and colour) charges, whose sum are the regular SM charges. This is necesarry to obtain gauge invariant Yukawa terms.
- In versions of the NCSM that arise from NC GUTs it is more natural to equip the neutrino (and other left-handed fermion fields) with left and right charges. The neutrino can in principle couple to photons in the presence of spacetime noncommutativity, even though its total charge is zero.
- Noncommutative gauge invariance implies the existence of many new couplings of gauge fields: Abelian gauge bosons self-interact via a star-commutator term that resembles the self-interaction of non-abelian gauge bosons and we find many new interaction terms that involve gauge fields as a consequence of the Seiberg-Witten maps.
- The perturbation theory is based on the free commutative action. Assymptotic states are the plane-wave eigenstates of the free commutative Hamiltonian. Both ordinary interaction terms and interactions due to noncommutative effects are

treated on equal footing. This makes it particularly simple to derive Feynman rules and compute the invariant matrix elements of fundamental processes. While there is no need to reinvent perturbation theory, care has to be taken nevertheless for a time-like θ-tensor to avoid problems with unitarity.

- Violation of spacetime symmetries and in particular of angular momentum conservation and discrete symmetries like P, CP, and possibly even CPT can be induced by spacetime noncommutativity. This symmetry breaking is spontaneous in the sense that it is with respect to a fixed θ-"vacuum". (As long as θ is also transformed as a tensor, everything is fully covariant.)

The physically interpretation of these violations of conservation laws is that angular momentum (and even energy-momentum) can be transferred to the noncommuative spacetime structure in much the same way as energy can be carried away from binary stars by gravitational waves.

3 Standard Model Forbidden Processes

A general feature of gauge theories on noncommutative spacetime is the appearance of many new interactions including Standard Model-forbidden processes. The origin of these new interactions is two-fold: One source are the star products that let abelian gauge theory on NC spacetime resemble Yang-Mills theory with the possibility of triple and quadruple gauge boson vertices. The other source are the gauge fields in the Seiberg-Witten maps for the gauge and matter fields. These can be pictured as a cloud of gauge bosons that dress the original 'commutative' fields and that have their origin in the interaction between gauge fields and the NC structure of spacetime. One of the perhaps most striking effects and a possible signature of spacetime noncommutativity is the spontaneous breaking of continuous and discrete spacetime symmetries.

3.1 Triple Gauge Boson Couplings

New anomalous triple gauge boson interactions that are usually forbidden by Lorentz invariance, angular moment conservation and Bose statistics (Yang theorem) can arise within the framework of the non-minimal noncommutative standard model [19, 20], and also in the alternative approach to the NCSM given in [15].

The new triple gauge boson (TGB) terms in the action have the following form [19, 20]:

$$
\begin{aligned}
S_{gauge} &= -\frac{1}{4}\int \mathrm{d}^4x\, f_{\mu\nu} f^{\mu\nu} \\
&- \frac{1}{2}\int \mathrm{d}^4x\, \mathrm{Tr}\left(F_{\mu\nu}F^{\mu\nu}\right) - \frac{1}{2}\int \mathrm{d}^4x\, \mathrm{Tr}\left(G_{\mu\nu}G^{\mu\nu}\right) \\
&+ g_s\, \theta^{\rho\tau}\!\int \mathrm{d}^4x\, \mathrm{Tr}\left(\frac{1}{4}G_{\rho\tau}G_{\mu\nu} - G_{\mu\rho}G_{\nu\tau}\right) G^{\mu\nu} \\
&+ g'^3 \kappa_1 \theta^{\rho\tau}\!\int \mathrm{d}^4x \left(\frac{1}{4} f_{\rho\tau} f_{\mu\nu} - f_{\mu\rho} f_{\nu\tau}\right) f^{\mu\nu}
\end{aligned}
$$

$$+ g' g^2 \kappa_2 \, \theta^{\rho\tau} \int d^4x \sum_{a=1}^{3} \left[\left(\frac{1}{4} f_{\rho\tau} F^a_{\mu\nu} - f_{\mu\rho} F^a_{\nu\tau} \right) F^{\mu\nu,a} + c.p. \right]$$

$$+ g' g_s^2 \kappa_3 \, \theta^{\rho\tau} \int d^4x \sum_{b=1}^{8} \left[\left(\frac{1}{4} f_{\rho\tau} G^b_{\mu\nu} - f_{\mu\rho} G^b_{\nu\tau} \right) G^{\mu\nu,b} + c.p. \right] , \qquad (24)$$

where *c.p.* means cyclic permutations. Here $f_{\mu\nu}$, $F^a_{\mu\nu}$ and $G^b_{\mu\nu}$ are the physical field strengths corresponding to the groups $\mathrm{U(1)_Y}$, $\mathrm{SU(2)_L}$ and $\mathrm{SU(3)_C}$, respectively. The constants κ_1, κ_2 and κ_3 are functions of $1/g_i^2$ $(i = 1, \ldots, 6)$:

$$\begin{aligned}
\kappa_1 &= -\frac{1}{g_1^2} - \frac{1}{4g_2^2} + \frac{8}{9g_3^2} - \frac{1}{9g_4^2} + \frac{1}{36g_5^2} + \frac{1}{4g_6^2} , \\
\kappa_2 &= -\frac{1}{4g_2^2} + \frac{1}{4g_5^2} + \frac{1}{4g_6^2} , \\
\kappa_3 &= +\frac{1}{3g_3^2} - \frac{1}{6g_4^2} + \frac{1}{6g_5^2} .
\end{aligned} \qquad (25)$$

The g_i are the coupling constants of the non-commutative electroweak sector up to first order in θ. The appearance of new coupling constants beyond those of the standard model reflect a freedom in the strength of the new TGB couplings. Matching the SM action at zeroth order in θ, three consistency conditions are imposed on (24):

$$\begin{aligned}
\frac{1}{g'^2} &= \frac{2}{g_1^2} + \frac{1}{g_2^2} + \frac{8}{3g_3^2} + \frac{2}{3g_4^2} + \frac{1}{3g_5^2} + \frac{1}{g_6^2} , \\
\frac{1}{g^2} &= \frac{1}{g_2^2} + \frac{3}{g_5^2} + \frac{1}{g_6^2} , \\
\frac{1}{g_s^2} &= \frac{1}{g_3^2} + \frac{1}{g_4^2} + \frac{2}{g_5^2} .
\end{aligned} \qquad (26)$$

From the action (24) we extract neutral triple-gauge boson terms which are not present in the SM Lagrangian. The allowed range of values for the coupling constants

$$\begin{aligned}
\mathrm{K}_{\gamma\gamma\gamma} &= \frac{1}{2}\, g g' (\kappa_1 + 3\kappa_2) , \\
\mathrm{K}_{Z\gamma\gamma} &= \frac{1}{2} \left[g'^2 \kappa_1 + \left(g'^2 - 2g^2 \right) \kappa_2 \right] , \\
\mathrm{K}_{Zgg} &= \frac{g_s^2}{2} \left[1 + (\frac{g'}{g})^2 \right] \kappa_3 ,
\end{aligned}$$

compatible with conditions (26) and the requirement that $1/g_i^2 > 0$ are plotted in Fig. 1. The remaining three coupling constants $\mathrm{K}_{ZZ\gamma}$, K_{ZZZ} and $\mathrm{K}_{\gamma gg}$, are uniquely fixed by the equations

$$\begin{aligned}
\mathrm{K}_{ZZ\gamma} &= \frac{1}{2} \left(\frac{g}{g'} - 3\frac{g'}{g} \right) \mathrm{K}_{Z\gamma\gamma} - \frac{1}{2} \left(1 - \frac{g'^2}{g^2} \right) \mathrm{K}_{\gamma\gamma\gamma} , \\
\mathrm{K}_{ZZZ} &= \frac{3}{2} \left(1 - \frac{g'^2}{g^2} \right) \mathrm{K}_{Z\gamma\gamma} - \frac{1}{2} \frac{g'^2}{g^2} \left(3 - \frac{g'^2}{g^2} \right) \mathrm{K}_{\gamma\gamma\gamma} , \\
\mathrm{K}_{\gamma gg} &= -\frac{g}{g'} \mathrm{K}_{Zgg} .
\end{aligned} \qquad (27)$$

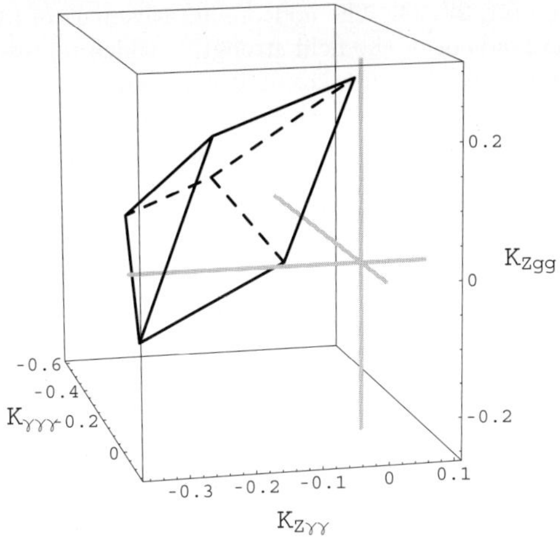

Fig. 1. The three-dimensional pentahedron that bounds possible values for the coupling constants $\mathrm{K}_{\gamma\gamma\gamma}$, $\mathrm{K}_{Z\gamma\gamma}$ and K_{Zgg} at the M_Z scale

We see that any combination of two TGB coupling constants does not vanish simultaneously due to the constraints set by the values of the SM coupling constants at the M_Z scale [20].

We conclude that the gauge sector is a possible place for an experimental search for noncommuative effects. The experimental discovery of the kinematically allowed $Z \to \gamma\gamma$ decay would indicate a violation of the Yang theorem and would be a possible signal of spacetime non-commutativity.

3.2 Electromagnetic Properties of Neutrinos

In the presence of spacetime noncommutativity, neutral particles can couple to gauge bosons via a $\star$-commutator

$$D_\mu^{\mathrm{NC}}\widehat{\psi} = \partial_\mu\widehat{\psi} - \mathrm{i}e\widehat{A}_\mu \star \widehat{\psi} + \mathrm{i}e\widehat{\psi} \star \widehat{A}_\mu \,. \tag{28}$$

Expanding the $\star$-product in (28) to first order in the antisymmetric (Poisson) tensor $\theta^{\mu\nu}$, we find the following covariant derivative on neutral spinor fields:

$$D_\mu^{\mathrm{NC}}\widehat{\psi} = \partial_\mu\widehat{\psi} + e\theta^{\nu\rho}\,\partial_\nu\widehat{A}_\mu\,\partial_\rho\widehat{\psi} \,. \tag{29}$$

We treat $\theta^{\mu\nu}$ as a constant background field of strength $|\theta^{\mu\nu}| = 1/\Lambda_{\mathrm{NC}}^2$ that models the non-commutative structure of spacetime in the neighborhood of the interaction region. As θ is not invariant under Lorentz transformations, the neutrino field can pick up angular momentum in the interaction. The gauge-invariant action for a neutral fermion that couples to an Abelian gauge boson via (29) is

$$S = \int \mathrm{d}^4x\,\bar{\psi}\left[\Big(\mathrm{i}\gamma^\mu\partial_\mu - m\Big) - \frac{e}{2}F_{\mu\nu}\left(\mathrm{i}\theta^{\mu\nu\rho}\partial_\rho - \theta^{\mu\nu}m\right)\right]\psi \,, \tag{30}$$

$$\theta^{\mu\nu\rho} = \theta^{\mu\nu}\gamma^\rho + \theta^{\nu\rho}\gamma^\mu + \theta^{\rho\mu}\gamma^\nu \,,$$

up to first order in θ [21, 22, 23]. The noncommutative part of (30) induces a *force*, proportional to the gradient of the field strengths, which represents an interaction of Stern-Gerlach type [24]. This interaction is non-zero even for $m_\nu = 0$ and in this case reduces to the coupling between the stress–energy tensor of the neutrino $T^{\mu\nu}$ and the symmetric tensor composed from θ and F [21]. The following is based on [22].

Neutrino Dipole Moments in the Mass-Extended Standard Model

Following the general arguments of [25, 26, 27, 28] only the Dirac neutrino can have a magnetic moment. However, the transition matrix elements relevant for $\nu_i \longrightarrow \nu_j$ may exist for both Dirac and Majorana neutrinos. In the neutrino-mass extended standard model [28], the photon–neutrino effective vertex is determined from the $\nu_i \longrightarrow \nu_j + \gamma$ transition, which is generated through 1-loop electroweak process that arise from the so-called "neutrino–penguin" diagrams via the exchange of $\lambda = e, \mu, \tau$ leptons and weak bosons, and is given by [25, 23]

$$\begin{aligned} J_\mu^{\rm eff}(\gamma\nu\bar{\nu})\epsilon^\mu(q) = &\left\{F_1(q^2)\bar{\nu}_j(p')(\gamma_\mu q^2 - q_\mu \not{q})\nu_i(p)_L \right.\\ &- \mathrm{i}F_2(q^2)\left[m_{\nu_j}\bar{\nu}_j(p')\sigma_{\mu\nu}q^\nu \nu_i(p)_L \right.\\ &\left.\left. + \; m_{\nu_i}\bar{\nu}_j(p')\sigma_{\mu\nu}q^\nu \nu_i(p)_R\right]\right\}\epsilon^\mu(q)\,. \end{aligned} \tag{31}$$

The above effective interaction is invariant under the electromagnetic gauge transformation. The first term in (31) vanishes identically for real photon due to the electromagnetic gauge condition.

From the general decomposition of the second term of the transition matrix element T (31),

$$\mathrm{T} = -\mathrm{i}\epsilon^\mu(q)\bar{\nu}(p')\left[A(q^2) - B(q^2)\gamma_5\right]\sigma_{\mu\nu}q^\nu \nu(p)\,, \tag{32}$$

we found the following expression for the electric and magnetic dipole moments

$$d_{ji}^{\rm el} \equiv B(0) = \frac{-e}{M^{*2}}\left(m_{\nu_i} - m_{\nu_j}\right)\sum_{\lambda=e,\mu,\tau} \mathrm{U}_{jk}^\dagger \mathrm{U}_{ki} \mathrm{F}\left(\frac{m_{\lambda_k}^2}{m_W^2}\right)\,, \tag{33}$$

$$\mu_{ji} \equiv A(0) = \frac{-e}{M^{*2}}\left(m_{\nu_i} + m_{\nu_j}\right)\sum_{\lambda=e,\mu,\tau} \mathrm{U}_{jk}^\dagger \mathrm{U}_{ki} \mathrm{F}\left(\frac{m_{\lambda_k}^2}{m_W^2}\right)\,, \tag{34}$$

where $i, j, k = 1, 2, 3$ denotes neutrino species, and

$$\mathrm{F}\left(\frac{m_{\lambda_k}^2}{m_W^2}\right) \simeq -\frac{3}{2} + \frac{3}{4}\frac{m_{\lambda_k}^2}{m_W^2}\,, \qquad \frac{m_{\lambda_k}^2}{m_W^2} \ll 1\,, \tag{35}$$

was obtained after the loop integration. In (33) and (34) $M^* = 4\pi\, v = 3.1$ TeV, where $v = (\sqrt{2}\,G_F)^{-1/2} = 246$ GeV represents the vacuum expectation value of the scalar Higgs field [29].

The neutrino mixing matrix U [30] is governing the decomposition of a coherently produced left-handed neutrino $\widetilde{\nu}_{L,\lambda}$ associated with charged-lepton-flavour $\lambda = e, \mu, \tau$ into the mass eigenstates $\nu_{L,i}$:

$$|\widetilde{\nu}_{L,\lambda};\boldsymbol{p}\rangle = \sum_i \mathrm{U}_{\lambda i}|\nu_{L,i};\boldsymbol{p},m_i\rangle\,, \tag{36}$$

For a Dirac neutrino $i=j$ [26, 31], and using $m_\nu = 0.05\,\mathrm{eV}$ [32], from (34), in units of [e cm] and Bohr magneton, we obtain

$$\begin{aligned}\mu_{\nu_i} &= \frac{3e}{2M^{*2}}m_{\nu_i}\left[1-\frac{1}{2}\sum_{\lambda=e,\mu,\tau}\frac{m_\lambda^2}{m_W^2}|\mathrm{U}_{\lambda i}|^2\right]\,,\\ &= 3.0\times 10^{-31}\,[\mathrm{e\;cm}] = 1.6\times 10^{-20}\mu_\mathrm{B}\,.\end{aligned} \tag{37}$$

From formula (37) it is clear that the chirality flip, which is necessary to induce the magnetic moment, arises only from the neutrino masses: Dirac neutrino magnetic moment (37) is still much smaller than the bounds obtained from astrophysics [33, 34]. More detailes about Dirac neutrinos can be found in [35, 36].

In the case of the off-diagonal transition moments, the first term in (35) vanishes in the summation over λ due to the orthogonality condition of U (GIM cancellation)

$$d^{\mathrm{el}}_{\bar{\nu}_j\nu_i} = \frac{3e}{2M^{*2}}\left(m_{\nu_i}-m_{\nu_j}\right)\sum_{\lambda=e,\mu,\tau}\frac{m_{\lambda_k}^2}{m_W^2}\mathrm{U}^\dagger_{jk}\mathrm{U}_{ki}\,, \tag{38}$$

$$\mu_{\bar{\nu}_j\nu_i} = \frac{3e}{2M^{*2}}\left(m_{\nu_i}+m_{\nu_j}\right)\sum_{\lambda=e,\mu,\tau}\frac{m_{\lambda_k}^2}{m_W^2}\mathrm{U}^\dagger_{jk}\mathrm{U}_{ki}\,. \tag{39}$$

In Majorana 4-component notation the Hermitian, neutrino-flavor antisymmetric, electric and magnetic dipole operators are

$$\begin{pmatrix}D_5\\ D\end{pmatrix}^{\mu\nu}_{ij} = \mathrm{e}\,\psi_\mathrm{i}^\top\left[\mathrm{C}\,\sigma^{\mu\nu}\begin{pmatrix}\gamma_5\\ \mathrm{i}1\end{pmatrix}\right]\psi_\mathrm{j}\,. \tag{40}$$

Majorana fields have the property that the particle is not distinguished from antiparticle. This forces us to use both charged lepton and antilepton propagators in the loop calculation of "neutrin-penguin" diagrams. This results in a complex antisymmetric transition matrix element T in lepton-flavour space:

$$\begin{aligned}\mathrm{T}_\mathrm{ji} &= -\mathrm{i}\epsilon^\mu\bar{\nu}_j\left[(A_{ji}-A_{ij})-(B_{ji}-B_{ij})\gamma_5\right]\sigma_{\mu\nu}q^\nu\nu_i\\ &= -\mathrm{i}\epsilon^\mu\bar{\nu}_j\left[2\mathrm{iIm}A_{ji}-2\mathrm{Re}B_{ji}\gamma_5\right]\sigma_{\mu\nu}q^\nu\nu_i\,.\end{aligned} \tag{41}$$

From this equation it is explicitly clear that for $i=j$, $d^{\mathrm{el}}_{\nu_i}=\mu_{\nu_i}=0$. Also, considering transition moment, only one of two terms in (41) is non-vanishing if the interaction respects the CP invariance, i.e. the first term vanishes if the relative CP of ν_i and ν_j is even, and the second term vanishes if odd [27]. Finally, dipole moments describing the transition from Majorana neutrino mass eigenstate-flavour ν_j to ν_k in the mass extended standard model reads:

$$d^{\mathrm{el}}_{\nu_i\nu_j} = \frac{3e}{2M^{*2}}\left(m_{\nu_i}-m_{\nu_j}\right)\sum_{\lambda=e,\mu,\tau}\frac{m_{\lambda_k}^2}{m_W^2}\mathrm{Re}\mathrm{U}^\dagger_{jk}\mathrm{U}_{ki}\,, \tag{42}$$

$$\mu_{\nu_i\nu_j} = \frac{3e}{2M^{*2}}\left(m_{\nu_i}+m_{\nu_j}\right)\sum_{\lambda=e,\mu,\tau}\frac{m_{\lambda_k}^2}{m_W^2}\mathrm{i}\,\mathrm{Im}\mathrm{U}^\dagger_{jk}\mathrm{U}_{ki}\,, \tag{43}$$

For the Majorana case the neutrino-flavour mixing matrix U is approximatively unitary, i.e. it is necessarily of the following form [29]

$$\sum_{i=1}^{3} \mathrm{U}^{\dagger}_{jk}\mathrm{U}_{ki} = \delta_{ji} - \varepsilon_{ji} \ , \tag{44}$$

where ε is a hermitian nonnegative matrix (i.e. with all eigenvalues nonnegative) and

$$\begin{aligned} |\varepsilon| = \sqrt{\mathrm{Tr}\ \varepsilon^2} &= \mathcal{O}\left(m_{\nu_{\rm light}}/m_{\nu_{\rm heavy}}\right) , \\ &\sim 10^{-22} \ \text{to} \ 10^{-21} \ . \end{aligned} \tag{45}$$

For the sum and difference of neutrino masses we assume hierarchical structure and take $|m_3 + m_2| \simeq |m_3 - m_2| \simeq |\Delta m^2_{32}|^{1/2} = 0.05\,\mathrm{eV}$ [32]. For the MNS matrix elements we set $|\mathrm{ReU}^*_{2\tau}\mathrm{U}_{\tau 3}| \simeq |\mathrm{ImU}^*_{2\tau}\mathrm{U}_{\tau 3}| \leq 0.5$. The electric and magnetic transition dipole moments of neutrinos $d^{\rm el}_{\nu_2\nu_3}$ and $\mu_{\nu_2\nu_3}$ are then denoted as $\left(d^{\rm el}_{\rm mag}\right)_{23}$ and given by

$$\begin{aligned} \left|\left(d^{\rm el}_{\rm mag}\right)_{23}\right| &= \frac{3e}{2M^{*2}}\frac{m^2_\tau}{m^2_W}|\Delta m^2_{32}|^{1/2}\begin{pmatrix} |\mathrm{ReU}^*_{2\tau}\mathrm{U}_{\tau 3}| \\ |\mathrm{ImU}^*_{2\tau}\mathrm{U}_{\tau 3}| \end{pmatrix} , \\ &\lesssim 1.95 \times 10^{-30}\,[\mathrm{e/eV}] = 3.8 \times 10^{-35}\,[\mathrm{e\ cm}], \\ &= 2.0 \times 10^{-24}\,\mu_B \ . \end{aligned} \tag{46}$$

The electric transition dipole moments of light neutrinos are smaller than the one of the d-quark. This is *the* order of magnitude of light neutrino transition dipole moments underlying the see–saw mechanism. It is by orders of magnitude smaller than in unprotected SUSY models.

Limits on the Noncommutativity Scale

Now we extract an upper limit on the $\star$-gradient interaction. The strength of the interaction (30) becomes $|m_\nu\, e\, \theta\, F|$. We compare it with the dipole interaction $|F\,\mu_{\nu_i}|$ for Dirac neutrino (37), and with the dipole transition interactions $|F\, d^{\rm el}_{\rm mag}|$ for Majorana case (42,43). Assuming that contributions from the neutrino-mass extended standard model are at least as large as those from noncommutativity, we derive the following two bounds on noncommutativity arising from the Dirac and Majorana nature of neutrinos, respectively:

$$\Lambda^{\rm Dirac}_{\rm NC} \gtrsim \left|\frac{e\, m_\nu}{\mu_\nu}\right|^{1/2} \simeq 1.7\,\mathrm{TeV} \ . \tag{47}$$

$$\Lambda^{\rm Majorana}_{\rm NC} \gtrsim \left|\frac{e\, m_\nu}{\left(d^{\rm el}_{\rm mag}\right)_{23}}\right|^{1/2} \simeq 150\,\mathrm{TeV} \ . \tag{48}$$

The fact that the neutrino mass extended standard model, as a consequence of (35), produces very different dipole moments for Dirac neutrinos (37) and Majorana neutrinos (46) respectively, manifests in two different scales of noncommutativity (47) and (48). The (m^2_λ/m^2_W) suppression of Majorana dipole moments (46) relative to the Dirac ones (37), is the main source for the different scales of noncommutativity. The bounds on noncommutativity thus obtained fix the scale $\Lambda_{\rm NC}$ at which the expected values of the neutrino electromagnetic dipole moments due to noncommutativity matches the standard model contributions.

References

1. W. Heisenberg, Letter to Peirls (1930). In: *Wolfgang Pauli, Scientific Correspondence*, vol 2, ed. by K. von Meyenn (Springer, Berlin Heidelberg New York 1985) p 15
2. H. S. Snyder: Phys. Rev. **71**, 38 (1947)
3. T. Filk: Phys. Lett. B **376**, 53 (1996)
4. A. Connes: Noncommutative Geometry, Academic Press London (1994)
5. A. Connes and J. Lott: Nucl. Phys. **18**, 29 (Proc. Suppl.) (1990)
6. M.R. Douglas, and C. Hull: JHEP **9802**, 008 (1998); Y.-K.E. Cheung and M. Krogh: Nucl. Phys. B **528**, 185 (1998); C.-S. Chu and P.-M. Ho: Nucl. Phys. B **550**, 151 (1999), Nucl. Phys. B **568**, 447 (2000)
7. N. Seiberg and E. Witten: JHEP **9909**, 032 (1999)
8. J. Madore, S. Schraml, P. Schupp and J. Wess: Eur. Phys. J. C **16**, 161 (2000)
9. B. Jurco, S. Schraml, P. Schupp and J. Wess: Eur. Phys. J. C **17**, 521 (2000)
10. B. Jurco, P. Schupp and J. Wess: Nucl. Phys. B **604**, 148 (2001)
11. B. Jurco, L. Moller, S. Schraml, P. Schupp and J. Wess: Eur. Phys. J. C **21**, 383 (2001)
12. X. Calmet, B. Jurco, P. Schupp, J. Wess and M. Wohlgenannt: Eur. Phys. J. C **23**, 363 (2002)
13. S. L. Glashow: Nucl. Phys. **22**, 579 (1961), S. Weinberg: Phys. Rev. Lett. **19**, 1264 (1967), A. Salam: Elementary Particle Physics, in Proceedings of the 8th Nobel Symposium (1968)
14. A. Armoni: Nucl. Phys. B **593**, 229 (2001)
15. M. Chaichian, P. Presnajder, M. M. Sheikh-Jabbari and A. Tureanu: Eur. Phys. J. C **29**, 413 (2003)
16. N. Seiberg and E. Witten: JHEP **9909**, 032 (1999)
17. P. Schupp: "Non-Abelian gauge theory on noncommutative spaces," hep-th/0111038.
18. P. Aschieri, B. Jurco, P. Schupp and J. Wess: Nucl. Phys. B **651**, 45 (2003)
19. W. Behr, N. G. Deshpande, G. Duplancic, P. Schupp, J. Trampetic and J. Wess: Eur. Phys. J. C **29**, 441 (2003)
20. G. Duplancic, P. Schupp and J. Trampetic: Eur. Phys. J. C **32**, 141 (2003)
21. P. Schupp, J. Trampetic, J. Wess and G. Raffelt: hep-ph/0212292, to be published in Eur. Phys. J. C
22. P. Minkowski, P. Schupp and J. Trampetic: hep-th/0302175.
23. J. Trampetić: Acta Phys. Pol. **B33**, 4317 (2002)
24. W. Gerlach and O. Stern: Ann. d. Phys. **74-4**, 673 (1924)
25. A. Fukugita and T. Yanagida: Physics of neutrinos in: Physics and Astrophysics of Neutrinos; A. Fukugita and A. Suzuki eds. (Springer-Verlag, 1994)
26. W.A. Bardeen, R. Gastmans and B. Lautrup: Nucl. Phys. **B46**, 319 (1972); S.Y. Lee: Phys. Rev. **D6**, 1701 (1972); B.W. Lee and R.E. Shrock: Phys. Rev. **D16**, 1444 (1977); W.J. Marciano and A.I. Sanda: Phys. Lett. **67B**, 303 (1977)

27. R.E. Shrock: Nucl. Phys. **B206**, 359 (1982)
28. C.A. Heusch and P. Minkowski: Nucl. Phys. **B416**, 3 (1994); P. Minkowski: Acta Phys. Polon. **B32**, 1935 (2001)
29. G. Duplančić, P. Minkowski and J. Trampetić: hep-ph/0304162, to be published in Eur. Phys. J. **C**
30. J. Maki, M. Nakagawa and S. Sakata: Prog. Theor. Phys. **28**, 870 (1962)
31. K. Fujikawa and R.E. Shrock: Phys. Rev. Lett. **45**, 963 (1980)
32. Super-Kamiokande Collaboration: Phys. Rev. Lett. **86**, 5651 (2001); Phys. Lett. **B539**, 179 (2002)
33. M. Fukugita and S. Yazaki: Phys. Rev. **D36**, 3817 (1987); M. Haft, G. Raffelt and A. Wiess: Astrophys. J. **425**, 222 (1994)
34. A. Ayala, J.C. D'Olivio and M. Torres: Phys. Rev. **D59**, 111901, (1999); E. Torrente-Lujan: hep-ph/0302082
35. M. B. Voloshin and M. I. Vysotsky: Sov. J. Nucl. Phys. **44**, 544 (1986)
36. M. B. Voloshin, M. I. Vysotsky and L.B. Okun Sov. J. Nucl. Phys. **64**, 446 (1986).

The Dressed Sliver in VSFT

Loriano Bonora[1], Carlo Maccaferri[2], and Predrag Prester[3]

[1] International School for Advanced Studies (SISSA/ISAS) Via Beirut 2–4, 34014 Trieste, Italy, and INFN, Sezione di Trieste
bonora@sissa.it

[2] International School for Advanced Studies (SISSA/ISAS) Via Beirut 2–4, 34014 Trieste, Italy, and INFN, Sezione di Trieste
maccafer@sissa.it

[3] Department of Theoretical Physics, Faculty of Science, University of Zagreb, Bijenička c. 32, p.p. 331, 10002 Zagreb, Croatia
pprester@phy.hr

1 Introduction

According to Sen's conjecture the minimum of the tachyonic potential in open string theory in D = 26 dimensions should correspond to an entirely different phase of string theory. At the tachyon condensation point the negative tachyonic potential is expected to exactly compensate for the D25-brane tension while no open string mode is expected to be excited, so that the BRST cohomology must be trivial. The only physics this scenario is likely to describe is that of a closed string theory vacuum. There are various ways to describe this phenomenon, but no doubt the most appropriate framework is that of Witten's Open String Field Theory (OSFT), [1]. Unfortunately so far we have been unable to find the exact solution describing tachyon condensation. It has been possible nevertheless to carry out explicit numerical calculations which have brought up evidence in favor of Sen's conjecture.

On the wake of this numerical progress, a new version of OSFT was proposed, [2], which is supposed to describe the theory at the minimum of the tachyonic potential, and, for this reason, was called vacuum string field theory (VSFT). VSFT is a simplified version of OSFT since the relevant BRST charge is made out only of ghost oscillators. This simplification induced a considerable progress. Many classical solutions were found, which are candidates to describe D-branes (the sliver, the butterfly, etc.), together with other classical solutions (lump solutions) which may represent lower dimensional D-branes [3, 4, 5, 6, 7, 8]. In some cases the spectrum around such solutions has been analyzed, and there is partial evidence that it provides the modes of a D-brane spectrum.

However the responses of VSFT are still far from being satisfactory. There are a series of nontrivial problems left behind. Let us consider for definiteness the sliver solution. To start with it has vanishing action for the matter part and infinite action for the ghost part, but it is impossible to make a finite number out of them, [18]. Moreover it is not at all clear, at least in the operator formalism, that the solutions to the linearized equations of motion around the sliver can accommodate all the open string modes one would expect if the sliver has to represent a D25-brane, [9, 10, 11, 12, 13, 14, 15, 16].

We believe the explanation for these drawbacks of the sliver solution is that the sliver is a too singular solution, not very fit to represent a D25-brane. In [21] it was shown that it is possible to find a more convenient solution to the VSFT equation of motion that is more appropriate than the sliver to represent the D25-brane of open string field theory. The solution was taylor–made in order to preserve most of the interesting and simplifying properties of the sliver. In fact one starts from the sliver defining matrix S and "perturbs it" by adding to CS a suitable rank one projector P. One can show not only that the corresponding squeezed state is a solution to the VSFT equations of motion, but that one can define infinite many independent such solutions. We call them *dressed slivers*.

However this is not the end of the story because these new solutions as well turn out to have an ill-defined *bpz*-norms and actions. To remedy this, we multiply the projector P by a real parameter ϵ, thus creating an interpolating family of states between the sliver ($\epsilon = 0$) and the dressed sliver ($\epsilon = 1$). This parameter plays the role of a regulator. On the other hand one can define a suitable regularization procedure for the determinants that appear in this kind of trade by means of the level truncation parameter L. By suitably tuning the latter to $1 - \epsilon$ we are able to show that the norm and the action corresponding to the dressed sliver can be made finite.

The purpose of the remaining part of this contribution is to give a simplified account of the derivation and properties of the dressed sliver solutions.

2 Dressing the Sliver

To start with we recall some formulas relevant to VSFT. The action is

$$\mathcal{S}(\Psi) = -\frac{1}{g_0^2}\left(\frac{1}{2}\langle\Psi|\mathcal{Q}|\Psi\rangle + \frac{1}{3}\langle\Psi|\Psi * \Psi\rangle\right) \tag{1}$$

where

$$\mathcal{Q} = c_0 + \sum_{n>0} (-1)^n \left(c_{2n} + c_{-2n}\right) \tag{2}$$

Notice that the action (1) does not contain any singular normalization constant, as opposed to [8, 9]. This important issue will be commented upon later on. The equation of motion is

$$\mathcal{Q}\Psi = -\Psi * \Psi \tag{3}$$

For nonperturbative solutions one makes the following factorized ansatz

$$\Psi = \Psi_m \otimes \Psi_g \tag{4}$$

where Ψ_g and Ψ_m depend purely on ghost and matter degrees of freedom, respectively. Then (3) splits into

$$\mathcal{Q}\Psi_g = -\Psi_g *_g \Psi_g \tag{5}$$

$$\Psi_m = \Psi_m *_m \Psi_m \tag{6}$$

where $*_g$ and $*_m$ refer to the star product involving only the ghost and matter part, respectively. The action for this type of solution becomes

$$\mathcal{S}(\Psi) = -\frac{1}{6g_0^2}\langle\Psi_g|\mathcal{Q}|\Psi_g\rangle\langle\Psi_m|\Psi_m\rangle \tag{7}$$

$\langle\Psi_m|\Psi_m\rangle$ is the ordinary inner product, $\langle\Psi_m|$ being the *bpz* conjugate of $|\Psi_m\rangle$ (see below).

To start with let us concentrate on the matter part (6). Since we are interested in a solution representing the D25-brane, which is translationally invariant, the $*_m$ product we need takes a simplified form

$$ {}_{123}\langle V_3|\Psi_1\rangle_1|\Psi_2\rangle_2 =_3 \langle \Psi_1 *_m \Psi_2| \tag{8} $$

where the three strings vertex V_3 is the reduced one (without momentum dependence)

$$ |V_3\rangle_{123} = \exp(-E)\,|0\rangle_{123}, \qquad E = \frac{1}{2}\sum_{a,b=1}^{3}\sum_{m,n\geq 1} \eta_{\mu\nu} a_m^{(a)\mu\dagger} V_{mn}^{ab} a_n^{(b)\nu\dagger} \tag{9} $$

Summation over the Lorentz indices $\mu,\nu = 0,\ldots,25$ is understood and η denotes the flat Lorentz metric. The operators $a_m^{(a)\mu}, a_m^{(a)\mu\dagger}$ denote the non-zero modes matter oscillators of the a-th string. They satisfy

$$ [a_m^{(a)\mu}, a_n^{(b)\nu\dagger}] = \eta^{\mu\nu}\delta_{mn}\delta^{ab}, \qquad m,n \geq 1 \tag{10} $$

while $|0\rangle_{123} \equiv |0\rangle_1 \otimes |0\rangle_2 \otimes |0\rangle_3$ is the tensor product of the Fock vacuum states relative to the three strings. The symbols V_{nm}^{ab} will denote the Neumann coefficients in the notation of Appendix A and B of [5].

Finally the bpz conjugation properties of the oscillators are

$$ bpz(a_n^{(a)\mu}) = (-1)^{n+1} a_{-n}^{(a)\mu} $$

Let us now return to (6). Its solutions are projectors of the $*_m$ algebra. We recall the simplest one, the sliver. It is defined by

$$ |\Xi\rangle = \mathcal{N} e^{-\frac{1}{2}a^\dagger S a^\dagger}|0\rangle, \qquad a^\dagger S a^\dagger = \sum_{n,m-1}^{\infty} a_n^{\mu\dagger} S_{nm} a_m^{\nu\dagger}\eta_{\mu\nu} \tag{11} $$

This state satisfies (6) provided the matrix S satisfies the equation

$$ S = V^{11} + (V^{12}, V^{21})(1 - \Sigma\mathcal{V})^{-1}\Sigma \begin{pmatrix} V^{21} \\ V^{12} \end{pmatrix} \tag{12} $$

where

$$ \Sigma = \begin{pmatrix} S & 0 \\ 0 & S \end{pmatrix}, \qquad \mathcal{V} = \begin{pmatrix} V^{11} & V^{12} \\ V^{21} & V^{22} \end{pmatrix}, \tag{13} $$

The proof of this fact is well-known. First one expresses (13) in terms of the twisted matrices $X = CV^{11}, X_+ = CV^{12}$ and $X_- = CV^{21}$, together with $T = CS = SC$, where $C_{nm} = (-1)^n\delta_{nm}$. The matrices X, X_+, X_- are mutually commuting. Then, requiring that T commute with them as well, one can show that (13) reduces to the algebraic equation

$$ XT^2 - (1+X)T + X = 0 \tag{14} $$

The interesting solution is

$$ T = \frac{1}{2X}(1 + X - \sqrt{(1+3X)(1-X)}) \tag{15} $$

The normalization constant $\mathcal{N}$ is given by

$$\mathcal{N} = (\mathrm{Det}(1 - \Sigma\mathcal{V}))^{\frac{D}{2}} \tag{16}$$

where $D = 26$. The contribution of the sliver to the matter part of the action (see (7)) is

$$\langle \Xi | \Xi \rangle = \frac{\mathcal{N}^2}{(\det(1 - S^2))^{\frac{D}{2}}} \tag{17}$$

Both (16) and (17) are ill-defined and need to be regularized, after which they both turn out to vanish (see below).

The dressed sliver is constructed by deforming the sliver with the addition of some special matrix to S. To this end we first introduce the infinite vector $\xi = \{\xi_n\}$ which is chosen to satisfy the condition

$$\rho_1 \xi = 0, \qquad \rho_2 \xi = \xi \,, \tag{18}$$

where ρ_1, ρ_2 are Fock space projectors each of which projects out half of the string modes, [6]. Next we set

$$\xi^T \frac{1}{1 - T^2} \xi = 1 \,, \qquad \xi^T \frac{T}{1 - T^2} \xi = \kappa \tag{19}$$

where ^{T} denotes matrix transposition. Our candidate for the *dressed sliver* solution is given by an ansatz similar to (42)

$$|\hat{\Xi}\rangle = \hat{\mathcal{N}} \mathrm{e}^{-\frac{1}{2} a^\dagger \hat{S} a^\dagger} |0\rangle \,, \tag{20}$$

with S replaced by

$$\hat{S} = S + R, \qquad R_{nm} = \frac{1}{\kappa + 1} \left(\xi_n (-1)^m \xi_m + \xi_m (-1)^n \xi_n \right) \tag{21}$$

As a consequence T is replaced by

$$\hat{T} = T + P, \qquad P_{nm} = \frac{1}{\kappa + 1} \left(\xi_m \xi_n + \xi_n (-1)^{m+n} \xi_m \right) \tag{22}$$

With a synthetic notation

$$P = \frac{1}{\kappa + 1} \left(|\xi\rangle\langle\xi| + |C\xi\rangle\langle C\xi| \right) \tag{23}$$

This operator is hermitean if ξ is real. We remark at this point that the conditions (19) are not very stringent. The only thing one has to worry is that the *lhs*'s are finite (this is the only true condition). Once this is guaranteed the rest follows from suitably rescaling ξ, so that the first equation is satisfied, and from the reality of ξ (see [21]).

We claim that $|\hat{\Xi}\rangle$ is a projector. To prove it one must show that

$$V^{11} + (V^{12}, V^{21})(1 - \hat{\Sigma}\mathcal{V})^{-1} \Sigma \begin{pmatrix} V^{21} \\ V^{12} \end{pmatrix} = \hat{S} \tag{24}$$

where

$$\hat{\Sigma} = \begin{pmatrix} \hat{S} & 0 \\ 0 & \hat{S} \end{pmatrix} \tag{25}$$

This is indeed the case, see [21].

We remark that, due to the arbitrariness of ξ, the result we have obtained brings into the game an infinite family of solutions to the equations of motion.

The normalization constant $\hat{\mathcal{N}}$ is given by

$$\hat{\mathcal{N}} = \mathrm{Det}(1 - \hat{\Sigma}\mathcal{V})^{\frac{D}{2}} = \mathrm{Det}(1 - \mathcal{T}\mathcal{M})^{\frac{D}{2}} \cdot \frac{1}{(\kappa+1)^D} \tag{26}$$

where $\mathcal{T} = C\Sigma$ and $\mathcal{M} = C\mathcal{V}$. However, if one tries to compute the norm of this state (which corresponds to its contribution to the action), i.e. $\langle\hat{\Xi}|\hat{\Xi}\rangle$, one finds an indeterminate result: the determinants involved in such calculations are in general not well-defined. It is evident that one has to introduce a regulator in order to end up with a finite action. Our idea is to deform the dressed sliver by introducing a parameter ϵ, so that we get the dressed sliver when $\epsilon = 1$. We notice that when $\epsilon \neq 1$ the state we obtain is in general not a $*$-algebra projector. We will use ϵ as a regulator and define the norm of the dressed sliver by means of the limit of a sequence of such states.

Therefore we introduce

$$|\hat{\Xi}_\epsilon\rangle = \hat{\mathcal{N}}_\epsilon \mathrm{e}^{-\frac{1}{2}a^\dagger \hat{S}_\epsilon a^\dagger}|0\rangle\,, \tag{27}$$

where

$$\hat{S}_\epsilon = S + \epsilon R, \tag{28}$$

As a consequence T is replaced by

$$\hat{T}_\epsilon = T + \epsilon P\,, \tag{29}$$

The consequences of this deformation will be worked out in the next section.

As for the ghost part of the equation of motion (5) the procedure is similar. The sliver-like solution to this equation takes the form

$$|\widetilde{\Xi}\rangle = \widetilde{\mathcal{N}} \exp\left(\sum_{n,m=1}^{\infty} c_n^\dagger \widetilde{S}_{nm} b_m^\dagger\right) c_1|0\rangle\,, \tag{30}$$

where the matrix $\widetilde{S}$ satisfies an equation similar to (12). The ghost dressed sliver solution can be constructed in the same way as above. We introduce two real vectors $\beta = \{\beta_n\}$ and $\delta = \{\delta_n\}$ which satisfy

$$\widetilde{\rho}_1\beta = \widetilde{\rho}_1\delta = 0, \qquad \widetilde{\rho}_2\beta = \beta, \qquad \widetilde{\rho}_2\delta = \delta\,. \tag{31}$$

We also set

$$\left\langle \beta | \frac{1}{1-\widetilde{T}^2} |\delta \right\rangle = 1\,, \qquad \left\langle \beta | \frac{\widetilde{T}}{1-\widetilde{T}^2} |\delta \right\rangle = \widetilde{\kappa} \tag{32}$$

where $\widetilde{\kappa}$ is a non-negative number. Now we dress the ghost part of the sliver by introducing the squeezed state

$$|\widehat{\widetilde{\Xi}}_{\widetilde{\epsilon}}\rangle = \widehat{\widetilde{\mathcal{N}}}_{\widetilde{\epsilon}}\, \mathrm{e}^{c^\dagger \widehat{\widetilde{S}}_{\widetilde{\epsilon}} b^\dagger} c_1|0\rangle \tag{33}$$

where instead of $\widetilde{S}$ we now have

$$\widehat{\widetilde{S}}_{\tilde{\epsilon}} = \widetilde{S} + \tilde{\epsilon}\widetilde{R}\,, \qquad \widetilde{R} = \frac{1}{\widetilde{\kappa}+1}(|C\delta\rangle\langle\beta| + |\delta\rangle\langle C\beta|)$$

It is easy to see that $\widehat{\widetilde{S}}^*_{\tilde{\epsilon}} = C\widehat{\widetilde{S}}_{\tilde{\epsilon}}C$ for β, δ real, which means that the string field is real. The idea is once again to recover the action contribution of the ghost via the limit $\tilde{\epsilon} \to 1$.

3 The Finite Dressed Sliver Action

In the previous section we have introduced for the matter part a state, depending on a parameter ϵ, that interpolates between the sliver $\epsilon = 0$ and the dressed sliver $\epsilon = 1$. Now we intent to show that by its means, we can give a precise definition of the norm of the dressed sliver, so that both its norm and its action can be made finite.

As already mentioned above, the determinants in (16), (17) relevant to the sliver are ill-defined. They are actually well defined for any finite truncation of the matrix X to level L and need a regulator to account for its behaviour when $L \to \infty$. A regularization that fits particularly our needs was introduced by Okuyama [18] and we will use it here. It consists in using an asymptotic expression for the eigenvalue density $\rho(k)$ of X (see also Sect. 3), $\rho(k) \sim \frac{1}{2\pi}\log L + \rho_{fin}(k)$, for large (continuous) L, where $\rho_{fin}(k)$ is a finite contribution when $L \to \infty$, see [19]. This leads to asymptotic expressions for the various determinants we need. In particular the scale of L can be chosen in such a way that

$$\begin{aligned} \det(1+T) &= h_+\, L^{-\frac{1}{3}} + \ldots \\ \det(1-T) &= h_-\, L^{\frac{1}{6}} + \ldots \\ \det(1-X) &= h_X\, L^{\frac{1}{9}} + \ldots \end{aligned} \tag{34}$$

where dots denote non-leading contribution when $L \to \infty$ and h_+, h_-, h_X are suitable numerical constants which arise due to the finite contribution in the eigenvalue density. Our strategy consists in tuning L with $1-\epsilon$ in such a way as to obtain finite results. It turns out however that the result we find depends very much on the way we take the limit $\epsilon \to 1$. The problem arises when simultaneous multiple limits are involved. Therefore the question is: do we have a criterion to select among all the different limits? The answer is affirmative. The criterion is the requirement that the equation of motion be satisfied, i.e. we must have

$$\lim_{\epsilon_1,\epsilon_2 \to 1} \langle \hat{\Xi}_{\epsilon_1} | \hat{\Xi}_{\epsilon_2} \rangle = \lim_{\epsilon_1,\epsilon_2,\epsilon_3 \to 1} \langle \hat{\Xi}_{\epsilon_1} | \hat{\Xi}_{\epsilon_2} * \hat{\Xi}_{\epsilon_3} \rangle \tag{35}$$

The analysis carried out in [21] tells us that a good procedure is defined by the nested limits, for it turns out that

$$\lim_{\epsilon_1 \to 1}\left(\lim_{\epsilon_2 \to 1} \langle \hat{\Xi}_{\epsilon_1} | \hat{\Xi}_{\epsilon_2} \rangle \right) = \lim_{\epsilon_1 \to 1}\left(\lim_{\epsilon_2 \to 1}\left(\lim_{\epsilon_3 \to 1} \langle \hat{\Xi}_{\epsilon_1} | \hat{\Xi}_{\epsilon_2} * \hat{\Xi}_{\epsilon_3} \rangle \right)\right) \tag{36}$$

Applying this criterion in calculating the norm of the dressed sliver one finds

$$\frac{1}{\langle 0|0\rangle} \lim_{\epsilon_1\to 1}\left(\lim_{\epsilon_2\to 1}\langle\hat{\Xi}_{\epsilon_1}|\hat{\Xi}_{\epsilon_2}\rangle\right) \tag{37}$$

$$= \lim_{\epsilon_1\to 1}\left(\frac{\mathrm{Det}(1-\Sigma\mathcal{V})}{\sqrt{\det(1-S^2)}}\right)^D \left(\frac{1}{4(\kappa+1)^2}\right)^{\frac{D}{2}}\left(\frac{4}{(1-\epsilon_1)^2}\right)^{\frac{D}{2}} + \dots$$

$$= \lim_{\epsilon_1\to 1}\left(\frac{h}{(\kappa+1)^2}\right)^{\frac{D}{2}}\left(\frac{L^{-\frac{5}{36}}}{1-\epsilon_1}\right)^D + \dots = \left(\frac{h}{(\kappa+1)^2 s_1^2}\right)^{\frac{D}{2}}$$

where dots denote non-leading terms and $h = \frac{h_X^2 h_+}{h_-}$. In the last passage we tuned the parameters as follows

$$1-\epsilon_1 = s_1 L^{-\frac{5}{36}} \tag{38}$$

Now we take the norm of $\hat{\Xi}$ to be defined by the limit in (37).

What we have achieved so far is to prove that it is possible to assign a finite positive number to the expression (norm) $\langle\hat{\Xi}|\hat{\Xi}\rangle$, in a way which is consistent with the matter equation of motion. It does not mean that a state exists in the Hilbert space which is the limit of $\hat{\Xi}_\epsilon$ when $\epsilon \to 1$. In fact it is possible to show that, while the regularization procedure defined above guarantees that we can associate a positive finite number to the symbol $\langle\hat{\Xi}|\hat{\Xi}\rangle$, it does not allow us to associate any Hilbert space state to $\hat{\Xi}$. *The state $\hat{\Xi}$ lives outside the Hilbert space.*

For the ghost part we proceed likewise. To define action terms we use the nested limits prescription. In particular

$$\lim_{\tilde{\epsilon}_1\to 1}\left(\lim_{\tilde{\epsilon}_2\to 1}\langle\hat{\tilde{\Xi}}_{\tilde{\epsilon}_1}|\mathcal{Q}|\hat{\tilde{\Xi}}_{\tilde{\epsilon}_2}\rangle\right) = \lim_{\tilde{\epsilon}_1\to 1}\left[1-\frac{\tilde{\epsilon}_1}{1+(1-\tilde{\epsilon}_1)\kappa}\right]^2 \frac{\det(1-\hat{\tilde{S}}^2)}{\left[\mathrm{Det}(1-\Sigma\mathcal{V})\right]^2}$$

$$= \lim_{\tilde{\epsilon}_1\to 1}\left[\frac{(\tilde{\kappa}+1)(1-\tilde{\epsilon}_1)}{1+(1-\tilde{\epsilon}_1)\tilde{\kappa}}\right]^2 \frac{L^{\frac{11}{18}}}{\tilde{h}} + \dots . \tag{39}$$

Therefore, if we assume that

$$1-\tilde{\epsilon}_1 = \tilde{s}L^{-\frac{11}{36}} \tag{40}$$

for some constant $\tilde{s}$, we have

$$\lim_{\tilde{\epsilon}_1\to 1}\left(\lim_{\tilde{\epsilon}_2\to 1}\langle\hat{\tilde{\Xi}}_{\tilde{\epsilon}_1}|\mathcal{Q}|\hat{\tilde{\Xi}}_{\tilde{\epsilon}_2}\rangle\right) = (\tilde{\kappa}+1)^2\frac{\tilde{s}^2}{\tilde{h}} . \tag{41}$$

which defines a finite value for the kinetic term in the action.

The calculation for the cubic term in the action goes along similar lines and one obtains that the nested limits preserve the equation of motion in the ghost case as well.

Now let us come to the conclusion concerning the regularized action for the full solution $\hat{\Psi} = \hat{\Xi}\otimes\hat{\tilde{\Xi}}$. Collecting the results (37,41) and plugging them into (7) one gets

$$-\frac{\mathcal{S}(\hat{\Psi})}{V^{(D)}} = \frac{1}{6g_0^2(2\pi)^D}\frac{(\tilde{\kappa}+1)^2\tilde{s}^2}{(\kappa+1)^D s^D}\frac{h^{\frac{D}{2}}}{\tilde{h}} \tag{42}$$

The value of the *rhs* can now be tuned to the physical value of the D25-brane tension. We stress that, apart from g_0, the parameters in the *rhs* are not present in the initial action, but arise from the regularization procedure.

4 Comments

In the present work we have considered a finitely normalized action

$$S[\hat{\psi}] = -\frac{1}{g_0^2}\left(\frac{1}{2}\langle\hat{\psi}\mathcal{Q}\hat{\psi}\rangle + \frac{1}{3}\langle\hat{\psi}\hat{\psi} * \hat{\psi}\rangle\right) \tag{43}$$

By means of the operator field redefinition [17]

$$\psi = \mathrm{e}^{-\frac{1}{4}\ln\epsilon(K_2-4)}\hat{\psi} \tag{44}$$

it can be brought to the form

$$S'[\psi] = -\frac{1}{g_0^2\epsilon^3}\left(\frac{1}{2}\langle\psi\mathcal{Q}\psi\rangle + \frac{1}{3}\langle\psi\psi * \psi\rangle\right) = -\frac{1}{g_0^2}\left(\frac{1}{2\epsilon}\langle\tilde{\psi}\mathcal{Q}\tilde{\psi}\rangle + \frac{1}{3}\langle\tilde{\psi}\tilde{\psi} * \tilde{\psi}\rangle\right) \tag{45}$$

where $\tilde{\psi} = \epsilon\psi$. Both forms of the action have been considered previously in the literature, [9, 8], in the limit $\epsilon \to 0$, implying a singular normalization of the action. What we have shown above is that free effective parameters appear in the process of regularizing the classical action so that a singular normalization of the latter can be avoided. This remark is of more consequence than it looks at first sight. The point is that the ridefiniton (44) can harmlessly be implemented only in $D = 26$, [20]. In noncritical dimensions as a consequence of such redefinition an anomaly appears. In the course of our derivation above the critical dimension has never featured, but this remark brings it back into the game. This has an important consequence: setting $\epsilon = g_0^{2/3}$ in the middle term of (45), it is evident that in critical dimensions we can make any parameter to completely disappear from the action by means of a field redefinition. So, in $D = 26$ the value of the brane tension is dinamically produced and not put in by hand. Although such a conclusion does not have the strength of a formal constraint, it marks the critical dimension as the privileged one.

This scenario in VSFT has to be contrasted with the situation in OSFT. Going back to (44), we recall that the family of operators $K_n = L_n - (-1)^n L_{-n}$ leaves the action cubic term invariant while it acts non trivially on the kinetic term as

$$[K_{2n}, \mathcal{Q}] = -4n(-1)^n\mathcal{Q} \tag{46}$$

In OSFT one cannot implement a redefinition like (44) keeping the ratio of the kinetic and interaction terms constant. Only in VSFT is this possible and therefore only in VSFT can we say that no free parameters appear in the action.

To conclude, in this paper we have shown that it is possible to find solutions of VSFT with finite *bpz*-norm and action. This result was achieved by first introducing a new kind of solutions of VSFT, which we called dressed slivers. The latter is a deformation of the well-known sliver solution by the addition of a rank one projector to the Neumann matrix. The dressed sliver, introduced in this naive way, has still an ill-defined norm (and action), but one can naturally introduce a regularization parameter ϵ (which interpolates between the sliver and the dressed sliver), and tune it to the level truncation parameter L. This leads to a finite *bpz*-norm and action. In [21] it was shown that in fact one can extend these conclusions to a large class of solutions, representing in particular parallel coincident and lower dimensional D-branes.

References

1. E. Witten, *Noncommutative Geometry and String Field Theory*, Nucl. Phys. B **268** (1986) 253.
2. L. Rastelli, A. Sen and B. Zwiebach, *"Vacuum string field theory,"* arXiv:hep-th/0106010.
3. V.A. Kostelecky and R. Potting, *Analytical construction of a nonperturbative vacuum for the open bosonic string*, Phys. Rev. D **63** (2001) 046007 [arXiv:hep-th/0008252].
4. L. Rastelli, A. Sen and B. Zwiebach, *String field theory around the tachyon vacuum*, Adv. Theor. Math. Phys. **5** (2002) 353 [arXiv:hep-th/0012251].
5. L. Rastelli, A. Sen and B. Zwiebach, *Classical solutions in string field theory around the tachyon vacuum*, Adv. Theor. Math. Phys. **5** (2002) 393 [arXiv:hep-th/0102112].
6. L. Rastelli, A. Sen and B. Zwiebach, *Half-strings, Projectors, and Multiple D-branes in Vacuum String Field Theory*, JHEP **0111** (2001) 035 [arXiv:hep-th/0105058].
7. T. Okuda, *The Equality of Solutions in Vacuum String Field Theory*, Nucl. Phys. B **641** (2002) 393 [arXiv:hep-th/0201149].
8. D. Gaiotto, L. Rastelli, A. Sen and B. Zwiebach, *Star Algebra Projectors*, JHEP **0204** (2002) 060 [arXiv:hep-th/0202151].
9. H. Hata and T. Kawano, *Open string states around a classical solution in vacuum string field theory*, JHEP **0111** (2001) 038 [arXiv:hep-th/0108150].
10. H. Hata and S. Moriyama, *Observables as Twist Anomaly in Vacuum String Field Theory*, JHEP **0201** (2002) 042 [arXiv:hep-th/0111034].
11. H. Hata, S. Moriyama and S. Teraguchi, *"Exact results on twist anomaly,"* JHEP **0202** (2002) 036 [arXiv:hep-th/0201177].
12. H. Hata and S. Moriyama, *Reexamining Classical Solution and Tachyon Mode in Vacuum String Field Theory*, Nucl. Phys. B **651** (2003) 3 [arXiv:hep-th/0206208].
13. L. Rastelli, A. Sen and B. Zwiebach, *A note on a Proposal for the Tachyon State in Vacuum String Field Theory*, JHEP **0202** (2002) 034 [arXiv:hep-th/0111153].
14. H. Hata and H. Kogetsu *Higher Level Open String States from Vacuum String Field Theory*, JHEP **0209** (2002) 027, [arXiv:hep-th/0208067].
15. J.R. David, *Excitations on wedge states and on the sliver*, JHEP **0107** (2001) 024 [arXiv:hep-th/0105184].
16. Y. Okawa, *Open string states and D-brane tension form vacuum string field theory*, JHEP **0207** (2002) 003 [arXiv:hep-th/0204012].
17. Y. Okawa, *"Some exact computations on the twisted butterfly state in string field theory,"* [arXiv:hep-tharXiv:hep-th/0310264.]
18. K. Okuyama, *Ghost Kinetic Operator of Vacuum String Field Theory*, JHEP **0201** (2002) 027 [arXiv:hep-th/0201015].
19. D.M. Belov, A. Konechny, *On spectral density of Neumann matrices*, Phys. Lett. B **558** (2003) 111-118 [arXiv:hep-th/0210169].
20. D.J. Gross and A. Jevicki, *Operator Formulation of Interacting String Field Theory*, Nucl. Phys. **B283** (1987) 1.
21. L. Bonora, C. Maccaferri, P. Prester, *Dressed Sliver solutions in Vacuum String Field Theory*, [arXiv:hep-th/0311198], to be published in JHEP.

M5-Branes and Matrix Theory

Martin Cederwall[1] and Henric Larsson[2]

[1] martin.cederwall@fy.chalmers.se
[2] solo@fy.chalmers.se
Department of Theoretical Physics, Göteborg University
and Chalmers University of Technology, 412 96 Göteborg, Sweden

1 Introduction

Supermembrane [1] theory [2] is a very promising candidate for a microscopic description of M-theory. Although it is not background invariant, it gives a completely new picture of the nature of space and time at small scales, together with a description of quantum-mechanical states that goes beyond local quantum field theory. These features are most clear in the matrix [3] truncation [2] of the membrane. It is widely appreciated that first-quantised supermembrane theory through its continuous spectrum [4] is capable of describing an entire ("multi-particle") Fock space. For reviews on the subject of membranes and matrices, see [5]. Due to the immense technical difficulties associated with actual calculations in the theory, which is non-linear and inherently non-perturbative, few quantitative features are known in addition to the general picture, which is supported by many qualitative arguments. Maybe the most important one is the proof that $su(N)$ matrix theory has a unique supersymmetric ground state [6, 7], which gives the relation to the massless degrees of freedom of $D = 11$ supergravity.

Many situations in M-theory backgrounds involve membranes that are not closed. Supermembranes may end on "defects", i.e., 5-branes and 9-branes [8, 9, 10, 11, 12, 13, 14]. It is urgent to have some mathematical formulation of these situations in order to understand the microscopic properties of physics in such backgrounds. One old enigma is the nature of the theory on multiple 5-branes, which we address in the present talk. There are several issues to be resolved. The membrane may be stretched between multiple 5-branes, and the truncation has to be consistent with this situation. In addition, the C-field, the 3-form potential of M-theory, may take some non-vanishing self-dual value on the 5-brane. The new results contained in this talk refer to the latter situation. There are several reasons to consider this specific situation. It should be connected to the theory on multiple $M5$-branes, which is some kind of non-abelian theory of self-dual tensors [15]. It should be possible to verify the decoupling limit of OM-theory [16] from microscopic considerations. There might also be information about the open membrane metric [17, 18] and maybe even some clue concerning the proper generalisation of the string endpoint non-commutativity to membranes [19, 18].

We start out by reviewing the consistent truncation of membranes to matrices via non-commutativity in Sect. 2. Section 3 describes how this construction is generalised to situations where the membrane has a boundary [20, 21, 14, 22]. Here we review the alternative constructions present in the literature, and discuss their relative applicability. In Sect. 4, we generalise the picture to include non-vanishing C-field, both light-like [23] and general. We identify the deformation of the 6-dimensional super-Yang-Mills theory whose dimensional reduction is the matrix theory associated to turning on the C-field. In Sect. 5, we discuss the possible applications and limitations of the model.

2 From Membranes to Matrices

We start from the action for the supermembrane coupled to an on-shell background of $D = 11$ supergravity,

$$S = -T \int \mathrm{d}^3\xi \sqrt{-g} + T \int C \,. \tag{1}$$

Here, the metric and C-field are pullbacks from superspace to the bosonic world-volume. In what follows, we will consider flat backgrounds, but allow for non-zero constant C.

Let us first remind of the consistent truncation to matrix theory of a closed membrane (we just display the bosonic degrees of freedom; fermions are straightforwardly included). Here, the C-field is irrelevant. In light-cône gauge, where reparametrisation invariance is used up except for area-preserving diffeomorphisms of the membrane "space-sheet". The light-cône hamiltonian p^- is given by

$$\frac{p^+p^-}{A} = \int \mathrm{d}^2\xi \left(P_I P_I + \frac{T^2}{4}\{X^I, X^J\}\{X^I, X^J\} \right) , \tag{2}$$

where A is the parametric area of the space-sheet, and $\{A, B\} = \epsilon^{ij}\partial_i A \partial_j B$ is the "Poisson bracket" on the space-sheet. The remaining gauge invariance is generated by the Poisson bracket as $\delta_f A = \{f, A\}$ [24]. Even though it is known that the algebra of area-preserving diffeomorphisms in a certain sense is $su(\infty)$ [2, 25], $su(N)$ is not contained as a subalgebra, and there is no way of getting to $su(N)$ matrix theory as a consistent truncation.

In order to obtain matrix theory as a consistent truncation, one introduces a non-commutativity on the membrane space-sheet (for simplicity, we consider a toroidal membrane), $[\xi^1, \xi^2] = \theta$, encoded in the Weyl-ordered star product $f \star g = f \exp(\frac{i}{2}\theta\epsilon^{ij}\overleftarrow{\partial_i}\overrightarrow{\partial_j})g$. Commutators between Fourier modes become

$$[\mathrm{e}^{\mathrm{i}k\cdot\xi}, \mathrm{e}^{\mathrm{i}k'\cdot\xi}] = -2i \sin\left(\frac{\theta}{2}\epsilon^{ij}k_i k'_j\right) \mathrm{e}^{\mathrm{i}(k+k')\cdot\xi} \,.$$

The Poisson bracket is recovered as $\{\cdot,\cdot\} = -i \lim_{\theta\to 0} \theta^{-1}[\cdot,\cdot]$. Choosing $\theta = \frac{2\pi}{N}$ implies that the functions $\mathrm{e}^{\mathrm{i}N\xi^1}, \mathrm{e}^{\mathrm{i}N\xi^2}$ are central. Their action on any function can consistently be modded out according to the equivalence relation $\mathrm{e}^{\mathrm{i}N\xi^{1,2}} \star f \approx f$. The remaining "square of functions" with mode numbers ranging from 0 to $N-1$ generate $u(N)$ [2, 25].

The model thus obtained as a consistent truncation of the supermembrane is an $su(N)$ supersymmetric matrix model, identical to the dimensional reduction to $D = 1$ of $D = 10$ super-Yang-Mills theory. This example sets the procedure we want to apply to other cases: deform by non-commutativity, replace Poisson brackets by commutators and perform a consistent truncation of the deformed theory. We would like to stress the importance of making a consistent truncation, in contrast to an approximation; the fact that the commutator obeys the same algebraic identities as the Poisson bracket means that one has control over the symmetries of the model, e.g., supersymmetry. The only symmetries that are lost in the matrix truncation are the super-Poincaré generators that are non-linearly realised in the light-cône gauge.

3 Matrix Theory for Membranes with Boundary

Let us now turn to the first modification of the previous situation, namely when the membrane has boundaries (we think of these as lying on $M5$-branes, but much if what is said applies to any possible boundary). It is expected that the "no-topology" theorem that applies for closed membranes persists for membranes with boundary, so that it is irrelevant e.g., whether a membrane ending on only one 5-brane is modeled as a half sphere, a half torus or some more complicated manifold. This is an assumption we make; a proof would be desirable.

We can distinguish between two classes of approaches to this kind of configuration: **A.** This approach was first physically motivated by double dimensional reduction to a $D4$-brane. The theory obtained after reduction is $D = 5$ super-Yang-Mills, and opening up the sixth direction should correspond to a strong coupling limit. In this limit, path integrals are dominated by saddle points at the moduli space of "instanton" solitons. The moduli space of N instantons in $\mathrm{U}(k)$ SYM has dimension $4kN$. The matrix theory should have this space as Higgs branch. It is the dimensional reduction of a $D = 6$ $\mathrm{U}(N)$ SYM with one adjoint and k fundamental hypermultiplets [26]. We will motivate this from the point of view of the supermembrane.

B. For a fixed membrane topology (a half torus, say), the boundary conditions may be solved, at least when $C = 0$ (see [14]). For the 5 directions transverse to the (flat) $M5$-brane one gets Dirichlet boundary conditions, which for torus topology means sine functions, and for the 4 transverse (2 have been eliminated when going to light-cône gauge) one gets Neumann boundary conditions, leading to cosine functions. The sine functions generate $\mathrm{SO}(N)$ [13], and the cosine function transform as the symmetric representation. The matrix model obtained is the dimensional reduction of a $D = 6$ $\mathrm{SO}(N)$ SYM with a hypermultiplet in the symmetric representation. A couple of comments can be made. The first one concerns the global symmetries of the matrix theory and of the M-theory configuration it describes. A $D = 6$ SYM theory with hypermultiplets has a lagrangian

$$\begin{aligned}\mathcal{L} = &-\frac{1}{4}F^A_{\mu\nu}F^{A\mu\nu} + \frac{1}{2}\Re(\lambda^{\dagger A}\gamma^\mu D_\mu\lambda^A) - \frac{1}{2}D^\mu\phi^I D_\mu\phi^*_I - \frac{1}{2}\Re(\psi^\dagger_I\tilde{\gamma}^\mu D_\mu\psi^I)\\ &- (\varrho^A)_I{}^J\Re(\lambda^{\dagger A}\psi^I\phi^*_J) + \frac{1}{8}(\varrho^A)_I{}^J(\varrho^A)_K{}^L\phi^I\phi^*_J\phi^K\phi^*_L\,. \end{aligned} \tag{3}$$

We use the isomorphism $\mathrm{Spin}(1,5) \approx \mathrm{SL}(2;\mathbb{H})$ and two-component quaternionic spinors. The scalars ϕ are quaternionic, $\phi = \phi_i e_i$, where $i = 1,\ldots,4$. Indices $I, J, \ldots$ label the representation of the hypermultiplet, and ϱ is the representation matrix. For real hypermultiplets (as in case **B** above) there is an $\mathrm{SU}(2)_L \times \mathrm{SU}(2)_R$ R-symmetry realised as multiplication by unit quaternions as

$$\begin{aligned} &A \to A\,, \qquad \lambda \to \lambda h_L\,, \\ &\phi \to h_L^* \phi h_R\,,\ \psi \to \psi h_R\,. \end{aligned} \tag{4}$$

When the hypermultiplet is complex, the right action is occupied by the gauge group. If there is an even number of hypermultiplets in the same representation, there will however be a flavour symmetry SU($2k$) The representations (specified by dimensions) of the fields under the Lorentz rotations and R-symmetry are thus

$$\begin{aligned} &A : (6, 1, 1)\,, \qquad \lambda : (4, 2, 1)\,, \\ &\phi : (1, 2, 2\,\text{or}\,1)\,,\ \psi : (\overline{4}, 1, 2\,\text{or}\,1) \end{aligned} \tag{5}$$

(the last possibility for $su(2)_R$ representations of the hypermultiplet is for the minimal content of a complex representation). The R-symmetry of the super-Yang-Mills theory is the rotation symmetry of the membrane/matrix theory in light-cône gauge, and the SO(5) rotation symmetry remaining on the super-Yang-Mills side after dimensional reduction is the R-symmetry of the membrane/matrices.

The supersymmetry transformation rules are

$$\begin{aligned} &\delta_\epsilon A_\mu^A = \Re(\epsilon^\dagger \gamma_\mu \lambda^A)\,,\ \delta_\epsilon \lambda^A = \tfrac{1}{2} F_{\mu\nu}^A \gamma^{\mu\nu} \epsilon + \epsilon W^A\,, \\ &\delta_\epsilon \phi^I = -\epsilon^\dagger \psi^I\,, \qquad \delta_\epsilon \psi^I = \gamma^\mu \epsilon D_\mu \phi^I\,, \end{aligned} \tag{6}$$

where ϵ is a spinor in the same representation as λ. Since we later want to identify the presence of a C-field with certain deformations of SYM that leave supersymmetry unbroken, we have written the transformation of the adjoint spinor using W^A, an imaginary quaternion (i.e., transforming in $(1, 3, 1)$) in the adjoint of the gauge group. In the undeformed case, $W^A = W_0^A = \frac{1}{2}(\varrho^A)_I{}^J \phi^I \phi_J^\star$. Note that the hypermultiplet potential is the square of W, $V(\phi) = \frac{1}{2} W^A W^{\star A}$. The deformations will be encoded in the form of W. The most convenient way of checking supersymmetry is to note that W is contained in the same supermultiplet as the hypermultiplet gauge current:

$$\begin{aligned} \delta_\epsilon W^A &= -\Im(\epsilon^\dagger \mu^A)\,, \\ \delta_\epsilon \mu^A &= J_\mu^A \gamma^\mu \epsilon + \gamma^\mu \epsilon D_\mu W^A\,, \end{aligned} \tag{7}$$

where

$$\begin{aligned} \mu^A &= \mu_0^A = (\varrho^A)_I{}^J \psi^I \phi_J^\star, \\ J_\mu^A &= \frac{1}{2}(\varrho^A)_I{}^J (D_\mu \phi^I \phi_J^\star - \phi^I D_\mu \phi_J^\star - \psi_J^\dagger \tilde{\gamma}_\mu \psi^I)\,. \end{aligned} \tag{8}$$

Before turning to the derivation of case **A** from the supermembrane, let us discuss the advantages and limitations of the two approaches and some aspects of their physical content. Both cases are defined as dimensional reductions of $D = 6$ SYM with matter. The expression for the potential is a sum of positive semidefinite terms, so the Higgs branch is determined by $W = 0$. In light of the correspondence with five-dimensional physics mentioned above, it is interesting to investigate the geometry of the Higgs branch. The low-energy limit of adiabatic motion on the Higgs branch is also the situation when bulk excitations (gravity) decouple. Counting the dimension of the Higgs branch as #(scalar matter fields)$-\#W-$dim(gauge group), one gets in case **A**: $4N^2 + 4kN - 3N^2 - N^2 = 4kN$, and in case **B**: $4\frac{N(N+1)}{2} - 3\frac{N(N-1)}{2} - \frac{N(N-1)}{2} = 4N$. Closer investigation reveals that the spaces agree for

$k = 1$, and that the Higgs branch then is $\mathbb{R}^4 \times (R^{4(N-1)}/P_N)$, interpreted as the space of loci of N indistinguishable partons/D0-branes. This is a flat hyper-Kähler space with conical singularities where partons coincide, which is where the Higgs branch intersects the Coulomb branch.

There is an index theorem [27] stating that the matrix theory has a unique supersymmetric ground state. The eight fermion zero modes lie in the representation $(4,1,2)$, so the ground state is the breaking to SO(5)×SU(2) of an SO(8) spinor $8_s \oplus 8_c$ when the vector decomposes as $8_v \to (4,2)$ (the "Hopf breaking"). Then $8_s \to (1,3) \oplus (5,1)$ and $8_c \to (4,2)$, giving the bosonic and fermionic fields of the self-dual $D = 6$ tensor multiplet in the light-cône gauge.

Note that approach **B** does not seem to accommodate multiple $M5$-branes in a natural way. On the other hand, approach **A**, as we will see, is less adaptable to incorporate the stringy nature of the membrane boundary. This is connected to the way it is derived from the membrane below; no boundary conditions are solved, the nature of the boundary is rather point-like. It is also unclear how **A** generalises to separated $M5$-branes. Concerning the incorporation of a non-vanishing C-field (following section), approach **A** has the advantage of being more or less directly applicable, while approach **B** encounters problems, due to the difficulty (impossibility?) of solving the boundary conditions in the presence of a C-field.

Let us sketch briefly how case **A** is derived as a consistent truncation from the supermembrane. As we already mentioned, no boundary conditions are solved before performing the matrix truncation. Instead we introduce the "boundary" through the truncated δ-function

$$\varDelta \equiv \sum_{n=0}^{N-1} \mathrm{e}^{\imath n\varrho} \; .$$

Here we consider a boundary located at $\varrho = 0$, where ϱ for simplicity is a coordinate on a torus. Due to the identities

$$\varDelta^2 = N\varDelta, \quad \varDelta \star f \star \varDelta = 0 \; ,$$

left and right star multiplication with $\varDelta/\sqrt{N}$ projects on two "boundary representations" N and $\bar{N}$ with opposite U(1) charge under adjoint action of $\varDelta$, e.g.,

$$[\varDelta, \varDelta \star f] = \varDelta \star \varDelta \star f - \varDelta \star f \star \varDelta = N\varDelta \star f \; .$$

Introduction of the "boundary" breaks $su(N)$ to $su(N-1) \oplus u(1)$. Higher rank δ-functions (sums of $\varDelta$'s with ϱ shifted by $\frac{2\pi}{N}$ times an integer) gives $su(N-k) \oplus su(k) \oplus u(1)$.

Let us also show how approach **A** generalises to a situation where the membrane is stretched between two separated parallel $M5$-branes (separation $L/2$) or where the membrane is wound on a non-contractible circle (length L) [14]. The mode expansion of a coordinate of a cylindrical membrane in the separation direction then contains a linear term in addition to the oscillators:

$$Y(\sigma, \varrho) = \frac{L}{2\pi}\varrho + \frac{1}{\pi}\sum_{n=-\infty}^{\infty}\sum_{m=1}^{\infty} y_{nm}\mathrm{e}^{in\sigma}\sin m\varrho \; .$$

The star-adjoint action of $\frac{\varrho}{2\pi}$ is identical to $-\frac{i}{N}\frac{\partial}{\partial_\sigma}$. Therefore, $\frac{\varrho}{2\pi}$ is an outer derivation on the algebra of functions, and its presence means that it is not consistent to truncate in the σ-direction. Truncating in the ϱ-direction only leads to an affine SO(N) algebra. The matrix theory is an "affine matrix theory", or a matrix string theory, which is the dimensional reduction to $D = 2$ of a $D = 6$ SYM theory with a hypermultiplet in the symmetric representation. Note that the coupling constant is relevant, since there is a dimensionless quotient between the eleven-dimensional Planck length and the brane separation.

4 Non-Vanishing C-Field

We now turn to the situation where there is a non-vanishing C-field on the $M5$-brane (the gauge invariant statement is in terms of the self-dual 3-form field strength $F_{(3)} = dB_{(2)} - C_{(3)}$ on the brane). In the process of choosing a light-cône gauge for the membrane, the same is done for the C-field. We choose $C_{-ij} = 0$. Then a self-dual C falls in either of the two classes, modulo choice of frame:
1. "Light-like": $C_{ijk} = 0$, $C_{+-i} = 0$, C_{+ij} self-dual in four dimensions. The transverse rotation are broken as $so(4) \approx su(2) \oplus su(2) \rightarrow su(2) \oplus u(1)$.
2. "Space-like": $C_{+ij} = 0$, $C_{ijk} = \epsilon_{ijkl} C_{+-l}$. Transverse rotations broken as $so(4) \approx su(2) \oplus su(2) \rightarrow (su(2))_{\text{diag}}$.
We now have to include the Wess–Zumino term of (1) in the canonical analysis. The light-cône membrane hamiltonian becomes

$$
\begin{aligned}
H \equiv \frac{p^+ p^-}{A} = \int \mathrm{d}^2\xi \Big[&\frac{1}{2} \Pi_I \Pi_I + \frac{T^2}{4} \{X^I, X^J\}\{X^I, X^J\} \\
&- \frac{T^2}{2} C_{+-J} C_{+-K} \{X^I, X^J\}\{X^I, X^K\} - \frac{T^2}{2} C_{IJK} C_{+-L} \{X^I, X^J\}\{X^K, X^L\} \\
&- \frac{p^+ T}{2A} C_{+IJ} \{X^I, X^J\} \Big] ,
\end{aligned} \tag{9}
$$

where $\Pi_I \,(= \dot{X}_I) = P_I - \frac{T}{2} C_{IJK} \{X^J, X^K\} - T C_{+-J} \{X_I, X^J\}$. In order to identify the connection with SYM, it is useful to form the lagrangian

$$
L = \int \mathrm{d}^2\xi \Big[\frac{1}{2} \dot{X}^I \dot{X}^I + \frac{T}{2} C_{IJK} \dot{X}^I \{X^J, X^K\} + T C_{+-J} \dot{X}^I \{X^I, X^J\} - V(X) \Big] \tag{10}
$$

Due to the difficulties with solving the non-linear boundary conditions in the presence of a C-field, we choose to work in the approach **A**. The light-like case is much simpler, and already well known (although not, to our knowledge, derived from the membrane). There, the last line in (9) represents the only deformation. We note that in the membrane hamiltonian a term

$$
\int \mathrm{d}\sigma \mathrm{d}\varrho \{A, B\} = \Big(\int_{\varrho=0} - \int_{\varrho=\pi} \Big) \mathrm{d}\sigma A \partial_\sigma B
$$

is a cocycle that is not well defined in the matrix truncation (since it is defined using the derivation ϱ. Any boundary term should be represented by a cocycle, defined by a derivation ∂ as $\mathrm{tr}(A[\partial, B])$. Since a finite-dimensional Lie algebra only has inner derivations, one may be lead to conclude that it is necessary to use the affine matrix theory mentioned earlier. This is however not true. The relevant derivation is the truncated δ-function $[\Delta, \cdot]$, which is inner, so that the cocycle $\mathrm{tr}(A, [\Delta, B])$ is exact in the space of functions. We get two equivalent pictures, one with a deformed algebra $[A,B]_k = [A,B] + k\mathrm{tr}(A,[\Delta,B])$, and one with an undeformed algebra (obtained from the deformed generators by a redefinition containing Δ) and a modified trace involving $tr\Delta \neq 0$. This gives a coupling containing the boundary representations N and $\bar{N}$. It amounts to the introduction in the SYM theory of a Fayet–Iliopoulos term [23] by $W^A = W_0^A + \zeta^A$, ζ being a fixed vector in the U(1) direction defined by Δ. It breaks the rotational $so(4)$ symmetry to $su(2) \oplus u(1)$ and leaves supersymmetry unbroken. Its effect is to resolve the singularities of the Higgs branch.

Turning to space-like C-field, we use the self-duality condition[1] on C to rewrite the terms in the lagrangian (10) linear in time derivatives as $\frac{1}{2}C_{+-J}(2\dot{X}^I\{X^I,X^J\}+\epsilon_{KLMJ}\dot{X}^K\{X^L,X^M\})$. Choosing a basis where $C_{+-4} = \gamma = C_{123}$ and splitting quaternions in real and imaginary parts with $X^4 = \Re X$, this can be rewritten as proportional to $f^{\mathcal{ABC}}\Re(\dot{X}^{\mathcal{A}}X^{\mathcal{B}}X^{\star\mathcal{C}})$, where indices $\mathcal{A},\mathcal{B},\ldots$ enumerate the truncated basis of functions. The only contribution from this term which is not a total derivative comes from the cocycle mentioned earlier, and the relevant part is then proportional to $\mathrm{tr}(\dot{\phi}W_0)$, which leads to the conclusion that space-like C-field corresponds to a deformation of the SYM theory given by

$$W = W_0 + \gamma\Im\dot{\phi}\ , \qquad \mu = \mu_0 + \gamma\dot{\psi}\ , \tag{11}$$

where the deformations take values in $u(1)$. Of course, also the potential terms have to be matched against the SYM theory. It is straightforward to show, using the supersymmetry transformations of (6), that this deformation preserves supersymmetry. The details of this are left for a future publication [28], where a fuller account will be given.

5 Conclusions

We have reviewed and constructed matrix theories describing situations where supermembranes end on M-theory 5-branes. Special emphasis has been put on non-vanishing C-field, which is also where the new results are found.

There are some potential applications of the results, that will be investigated in a future publication. One is to obtain the decoupling from gravity in the limit of maximal C-field, the OM limit. For any value of the C-field, we should be able to use our formulation to derive the open membrane metric, which should arise naturally after certain rescalings in the process of matching the truncated membrane hamiltonian to the SYM one.

One of our motivations for initiating this work was the prospect of treating membrane boundary conditions in the presence of non-vanishing C. In order for this to work, and to get information of the generalisation of the string end-point non-commutativity to membrane end-strings, one would need to find a generalisation of the approach **B** described above, so that the string nature of the boundary is preserved. We have not been able to do this. An intriguing observation is that there

[1] We use a linear self-duality condition, although the self-duality on an $M5$-brane should really be non-linear. It is not obvious to us why a linear relation seems to produce the right result.

are two inequivalent cocycles extending an $su(N)$ loop algebra to an affine algebra – the untwisted and the twisted one. The zero-modes of the twisted affine algebra form an $so(N)$ algebra, which certainly indicates a connection to approach **B**. Further investigations along this line of thought might provide interesting results.

References

1. E. Bergshoeff, E. Sezgin and P.K. Townsend: Phys. Lett. **B189**, 75 (1987); Ann. Phys. **185**, 330 (1988).
2. B. de Wit, J. Hoppe and H. Nicolai: Nucl. Phys. **B305**, 545 (1988).
3. T. Banks, W. Fischler, S.H. Shenker and L. Susskind: Phys. Rev. **D55**, 5112 (1997).
4. B. de Wit, M Lüscher and H. Nicolai: Nucl. Phys. **B320**, 135 (1989).
5. M.J. Duff: hep-th/9611203; H. Nicolai and R. Helling: hep-th/9809103; B. de Wit: hep-th/9902051; T. Banks: hep-th/9911068; W. Taylor: hep-th/0101126.
6. S. Sethi and M. Stern: Commun. Math. Phys. **194**, 675 (1998).
7. V.G. Kac and A.V. Smilga: Nucl. Phys. **B571**, 515 (2000).
8. P. Hořava and E. Witten: Nucl. Phys. **B460**, 506 (1996); Nucl. Phys. **B475**, 94 (1996).
9. M. Cederwall: Mod. Phys. Lett. **A12**, 2641 (1997).
10. Ph. Brax and J. Mourad: Phys. Lett. **B408**, 142 (1997); Phys. Lett. **B416**, 295 (1998).
11. C.-S. Chu and E. Sezgin: JHEP **12**, 001 (1997).
12. K. Becker and M. Becker: Nucl. Phys. **B472**, 221 (1996).
13. N. Kim and S.-J. Rey: Nucl. Phys. **B504**, 189 (1997).
14. M. Cederwall: JHEP **12**, 008 (2002).
15. A. Gustavsson and M. Henningson: JHEP **06**, 054 (2001).
16. R. Gopakumar, S. Minwalla, N. Seiberg and A. Strominger: JHEP **08**, 008 (2000); E. Bergshoeff, D.S. Berman, J.P. van der Schaar and P. Sundell: Phys. Lett. **B492**, 193 (2000).
17. J.P. van der Schaar: JHEP **08**, 048 (2001).
18. D.S. Berman, M. Cederwall, U. Gran, H. Larsson, M. Nielsen, B.E.W. Nilsson and P. Sundell: JHEP **02**, 012 (2002).
19. E. Bergshoeff, D.S. Berman, J.P. van der Schaar and P. Sundell: Nucl. Phys. **B590**, 173 (2000).
20. B. de Wit, K. Peeters and J. Plefka: Phys. Lett. **B409**, 117 (1997).
21. K. Ezawa, Y. Matsuo and K. Murakami: Phys. Rev. **D57**, 5118 (1998).
22. Y. Sekino and T. Yoneya: Nucl. Phys. **B619**, 22 (2001).
23. Ori J. Ganor and Joanna L. Karczmarek, JHEP **10**, 024 (2000); M. Berkooz: hep-th/0010158.
24. E. Bergshoeff, E. Sezgin, Y. Tanii and P.K. Townsend: Ann. Phys. **199**, 340 (1990).
25. D.B. Fairlie and C.K. Zachos: Phys. Lett. **B224**, 101 (1989).
26. O. Aharony, M. Berkooz, S. Kachru, N. Seiberg and E. Silverstein: Adv. Theor. Math. Phys. **1**, 148 (1998); O. Aharony, M. Berkooz and N. Seiberg: Adv. Theor. Math. Phys. **2**, 119 (1998).
27. S. Sethi and M. Stern: Nucl. Phys. **B578**, 163 (2000).
28. M. Cederwall and H. Larsson, to appear.

Brane Gravity

Merab Gogberashvili

Andronikashvili Institute of Physics
6 Tamarashvili Str., Tbilisi 0177, Georgia
gogber@ictp.trieste.it

1 Introduction

The scenario where our world is associated with a brane embedded in a higher dimensional space-time with non-factorizable geometry has attracted a lot of interest since the appearance of the papers [1, 2, 3]. Here we shell concentrate on two models of brane gravity considered in the papers [4] and [5, 6].

In brane approach gravity usually is described by multidimensional Einstein equations. However, the difficulties of General Relativity are well known even in four dimensions. As it is known in brane models the gravitational constant can be constructed with the fundamental scale and the brane width [1]. Possibly on the brane not only gravitational constant but Einstein equations can be effective as well. In multi dimensions the equations describing gravity can be quite different from Einstein's equations.

The second section (Sect. 2) devoted to the model where a multi-dimensional vector field is used to describe gravity on the brane [4]. It can be shown that when a brane is embedded in pseudo-Euclidean space, multi-dimensional vector fields together with the brane geometry can imitate Einstein gravity on the brane and solutions of 4-dimensional Einstein's equations could be constructed with the solutions of multidimensional Maxwell's equations. In this picture gravity exhibits tensor character only on the brane and graviton appears to be the combination of two spin-1 massless particles.

Another aspect of brane models considered here is difficulties with gravitational trapping mechanism for matter fields. In the existing (1+4)-dimensional models spin 0 and spin 2 fields can be localized on the brane with an exponentially decreasing gravitational warp factor, spin 1/2 field with an increasing factor [7], and spin 1 fields are not localized at all [8]. For the case of (1+5)-dimensions it was found that spin 0, spin 1 and spin 2 fields are localized on the brane with a decreasing warp factor and spin 1/2 fields again are localized with an increasing factor [9]. So in both (1+4)-, or (1+5)-space models with warped geometry one is required to introduce some non-gravitational interaction in order to localize all the Standard Model particles. For reasons of economy and to avoid charge universality obstruction [10] one would like to have a universal gravitational trapping mechanism for all fields.

In the last section (Sect. 3) the solution of 6-dimensional Einstein equations which localized all kind of bulk fields on the brane is considered [5, 6]. In contrast with the standard approach [3], this solution contain non-exponential scale factor, which increase from the brane, and asymptotically approach a finite value at infinity.

2 Vector Gravity

To show that Einstein equations on the brane can be received from multidimensional vector field equations we start with reminding that any n-dimensional Riemannian space can be embedded into N-dimensional pseudo-Euclidean space with $n \leq N \leq n(n+1)/2$ [11]. Thus, no more than ten dimensions are required to embed any 4-dimensional solution of Einstein's equations with arbitrary energy-momentum tensor. Embedding the space-time with the coordinates x^α and metric $g_{\alpha\beta}$ into pseudo-Euclidean space with Cartesian coordinates ϕ^A and Minkowskian metric η_{AB} is given by

$$\mathrm{d}s^2 = g_{\alpha\beta}\mathrm{d}x^\alpha \mathrm{d}x^\beta = \eta_{AB}h^A_\alpha h^B_\beta \mathrm{d}x^\alpha \mathrm{d}x^\beta = \eta_{AB}\mathrm{d}\phi^A \mathrm{d}\phi^B \; . \tag{1}$$

Capital Latin letters $A, B, \ldots$ labels coordinates of embedded space, while Greek indices $\alpha, \beta, \ldots$ enumerate coordinates in four dimensions. Existence of the embedding (1) demonstrates that the multi-dimensional 'tetrad' fields h^A_α can be expressed as a derivatives of some vector

$$h^A_\alpha = \partial_\alpha \phi^A \; . \tag{2}$$

In four dimensions when tetrad index run over only four values such relation is impossible in general and according to (1) it could be always written in multi dimensions.

Let's suppose that in multi-dimensional flat space-time there exists (1+3)-brane with arbitrary geometry. In order to simplify demonstration of the idea, let us first consider the case of only one extra space-like dimension. Generalization for arbitrary dimensions and signature is obvious.

Let's say that the equation of the branes surface in the Cartesian 5-dimensional coordinates X^A has the form:

$$W(X^A) = 0 \; . \tag{3}$$

Introducing the function

$$\xi(X^A) = \frac{W(X^A)}{\sqrt{|\partial_B W \partial^B W|}} \; , \tag{4}$$

the metric of pseudo-Euclidean bulk (1+4)-space can be transformed to the Gaussian normal coordinates

$$\mathrm{d}s^2 = -d\xi^2 + g_{\alpha\beta}(\xi, x^\nu)\mathrm{d}x^\alpha \mathrm{d}x^\beta \; . \tag{5}$$

Since $\xi = 0$ is the equation of the hyper-surface, the induced metric $g_{\alpha\beta}(0, x^\nu)$, which determines the geometry on the brane, is the same 4-dimensional metric as used in (1) for the embedding.

Introducing unit normal vector to the brane

$$n^A = \partial^A \xi \;|_{\xi=0} \; , \tag{6}$$

one can decompose tensors of bulk space in a standard way (see e.g. [12]).

In the Gaussian system of coordinates (5) the Christoffel symbols on the brane are:

$$\Gamma^\alpha_{\nu\lambda} = h^{A\alpha}\partial_\lambda h_{A\nu} \; . \tag{7}$$

Raising and lowering of Greek indices is made with the induced metric tensor $g_{\alpha\beta}$ and Latin indices with 5-dimensional Minkowskian metric tensor η_{AB}. The Christoffel symbols containing two or three indices ξ are equal to zero. The connections containing just one index ξ are forming outer curvature tensor, which, after using (2), can be written as:

$$K_{\alpha\beta} = n_A D_\beta h^A_\alpha = -\Gamma^\xi_{\alpha\beta} = \partial_\alpha \partial_\beta \phi^\xi \ , \tag{8}$$

where D_β denotes covariant derivatives in Gaussian coordinates (5) and ϕ^ξ is the transversal component of the embedding function.

Since bulk 5-dimensional space-time is pseudo-Euclidean, its scalar curvature is zero:

$$^5R = R + K^2 - K_{\alpha\beta}K^{\alpha\beta} = 0 \ . \tag{9}$$

From this relation the 4-dimensional scalar curvature R can be expressed with the quadratic combinations of the extrinsic curvature $K_{\alpha\beta}$. Thus, using (8) Hilbert's 4-dimensional gravitational action (after removing of boundary terms) can be written in the form:

$$\begin{aligned} S_g &= -M^2_{Pl} \int R\sqrt{-g}\mathrm{d}^4x = \\ &= M^2_{Pl} \int \left(\Delta\phi^\xi \Delta\phi^\xi - \partial_\alpha\partial_\beta\phi^\xi \partial^\alpha\partial^\beta\phi^\xi \right) \sqrt{-g}\mathrm{d}^4x \ , \end{aligned} \tag{10}$$

where $\Delta = \partial_\alpha\partial^\alpha$ is the 4-dimensional wave operator and g is the determinant in Gaussian coordinates. It is clear now that embedding theory allows us to rewrite 4-dimensional gravitational action in terms of derivatives of the normal components of some multi-dimensional vector.

Now let us consider the bulk massless vector field A^B that obeys 5-dimensional Maxwell's equations

$$\partial_A F^{AB} = 0 \ , \tag{11}$$

where $F_{AB} = \partial_A A_B - \partial_B A_A$ is the ordinary field strength. We avoid connection of the functions A^B with the bulk coordinates X^A, not to restrict ourselves with pure geometrical interpretation.

The action for Maxwell's field can be written in the general form:

$$\begin{aligned} S_A &= -\frac{1}{4} \int F_{AB}F^{AB}\mathrm{d}^5X = \\ &= -\frac{1}{2} \int [\partial_A A_B(\partial^A A^B + \partial^B A^A) - 2\partial_A A^A \partial_B A^B]\mathrm{d}^5X \ . \end{aligned} \tag{12}$$

We shall demonstrate below that on the brane this action can be reduced to the 4-dimensional gravity action (10).

Note that so called vacuum gauge fields – the solutions of the equations

$$F_{AB} = 0 \ , \tag{13}$$

are always present in the space-time. These fields are solutions of Maxwell's equations (11) as well. If there are no topological defects in space, the solutions of (13) are pure gauges. The example of non-trivial solution of (13) in space with the linear defect is Aharonov-Bohm field (see e.g. [13]). For the brane the normal to its surface components of vacuum gauge fields also are non-trivial and we shall show that they can resemble gravity on the brane.

In the Gaussian coordinates (5) the ansatz which satisfies (13) and has the symmetries of the brane, with the accuracy of constants, can be written in the form:

$$A^\alpha = G(\xi)\partial^\alpha\phi(x^\beta) , \qquad A^\xi = \phi(x^\beta)\partial^\xi G(\xi) , \tag{14}$$

where $G(\xi)$ and $\phi(x^\beta)$ are some functions depending on the fifth coordinate ξ, and four-coordinates x^β respectively. We assume that $G(\xi)$ is an even function of ξ and the integrals from $G(\xi)$ and from its derivative are convergent. Simple example of such a function is $\exp(\xi^2)$. So, we choose the ansatz:

$$A^\alpha = \frac{c}{\epsilon^{3/2}}\partial^\alpha\phi^\xi(x^\beta)\exp\left(-\frac{\xi^2}{\epsilon^2}\right) , \qquad A^\xi = -\frac{2c}{\epsilon^{7/2}}\,\xi\phi^\xi(x^\beta)\exp\left(-\frac{\xi^2}{\epsilon^2}\right) , \tag{15}$$

where ϵ is the brane width and c is some dimensionless constant. We assume that the width of the brane ϵ is constant all along the surface.

Inserting the ansatz (15) into action integral (12), and integrating by the normal coordinate ξ we receive the induced action on the brane:

$$S_A = \sqrt{\frac{\pi}{2}}\frac{c^2}{\epsilon^2}\int\left(\Delta\phi^\xi\Delta\phi^\xi - \partial_\alpha\partial_\beta\phi^\xi\partial^\alpha\partial^\beta\phi^\xi\right)\sqrt{-g}\mathrm{d}^4x ,$$

where summing is made with the intrinsic metric $g_{\alpha\beta}(x^\nu)$. If we put

$$M_{Pl}^2 = \sqrt{\frac{\pi}{2}}\frac{c^2}{\epsilon^2} , \tag{16}$$

the effective action of 5-dimensional vector field (16) becomes equivalent to Hilbert's action for 4-dimensional gravity (10). So, 4-dimensional Einstein's equations on the brane can be received from Maxwell's multi-dimensional equations in flat space-time.

The same result can be obtained in the case $N > 5$. Now ξ and ϕ^ξ in (6), (8) and (15) must be replaced by ξ^i and ϕ^i, and action integrals (10) and (16) transform to the sum:

$$S_\Sigma = M_{Pl}^2\eta_{ij}\int\left(\Delta\phi^i\Delta\phi^j - \partial_\alpha\partial_\beta\phi^i\partial^\alpha\partial^\beta\phi^j\right)\sqrt{-g}\mathrm{d}^4x , \tag{17}$$

where η_{ij} is the Minkowskian metric of the normal space to the brane. Small Latin indices $i, j, \ldots$ enumerates extra $(N-4)$ coordinates. If we assume that all the brane widths are equal to ϵ for the scale in (17) we have:

$$M_{Pl}^2 = \frac{c^2}{\epsilon^2}\left(\frac{\pi}{2}\right)^{(N-4)/2} . \tag{18}$$

Hence, the equivalence of the descriptions of 4-dimensional gravity both with the intrinsic metric, and with the multi-dimensional vector field was demonstrated [4]. We have not considered here models for the field that forms the brane, as well as the question how the brane geometry is changed because of couplings with 4-dimensional matter was not raised.

3 Localization Problem

Now we consider another model. We want to show that when gravity in multi dimensions is described by Einstein equations there possible to find there solution in six dimensions, which provides trapping of all kind of physical fields on the brane [5, 6].

The general form of action of the gravitating system in six dimensions is

$$S = \int d^6x\sqrt{-{}^6g}\left[\frac{M^4}{2}({}^6R + 2\Lambda) +{}^6 L\right] , \tag{19}$$

where $\sqrt{-{}^6g}$ is the determinant, M is the fundamental scale, ${}^6R$ is the scalar curvature, Λ is the cosmological constant and ${}^6L$ is the Lagrangian of matter fields. All of these quantities are six dimensional.

The 6-dimensional Einstein equations with stress-energy tensor T_{AB} are

$${}^6R_{AB} - \frac{1}{2}g_{AB}\ {}^6R = \frac{1}{M^4}\left(\Lambda g_{AB} + T_{AB}\right) . \tag{20}$$

Capital Latin indices run over $A, B, \ldots = 0, 1, 2, 3, 5, 6$.

For the metric of the 6-dimensional space-time we choose the ansatz [5, 6]

$$ds^2 = \phi^2(r)\eta_{\alpha\beta}(x^\nu)dx^\alpha dx^\beta - \lambda(r)(dr^2 + r^2d\theta^2) , \tag{21}$$

where the Greek indices $\alpha, \beta, \ldots = 0, 1, 2, 3$ refer to 4-dimensional coordinates. The metric of ordinary 4-space, $\eta_{\alpha\beta}(x^\nu)$, has the signature $(+, -, -, -)$. The functions $\phi(r)$ and $\lambda(r)$ depend only on the extra radial coordinate, r, and thus are cylindrically symmetric in the transverse polar coordinates $(0 \le r < \infty, 0 \le \theta < 2\pi)$.

The ansatz (21) is different from the metric investigated in other (1+5)-space models with warped geometry [9, 14, 15]

$$ds^2 = \phi^2(r)\eta_{\alpha\beta}(x^\nu)dx^\alpha dx^\beta - dr^2 - \lambda(r)d\theta^2 . \tag{22}$$

In (21) the independent metric function of the extra space, $\lambda(r)$, serves as a conformal factor for the Euclidean 2-dimensional metric of the transverse space, just as the function $\phi^2(r)$ does for the 4-dimensional part. However, in (22) the function $\lambda(r)$ multiples only the angular part of the metric and corresponds to a cone-like geometry of a string-like defect with a singularity on the brane at $r = 0$.

The stress-energy tensor T_{AB} is assumed to have the form

$$T_{\mu\nu} = -g_{\mu\nu}F(r), \qquad T_{ij} = -g_{ij}K(r), \qquad T_{i\mu} = 0 . \tag{23}$$

Using the ansatz (21), the energy-momentum conservation equation gives a relationship between the two source functions $F(r)$ and $K(r)$ from (23)

$$K' + 4\frac{\phi'}{\phi}(K - F) = 0 . \tag{24}$$

The source functions $F(r)$ and $K(r)$, which satisfy restriction (24) are

$$F(r) = \frac{f_1}{2\phi^2} + \frac{3f_2}{4\phi} , \qquad K(r) = \frac{f_1}{\phi^2} + \frac{f_2}{\phi} , \tag{25}$$

where f_1, f_2 are constants. Note that these source functions do not have a vanishing value at $r \to \infty$, due to the asymptotic behavior of ϕ given in (32).

We require that the 4-dimensional Einstein equations have the ordinary form without a cosmological term

$$R_{\mu\nu} - \frac{1}{2}\eta_{\mu\nu}R = 0 \,. \tag{26}$$

The Ricci tensor in four dimensions $R_{\alpha\beta}$ is constructed from the 4-dimensional metric tensor $\eta_{\alpha\beta}(x^\nu)$ in the standard way. Then with the ansätze (21) and (23) the Einstein field equations (20) become

$$\begin{aligned}
3\frac{\phi''}{\phi} + 3\frac{\phi'}{r\phi} + 3\frac{(\phi')^2}{\phi^2} + \frac{1}{2}\frac{\lambda''}{\lambda} - \frac{1}{2}\frac{(\lambda')^2}{\lambda^2} + \frac{1}{2}\frac{\lambda'}{r\lambda} &= \frac{\lambda}{M^4}[F(r) - \Lambda] \,, \\
\frac{\phi'\lambda'}{\phi\lambda} + 2\frac{\phi'}{r\phi} + 3\frac{(\phi')^2}{\phi^2} &= \frac{\lambda}{2M^4}[K(r) - \Lambda] \,, \\
2\frac{\phi''}{\phi} - \frac{\phi'\lambda'}{\phi\lambda} + 3\frac{(\phi')^2}{\phi^2} &= \frac{\lambda}{2M^4}[K(r) - \Lambda] \,,
\end{aligned} \tag{27}$$

where the prime $= \partial/\partial r$. These equations are for the $\alpha\alpha$, rr, and $\theta\theta$ components respectively.

Subtracting the rr from the $\theta\theta$ equation and multiplying by ϕ/ϕ' we arrive at

$$\frac{\phi''}{\phi'} - \frac{\lambda'}{\lambda} - \frac{1}{r} = 0 \,. \tag{28}$$

This equation has the solution

$$\lambda(r) = \frac{\rho^2\phi'}{r} \,, \tag{29}$$

where ρ is an integration constant with units of length.

System (27), after the insertion of (29), reduces to only one independent equation. Taking either the rr, or $\theta\theta$ component of these equations and multiplying it by $r\phi^4$ gives

$$r\phi^3\phi'' + \phi^3\phi' + 3r\phi^2(\phi')^2 = \frac{\rho^2\phi^4\phi'}{2M^4}[K(r) - \Lambda] \,. \tag{30}$$

We require for ϕ the following boundary conditions near the origin $r = 0$

$$\phi(r \to 0) \approx 1 + \mathrm{d}r^2 \,, \quad \phi'(r \to 0) \approx 2\mathrm{d}r \,, \tag{31}$$

where d is some constant. At infinity we want $\phi(r)$ to behave as

$$\phi(r \to \infty) \to a \,, \qquad \phi'(r \to \infty) \to 0 \,, \tag{32}$$

where $a > 1$ is some constant.

Substituting (25) into (30), taking its first integral and setting the integration constant to zero yields

$$r\phi' = \frac{\rho^2\Lambda}{10M^4}\left(\frac{5f_1}{3\Lambda} + \frac{5f_2}{4\Lambda}\phi - \phi^2\right) . \tag{33}$$

By introducing the parameters A and a such that

$$\frac{\rho^2 \Lambda}{10M^4} = A\ , \qquad f_1 = -\frac{3\Lambda}{5} a\ , \qquad f_2 = \frac{4\Lambda}{5}(a+1)\ , \tag{34}$$

equation (33) becomes

$$r\phi' = A[-a + (a+1)\phi - \phi^2]\ . \tag{35}$$

Equation (35) is easy to integrate and using boundary conditions (31) and (32) the solution corresponding to a non-singular transverse gravitational potential has the form

$$\phi = \frac{3\epsilon^2 + ar^2}{3\epsilon^2 + r^2}\ . \tag{36}$$

From the condition that we have a 6-dimensional Minkowski metric on the brane, $\lambda(r=0) = 1$, (any other value corresponds only to a re-scaling of the extra coordinates) we can fix also the integration constant in (29)

$$\rho^2 = \frac{3\epsilon^2}{2(a-1)}\ . \tag{37}$$

The brane width can be expressed in terms of the bulk cosmological constant and fundamental scale

$$\epsilon^2 = \frac{40M^4}{3\Lambda}\ . \tag{38}$$

Now the metric tensor of the transverse space (29) is not dependent on a and has the form

$$\lambda = \frac{9\epsilon^4}{(3\epsilon^2 + r^2)^2}\ . \tag{39}$$

Using solutions (36), (39) and the relationship (29) to integrate the gravitational part of the action integral (19) over the extra coordinates we find

$$\begin{aligned} S_g &= \frac{M^4}{2}\int \mathrm{d}x^6 \sqrt{-{}^6 g}\ {}^6 R = \frac{M^4}{2}\int_0^{2\pi} \mathrm{d}\theta \int_0^\infty \mathrm{d}r\ r\phi^2\lambda \int \mathrm{d}x^4 \sqrt{-\eta}R = \\ &= \rho^2 \pi M^4 \int_1^a \mathrm{d}\phi\ \phi^2 \int \mathrm{d}x^4\sqrt{-\eta}R = \frac{M^4}{2}\epsilon^2\pi(a^2+a+1)\int \mathrm{d}x^4\sqrt{-\eta}R\ , \end{aligned} \tag{40}$$

where R and η are respectively the scalar curvature and determinant, in four dimensions.

The formula for the effective Planck scale in the model, which is two times the numerical factor in front of the last integral in (40)

$$m_{Pl}^2 = M^4 \pi \epsilon^2 (a^2 + a + 1)\ , \tag{41}$$

is similar to those from the "large" extra dimensions model [1]. The differences are, the presence of the value of gravitational potential at extra infinity, a, in (41), and that the radius of the extra dimensions is replaced by the brane width ϵ, which, as seen from (38), is expressed by the ratio of the fundamental scale M and the cosmological constant Λ.

The normalization condition for a physical field, that its action integral over the extra coordinates r, θ converges, is also the condition for its localization. As was shown in [6] Newtonian gravity is localized on the brane, since the action integral for gravity, (40), is convergent over the extra space.

When wave-functions of matter fields in six dimensions are peaked near the brane in the transverse dimensions there wave-functions on the brane can be factorized as

$$\Xi(x^A) = \frac{\xi(x^\nu)}{\kappa} , \tag{42}$$

where the parameter κ is the value of the constant zero mode with the dimension of length. These parameters can be found from the normalization condition for zero modes

$$\int_0^{2\pi} \mathrm{d}\theta \int_0^\infty \mathrm{d}r \sqrt{-{}^6g}\frac{1}{\kappa^2} = \sqrt{-\eta} , \tag{43}$$

which also guarantees the validity of the equivalence principle for different kinds of particles.

Let us consider the situation with the localization of particular matter fields. If we assume that the zero mode of a spin-0 field, Φ, is independent of the extra coordinates its action can be brought to the form [6]

$$S_\Phi = \int \mathrm{d}^6x\sqrt{-{}^6g}\ {}^6L_\Phi(x^A) = \frac{2\pi}{\kappa_\Phi^2}\int_0^\infty \mathrm{d}r\ r\phi^2\lambda \int \mathrm{d}^4x\sqrt{-\eta}L_\Phi(x^\nu) =$$
$$= \frac{\epsilon^2\pi(a^2+a+1)}{\kappa_\Phi^2}\int \mathrm{d}^4x\sqrt{-\eta}L_\Phi(x^\nu) , \tag{44}$$

where $L_\Phi(x^\nu)$ is the ordinary 4-dimensional Lagrangian of the spin-0 field and κ_Φ is value of the constant zero mode. The integral over r, θ in (44) is finite and the spin-0 field is localized on the brane.

The action for a vector field in the case of constant extra components ($A_i =$ const) also reduces to the 4-dimensional Yang-Mills action multiplied an integral over the extra coordinates [6]

$$S_A = \int \mathrm{d}^6x\sqrt{-{}^6g}\ {}^6L_A(x^B) = \frac{2\pi}{\kappa_A^2}\int_0^\infty \mathrm{d}r\ r\lambda \int \mathrm{d}^4x\sqrt{-\eta}L_A(x^\nu) =$$
$$= \frac{3\epsilon^2\pi}{\kappa_A^2}\int \mathrm{d}^4x\sqrt{-\eta}L_A(x^\nu) , \tag{45}$$

where κ_A is the value of the zero mode of the vector field. The extra integral in (45) is also finite and the gauge field is localized on the brane.

The factorization of the zero mode of a 6-dimensional spinor field in the ansatz (21) is different from the definition (42), having instead the form [6]

$$\Psi(x^A) = \frac{\psi(x^\nu)}{\kappa_\Psi\phi^2\left(r\phi'\right)^{1/4}} , \tag{46}$$

where κ_Ψ is the value of the constant zero mode. Integrating the action of fermions over the extra coordinates, using the explicit form (36), yields

$$S_\Psi = \int \mathrm{d}^6x\sqrt{-{}^6g}\ {}^6L_\Psi(x^A) = \frac{3\pi^2\epsilon^2}{\kappa_\Psi^2\sqrt{2a(a-1)}}\int \mathrm{d}^4x\sqrt{-\eta}L_\Psi(x^\nu) , \tag{47}$$

where L_Ψ is the 4-dimensional Dirac Lagrangian. The integral in (47) over r and θ is finite and Dirac fermions are localized on the brane.

Equating the coefficients of action integrals (44), (45) and (47) to 1, so as to satisfy the normalization condition (43), and to guarantee the equivalence principle for gravity, we find the values of the zero modes for spin 0, spin 1 and spin 1/2 fields

$$\kappa_{\Phi}^2 = \pi\epsilon^2(a^2 + a + 1) , \qquad \kappa_A^2 = 3\pi\epsilon^2 , \qquad \kappa_{\Psi}^2 = \frac{3\pi^2\epsilon^2}{\sqrt{2a(a-1)}} , \tag{48}$$

which are used to parameterize the 4-dimensional fields in the Lagrangians.

To summarize, it is shown that for a realistic form of the brane stress-energy, there exists a static, non-singular solution of the 6-dimensional Einstein equations, which provides a gravitational trapping of 4-dimensional gravity and matter fields on the brane [5, 6].

References

1. N. Arkani-Hamed, S. Dimopoulos and G. Dvali, Phys. Lett., **B 429**, 263 (1998); I. Antoniadis, N. Arkani-Hamed, S. Dimopoulos and G. Dvali, Phys. Lett., **B 436**, 257 (1998).
2. M. Gogberashvili, Int. J. Mod. Phys., **A 14**, 2024 (1999); ibid. **D 11**, 1635 (2002); ibid. **D 11**, 1639 (2002).
3. L. Randall and R. Sundrum, Phys. Rev. Lett., **83**, 3370 (1999); ibid. **83**, 4690 (1999).
4. M. Gogberashvili, Phys. Lett., **B 553**, 284 (2003).
5. M. Gogberashvili and P. Midodashvili, Phys. Lett., **B 515**, 447 (2001); Europhys. Lett., **61**, 308 (2003).
6. M. Gogberashvili and D. Singleton, Phys. Rev., **D**, (2004) Accepted; Phys. lett., **B**, (2004) Accepted.
7. B. Bajc and G. Gabadadze, Phys. Lett., **B 474**, 282 (2000).
8. A. Pomarol, Phys. Lett., **B 486**, 153 (2000).
9. I. Oda, Phys. Rev., **D 62**, 126009 (2000).
10. S.L. Dubovsky, V.A. Rubakov and P.G. Tinyakov, JHEP, **0008**, 041 (2000).
11. L. P. Eisenhart, *Riemannian Geometry* (New Jersey: Princeton University Press, 1949); A. Friedman, J. Math. Mech., **10**, 625 (1961); Rev. Mod. Phys., **37**, 201 (1965).
12. C. W. Misner, K. S. Thorne and J. A. Wheeler, *Gravitation* (San Francisco: Freeman, 1973).
13. M. Peskin and A. Tonomura, *The Aharonov-Bohm Effect* (Berlin: Springer-Verlag, 1989).
14. R. Gregory, Phys. Rev. Lett., **84**, 2564 (2000).
15. T. Gherghetta and M. Shaposhnikov, Phys. Rev. Lett., **85**, 240 (2000).

Stringy de Sitter Brane-Worlds

Tristan Hübsch

Department of Physics and Astronomy, Howard University,
Washington DC, USA
thubsch@howard.edu

Dedicated to the memory of Pavao Senjanović

The possibility that our $3+1$-dimensional world might be a cosmic defect (brane-world) within a higher-dimensional spacetime[1] has recently attracted much interest, owing to the proof [7] that gravity may be localized on such brane-worlds. Randall and Sundrum showed that such geometries may also solve the hierarchy problem [8]. However, it remained unclear whether these and other desirable properties can be achieved within the *same* model.

Herein, we describe a family of stringy toy model brane-worlds [1, 2, 3, 4, 5, 6], which generalize the concept of spacetime variable cosmic strings [9, 10] and exhibits *simultaneously*:

1. exponential hierarchy of Plank mass scales,
2. localized gravity on the brane-world,
3. an induced de Sitter metric on the brane-world,
4. a phenomenologically acceptable value for the cosmological constant,
5. a dynamical mechanism for either trapping the bulk-roaming degrees of freedom to the brane-world, or decoupling them from it,

and where the spacetime geometry is driven by the anisotropy of the axion-dilaton moduli field. Furthermore, the axion-dilaton background configuration possesses crucial stringy $SL(2, \mathbf{Z})$ monodromy, and many of the features are a direct and quantifiable consequence of supersymmetry breaking.

1 A Stringy Family of Toy Models

We begin with a higher-dimensional string [12, 13] or F-theory [11] compactified on a Calabi-Yau (complex) n-fold, some moduli (ϕ^α) of which are allowed to vary over (the "transversal") part of the non-compact space. Following [9, 10], the effective action describing the coupling of the moduli to gravity of the observable spacetime is derived by dimensionally reducing the higher dimensional Einstein-Hilbert action. The relevant part of the low-energy effective D-dimensional action of the moduli, ϕ^α, of the Calabi-Yau n-fold coupled to gravity then reads:

$$S_0 + S^b_{\text{eff}} = \frac{1}{2\kappa^2} \int \mathrm{d}^D x \sqrt{-g}(R - \mathcal{G}_{\alpha\beta} g^{\mu\nu} \partial_\mu \phi^\alpha \partial_\nu \phi^\beta + \ldots) + S^b_{\text{eff}} \,. \tag{1}$$

[1] For a fairly complete bibliography on the subject, see [1, 2, 3, 4, 5, 6].

Here $\mu, \nu = 0, \cdots, D-1$, $2\kappa^2 = 16\pi G_N^D$, where G_N^D is the D-dimensional Newton constant, and $\mathcal{G}_{\alpha\bar{\beta}}$ is the metric on $\mathcal{M}_\phi$, the space of moduli ϕ^α. Higher derivative terms and all other fields in the theory are neglected. We also restrict the moduli to depend on the *"transversal"* coordinates, x_i, $i = D-2, D-1$, and so have a vanishing *"longitudinal"* gradient: $\partial_a \phi = 0$, $a = 0, \cdots, D-3$. S^b_{eff} is a purely $(D-1)$-dimensional effective action describing the (our?) brane-world implied by the explicit form of the solution described below. The moduli obey the equation of motion:

$$g^{ij}\Big(\nabla_i \nabla_j \phi^\alpha + \Gamma^\alpha_{\beta\gamma}(\phi, \bar{\phi}) \partial_i \phi^\beta \partial_j \phi^\gamma\Big) = 0\,, \tag{2}$$

where $\Gamma^\alpha_{\beta\gamma}$ is the Christoffel connection on $\mathcal{M}_\phi$. Note that S^b_{eff} does not depend on the moduli. The Einstein equations are:

$$R_{\mu\nu} - \tfrac{1}{2} 2 g_{\mu\nu} R = T_{\mu\nu}(\phi, \bar{\phi}) + T^b_{\mu\nu}\,, \tag{3}$$

$$T_{\mu\nu} = \mathcal{G}_{\alpha\bar{\beta}}\Big(\partial_\mu \phi^\alpha \partial_\nu \phi^{\bar{\beta}} - \tfrac{1}{2} 2 g_{\mu\nu}\, g^{\rho\sigma} \partial_\rho \phi^\alpha \partial_\sigma \phi^{\bar{\beta}}\Big)\,, \tag{4}$$

where $T^b_{\mu\nu}$ is a delta-function source, as shown below.

1.1 Matter

For our family of toy models, we *choose*: $\phi^\alpha = \tau := \mathrm{a} + i e^{-\Phi}$, representing the axion-dilaton system of the $D = 10$ Type IIB string theory, thought of as a T^2-compactification of F-Theory, so $\mathcal{G}_{\tau\bar{\tau}} = [\Im m(\tau)]^{-2}$ is the Teichmüller metric [11]. Now, we *assume* that $\tau = \tau(\theta)$, where $\theta = \arctan(\frac{x_{D-1}}{x_{D-2}})$ is the "polar" angle in the transversal (x_{D-2}, x_{D-1})-plane. With this, (2) becomes:

$$\tau'' + \frac{2}{\bar{\tau} - \tau} = 0\,, \tag{5}$$

and is solved by:

$$\tau_I(\theta) = \mathrm{a}_0 + i\, g_s^{-1}\, e^{\omega(\theta - \theta_0)}\,, \qquad \mathrm{a}_0, g_s, \omega, \theta_0 = \text{const}\,, \tag{6}$$

$$\tau_{II}(\theta) = \mathrm{a}_0 \pm g_s^{-1} \frac{\sinh[\omega(\theta - \theta_0)] + i}{\cosh[\omega(\theta - \theta_0)]}\,, \tag{7}$$

which satisfies our *requirement* that its energy-momentum tensor be a constant[2], $\propto \omega^2$. Both solutions are discontinuous across the branch-cut, $(\theta - \theta_0) = \pm\pi$, but the constants $\mathrm{a}_0, g_s, \omega$ may be chosen so that $\tau(\theta_0 + \pi) = M{\cdot}\tau(\theta_0 - \pi)$, where $M \in SL(2, \mathbf{Z})$. That is, both τ_I and τ_{II} exhibit (different) non-trivial $SL(2, \mathbf{Z})$ *monodromy* [1, 2].

The absence (in the limit of exact supersymmetry) of a potential for τ, and their nontrivial $SL(2, \mathbf{A})$ monodromy enforces the conclusion:

The metric-moduli system (1), (6-7) can only stem from a string theory.

---

[2] Requiring that $\tau = \tau(\theta)$ and that its energy-momentum tensor be constant permits solving for the metric as independent of θ by means of separation of variables.

1.2 Minkowski Metric

With a phenomenologically interesting $K3$ compactification of the $D = 10$ solution in the back of our minds (upon which the metric receives α' corrections), we however keep D unspecified for the sake of generality. The metric that interpolates between the two solutions of [1, 2], with $z = \log(r/\lambda)$, is:

$$\mathrm{d}s^2 = A(z)\eta_{ab}\mathrm{d}x^a\mathrm{d}x^b + \lambda^2 B(z)(\mathrm{d}z^2 + \mathrm{d}\theta^2)\ , \tag{8}$$

$$A(z) = Z^{\frac{2}{D-2}}\ , \qquad Z(z) := 1 + a_0|z|\ , \tag{9}$$

$$B(z) = Z^{-\frac{D-3}{D-2}}\,\mathrm{e}^{\frac{\xi}{a_0}(\beta - Z^2)}\ , \tag{10}$$

Here ξ, a_0 and β are free parameters, $\lambda \sim O(M_D^{-1})$ sets the transversal length scale, and η_{ab} is the Minkowski metric along the $(D-2)$-dimensional brane-world. The dependence on $|z|$ (in place of just z in [1, 2]) induces the δ-function terms in (3) (with $\varrho := a_0\mathrm{e}^{-\frac{\xi}{a_0}(\beta-1)}[\frac{D-3}{D-2} - 2\frac{\xi}{a_0}]$)

$$-\eta_{ab}\,\lambda^{-2}\Big[a_0\xi\,\mathrm{sign}^2(z)\,Z^{\frac{D-1}{D-2}}\,\mathrm{e}^{-\frac{\xi}{a_0}(\beta-Z^2)} - \varrho\,\delta(z)\Big] = T_{ab} + T^b_{ab}\ , \tag{11}$$

$$-a_0\xi\,\mathrm{sign}^2(z) = T_{zz} + T^b_{zz}\ , \tag{12}$$

$$+a_0\xi\,\mathrm{sign}^2(z) + 2a_0\,\delta(z) = T_{\theta\theta} + T^b_{\theta\theta}\ . \tag{13}$$

Hereafter, we refer to the brane at $z = 0$ as the *brane-world*: there, $\mathrm{sign}^2(z) = 0$ and so $T_{\mu\nu} = 0$ also. On the other hand, the δ-function terms in the left-hand side of the Einstein equations (3) are now non-zero and read:

$$T^b_{\mu\nu} = \lambda^{-2}\,\mathrm{diag}\Big[-\varrho, \varrho, \ldots, \varrho, 0, 2a_0\Big]\,\delta(z)\ . \tag{14}$$

Since τ depends on ω (which (11–13) fix to $\omega^2 \equiv 8a_0\xi \geq 0$), we see that $\varrho \geq 0$ and $\mathrm{sign}(T^b_{00}) = -\mathrm{sign}(a_0)$ for $|\xi| \leq |\xi_c| := \frac{1}{2}\frac{D-3}{D-2}|a_0|$. In particular, $T^b_{00} \geq 0$ in the $a_0 \leq 0$ case, when z is restricted between the naked (null) singularities at $z = \pm 1/a_0$. When $\xi = 0$ this form of the stress tensor is similar to that of spatial domain walls [14, 15], in which the surface energy density, σ, is equal to the surface tension, $-p$, where p is the pressure along the domain wall. In our case, however $|T^b_{\theta\theta}| > T^b_{00}$. From this it follows that the weak energy condition holds except for $T^b_{\theta\theta}$, i.e., $T^b_{\mu\nu}\zeta^\mu\zeta^\nu < 0$ only for the null vector $\zeta^\mu = (1, 0, \cdots, 0, \sqrt{A/B})$ representing a vortex in the transversal (z, θ)-plane. (For a related discussion of this feature of co-dimension two solutions consult for example [16].) Still, we *assume* that it is possible to associate an effective action for the source at $z = 0$,

$$S^b_{\mathrm{eff}} = \int \mathrm{d}^{D-2}x\,\mathrm{d}z\,\mathrm{d}\theta\,\sqrt{-g}\,\delta(z)\,\lambda\,\mathcal{L}^b\ , \tag{15}$$

depending on all matter localized[3] to this (our?) brane-world. Equating the $T^b_{\mu\nu}$ calculated from (15) with the δ-function contribution of the Einstein equations from (11–13), we obtain that

$$\lambda \sim -a_0\lambda^{-2}\mathrm{e}^{-\frac{\xi}{a_0}(\beta-1)} \qquad |\xi| \ll |\xi_c|\ . \tag{16}$$

[3] Localization of matter is a generic feature in superstring theories [17].

Note that the vacuum energy which couples to gravity is $\lambda\,\mathcal{L}^b = -\lambda$. Analogous results hold also in the $a_0 > 0$ and $\xi > \xi_c$ case. However, now $T^b_{\mu\nu}\zeta^\mu\zeta^\nu > 0$ for *all* null vectors. We thus have two subfamilies of solutions:

1. $a_0, (|\xi| - |\xi_c|) < 0$, where $T^b_{\mu\nu}\zeta^\mu\zeta^\nu < 0$ only for $\zeta^\mu = (1, 0, \cdots, 0, \sqrt{A/B})$, the brane-world is encircled by naked singularities at $z = \pm 1/a_0$;
2. $a_0, (|\xi| - |\xi_c|) > 0$, where $T^b_{\mu\nu}\zeta^\mu\zeta^\nu > 0$ for *all* null vectors, and the transverse space is infinite, with temporal singularities at $z = \pm\infty$.

Incidentally, replacing η_{ab} with *any Ricci-flat metric* (e.g., the Schwarzschild geometry), leaves the above solutions unchanged.

1.3 de Sitter Metric

Now modify (8) into:

$$\mathrm{d}s^2 = \tilde{A}^2(z)\,\tilde{g}_{ab}\,\mathrm{d}x^a\mathrm{d}x^b + \lambda^2\tilde{B}^2(z)\,(\mathrm{d}z^2 + \mathrm{d}\theta^2)\,, \tag{17}$$

$$[\tilde{g}_{ab}] = \mathrm{diag}[-1, \mathrm{e}^{2\sqrt{\Lambda}x^0}, \cdots, \mathrm{e}^{2\sqrt{\Lambda}x^0}]\,, \tag{18}$$

where Λ is the cosmological constant for the brane-world spacetime. (Note that there is no cosmological constant in D-dimensional spacetime of the original string or F-theory!)

The closed-form solutions (9–10) no longer apply. Instead, the purely longitudinal part of (3) reduces to a single equation, giving:

$$\tilde{B}^2 = \lambda^{-2}\Lambda^{-1}\frac{h''h^{-\frac{D-4}{D-2}}}{(D-2)}\,, \qquad h(z) := \tilde{A}(z)^{D-2}\,, \tag{19}$$

which determines $\tilde{B}(z)$ in terms of $h(z)$ and so $\tilde{A}(z)$. Upon this substitution, the remaining components of (3) produce the following single equation[4]:

$$\frac{1}{2(D-2)}\frac{h'^2}{h^2} - \frac{h''}{2h} + \frac{h'h'''}{2hh''} + \frac{1}{8}\omega^2 = 0\,. \tag{20}$$

This implies that $\Lambda > 0$, and that the Ansatz (17–18) does not permit a double Wick rotation into an anti-de Sitter spacetime, and conversely that our solution cannot be obtained from any anti-de Sitter solution of string theory. To see this, note that (20) determines $h(z)$, and hence $\tilde{A}(z)$, to be independent of Λ. But then, $\Lambda \to -\Lambda$ in (19) would imply $\tilde{B}(z)^2 < 0$, making the entire plane transverse to the cosmic brane also time-like.

Furthermore, with $h(z) = (1 - z/z_0)^{D-2}$, and so with

$$\tilde{A}_0(z) = \tilde{Z}(z) := (1 - z/z_0)\,, \quad \text{and} \quad \tilde{B}_0(z) := \frac{1}{\lambda z_0\sqrt{\Lambda}}\,, \tag{21}$$

the metric (17) satisfies the Einstein equations (3) for $\omega^2 = 0$, i.e., when $\tau = \text{const.}$ This solution describes the familiar Rindler space [18].

[4] It is straightforward to show that R_{zz} and $R_{\theta\theta}$ can be written as certain linear combinations of the left-hand side of the differential equation (20) and its derivatives.

For $\omega \neq 0$ ($\tau \neq$ const.), (20) has a perturbative solution[5] by expanding around the horizon, $\tilde{Z}(z) = 0$:

$$\tilde{A}(z) = \tilde{Z}(z)\Big(1 - \frac{\omega^2 z_0^2 (D-3)}{24(D-1)(D-2)}\tilde{Z}(z)^2 + O(\omega^4)\Big) , \tag{22}$$

$$\tilde{B}(z) = \frac{1}{\lambda z_0 \sqrt{\Lambda}}\Big(1 - \frac{\omega^2 z_0^2}{8(D-1)}\tilde{Z}(z)^2 + O(\omega^4)\Big) . \tag{23}$$

Notice that, depending on $\tilde{A}(z)^2$ and $\tilde{B}(z)^2$, the metric (17) is well-defined for all values of z, with merely a *horizon* [19, 20] at $z = z_0$. It is easy to check that for our solution (22–23) both the Ricci scalar and tensor vanish at $z = z_0$, as does the whole Riemann tensor. In fact, these tensors as well as the $R_{\mu\nu}R^{\mu\nu}$ and $R_{\mu\nu\rho\sigma}R^{\mu\nu\rho\sigma}$ curvature scalars all remain bounded for all finite z. So, close to the horizon spacetime is asymptotically flat in agreement with the behavior of Rindler space, see (21)–(23) [18]. However, the horizon does provide an effective cut-off of spacetime and, as usual in de Sitter space, we will only consider the degrees of freedom inside this horizon.

In contrast, when $\Lambda = 0$, the solution (9–10) with $a_0 < 0$ exhibits a naked singularity, at $z = \pm a_0^{-1}$ ($Z = 0$), for the global cosmic brane and the region $|z| > |a_0^{-1}|$ ($Z < 0$) is unphysical: the metric becomes complex. In comparing (9) with the de Sitter solution (22), *the singularity is effectively removed* by introducing a non-zero longitudinal cosmological constant. Note that in solving (20), $h'' \neq 0$ was assumed. But $\Lambda \to 0$ implies that $h'' \to 0$, which gives rise to the solution of $A(z)$ in (9), with (19) no longer valid.

While the naked singularity of the Minkowski solution (9–10) has been removed by the non-zero Λ, away from the horizon this Minkowski solution is still a good approximation to (23). To compare, we first obtain a power series solution of (23), expanding around the core, at $z = 0$. From this we determine the lowest order terms[6] in $h(z) = \sum_{n=0} h_n z^n$. Finally, we expand $\tilde{A}(z)$ and $\tilde{B}(z)$, expressed as functions of $h(z)$ and $h''(z)$ to lowest order in z,

$$\tilde{A}(z) = \left(1 + z\frac{h_1}{(D-2)}\right) , \; \tilde{B}(z) = \sqrt{\frac{2h_2}{(D-2)\Lambda\lambda^2}}\left[1 + z\left(\frac{3h_3}{2h_2} - \frac{h_1(D-4)}{2(D-2)}\right)\right] . \tag{24}$$

Here, the coefficients h_i for $i > 2$ are determined in terms of h_0, h_1, h_2 by (20), the overall rescaling of $\tilde{A}(z)$ and $\tilde{B}(z)$ is absorbed in a rescaling of x^a and λ, respectively, and the numerical values of h_1, h_2 are determined by comparison with the expansions (23). Comparing now (24) with (9–10), expanded to first order in z, leads to

$$a_0 \approx 0.9\frac{(D-2)}{\rho_0} , \quad \xi \approx \frac{1}{0.9}\frac{\omega^2 \rho_0}{8(D-2)} , \quad \text{and} \quad \frac{\omega^2}{2(D-2)\Lambda\lambda^2} = 1 . \tag{25}$$

[5] This solution is of the same form as that discussed by Gregory [19, 20] for the $U(1)$ vortex solution.

[6] This requires an initial guess for the value of $\omega^2\rho_0^2$ and that the higher order corrections in the expansion of $\tilde{A}(z)$ in terms of $\tilde{Z}(z)$ fall off fast enough. Indeed, we have computed the expansion of $\tilde{A}(z)$ to $O(\tilde{Z}^{12}(z))$, and determined $\omega^2\rho_0^2$ and the corresponding numerical values of the coefficients h_i recursively.

The last of the identifications (25) implies:

$$\Lambda = \frac{\omega^2}{2(D-2)\lambda^2}\,, \tag{26}$$

thus *expressing the cosmological constant in terms in the brane-world of the transversal anisotropy of the axion-dilaton system*! This gives a very non-trivial relation between the stringy moduli, and hence string theory itself, and a *positive* cosmological constant Λ. Since the dilaton is $\Phi = -\omega\theta$ it also follows from (26) that we have a strongly coupled theory[7].

Note also that $\Lambda \sim \omega^2/\lambda^2$ implies that supersymmetry breaking and a non-zero cosmological constant are related: In our family of toy models, supersymmetry is explicitly broken by $\omega^2 \neq 0$. But since $\Lambda \sim \omega^2/\lambda^2$, supersymmetry breaking by $\omega^2 \neq 0$ also induces a positive cosmological constant, which then can vanish only in the decompactifying limit, $\lambda \to \infty$. In the limit $\omega^2 = 0$ we recover supersymmetry and thus have a possible (supersymmetric) F-theory [11] background.

The cosmological constant on the brane-world is thus induced by the supersymmetry breaking caused by the anisotropy of the axion-dilaton system.

2 Localization of Gravity and Planck Mass

Unlike in the original Randall-Sundrum models [7, 8] (and, to the best of my knowledge, also any other brane-world model), for a suitable choice of parameters, the above family of toy models exhibits *both* an exponentially large hierarchy *and* localized gravity [4].

2.1 Exponential Hierarchy

The large hierarchy between the $(D-2)$- and D-dimensional Planck scales is the same as in [1, 2]:

$$M_{D-2}^{D-4} = M_D^{D-2} \int_{M_\perp} \mathrm{d}v\,\mathrm{d}\theta\; \psi_0^2(v) = M_D^{D-2}\frac{2\pi\lambda^2}{|a_0|}\mathrm{e}^{\frac{\beta\xi}{a_0}}\left(\frac{a_0}{\xi}\right)^{\frac{D-3}{2(D-2)}} I^D_{a_0,\xi}\,, \tag{27}$$

$$I^D_{a_0,\xi} = \begin{cases} \left[\Gamma\left(\frac{D-3}{2(D-2)}\right) - \gamma\left(\frac{D-3}{2(D-2)};\frac{\xi}{a_0}\right)\right] & \text{for } a_0 > 0,\\ \gamma\left(\frac{D-3}{2(D-2)};\frac{\xi}{a_0}\right) & \text{for } a_0 < 0. \end{cases} \tag{28}$$

where $M_\perp$ denotes the hyperbolic transverse space [2]. Note that the large hierarchy is controlled by the product of β and the ratio $\frac{\xi}{a_0} > 0$, where the positivity of the latter is due to the presence of the non-trivial stringy moduli. It is therefore possible to choose $\frac{\beta\xi}{a_0}$, so as to have a large hierarchy between M_D and M_{D-2}.

Following the discussion of Randall and Sundrum [8] we compute the coupling of gravity to the fields on the brane. Writing, $\bar{g}_{\mu\nu} := g_{\mu\nu}|_{z=0}$ for the metric on

---

[7] Recall: with $\tau = a_0 + i g_s^{-1}\exp(\omega\theta)$, the $SL(2,\mathbf{Z})$ monodromy sets $g_s^D \sim O(1)$ in D dimensions. However, in the $D-2$-dimensional brane-world, $g_s^{D-2} = g_s^D\sqrt{\alpha'/V_\perp}$, and since the volume of the transverse space, $V_\perp$, is large (27), $g_s^{D-2} \ll 1$.

the brane-world, there is a non-trivial contribution from $\sqrt{-g}$, i.e., $\sqrt{-g}|_{z=0} = \sqrt{-\bar{g}}\,\lambda^2 e^{(\beta-1)\xi/a_0}$. Hence, (15) becomes

$$S^b_{\rm eff} \sim -a_0 \int \mathrm{d}^{D-2}x \mathrm{d}\theta \,\sqrt{-\bar{g}}\,\mathcal{L}^b\,, \tag{29}$$

where we have taken into account the tension for the brane-world according to (16). Thus, unlike in [8], here the fields, masses, couplings and vev's in $\mathcal{L}^b$ retain their fundamental, D-dimensional value, $O(M_D)$. Also, using (8), the kinetic terms of a typical field, Ψ, expand

$$|\partial_\mu \Psi|^2 = |\partial_{\|}\Psi|^2 + \lambda^{-2} e^{-\frac{\xi}{a_0}(\beta-1)} |\partial_{\perp}\Psi|\,, \tag{30}$$

so that the transverse excitations of Ψ are exponentially suppressed.

2.2 Localization of Gravity

To understand the localization of gravity, we look at small gravitational fluctuations $\delta\eta_{ab} = h_{ab}$ of the longitudinal part of the metric[8]. From the Einstein equations, h_{ab} satisfies a wave equation of the form [21]:

$$\Box h_{ab} = \frac{1}{\sqrt{-g}}\,\partial_\mu(\sqrt{-g} g^{\mu\nu}\partial_\nu h_{ab}) = 0\,. \tag{31}$$

Following [22], we change coordinates:

$$\mathrm{d}v = \lambda\, Z^{-\frac{(D-1)}{2(D-2)}}\, e^{\frac{\xi}{2a_0}(\beta - Z^2)}\, \mathrm{d}z\,, \tag{32}$$

$$\mathrm{d}s^2 = A(v)\eta_{ab}\mathrm{d}x^a \mathrm{d}x^b + A(v)\mathrm{d}v^2 + B(v)\lambda^2 \mathrm{d}\theta^2\,, \tag{33}$$

and use the following Ansatz

$$h_{ab} = \epsilon_{ab} e^{\mathrm{i}p\cdot x} e^{\mathrm{i}n\theta}\, \frac{\phi}{\psi_0}\,, \tag{34}$$

dictated by the isometrics of the metric and where

$$\psi_0 := \sqrt{A^{-1}\sqrt{-g}} = \sqrt{A^{\frac{D-3}{2}} B^{\frac{1}{2}}} = Z^{\frac{D-3}{4(D-2)}}\, e^{\frac{\xi}{4a_0}(\beta - Z^2)}\,. \tag{35}$$

With these variables [1, 2, 22], (31) becomes a Schrödinger-like equation:

$$-\phi'' + \left(\frac{\psi_0''}{\psi_0} + \frac{A}{B} n^2\right)\phi = m^2\phi\,. \tag{36}$$

For simplicity, set $n = 0$. Integrating (32) gives

[8] Owing to the dearth of solutions in closed form, the analogous discussion of the de Sitter case (17–18), (22–23) is technically rather more involved, albeit just as straightforward conceptually.

$$v - v_0 = \mathrm{sign}(z) v_* \Big[\gamma_0^{-1}\gamma\Big(\frac{D-3}{4(D-2)}; \frac{\xi Z(z)^2}{2a_0}\Big) - 1\Big] ,$$

$$v_* = \frac{\lambda}{2a_0}\, \mathrm{e}^{\frac{\beta\xi}{2a_0}} \Big(\frac{2a_0}{\xi}\Big)^{\frac{D-3}{4(D-2)}} \gamma_0 , \qquad \gamma_0 = \gamma\Big(\frac{D-3}{4(D-2)}; \frac{\xi}{2a_0}\Big) . \tag{37}$$

The change of variables $z \to v$ is single-valued, continuous and smooth across $z = 0$, and $\mathrm{sign}(z) = \mathrm{sign}(v - v_0)$. However, the appearance of the "incomplete gamma function." $\gamma(a; x)$ prevents an *explicit* inversion of $v = v(z)$, and evaluation of ψ_0''/ψ_0 in (36). Nevertheless, in the $\xi \to 0$ limit:

$$\tilde{v} - v_0 = \mathrm{sign}(z)\, \tilde{v}_* \Big[Z(z)^{\frac{D-3}{2(D-2)}} - 1\Big] , \qquad \tilde{v}_* := \frac{2(D-2)}{D-3} \frac{\lambda}{a_0} , \tag{38}$$

which is easy to invert *explicitly*. Hereafter, we set $v_0 = 0$, focus on small but nonzero $|\xi|$ and drop the tilde. (Equivalently, we could consider the case in which $a_0, \xi > 0$ and expand around $v_\infty = \lim_{|z|\to\infty} v$, where the result is exactly the same as in the situation considered here [4].)

Equation (36) can now be written as

$$-\Big[\frac{\mathrm{d}^2}{\mathrm{d}^2 v} + \frac{\mathrm{sign}^2(v)}{4(|v_*| - |v|)^2} + \frac{1}{|v_*|}\delta(v)\Big]\phi_m = m^2 \phi_m . \tag{39}$$

Away from $v = 0$, this becomes the Bessel equation, so

$$\phi_m = a_m \sqrt{|v_*| - |v|}\, J_0\Big(m(|v_*| - |v|)\Big) + b_m \sqrt{|v_*| - |v|}\, Y_0\Big(m(|v_*| - |v|)\Big) , \tag{40}$$

which must satisfy the δ-function matching conditions at $v = 0$:

$$\Big[2\frac{\mathrm{d}\phi_m}{\mathrm{d}v} + \frac{1}{|v_*|}\phi_m\Big]_{v=0} = 0 . \tag{41}$$

Evaluating (40) for small values of m,

$$\phi_m \sim a_m \sqrt{|v_*| - |v|}\, 1 + b_m \sqrt{|v_*| - |v|}\, 2 \log\Big(\frac{1}{2} m(|v_*| - |v|)\Big) , \tag{42}$$

it is clear that (41) is satisfied only if $b_m = 0$.

It remains to determine a_m such that the normalization integral of ϕ_m is m-independent

$$\langle\phi_m|\phi_m\rangle = a_m^2 \int_{-v_*}^{v_*} \mathrm{d}v\ (|v_*| - |v|) J_0^2(m(|v_*| - |v|)) . \tag{43}$$

This integral can in fact be computed exactly, and turns out to be dominated by the plane wave approximation, i.e.,

$$J_0^2(m|v_*|) \sim \frac{\cos^2(m|v_*| - \pi/4)}{m|v_*|} , \qquad m|v_*| \gg 1 . \tag{44}$$

and is "regularized" by $|\xi| > 0$ [4]. The zero-mode wave function can be expressed in terms of v, and the normalization integral for ψ_0 becomes:

$$\langle\psi_0|\psi_0\rangle = 2\pi \int \mathrm{d}v\ |\psi_0|^2 = \frac{2(D-2)}{D-3} \frac{2\pi\lambda^2}{|a_0|}\, \mathrm{e}^{\frac{\beta|\xi|}{|a_0|}} . \tag{45}$$

When we compare this expression with the exact result (28), the only discrepancy occurs in the power of a_0/ξ and the overall $O(1)$ numerical factor. Similar arguments apply for the ϕ_m when $\xi \neq 0$. For the normalization $\langle\phi_m|\phi_m\rangle$ we get,

$$\langle\phi_m|\phi_m\rangle_{\xi\neq 0} = \mathrm{e}^{\frac{\beta|\xi|}{|a_0|}}\langle\phi_m|\phi_m\rangle_{\xi=0} \ . \tag{46}$$

The plane wave approximation is valid for $\xi \neq 0$ because $v_* \sim \lambda/|a_0|\exp(\frac{\beta|\xi|}{|a_0|})$ and hence when m is large, $mv_* \gg 1$. Since large m's are limited by λ^{-1}, we can compute $\langle\phi_m|\phi_m\rangle$ by looking at large mv_* for which the Bessel function, J_0, looks like a plane wave (44). This means that $\langle\phi_m|\phi_m\rangle \sim a_m^2 m^{-1} v_*$ and we have to choose $a_m \sim \sqrt{m}$. Since $v_* \sim \lambda$, then $\phi_m \sim \sqrt{m\lambda}$.

Thus, $\psi_0 \neq \lim_{m\to 0}\phi_m$, i.e., the *non-trivial stringy moduli guarantee localized gravity at* $z=0$ through the existence of the isolated zero mode.

With these, the Newton potential takes the following form [1, 2, 4]:

$$U(r) = \frac{1}{M_D^{D-2}}\frac{M_1 M_2}{r}\left(1 + \frac{\lambda^3}{r^3} + \cdots\right) , \tag{47}$$

where the correction term does not depend on a_0, β or ξ, and is very small. For example, $M_D \sim$ TeV, since $\lambda \sim (M_D)^{-1}$. The Newton potential has only been checked down to $r_e \sim 1\,\mathrm{mm} \sim 10^{-12}\,\mathrm{GeV}^{-1}$, so that $\lambda/r < \lambda/r_e \sim 10^{-15}$.

3 Dynamical Decoupling From, or Trapping of Bulk-Roaming Modes

In addition to the degrees of freedom discussed above, typical higher-dimensional models also include degrees of freedom of various spatial extendedness (many of which describable as D-brane probes) and Yang-Mills gauge fields. For the latter, we assume that a variation of the argument shown above for gravity will similarly localize the Coulomb forces, and it remains to discuss bulk-roaming D-brane probes.

In any brane-world cosmological model, "matter" degrees of freedom that are not localized to the brane-world through a "topological" mechanism [17], inevitably are permitted to roam the higher-dimensional bulk of the spacetime. Since the brane-world is embedded in the bulk spacetime, this bulk-roaming matter will pass through the brane-world. Unless its interactions with *all* of the brane-world matter and *all* localized gauge fields (including gravity) are for some reason negligibly small, this will violate brane-world conservation laws. Surprisingly, our family of toy models includes an automatic dynamical mechanism for "stabilization" in this respect.

References [2, 3] have analyzed the dynamics of D-brane probes in the vicinity of the naked singularities using the appropriate Born-Infeld action [23, 24, 25, 26, 27]:

$$S_{\mathrm{BI}} = 2\pi(2\pi\sqrt{\alpha'})^{-(p+1)}\int \mathrm{d}^{p+1}x\,\left[C_{p+1} - \mathrm{e}^{-\Phi}\sqrt{-\det G^s_{ab}}\right] , \tag{48}$$

where G^s_{ab} is the metric on the brane-probe (of p-dimensional spatial extent) induced from the background string frame metric by embedding the brane coordinates along the spacetime ones. C_{p+1} is the potential whose field strength is dual to

$F := \mathrm{d}a$, where a is the axion. The induced string frame metric on the brane-probe is (v is now the speed of the brane-probe!)

$$[G^s_{ab}] = \mathrm{e}^{\Phi/2} g_s^{-1/2} \mathrm{diag}[\,(\mathrm{e}^{2B} v^2 - \mathrm{e}^{2A})\,,\, \underbrace{\mathrm{e}^{2A}, \cdots, \mathrm{e}^{2A}}_{p}\,]\,, \tag{49}$$

whereby the action (48) may be formally identified with that of a relativistic particle, in which the rôle of "mass" and "speed of light" are played by rather complicated functions of the dilaton, Φ, and the metric warp factors A, B.

Unlike the supersymmetric case [27] where the effective potential (the negative of the Lagrangian evaluated at $v = 0$) vanishes, in our case V_{eff} turns out to be a linear function of $Z(|z|)$ [2, 3]: another consequence of supersymmetry breaking. In the two subfamilies of toy models described in Sect. 1.2, this potential has the form depicted in Fig. 1.

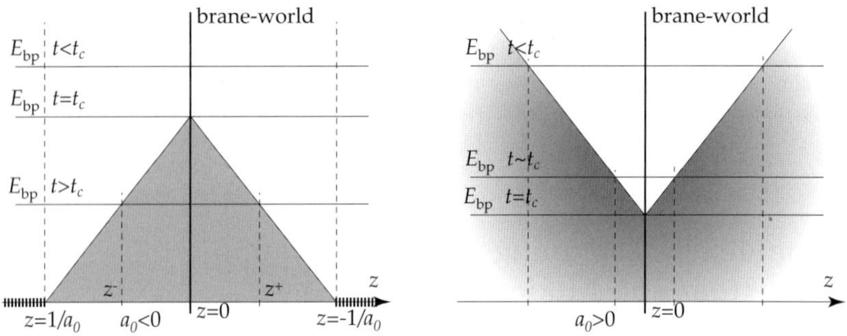

Fig. 1. The two scenarios of Sect. 1.2: *On the left*, naked singularities encircle the brane-world and the effective potential for the brane-probes falls off linearly. Thus, as a brane-probe looses its energy through *gravischtrahlung*, the region of the transversal plane with the brane-world becomes inaccessible to it: it *dynamically decouples from the brane-world. On the right*, the transversal plane extends to $z = \pm\infty$, and the effective potential for the brane-probe rises linearly. Now, as a brane-probe looses its energy, it becomes confined to a diminishing extent of the transversal plane, and eventually becomes *dynamically trapped to the brane-world*

Thus, *owing to supersymmetry breaking caused by the anisotropy of the axion-dilaton system, brane-probes either decouple (when $a_0 < 0$) from the brane-world, or become trapped (when $a_0 > 0$) in it.*

In the first case, all bulk-roaming modes *eventually* decouple from the brane-world at $z = 0$. In the second, all modes *eventually* become trapped, i.e., localized to the brane-world. In fact, it is amusing to realize that the latter process of localization would, from the point of view of a brane-world observer, seem as *creation of matter from nothing* – indeed, conceivably, of all of the brane-world matter.

Acknowledgements

This article is an abridged and annotated summary of [1, 2, 3, 4, 5, 6], and I am indebted to my co-authors, P. Berg lund and Đ. Minić, for all they have taug ht me; the errors however are entirely mine. I also wish to thank The US Department of Energy for their generous support under grant number DE-FG02-94ER-40854.

References

1. P. Berglund, T. Hübsch and D. Minic: *JHEP* **09**(2000)015 (hep-th/0005162).
2. P. Berglund, T. Hübsch and D. Minic: *JHEP* **02**(2001)010 (hep-th/0012042).
3. P. Berglund, T. Hübsch and D. Minic: *JHEP* **01**(2001)041 (hep-th/0012180).
4. P. Berglund, T. Hübsch and D. Minic: *Phys. Lett.* **512**(2001)155 (hep-th/0104057).
5. P. Berglund, T. Hübsch and D. Minic: *Phys. Lett.* **B534**(2002)147 (hep-th/0112079).
6. P. Berglund, T. Hübsch and D. Minic: *Phys. Rev.* **D67**(2003)041901 (hep-th/0201187).
7. L. Randall and R. Sundrum: *Phys. Rev. Lett.* **83**(1999)4690–4693 (hep-th/9906064).
8. L. Randall and R. Sundrum: *Phys. Rev. Lett.* **83**(1999)3370–3373 (hep-ph/9905221).
9. B.R. Greene, A. Shapere, C. Vafa and S.-T. Yau: *Nucl. Phys.* **B337**(1990)1.
10. P.S. Green and T. Hübsch: *Int. J. Mod. Phys.* **A9**(1994)3203–3227.
11. C. Vafa: *Nucl. Phys.* **B469**(1996)403–418 (hep-th/9602022).
12. M.B. Green, J.H. Schwarz and E. Witten: *Superstring Theory* (Cambridge University Press, Cambridge, 1987).
13. J. Polchinski: *String Theory* (Cambridge University Press, Cambridge, 1998).
14. A. Vilenkin: *Phys. Rev.* **D23** (1981)852.
15. J. Ipser and P. Sikivie: *Phys. Rev.* **D30**(1994)712.
16. P. Tinyakov and K. Zuleta: *Phys. Rev.* **D64**(2001)025022 (hep-th/0103062).
17. P. Berglund and T. Hübsch: *Int. J. Mod. Phys.* **A10**(1995)3381–3430 (hep-th/9411131).
18. N. Kaloper: *Phys. Rev.* **D60**(1999)123506 (hep-th/9905210).
19. R. Gregory: *Phys. Rev. Lett.* **84**(2000)2564–2567 (hep-th/9911015).
20. R. Gregory: *Phys. Rev.* **D54**(1996)4995–4962 (gr-qc/9606002).
21. C. Csáki, J. Erlich, T.J. Hollowood and Y. Shirman: *Nucl. Phys.* **B581**(2000)309 (hep-th/0001033).
22. A.G. Cohen and D.B. Kaplan: *Phys. Lett* **B470**(1999)52–58 (hep-th/9910132); *Phys. Lett* **B215**(1988)663.
23. E. S. Fradkin and A. Tseytlin: *Phys. Lett* **B163**(1985)123.
24. A. Abouelsaood, C.G. Callan, C.R. Nappi and S.A. Yost: *Nucl. Phys.* **B280**(1987)599.
25. R. Leigh: *Mod. Phys. Lett.* **A4**(1989)2767.
26. For general reference on D-branes consult J. Polchinski: TASI'96 lectures, hep-th/9611050.
27. C.V. Johnson: *D-Branes* (Cambridge University Press, Cambridge, 2003); see also hep-th/0007170.

Finite Unified Theories and the Higgs Mass Prediction

Abdelhak Djouadi[1], Sven Heinemeyer[2], Myriam Mondragón[3] and George Zoupanos[4]

[1] Laboratoire de Physique Mathématique et Théorique, Université de Montpellier II, France
abdelhak.djouadi@cern.ch

[2] Dept. of Physics, CERN, TH Division, 1211 Geneva 23, Switzerland
sven.heinemeyr@cern.ch

[3] Instituto de Física, UNAM, Apdo. Postal 20-364, México 01000, D.F., México
myriam@fisica.unam.mx

[4] Physics Dept., Nat. Technical University, 157 80 Zografou, Athens, Greece
george.zoupanos@cern.ch

1 Introduction

Finite Unified Theories are $N = 1$ supersymmetric Grand Unified Theories (GUTs) which can be made finite even to all-loop orders, including the soft supersymmetry breaking sector. The method to construct GUTs with reduced independent parameters [1, 2] consists of searching for renormalization group invariant (RGI) relations holding below the Planck scale, which in turn are preserved down to the GUT scale. Of particular interest is the possibility to find RGI relations among couplings that guarantee finitenes to all-orders in perturbation theory [3, 4]. In order to achieve the latter it is enough to study the uniqueness of the solutions to the one-loop finiteness conditions [3, 4]. The constructed *finite unified* $N = 1$ supersymmetric SU(5) GUTs using the above tools, predicted correctly from the dimensionless sector (Gauge-Yukawa unification), among others, the top quark mass [5]. The search for RGI relations and finiteness has been extended to the soft supersymmetry breaking sector (SSB) of these theories [6, 7], which involves parameters of dimension one and two. Eventually, the full theories can be made all loop finite and their predictive power is extended to the Higgs sector and the supersymmetric spectrum (s-spectrum). The purpose of the present article is to start an exhaustive search of these latter predictions, as well as to provide a rather dense review of the subject.

2 Reduction of Couplings and Finiteness in $N = 1$ SUSY Gauge Theories

Here let us review the main points and ideas concerning the *reduction of couplings* and *finiteness* in $N = 1$ supersymmetric theories. A RGI relation among couplings g_i, $\Phi(g_1, \cdots, g_N) = 0$, has to satisfy the partial differential equation $\mu\, d\Phi/d\mu = \sum_{i=1}^{N} \beta_i\, \partial\Phi/\partial g_i = 0$, where β_i is the β-function of g_i. There exist $(N-1)$ independent Φ's, and finding the complete set of these solutions is equivalent to solve the so-called reduction equations (REs) [2], $\beta_g\, (dg_i/dg) = \beta_i$, $i = 1, \cdots, N$, where

g and β_g are the primary coupling and its β-function. Using all the $(N-1)\,\Phi$'s to impose RGI relations, one can in principle express all the couplings in terms of a single coupling g. The complete reduction, which formally preserves perturbative renormalizability, can be achieved by demanding a power series solution, whose uniqueness can be investigated at the one-loop level.

Finiteness can be understood by considering a chiral, anomaly free, $N = 1$ globally supersymmetric gauge theory based on a group G with gauge coupling constant g. The superpotential of the theory is given by

$$W = \frac{1}{2}\, m^{ij}\, \Phi_i\, \Phi_j + \frac{1}{6}\, C^{ijk}\, \Phi_i\, \Phi_j\, \Phi_k \,, \tag{1}$$

where m^{ij} (the mass terms) and C^{ijk} (the Yukawa couplings) are gauge invariant tensors and the matter field Φ_i transforms according to the irreducible representation R_i of the gauge group G.

The one-loop β-function of the gauge coupling g is given by

$$\beta_g^{(1)} = \frac{\mathrm{d}g}{\mathrm{d}t} = \frac{g^3}{16\pi^2}\left[\sum_i l(R_i) - 3\,C_2(G)\right] \,, \tag{2}$$

where $l(R_i)$ is the Dynkin index of R_i and $C_2(G)$ is the quadratic Casimir of the adjoint representation of the gauge group G. The β-functions of C^{ijk}, by virtue of the non-renormalization theorem, are related to the anomalous dimension matrix γ_i^j of the matter fields Φ_i as:

$$\beta_C^{ijk} = \frac{\mathrm{d}}{\mathrm{d}t}\, C^{ijk} = C^{ijp} \sum_{n=1} \frac{1}{(16\pi^2)^n}\, \gamma_p^{k(n)} + (k \leftrightarrow i) + (k \leftrightarrow j) \tag{3}$$

At one-loop level γ_i^j is given by

$$\gamma_i^{j(1)} = \frac{1}{2} C_{ipq}\, C^{jpq} - 2\, g^2\, C_2(R_i)\delta_i^j \,, \tag{4}$$

where $C_2(R_i)$ is the quadratic Casimir of the representation R_i, and $C^{ijk} = C^*_{ijk}$.

All the one-loop β-functions of the theory vanish if the β-function of the gauge coupling $\beta_g^{(1)}$, and the anomalous dimensions of the Yukawa couplings $\gamma_i^{j(1)}$, vanish, i.e.

$$\sum_i l(R_i) = 3C_2(G), \frac{1}{2} C_{ipq} C^{jpq} = 2\delta_i^j g^2 C_2(R_i) \,, \tag{5}$$

where $l(R_i)$ is the Dynkin index of R_i, and $C_2(G)$ is the quadratic Casimir invariant of the adjoint representation of G.

A very interesting result is that the conditions (5) are necessary and sufficient for finiteness at the two-loop level [8, 9].

The one- and two-loop finiteness conditions (5) restrict considerably the possible choices of the irreps. R_i for a given group G as well as the Yukawa couplings in the superpotential (1). Note in particular that the finiteness conditions cannot be applied to the supersymmetric standard model (SSM), since the presence of a $U(1)$ gauge group is incompatible with the condition (5), due to $C_2[U(1)] = 0$. This leads to the expectation that finiteness should be attained at the grand unified level only, the SSM being just the corresponding, low-energy, effective theory.

The finiteness conditions impose relations between gauge and Yukawa couplings. Therefore, we have to guarantee that such relations leading to a reduction of the couplings hold at any renormalization point. The necessary, but also sufficient, condition for this to happen is to require that such relations are solutions to the reduction equations (REs) to all orders. The all-loop order finiteness theorem of [3] is based on: (a) the structure of the supercurrent in $N = 1$ SYM and on (b) the non-renormalization properties of $N = 1$ chiral anomalies [3]. Alternatively, similar results can be obtained [4, 10] using an analysis of the all-loop NSVZ gauge beta-function [11].

3 Soft Supersymmetry Breaking and Finiteness

The above described method of reducing the dimensionless couplings has been extended [6, 7] to the soft supersymmetry breaking (SSB) dimensionful parameters of $N = 1$ supersymmetric theories. More recently a very interesting progress has been made [9]–[20] concerning the renormalization properties of the SSB parameters based conceptually and technically on the work of [14]. In this work the powerful supergraph method [17] for studying supersymmetric theories has been applied to the softly broken ones by using the "spurion" external space-time independent superfields [18]. In the latter method a softly broken supersymmetric gauge theory is considered as a supersymmetric one in which the various parameters such as couplings and masses have been promoted to external superfields that acquire "vacuum expectation values". Based on this method the relations among the soft term renormalization and that of an unbroken supersymmetric theory have been derived. In particular the β-functions of the parameters of the softly broken theory are expressed in terms of partial differential operators involving the dimensionless parameters of the unbroken theory. The key point in the strategy of [12]–[20] in solving the set of coupled differential equations so as to be able to express all parameters in a RGI way, was to transform the partial differential operators involved to total derivative operators [12]. This is indeed possible to be done on the RGI surface which is defined by the solution of the reduction equations. In addition it was found that RGI SSB scalar masses in Gauge-Yukawa unified models satisfy a universal sum rule at one-loop [16]. This result was generalized to two-loops for finite theories [20], and then to all-loops for general Gauge-Yukawa and Finite Unified Theories [13].

In order to obtain a feeling of some of the above results, consider the superpotential given by (1) along with the Lagrangian for SSB terms

$$-\mathcal{L}_{\rm SB} = \frac{1}{6}\, h^{ijk}\, \phi_i\phi_j\phi_k + \frac{1}{2}\, b^{ij}\, \phi_i\phi_j + \frac{1}{2}\, (m^2)^{j}_{i}\, \phi^{*\,i}\phi_j + \frac{1}{2}\, M\, \lambda\lambda + \text{H.c.}\,, \tag{6}$$

where the ϕ_i are the scalar parts of the chiral superfields Φ_i , λ are the gauginos and M their unified mass. Since only finite theories are considered here, it is assumed that the gauge group is a simple group and the one-loop β-function of the gauge coupling g vanishes. It is also assumed that the reduction equations admit power series solutions of the form

$$C^{ijk} = g \sum_{n=0} \rho^{ijk}_{(n)} g^{2n} \,. \tag{7}$$

According to the finiteness theorem [3], the theory is then finite to all-orders in perturbation theory, if, among others, the one-loop anomalous dimensions $\gamma_i^{j(1)}$ vanish. The one- and two-loop finiteness for h^{ijk} can be achieved by [9]

$$h^{ijk} = -MC^{ijk} + \cdots = -M\rho^{ijk}_{(0)}\, g + O(g^5)\,. \tag{8}$$

An additional constraint in the SSB sector up to two-loops [20], concerns the soft scalar masses as follows

$$\frac{(m_i^2 + m_j^2 + m_k^2)}{MM^\dagger} = 1 + \frac{g^2}{16\pi^2}\,\Delta^{(2)} + O(g^4) \tag{9}$$

for i, j, k with $\rho^{ijk}_{(0)} \neq 0$, where $\Delta^{(2)}$ is the two-loop correction

$$\Delta^{(2)} = -2\sum_l \left[\left(m_l^2/MM^\dagger\right) - (1/3)\right]\, T(R_l), \tag{10}$$

which vanishes for the universal choice [9], i.e. when all the soft scalar masses are the same at the unification point.

If we know higher-loop β-functions explicitly, we can follow the same procedure and find higher-loop RGI relations among SSB terms. However, the β-functions of the soft scalar masses are explicitly known only up to two loops. In order to obtain higher-loop results, we need something else instead of knowledge of explicit β-functions, e.g. some relations among β-functions.

The recent progress made using the spurion technique [17, 18] leads to the following all-loop relations among SSB β-functions, [12]–[20]

$$\beta_M = 2\mathcal{O}\left(\frac{\beta_g}{g}\right)\,, \tag{11}$$

$$\begin{aligned}\beta_h^{ijk} &= \gamma^i{}_l h^{ljk} + \gamma^j{}_l h^{ilk} + \gamma^k{}_l h^{ijl} \\ &\quad -2\gamma_1^i{}_l C^{ljk} - 2\gamma_1^j{}_l C^{ilk} - 2\gamma_1^k{}_l C^{ijl}\,,\end{aligned} \tag{12}$$

$$(\beta_{m^2})^i{}_j = \left[\Delta + X\frac{\partial}{\partial g}\right]\gamma^i{}_j\,, \tag{13}$$

$$\mathcal{O} = \left(Mg^2\frac{\partial}{\partial g^2} - h^{lmn}\frac{\partial}{\partial C^{lmn}}\right)\,, \tag{14}$$

$$\Delta = 2\mathcal{O}\mathcal{O}^* + 2|M|^2 g^2\frac{\partial}{\partial g^2} + \tilde{C}_{lmn}\frac{\partial}{\partial C_{lmn}} + \tilde{C}^{lmn}\frac{\partial}{\partial C^{lmn}}\,, \tag{15}$$

where $(\gamma_1)^i{}_j = \mathcal{O}\gamma^i{}_j$, $C_{lmn} = (C^{lmn})^*$, and

$$\tilde{C}^{ijk} = (m^2)^i{}_l C^{ljk} + (m^2)^j{}_l C^{ilk} + (m^2)^k{}_l C^{ijl}\,. \tag{16}$$

It was also found [19] that the relation

$$h^{ijk} = -M(C^{ijk})' \equiv -M\frac{\mathrm{d}C^{ijk}(g)}{\mathrm{d}\ln g}\,, \tag{17}$$

among couplings is all-loop RGI. Furthermore, using the all-loop gauge β-function of Novikov et al. [11] given by

$$\beta_g^{\mathrm{NSVZ}} = \frac{g^3}{16\pi^2}\left[\frac{\sum_l T(R_l)(1-\gamma_l/2) - 3C(G)}{1-g^2C(G)/8\pi^2}\right], \tag{18}$$

it was found the all-loop RGI sum rule [13],

$$m_i^2 + m_j^2 + m_k^2 = |M|^2\left\{\frac{1}{1-g^2C(G)/(8\pi^2)}\frac{\mathrm{d}\ln C^{ijk}}{\mathrm{d}\ln g} + \frac{1}{2}\frac{\mathrm{d}^2\ln C^{ijk}}{\mathrm{d}(\ln g)^2}\right\}$$
$$+\sum_l \frac{m_l^2 T(R_l)}{C(G)-8\pi^2/g^2}\frac{\mathrm{d}\ln C^{ijk}}{\mathrm{d}\ln g}\,. \tag{19}$$

In addition the exact-β-function for m^2 in the NSVZ scheme has been obtained [13] for the first time and is given by

$$\beta_{m_i^2}^{\mathrm{NSVZ}} = \left[|M|^2\left\{\frac{1}{1-g^2C(G)/(8\pi^2)}\frac{\mathrm{d}}{\mathrm{d}\ln g} + \frac{1}{2}\frac{\mathrm{d}^2}{\mathrm{d}(\ln g)^2}\right\}\right.$$
$$\left.+\sum_l \frac{m_l^2 T(R_l)}{C(G)-8\pi^2/g^2}\frac{\mathrm{d}}{\mathrm{d}\ln g}\right]\gamma_i^{\mathrm{NSVZ}}\,. \tag{20}$$

4 Finite Unified Theories

In this section we examine two concrete $SU(5)$ finite models, where the reduction of couplings in the dimensionless and dimensionful sector has been achieved. A predictive Gauge-Yukawa unified $SU(5)$ model which is finite to all orders, in addition to the requirements mentioned already, should also have the following properties:

1. One-loop anomalous dimensions are diagonal, i.e., $\gamma_i^{(1)\,j} \propto \delta_i^j$.
2. Three fermion generations, in the irreducible representations $\overline{\mathbf{5}}_i, \mathbf{10}_i$ $(i = 1, 2, 3)$, which obviously should not couple to the adjoint $\mathbf{24}$.
3. The two Higgs doublets of the MSSM should mostly be made out of a pair of Higgs quintet and anti-quintet, which couple to the third generation.

In the following we discuss two versions of the all-order finite model. The model of [5], which will be labeled **A**, and a slight variation of this model (labeled **B**), which can also be obtained from the class of the models suggested by Kazakov et al. [12] with a modification to suppress non-diagonal anomalous dimensions[1].

The superpotential which describes the two models takes the form [5, 20]

$$W = \sum_{i=1}^{3}\left[\frac{1}{2}g_i^u\,\mathbf{10}_i\mathbf{10}_i H_i + g_i^d\,\mathbf{10}_i\overline{\mathbf{5}}_i\,\overline{H}_i\right] + g_{23}^u\,\mathbf{10}_2\mathbf{10}_3 H_4 \tag{21}$$
$$+g_{23}^d\,\mathbf{10}_2\overline{\mathbf{5}}_3\,\overline{H}_4 + g_{32}^d\,\mathbf{10}_3\overline{\mathbf{5}}_2\,\overline{H}_4 + \sum_{a=1}^{4} g_a^f\,H_a\,\mathbf{24}\,\overline{H}_a + \frac{g^\lambda}{3}(\mathbf{24})^3\,,$$

where H_a and $\overline{H}_a$ $(a = 1, \ldots, 4)$ stand for the Higgs quintets and anti-quintets.

[1] An extension to three families, and the generation of quark mixing angles and masses in Finite Unified Theories has been addressed in [21], where several realistic examples are given. These extensions are not considered here.

The non-degenerate and isolated solutions to $\gamma_i^{(1)} = 0$ for the models $\{\mathbf{A}, \mathbf{B}\}$ are:

$$(g_1^u)^2 = \left\{\frac{8}{5}, \frac{8}{5}\right\} g^2, (g_1^d)^2 = \left\{\frac{6}{5}, \frac{6}{5}\right\} g^2, (g_2^u)^2 = (g_3^u)^2 = \left\{\frac{8}{5}, \frac{4}{5}\right\} g^2, \tag{22}$$
$$(g_2^d)^2 = (g_3^d)^2 = \left\{\frac{6}{5}, \frac{3}{5}\right\} g^2, (g_{23}^u)^2 = \left\{0, \frac{4}{5}\right\} g^2, (g_{23}^d)^2 = (g_{32}^d)^2 = \left\{0, \frac{3}{5}\right\} g^2,$$
$$(g^\lambda)^2 = \frac{15}{7} g^2, (g_2^f)^2 = (g_3^f)^2 = \left\{0, \frac{1}{2}\right\} g^2, (g_1^f)^2 = 0, (g_4^f)^2 = \{1, 0\} g^2.$$

According to the theorem of [3] these models are finite to all orders. After the reduction of couplings the symmetry of W is enhanced [5, 20].

The main difference of the models **A** and **B** is that three pairs of Higgs quintets and anti-quintets couple to the **24** for **B** so that it is not necessary to mix them with H_4 and $\overline{H}_4$ in order to achieve the triplet-doublet splitting after the symmetry breaking of $SU(5)$.

In the dimensionful sector, the sum rule gives us the following boundary conditions at the GUT scale [20]:

$$m_{H_u}^2 + 2m_{\mathbf{10}}^2 = m_{H_d}^2 + m_{\overline{\mathbf{5}}}^2 + m_{\mathbf{10}}^2 = M^2 \quad \text{for} \quad \mathbf{A}; \tag{23}$$
$$m_{H_u}^2 + 2m_{\mathbf{10}}^2 = M^2, \; m_{H_d}^2 - 2m_{\mathbf{10}}^2 = -\frac{M^2}{3},$$
$$m_{\overline{\mathbf{5}}}^2 + 3m_{\mathbf{10}}^2 = \frac{4M^2}{3} \quad \text{for} \quad \mathbf{B}, \tag{24}$$

where we use as free parameters $m_{\overline{\mathbf{5}}} \equiv m_{\overline{\mathbf{5}}_3}$ and $m_{\mathbf{10}} \equiv m_{\mathbf{10}_3}$ for the model **A**, and $m_{\mathbf{10}} \equiv m_{\mathbf{10}_3}$ for **B**, in addition to M.

5 Predictions of Low Energy Parameters

Since the gauge symmetry is spontaneously broken below M_{GUT}, the finiteness conditions do not restrict the renormalization property at low energies, and all it remains are boundary conditions on the gauge and Yukawa couplings (22), the $h = -MC$ relation, and the soft scalar-mass sum rule (9) at M_{GUT}, as applied in the various models. Thus we examine the evolution of these parameters according to their RGEs up to two-loops for dimensionless parameters and at one-loop for dimensionful ones with the relevant boundary conditions. Below M_{GUT} their evolution is assumed to be governed by the MSSM. We further assume a unique supersymmetry breaking scale M_s and therefore below that scale the effective theory is just the SM.

The predictions for the top quark mass M_t are ~183 and ~174 GeV in models **A** and **B** respectively. Comparing these predictions with the most recent experimental value $M_t^{exp} = (177.9 \pm 4.4)$ GeV [22], and recalling that the theoretical values for M_t may suffer from a correction of ~4% [23], we see that they are consistent with the experimental data. In addition the value of $\tan\beta$ is found to be $\tan\beta$ ~54 and ~48 for models **A** and **B** respectively.

In the SSB sector, besides the constraints imposed by finiteness there are further restrictions imposed by phenomenology. In the case where all the soft scalar masses are universal at the unfication scale, there is no region of $M_s = M$ below $O(few\,\text{TeV})$ in which $m^2_{\tilde{\tau}} > m^2_{\chi^0}$ is satisfied (where $m_{\tilde{\tau}}$ is the lightest $\tilde{\tau}$ mass, and m_{χ^0} the lightest neutralino mass, which is the lightest supersymmetric particle). But once the universality condition is relaxed this problem can be solved naturally (thanks to the sum rule). More specifically, using the sum rule (9) and imposing the conditions a) successful radiative electroweak symmetry breaking, b) $m^2_{\tilde{\tau}} > 0$ and c) $m^2_{\tilde{\tau}} > m^2_{\chi^0}$, a comfortable parameter space for both models (although model **B** requires large $M \sim 1\,\text{TeV}$) is found.

As an additional constraint, we take into account the $BR(b \to s\gamma)$ [24]. We do not take into account, though, constraints coming from the muon anomalous magnetic moment (g-2) in this work, which excludes a small region of the parameter space. In the graphs we show the **FUTA** results concerning m_h (including the large corrections due to $\tan\beta$), m_{χ^0}, and M_A, for different values of M, for the case $\mu < 0$ and the LSP is a neutralino χ^0. The results for $\mu > 0$ are slightly different: the spectrum starts around 500 GeV. The main difference, though, is in the value of the running bottom mass $m_{bot}(m_{bot})$, where we have included the corrections coming from bottom squark-gluino loops and top squark-chargino loops [25]. In the $\mu < 0$ case, $m_{bot} \sim 3.5 - 4.0\,\text{GeV}$ is just below the experimental value $m^{exp}_{bot} \sim 4.0 - 4.5\,\text{GeV}$ [11], while in the $\mu > 0$ case, $m_{bot} \sim 4.8 - 5.3\,\text{GeV}$, i.e. above the experimental value.

The Higgs mass prediction of the two models is, although the details of each of the models differ, in the following range

$$m_h = \quad \sim 112 - 132\,\text{GeV}\,, \tag{25}$$

where the uncertainty comes from variations of the gaugino mass M and the soft scalar masses, and from finite (i.e. not logarithmically divergent) corrections in

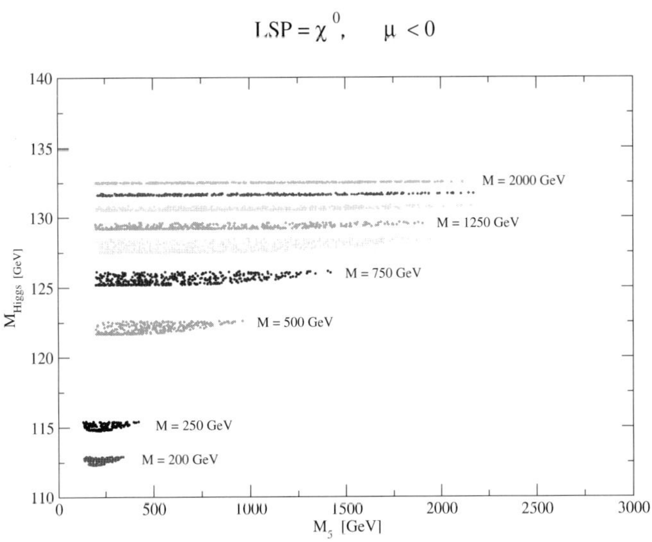

Fig. 1. m_h as function of m_5 for different values of M for model **FUTA**, for $\mu < 0$

Table 1. A representative example of the bottom (*running*) and top (*pole*) masses, plus the supersymmetric spectrum for Model FUTA, with $m_5 = 697\,\mathrm{GeV}$, $m_{10} = 806\,\mathrm{GeV}$, $M_{susy} = 1681\,\mathrm{GeV}$, $\mu < 0$. All masses in the Table are in GeV

M_{top}	183	m_{bot}	3.9
$\tan\beta =$	54.4	α_s	.118
$m_{\chi 1}$	452	$m_{\tilde{\tau}_2}$	916
$m_{\chi 2}$	843	$m_{\tilde{\nu}_3}$	883
$m_{\chi 3}$	850	μ	−1494
$m_{\chi 4}$	1500	B	3543
$m_{\chi_1^\pm}$	843	m_A	555
$m_{\chi_2^\pm}$	1500	$m_{H^\pm}$	560
$m_{\tilde{t}_1}$	1578	m_H	555
$m_{\tilde{t}_2}$	1776	m_h	127.5
$m_{\tilde{b}_1}$	1580	M_1	452
$m_{\tilde{b}_2}$	1766	M_2	846
$m_{\tilde{\tau}_1}$	654	M_3	2210

Table 2. A representative example of the bottom (*running*) and top (*pole*) masses, plus the supersymmetric spectrum for Model FUTB, with $m_{10} = 945\,\mathrm{GeV}$, $M_{susy} = 2278\,\mathrm{GeV}$, $\mu < 0$. All masses in the Table are in GeV

M_{top}	173	m_{bot}	4.2
$\tan\beta =$	48	α_s	.116
$m_{\chi 1}$	669	$m_{\tilde{\tau}_2}$	970
$m_{\chi 2}$	912	$m_{\tilde{\nu}_3}$	916
$m_{\chi 3}$	1289	μ	-1900
$m_{\chi 4}$	1909	B	4010
$m_{\chi_1^\pm}$	1289	m_A	1106
$m_{\chi_2^\pm}$	909	$m_{H^\pm}$	1109
$m_{\tilde{t}_1}$	2236	m_H	1106
$m_{\tilde{t}_2}$	2519	m_h	123.5
$m_{\tilde{b}_1}$	2163	M_1	700
$m_{\tilde{b}_2}$	2501	M_2	1293
$m_{\tilde{\tau}_1}$	766	M_3	3256

changing renormalization scheme. The value of M was varied from $200-2000\,\mathrm{GeV}$ for **FUTA**. In the analysis we have also included a small variation, due to threshold corrections at the GUT scale, of $1-2\%$ of the FUT boundary conditions. This small variation does not give a noticeable effect in the results at low energies. From Fig. 1 we already see that the requirement $m_h > 114.4\,\mathrm{GeV}$ [27] (neglecting the theoretical uncertainties) excludes the possibility of $M = 200\,\mathrm{GeV}$ for FUTA, which is taken into account in the presentation of the next graphs.

A more detailed numerical analysis, where the results of our program and of the known programs FeynHiggs [28] and Suspect [29] are combined, is currently in progress [30].

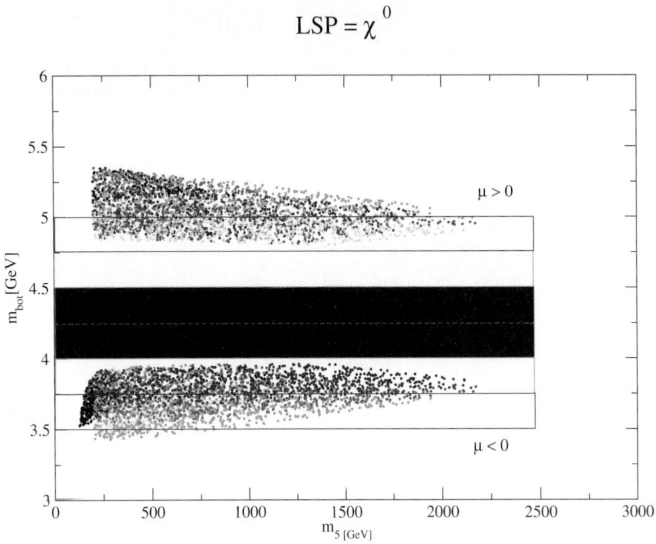

Fig. 2. $m_{bot}(m_{bot})$ as function of m_5 for different values of M for model **FUTA**, for $\mu < 0$ and $\mu > 0$. The shaded region shows the experimentally accepted value of $m_{bot}(m_{bot})$ according to [11]

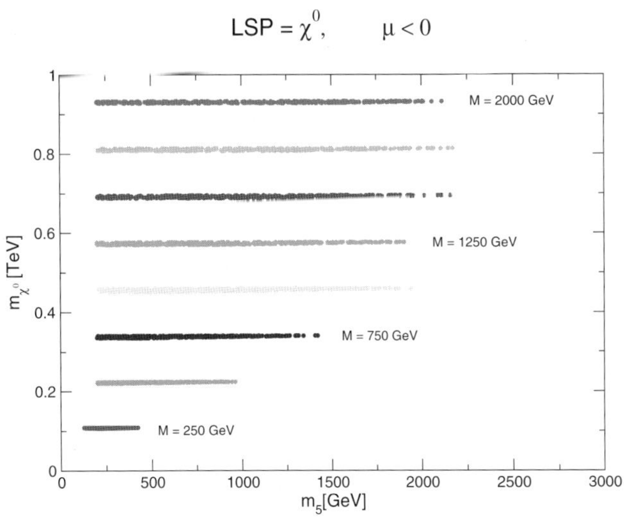

Fig. 3. m_{χ^0} as function of m_5 for different values of M for model **FUTA**, for $\mu < 0$

In Tables 1 and 2 we present representative examples of the values obtained for the s-spectra for each of the models. The value of the lightest Higgs physical mass m_h has already the one-loop radiative corrections included, evaluated at the appropriate scale [31].

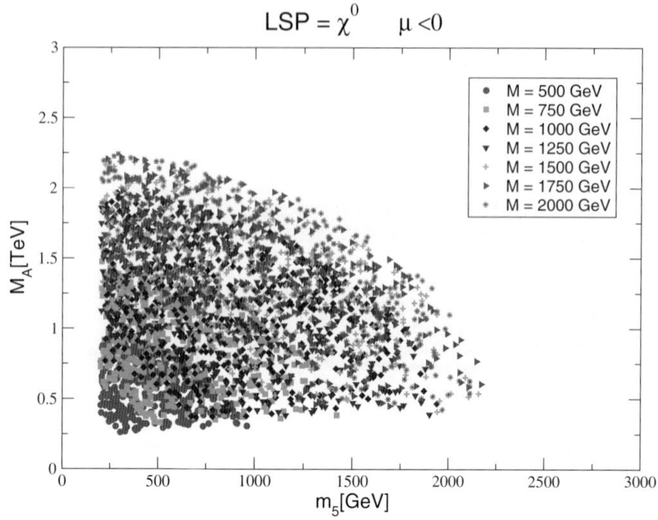

Fig. 4. M_A as function of m_5 for different values of M for model **FUTA**, for $\mu < 0$

Acknowledgements

It is a pleasure to thank the Organising Committee for the very warm hospitality offered to one of us (G.Z.). Supported by the projects PAPIIT-IN116206 and Conacyt grant 42026-F and partially by the RTN contract HPRN-CT-2000-00148, the Greek-German Bilateral Programme IKYDA-2001 and by the NTUA Programme for Fundamental Research "THALES".

References

1. Kubo, J., Mondragón, M. and Zoupanos, G. (1994) *Nucl. Phys.* **B424** 291.
2. Zimmermann, W., (1985) *Com. Math. Phys.* **97** 211; Oehme, R. and Zimmermann, W. (1985) *Com. Math. Phys.* **97** 569; Ma, E. (1978) *Phys. Rev* **D 17** 623;ibid (1985) **D31** 1143.
3. Lucchesi, C., Piguet, O. and Sibold, K. (1988) *Helv. Phys. Acta* **61** 321; Piguet, O. and Sibold, K. (1986) *Intr. J. Mod. Phys.* **A1** 913; (1986) *Phys. Lett.* **B177** 373; see also Lucchesi, C. and Zoupanos, G. (1997) *Fortsch. Phys.* **45** 129.
4. Ermushev, A.Z., Kazakov, D.I. and Tarasov, O.V. (1987) *Nucl. Phys.* **281** 72; Kazakov, D.I. (1987) *Mod. Phys. Lett.* **A9** 663.
5. Kapetanakis D., Mondragón, M. and Zoupanos, G., (1993) *Zeit. f. Phys.* **C60** 181; Mondragón, M. and Zoupanos, G. (1995) *Nucl. Phys. B* (Proc. Suppl.) **37C**) 98.
6. Kubo, J., Mondragón, M. and Zoupanos, G. (1996) *Phys. Lett.* **B389** 523.
7. Jack, I. and Jones, D.R.T. (1995) *Phys. Lett.* **B349** 294.
8. D.R.T. Jones, L. Mezincescu and Y.-P. Yao, *Phys. Lett.* **B148** 317 (1984).
9. Jack, I. and Jones, D.R.T. (1994) *Phys. Lett.* **B333** 372.

10. Leigh, R.G., and Strassler, M.J. (1995) *Nucl. Phys.* **B447** 95.
11. Novikov, V., Shifman, M., Vainstein, A., and Zakharov, V. (1983) *Nucl. Phys.* **B229** 381; (1986)*Phys. Lett.* **B166** 329; Shifman, M. (1996) *Int.J. Mod. Phys.* **A11** 5761 and references therein.
12. Avdeev, L.V., Kazakov, D.I. and Kondrashuk, I.N. (1998) *Nucl. Phys.* **B510** 289; Kazakov, D.I. (1999) *Phys.Lett.* **B449** 201.
13. T. Kobayashi, J. Kubo and G. Zoupanos, (1998) Phys. Lett. **B427** 291; T. Kobayashi et al., in proc. of "Supersymmetry, Supergravity and Superstrings", pp. 242-268, Seoul, 1999; T.Kobayashi et al. (2001) *Surveys High Energy Phys.* **16** 87-129.
14. Yamada, Y. (1994) *Phys. Rev.* **D50** 3537.
15. J. Hisano and M. Shifman, *Phys. Rev.* **D56** (1997) 5475.
16. Kawamura, T., Kobayashi, T. and Kubo, J. (1997) *Phys. Lett.* **B405** 64.
17. Delbourgo, R. (1975) *Nuovo Cim* **25A** 646; Salam, A. and Strathdee, J. (1975) *Nucl. Phys.* **B86** 142; Fujikawa, K. and Lang, W. (1975) *Nucl. Phys.* **B88** 61; Grisaru, M.T., Rocek, M. and Siegel, W. (1979) *Nucl. Phys.* **B59** 429.
18. Girardello, L. and Grisaru, M.T. (1982) *Nucl. Phys.* **B194** 65; Helayel-Neto, J.A. (1984) *Phys. Lett.* **B135** 78; Feruglio, F., Helayel-Neto, J.A. and Legovini, F. (1985) *Nucl. Phys.* **B249** 533; Scholl, M. (1985) *Zeit. f. Phys.* **C28** 545.
19. I. Jack, D.R.T. Jones and A. Pickering, *Phys. Lett.* **B426** (1998) 73.
20. Kobayashi, T., Kubo, J., Mondragón, M. and Zoupanos, G. (1998) *Nucl. Phys.* **B511** 45; in proc. of ICHEP 1998, vol. 2, p. 1597 (Vancouver 1998); Acta Phys. Polon. **B30** (1999) 2013; in proc. of HEP99, p. 804 (Tampere 1999);
M. Mondragon and G. Zoupanos, Acta Phys. Polon. B **34** (2003) 5459.
21. K. S. Babu, T. Enkhbat and I. Gogoladze (2003) Phys. Lett. B **555** 238 [arXiv:hep-ph/0204246];
22. J. Erler, to appear in the next edition of the PDG, private communication.
23. For an extended discussion and a complete list of references see: Kubo, J., Mondragón, M. and Zoupanos, G. (1997) *Acta Phys. Polon.* **B27** 3911.
24. P. Cho, M. Misiak and D. Wyler, *Phys. Rev.* **D 54**, 3329 (1996); A. Kagan and M. Neubert, *Eur. Phys. J.* **C 7** (1999) 5, K. Chetyrkin, M. Misiak and M. Munz, *Phys. Lett.* **B 400**, (1997) 206, [Erratum-ibid. **B 425** (1998) 414] P. Gambino and M. Misiak, *Nucl. Phys.* **B 611** (2001) 338, R. Barate et al. [ALEPH Collaboration], *Phys. Lett.* **B 429** (1998) 169; S. Chen et al. [CLEO Collaboration], *Phys. Rev. Lett.* **87** (2001) 251807, K. Abe et al. [Belle Collaboration], *Phys. Lett.* **B 511** (2001) 151, B. Aubert et al. [BABAR Collaboration];
25. L. J. Hall, R. Rattazzi and U. Sarid, Phys. Rev. D **50**, (1994) 7048 [arXiv:hep-ph/9306309]; R. Hempfling, Z. Phys. C **63**, (1994) 309 [arXiv:hep-ph/9404226]; F. M. Borzumati, M. Olechowski and S. Pokorski, Phys. Lett. B **349**, (1995) 311 [arXiv:hep-ph/9412379]; T. Blazek, S. Raby and S. Pokorski, Phys. Rev. D **52**, (1995) 4151 [arXiv:hep-ph/9504364]. M. Carena, M. Olechowski, S. Pokorski and C. E. M. Wagner, Nucl. Phys. B **426** (1994) 269 [arXiv:hep-ph/9402253].
26. K. Hagiwara et al., *Phys. Rev. D66*, 010001 (2002).
27. LEP Higgs working group, *Phys. Lett.* **B 565** (2003) 61, hep-ex/0306033.
28. S. Heinemeyer, W. Hollik and G. Weiglein, Comput. Phys. Commun. **124** (2000) 76 [arXiv:hep-ph/9812320]; ibid Eur. Phys. J. C **9** (1999) 343 [arXiv:hep-ph/9812472].

29. A. Djouadi, J. L. Kneur and G. Moultaka, arXiv:hep-ph/0211331.
30. Djouadi, A., Heinemeyer, S., Mondragón, M. and Zoupanos, G., work in progress.
31. Gladyshev, A.V., Kazakov, D.I., de Boer, W., Burkart, G. and Ehret, R. (1997) *Nucl. Phys.* **B498** 3; Carena, M. et al., (1995) *Phys. Lett.* **B355** 209.

Non-Commutative GUTs, Standard Model and C, P, T Properties from Seiberg-Witten Map

Paolo Aschieri

Dipartimento di Scienze e Tecnologie Avanzate, Universitá del Piemonte Orientale, and INFN, Piazza Ambrosoli 5, 15100, Alessandria, Italy
Max-Planck-Institut für Physik Föhringer Ring 6, 80805 München
Sektion Physik, Universität München Theresienstraße 37, 80333 München
aschieri@theorie.physik.uni-muenchen.de

1 Introduction

There are different examples of noncommutative theories, we here concentrate on the case where noncommutativity is described by a constant parameter $\theta^{\mu\nu}$. The commutation relations among the coordinates read $[x^\mu \overset{\star}{,} x^\nu] \equiv x^\mu \star x^\nu - x^\nu \star x^\mu = \mathrm{i}\theta^{\mu\nu}$, where the star product between functions f, g is given by $f \star g = f\, \mathrm{e}^{\frac{i}{2}\theta^{\mu\nu}\overleftarrow{\partial_\mu}\overrightarrow{\partial_\nu}}\, g$. We do not claim that spacetime has exactly this noncommutativity, rather we are interested in investigating a mathematically sound gauge theory based on this easiest noncommutative structure. General aspects of this noncommutative theory will then probably be in common with more refined choices of θ. In particular the choice $\theta^{\mu\nu}$ = constant breakes the Lorentz group in a spontaneous way; in a bigger theory we would like to consider $\theta^{\mu\nu}$ (or the related B field) dynamical and not frozen to a constant value, thus recovering Lorentz covariance. One can also consider gauge theories with θ nondynamical but frozen to a particular nonconstant value, linear in the coordinates, such that one has a (kappa) deformed Poincaré symmetry, see [1].

Using Seiberg-Witten map [8], that relates commutative gauge fields to noncommutative ones in such a way that commutative gauge transformations are mapped in NC gauge transformations, one can construct NC gauge theories with arbirary gauge groups [6, 4], see also B. Jurčo contibution to the proceedings. These theories are invariant under both commutative and noncommutative gauge transformations. Along these lines noncommutative generalizations of the standard model and GUT theories have been studied [5, 6]. The SW map and the $\star$ product allow us to expand these noncommutative actions order by order in θ and to express them in terms of ordinary commutative fields so that one can then study the physics properties of these θ-expanded commutative actions, see for [7], see also P. Schupp contribution to the proceedings.

It turns out that given a commutative YM theory, SW map and commutative/noncommutative gauge invariance are in general not enough in order to single out a unique noncommutative generalization of the original YM theory. One can follow different criteria in order to select a specific noncommutative generalization. We here focus on a classical analysis, in particular imposing the constraint that the noncommutative generalization of the Standard Model should be compatible with noncommutative GUT theories. Another issue would be to single out a noncommutative SM or GUT that is well behaved at the quantum level. We refer to the problems relative to renormalization, see for [8]. On the other hand chiral gauge anomalies are absent in these models [9], see also F. Brandt contribution to the proceedings.

In this talk, following [6], we present a general study of the ambiguities that appear when constructing NCYM theories. We then see that at first order in θ there is no ambiguity in $SO(10)$ NCYM theory. In particular no triple gauge bosons coupling of the kind θFFF is present. We further study the noncommutative SM compatible with $SO(10)$.

We next study the reality, hermiticity, charge conjugation, parity and time reversal properties of the SW map and of θ-expanded NCYM theories. This constraints the possible freedom in the choice of a "good" SW map. In [10] the C, P, T properties of NCQED were studied assuming the usual C, P and T transformations also for noncommutative fields. We here show, that the usual C, P, T transformation on commutative spinors and nonabelian gauge potentials imply, via SW map, the same C, P, T transformations for the noncommutative spinors and gauge potentials. We also see that CPT is always a symmetry of noncommutative actions. In [11] CPT is studied more axiomatically.

The reality property of the SW map is then used to analyze the difference between the SM in [5] and the GUT inspired SM proposed here. It is a basic one, and can be studied also in a QED model. While in [5], and in general in the literature, left and right handed components of a noncommutative spinor field are built with the same SW map, we here use and advocate a different choice: if noncommutative left handed fermions are built with the $+\theta$ SW map then their right handed companions should be built with the $-\theta$ SW map; this implies that both noncommutative ψ_L and $\psi^C_L \equiv -i\sigma_2\, {\psi_R}^*$ are built with the $+\theta$ SW map. In other words, with this choice, noncommutativity does not distinguish between a left handed fermion and a left handed antifermion, but does distinguish between fermions with different chirality. This appears to be the only choice compatible with GUT theories.

2 Seiberg-Witten Map and NC Particle Models

Consider an ordinary "commutative" YM action with gauge group G, and one fermion multiplet, $\int \mathrm{d}^4x \,\frac{-1}{2g^2}\,\mathrm{Tr}(F_{\mu\nu}F^{\mu\nu}) + \overline{\Psi} i\not{D}\Psi$. This action is gauge invariant under $\delta\Psi = \mathrm{i}\rho_\Psi(\Lambda)\Psi$ where ρ_Ψ is the representation of G determined by Ψ. Following [4] the noncommutative generalization of this action is given by

$$\widehat{S} = \int \mathrm{d}^4x \,\frac{-1}{2g^2}\,\mathrm{Tr}(\widehat{F}_{\mu\nu} \star \widehat{F}^{\mu\nu}) + \overline{\widehat{\Psi}} \star i\widehat{\not{D}}\widehat{\Psi} \tag{1}$$

where the noncommutative field strength $\widehat{F}$ is defined by $\widehat{F}_{\mu\nu} = \partial_\mu\widehat{A}_\nu - \partial_\nu\widehat{A}_\mu - \mathrm{i}\sigma[\widehat{A}_\mu, \widehat{A}_\nu]$. The covariant derivative is given by

$$\widehat{D}_\mu\widehat{\Psi} = \partial_\mu\widehat{\Psi} - \mathrm{i}\rho_\Psi(\widehat{A}_\mu) \star \widehat{\Psi} \;. \tag{2}$$

The action (1) is invariant under the noncommutative gauge transformations

$$\hat{\delta}\widehat{\Psi} = \mathrm{i}\rho_\Psi(\widehat{\Lambda}) \star \widehat{\Psi} \quad , \quad \hat{\delta}\widehat{A}_\mu = \partial_\mu\widehat{\Lambda} + \mathrm{i}\sigma[\widehat{\Lambda}, \widehat{A}_\mu] \;. \tag{3}$$

The fields $\widehat{A}$, $\widehat{\Psi}$ and $\widehat{\Lambda}$ are functions of the commutative fields A, Ψ, Λ and the noncommutativity parameter θ via the SW map [8]. At first order in θ we have

$$\widehat{A}_\xi[A,\theta] = A_\xi + \frac{1}{4}\theta^{\mu\nu}\{A_\nu, \partial_\mu A_\xi\} + \frac{1}{4}\theta^{\mu\nu}\{F_{\mu\xi}, A_\nu\} + \mathcal{O}(\theta^2) \tag{4}$$

$$\widehat{\Lambda}[\Lambda, A, \theta] = \Lambda + \frac{1}{4}\theta^{\mu\nu}\{\partial_\mu \Lambda, A_\nu\} + \mathcal{O}(\theta^2) \tag{5}$$

$$\widehat{\Psi}[\Psi, A, \theta] = \Psi + \frac{1}{2}\theta^{\mu\nu}\rho_\Psi(A_\nu)\partial_\mu\Psi + \frac{i}{8}\theta^{\mu\nu}[\rho_\Psi(A_\mu), \rho_\Psi(A_\nu)]\Psi + \mathcal{O}(\theta^2) \tag{6}$$

In terms of the commutative fields the action (1) is also invariant under the ordinary gauge transformation $\delta A_\mu = \partial_\mu\Lambda + \mathrm{i}[\Lambda, A_\mu]$, $\delta\Psi = \mathrm{i}\rho_\Psi(\Lambda)\Psi$.

In (1) the information on the gauge group G is through the dependence of the noncommutative fields on the commutative ones. The commutative gauge potential A and gauge parameter Λ are valued in the G Lie algebra, $A = A^a T^a, \Lambda = \Lambda^a T^a$; and from (4), (5) it follows that $\widehat{A}$ and $\widehat{\Lambda}$ are valued in the universal enveloping algebra of the G Lie algebra. However, due to the SW map, the degrees of freedom of $\widehat{A}$ are the same as that of A. Similarly to $\widehat{A}$, also $\widehat{F}$ is valued in the universal enveloping algebra of G. Now expression (1) is ambiguous because in $\mathrm{Tr}(\widehat{F}_{\mu\nu} \star \widehat{F}^{\mu\nu})$ we have not specified the representation $\rho(T^a)$. We can render explicit the ambiguity in (1) by writing

$$\frac{1}{g^2}\mathrm{Tr}(\widehat{F}_{\mu\nu} \star \widehat{F}^{\mu\nu}) = \sum_\rho c_\rho \mathrm{Tr}(\rho(\widehat{F}_{\mu\nu}) \star \rho(\widehat{F}^{\mu\nu})) \tag{7}$$

where the sum is extended over all unitary irreducible and inequivalent representations ρ of G. The real coefficients c_ρ parametrize the ambiguity in (7). They are constrained by requiring that in the commutative limit, $\theta \to 0$, (7) becomes the correctly normalized commutative gauge kinetic term.

The ambiguity (7) in the action (1) can also be studied by expanding (7) in terms of the commutative fields Ψ, Λ, F. At first order in θ we have

$$\begin{aligned}\widehat{S}_{gauge} = & -\frac{1}{4g^2}\int \mathrm{d}^4x \sum_{a=1}^{\dim G} F^a_{\mu\nu}F^{a\,\mu\nu} \\ & + \left(\textstyle\sum_\rho c_\rho D^{abc}_\rho\right) \frac{\theta^{\mu\nu}}{4}\int \mathrm{d}^4x\ \frac{1}{4}F^a_{\mu\nu}F^b_{\rho\sigma}F^{c\,\rho\sigma} - F^a_{\mu\rho}F^b_{\nu\sigma}F^{c\,\rho\sigma}\end{aligned} \tag{8}$$

where

$$\frac{1}{2}D^{abc}_\rho \equiv \mathrm{Tr}(\rho(T^a)\{\rho(T^b), \rho(T^c)\}) = \mathcal{A}(\rho)\mathrm{Tr}(t^a\{t^b, t^c\}) \equiv \frac{1}{2}\mathcal{A}(\rho)d^{abc}\,. \tag{9}$$

Here t^a denotes the fundamental representation, and we are using that the completely symmetric D^{abc}_ρ tensor in the representation ρ is proportional to the d^{abc} one defined by the fundamental representation. In particular for all simple Lie groups, except $SU(N)$ with $N \geq 3$, we have $D^{abc}_\rho = 0$ for any representation ρ. Thus from (8) we see that at first order in θ the ambiguity (7) is present just for $SU(N)$ Lie groups.

Among the possible representations that one can choose in (7) there are two natural ones. The fermion representation and the adjoint representation. The adjoint representation is particularly appealing if we just have a pure gauge action, then, since only the structure constants appear in the commutative gauge kinetic term $\sum_a F^a_{\mu\nu} F^{a\,\mu\nu}$, a possible choice is indeed to consider only the adjoint representation. This is a minimal choice in the sense that in this case only structure constants enter (7). (It can be shown [6] that in this case the gauge action is even in θ). If we also have matter fields then from (2) we see that we must consider the particle representation ρ_Ψ given by the multiplet Ψ (and inherited by $\widehat{\Psi}$). In (7) one could then make the minimal choice of selecting just the ρ_Ψ representation.

Along the lines of the above NCYM theories framework we now examine the $SO(10)$, the $SU(5)$ and the Standard Model noncommutative gauge theories.

Noncommutative *SO*(10) We consider only one fermion generation: the 16-dimensional spinor representation of $SO(10)$ usually denoted 16^+ (no relevant new effects appear considering all three families). We write the left handed multiplet as

$$\Psi_L^+ = (u^i, d^i\,,\, -u_i^C\,,\, d_i^C\,,\, \nu, e^-\,,\, e^+\,,\, -\nu^C)_L \tag{10}$$

where i is the $SU(3)$ color index and $\nu^C{}_L = -\mathrm{i}\sigma_2\, {\nu_R}^*$ is the charge conjugate of the neutrino particle ν_R (not present in the Standard Model). The gauge and fermion sector of noncommutative $SO(10)$ is then simply obtained by replacing $\widehat{\Psi}$ with $\widehat{\Psi}_L^+$ in (1). Notice that no linear term in θ, i.e. no cubic term in F can appear. This is so because $SO(10)$ is anomaly free: $D_\rho^{abc} = 0$ forall ρ. In other words, at first order in θ, noncommutative $SO(10)$ gauge theory is unique.

Noncommutative *SU*(5) The fermionic sector of $SU(5)$ has the $\psi^C{}_L$ multiplet that transforms in the $\overline{5}$ of $SU(5)$ and the χ_L multiplet that transforms according to the 10 of $SU(5)$. In this case we expect that the adjoint, the $\overline{5}$ and the 10 representations enter in (7). In principle one can consider the coefficients $c_{\overline{5}} \neq c_{10}$, i.e. while the $(\psi^C{}_L, \chi_L)$ fermion rep. is $\overline{5} \oplus 10$, in (7) the weights c_ρ of the $\overline{5}$ and the 10 can possibly be not the same. It turns out that only if $c_{\overline{5}} \neq c_{10}$ then $\sum_\rho c_\rho D_\rho^{abc} \neq 0$ in (8). We see that, already at first order in the noncommutativity parameter θ, noncommutative $SU(5)$ gauge theory is not uniquely determined by the gauge coupling constant g, but also by the value of $\sum_\rho c_\rho D_\rho^{abc}$. Thus $SU(5)$ is *not* a truly unified theory in a noncommutative setting. It is tempting to set $c_{\overline{5}} = c_{10}$ so that exactly the fermion representation $\overline{5} \oplus 10$ enters (8). We then have $\sum_\rho c_\rho D_\rho^{abc} = 0$, (however this relation is not protected by symmetries).

(GUT Inspired) Noncommutative Standard Model One proceeds similarly for the SM gauge group. The full ambiguity of the gauge kinetic term is given in [6]. About the fermion kinetic term, the fermion vector $\widehat{\Psi}_L$ is constructed from $\Psi_L = (u^i, d^i\,,\, -u_i^C\,,\, d_i^C\,,\, \nu, e^-\,,\, e^+)_L$. The covariant derivative is as in (2), with $\Psi \to \Psi_L$ and with $A_\mu = A_\mu^A T^A$, where $\{T^A\} = \{Y, T_L^a, T_S^\lambda\}$ are the generators of $U(1) \otimes SU(2) \otimes SU(3)$. The fermion kinetic term is then as in (1). This Standard Model is built using only left handed fermions and antifermions. We call it GUT inspired because its noncommutative structure can be embedded in $SO(10)$ GUT. Indeed Ψ_L and Ψ_L^+ differ just by the extra neutrino $\nu^C{}_L = -\mathrm{i}\sigma_2\, {\nu_R}^*$; moreover under an infinitesimal gauge transformation $\widehat{\Lambda}$, all fermions in Ψ_L transform with $\widehat{\Lambda}$ on the left. This GUT inspired Standard Model differs from the one considered in [5]; indeed here we started from the chiral vector Ψ_L, while there the vector $\Psi' = (u_L^i, d_L^i\,,\, u_R^i\, d_R^i\,,\, \nu_L, e_L^-\,,\, e_R^-)$ is considered. In the commutative case

$\int \overline{\Psi'} \not{D} \Psi' = \int \overline{\Psi_L} \not{D} \Psi_L$ but in the noncommutative case (see later) this is no more true: $\int \widehat{\overline{\Psi'}} \star \widehat{\not{D}} \widehat{\Psi'} \neq \int \widehat{\overline{\Psi_L}} \star \widehat{\not{D}}\, \widehat{\Psi_L}$; if we change θ into $-\theta$ in the right handed sector of $\int \widehat{\overline{\Psi'}} \star \widehat{\not{D}} \widehat{\Psi'}$, then the two expressions coincide.

Finally it is a natural choice to consider in the SM gauge kinetic term only the adjoint rep. and the fermion rep., we then have that at first order in θ there are no modifications to the SM gauge kinetic term. This is so because the fermion rep. is anomaly free: $D^{AA'A''}_{\rho_{\mathrm{fermion}}} = 0$ and because for $U(1)$ the adjoint rep. is trivial.

Higgs Sectors While the noncommutative Higgs kinetic and potential terms are given by

$$(\widehat{D}_\mu \widehat{\phi})^\dagger \star \widehat{D}^\mu \widehat{\phi} + \mu^2 \widehat{\phi}^\dagger \star \widehat{\phi} - \lambda\, \widehat{\phi}^\dagger \star \widehat{\phi} \star \widehat{\phi}^\dagger \star \widehat{\phi}\,, \tag{11}$$

a noncommutative version of the SM and GUT Yukawa terms is not straighforward and requires the introduction of the hybrid Seiberg-Witten maps $\widehat{\ }^H$ on fermions. A typical noncommutative Yukawa term then reads

$$\widehat{\phi}^\dagger \star \widehat{L_L}^H \star \widehat{e^*_R} + \mathit{herm.\ conj}\,. \tag{12}$$

where $L_L = \begin{pmatrix} \nu_L \\ \mathrm{e}_L \end{pmatrix}$. Under an infinitesimal $U(1)\otimes SU(2)\otimes SU(3)$ gauge transformation Λ, $\widehat{L_L}^H$ transforms as $\delta\, \widehat{L_L}^H = \mathrm{i}\rho_\phi(\widehat{\Lambda}) \star \widehat{L_L}^H - i \widehat{L_L}^H \star \rho_{e^*_R}(\widehat{\Lambda})$. We see that in the hybrid SW map $\widehat{\Lambda}$ appears both on the left and on the right of the fermions, moreover the representation of $\widehat{\Lambda}$ is inherited from the Higgs and fermions that sandwich $\widehat{L_L}^H$. The Yukawa term (12) is thus invariant under noncommutative gauge transformations. Of course in the $\theta \to 0$ limit we recover the usual gauge transformation for the leptons. An explicit formula for the hybrid SW map at first order in θ is in [5, 6]. The Yukawa terms (12) differ from those studied in [5]. There the hybrid SW map is considered on ϕ, in particular there $\widehat{\phi}^H$ is not invariant under $SU(3)$ gauge transformations (and this implies that in [5] gluons couple directly to the Higgs field).

The Higgs sector in the $SO(10)$ and $SU(5)$ models can be constructed with similar techniques [6].

3 Hermiticity and Reality Properties of SW Map

From (4) we see that if A is hermitian then $\widehat{A}$ is also hermitian. Actually, to all orders in θ, $\widehat{A}$ and $\widehat{\Lambda}$ can be chosen hermitian if A and Λ are hermitian. Otherwise stated, SW map can be chosen to be compatible with hermitian conjugation. Compatibility of SW map with complex conjugation reads,

$$\widehat{\Psi^*} = \widehat{\Psi}^* \tag{13}$$

where $\widehat{\Psi} = \mathrm{SW}[\Psi, \rho_\Psi(A), \theta]$ denotes the SW map of Ψ constructed with the representation ρ_Ψ of the potential A, and the SW map of the complex conjugate spinor Ψ^* is defined by

$$\widehat{\Psi^*} \equiv \mathrm{SW}[\Psi^*, \rho_{\Psi^*}(A), -\theta] \tag{14}$$

where ρ_{Ψ^*} is the representation conjugate to ρ_Ψ[1]. Notice that in (14) the noncommutativity parameter θ appears with opposite sign w.r.t. the θ in the SW map of Ψ. Similarly to (13) we have $\widehat{\rho_{\Psi^*}(A)} = -\overline{\widehat{\rho_\Psi(A)}}$ and $\widehat{\rho_{\Psi^*}(\Lambda)} = -\overline{\widehat{\rho_\Psi(\Lambda)}}$ where $\widehat{\rho_{\Psi^*}(A)} \equiv \widehat{A}[\rho_{\Psi^*}(A), -\theta]$ and $\widehat{\rho_{\Psi^*}(\Lambda)} \equiv \widehat{\Lambda}[\rho_{\Psi^*}(A), \rho_{\Psi^*}(\Lambda), -\theta]$. The proof of (13) relies on showing that the SW differential equations [8] (obtained by requiring that gauge equivalence classes of the gauge theory with noncommutativity $\theta + \delta\theta$, correspond to gauge equivalence classes of the gauge theory with noncommutativity θ) are themselves compatible with complex conjugation [6]. One proceeds similarly for the case of hermitian conjugation.

Noncommutativity and Chirality We can now discuss a further ambiguity of noncommutative gauge theories, and resolve it by requiring compatibility with grand unified theories. For simplicity we consider noncommutative QED. Let ψ be a 4-component Dirac spinor, and decompose it into its Weil spinors ψ_L and ψ_R. Their charge conjugate spinors are ${\psi_L}^C = {\psi^C}_L = -\mathrm{i}\sigma_2\psi_R^*$ and ${\psi_R}^C = {\psi^C}_R = \mathrm{i}\sigma_2\psi_L^*$. Consider the noncommutative left-handed spinor $\widehat{\psi_L} = \mathrm{SW}[\psi_L, \rho_{\psi_L}(A), \theta]$, we then have the $\pm\theta$ choice

$$\widehat{\psi_R} = \mathrm{SW}[\psi_R, \rho_{\psi_R}(A), \pm\theta] \tag{15}$$

for the right handed one. In the literature the choice $+\theta$ is usually considered so that for the 4-component Dirac spinor ψ we can write $\widehat{\psi} = \mathrm{SW}[\psi, A, \theta, \partial, i]$, $\delta\widehat{\psi} = \mathrm{i}\widehat{\Lambda}\star\widehat{\psi}$. We here advocate the opposite choice $(-\theta)$ in (15). Indeed from (13) we have that $\widehat{{\psi_L}^C} = -\mathrm{i}\sigma_2\,\widehat{\psi_R}^{\,*}$ and therefore

$$\widehat{\psi_R} = \mathrm{SW}[\psi_R, \rho_{\psi_R}(A), -\theta, \partial, i] \quad \Longleftrightarrow \quad \widehat{{\psi_L}^C} = \mathrm{SW}[{\psi_L}^C, \rho_{{\psi_L}^C}(A), +\theta, \partial, i] \tag{16}$$

so that with the $-\theta$ choice in (15), both left handed fermions $\widehat{\psi_L}$, $\widehat{{\psi_L}^C}$ are associated with θ while the right handed ones $\widehat{\psi_R}$, $\widehat{{\psi_R}^C}$ are associated with $-\theta$. In GUT theories we have multiplets of definite chirality and therefore this is the natural choice to consider in this setting.

These observations allow us to compare QED$_+$ with QED$_-$, the two different QED theories obtained with the two different $\pm\theta$ choices (15). This difference immediately extends to the fermion kinetic terms of nonabelian gauge theories and allows us to compare the NCSM discussed in [5] with the present GUT compatible one. We have (up to gauge kinetic terms)

$$S_{\mathrm{QED}_+} = \int \widehat{\psi_L}^\dagger \star \mathrm{i}\widehat{\not{D}}\,\widehat{\psi_L} + \widehat{\psi_R}^\dagger \star \mathrm{i}\widehat{\not{D}}\,\widehat{\psi_R} \quad , \quad S_{\mathrm{QED}_-} = \int \widehat{\psi_L}^\dagger \star \mathrm{i}\widehat{\not{D}}\,\widehat{\psi_L} + \widehat{{\psi_L}^C}^\dagger \star \mathrm{i}\widehat{\not{D}}\,\widehat{{\psi_L}^C}$$

where the GUT inspired QED$_-$ is obtained using the left handed spinor $\begin{pmatrix}\psi_L\\ \psi_L^C\end{pmatrix}$ so that $\widehat{\psi_L^C} = \mathrm{SW}[\psi_L^C, \rho_{\psi_L^C}(A), \theta]$. Now from $\widehat{{\psi_L}^C} = -\mathrm{i}\sigma_2\,\widehat{\psi_R}^{\,*}$ and from σ matrix algebra we have $\int \widehat{{\psi_L}^C}^\dagger \star \mathrm{i}\widehat{\not{D}}\,\widehat{{\psi_L}^C} = \int \widehat{\psi_R}^{op\,\dagger}\,\mu * \mathrm{i}\widehat{\not{D}}^{op}\,\widehat{\psi_R}^{op}$, where we have emphasized that we are using the $-\theta$ convention in the SW map by writing $\widehat{\ }^{op}$ instead of $\widehat{\ }$. We conclude that in order to obtain QED$_-$ from QED$_+$ we just need to change θ into $-\theta$ in the right handed fermion sector of QED$_+$.

[1] Given the group element $g = \mathrm{e}^{\mathrm{i}\Lambda} = \mathrm{e}^{\mathrm{i}\Lambda^a T^a}$ we have $\rho_{\Psi^*}(g) \equiv \overline{\rho_\Psi(g)}$ and, since Λ^a, A^a are real, we have $\rho_{\Psi^*}(\Lambda) = -\overline{\rho_\Psi(\Lambda)}$, $\rho_{\Psi^*}(A) = -\overline{\rho_\Psi(A)}$.

4 C, P, T Properties of SW Map and of NCYM Actions

Using compatibility of SW map with complex conjugation and the tensorial properties of SW map (i.e. that SW map preserves the space-time index) one can study the properties of SW map with respect to the C, P and T operations. In particular these same expressions as in the commutative case holds:

$$\widehat{\Psi_L}^{\,T} = -\mathrm{i}\sigma_1\sigma_3\widehat{\Psi_L} \quad , \quad \widehat{\Psi_R}^{\,T} = -\mathrm{i}\sigma_1\sigma_3\widehat{\Psi_R} \, , \quad \widehat{A}_\mu^{\,T} = (\widehat{A}_0, -\widehat{A}_i) \tag{17}$$

$$\widehat{\Psi_L}^{\,CP} = \mathrm{i}\sigma_2\widehat{\Psi_L}^{\,*} \, , \quad \widehat{\Psi_R}^{\,CP} = -\mathrm{i}\sigma_2\widehat{\Psi_R}^{\,*} \quad , \quad \widehat{A}_\mu^{\,CP} = (-\overline{\widehat{A}_0}, \overline{\widehat{A}_i}) \tag{18}$$

where the action of the P and C operators on spinors is given by

$$\widehat{\Psi_L}^{\,P} = \mathrm{SW}[\Psi_L^{\,P}, \rho_{\Psi_L}(A^P), \theta^P, \partial^P, i] \quad , \quad \widehat{\Psi_L}^{\,C} = \mathrm{SW}[\Psi_L^{\,C}, (\rho_{\Psi_L}(A))^C, \theta^C, \partial, i] \; ,$$

while the time inversion is given by $\widehat{\Psi_L}^{\,T} = \mathrm{SW}[\Psi_L^{\,T}, \rho_{\Psi_L^T}(A^T), \theta^T, \partial^T, -i]$. In these expressions we have written explicitly the dependence on the partial derivatives, and the imaginary unit i in the last slot marks that the coefficients in the SW map are in general complex coefficients. The $-i$ in the last expression means that we are considering the complex conjugates of the coefficients in the SW map, this is so because T is antilinear and multiplicative. Relations (17) and (18) hold provided that $\theta^{\mu\nu}$ transforms under C, P, T as a $U(1)$ field strenght $F_{\mu\nu}$. If we choose $+\theta$ in (15) then parity and charge conjugation sepatately assume the same expression as in the commutative case.

Now we discuss the transformations properties of NCYM actions under C, P and T. In the $+\theta$ choice (15) we have that NCYM actions are invariant under C, P and T iff in the commutative limit they are invariant. On the other hand, in the $-\theta$ choice NCYM actions are invariant under CP and T iff in the commutative limit they are invariant. For the fermion kinetic term these statements are a straighforward consequence of $\int \widehat{\Psi_L}^{\dagger} \star \partial\!\!\!/ \widehat{\Psi_L} = \int \widehat{\Psi_L}^{\dagger} \partial\!\!\!/ \widehat{\Psi_L}$. Since $\widehat{F}$ transforms like F under CP and T, and in the $+\theta$ case also under C and P separately, the C, P, T properties of the gauge kinetic term $\int \mathrm{Tr}(\widehat{F} \star \widehat{F}) = \int \mathrm{Tr}(\widehat{F}\widehat{F})$ easily follow. Inspection of the fermion gauge bosons interaction term leads also to the same conclusion.

We have studied the C, P and T symmetry properties of NCYM actions where θ transforms under C, P and T as a field strenght. Viceversa, *if we keep θ fixed* under C, P and T transformations, we in general have that NCYM theories *break* C, P and T symmetries.

Finally a $U(1)$ field strenght is invariant under the combined CPT transformation, and therefore θ does not change. This implies that CPT is always a symmetry of NCYM actions.

Acknowledgements

It is a pleasure to thank the organizers for the nice and stimulating atmosphere at the conference and its efficient organization.

References

1. M. Dimitrijevic, F. Meyer, L. Moller and J. Wess, "Gauge theories on the kappa-Minkowski spacetime," arXiv:hep-th/0310116.
2. N. Seiberg and E. Witten, JHEP **9909**, 032 (1999) [hep-th/9908142].
3. J. Madore, S. Schraml, P. Schupp and J. Wess, Eur. Phys. J. C **16**, 161 (2000) [hep-th/0001203].
   B. Jurco, S. Schraml, P. Schupp and J. Wess, Eur. Phys. J. C **17**, 521 (2000) [hep-th/0006246].
4. B. Jurco, L. Moller, S. Schraml, P. Schupp and J. Wess, Eur. Phys. J. C **21**, 383 (2001) [hep-th/0104153].
5. X. Calmet, B. Jurco, P. Schupp, J. Wess and M. Wohlgenannt, Eur. Phys. J. C **23** (2002) 363 [arXiv:hep-ph/0111115].
6. P. Aschieri, B. Jurco, P. Schupp and J. Wess, Nucl. Phys. B **651** (2003) 45 [arXiv:hep-th/0205214].
7. P. Schupp, J. Trampetic, J. Wess and G. Raffelt, "The photon neutrino interaction in non-commutative gauge field theory and astrophysical bounds," arXiv:hep-ph/0212292.
8. M. Buric and V. Radovanovic, JHEP **0402** (2004) 040 [arXiv:hep-th/0401103].
9. F. Brandt, C. P. Martin and F. R. Ruiz, JHEP **0307** (2003) 068 [arXiv:hep-th/0307292].
10. M. M. Sheikh-Jabbari, Phys. Rev. Lett. **84** (2000) 5265 [arXiv:hep-th/0001167].
11. M. Chaichian, K. Nishijima and A. Tureanu, Phys. Lett. B **568**, 146 (2003) [arXiv:hep-th/0209008]
    L. Alvarez-Gaume and M. A. Vazquez-Mozo, Nucl. Phys. B **668** (2003) 293 [arXiv:hep-th/0305093]
    N. Mahajan, Phys. Lett. B **569** (2003) 85 [arXiv:hep-th/0305105].

Noncommutative Gauge Theory on the q-Deformed Euclidean Plane

Frank Meyer[1,2] and Harold Steinacker[2]

[1] Max-Planck-Institute for Physics (Werner-Heisenberg-Institut), Föhringer Ring 6, 80805 München, Germany
meyerf@mppmu.mpg.de
[2] University of Munich, Theresienstr. 37, 80333 München, Germany
hsteinac@theorie.physik.uni-muenchen.de

1 Introduction

The physical nature of space at very short distances is still not known. Already Heisenberg proposed in a letter to Peierls [1] that spacetime is quantized below some scale, suggesting that this could help to resolve the problem of infinities in quantum field theories. With this motivation, there has been a lot of work and progress in the formulation of quantum field theory on quantized or noncommutative spaces. Noncommutativity is implemented by replacing a differentiable space-time manifold by an algebra of noncommutative coordinates

$$[x^i, x^j] = \theta^{ij}(x) \neq 0 \,. \tag{1}$$

The simplest case is the so-called canonical quantum plane $\mathbb{R}^n_\theta$, where θ^{ij} is a constant tensor independent of x. This is the space which is usually considered in the literature [2]. However, most of the rotational symmetry is lost on $\mathbb{R}^n_\theta$. On the other hand, there exist quantum spaces which admit a generalized notion of symmetry, being covariant under a quantum group. Not much is known about field theory on this type of spaces.

One of the simplest spaces with quantum group symmetry is the Euclidean quantum plane $\mathbb{R}^2_q$. It is covariant with respect to the q-deformed two-dimensional Euclidean group $E_q(2)$. We report here on our work [3], proposing a formulation of gauge theory based on the natural algebraic structures on this spaces and using a suitable star product.

2 The $E_q(2)$-Symmetric Plane

The $E_q(2)$-Symmetric Plane is generated by the complex coordinates $z, \bar{z}$ with the commutation relation

$$z\bar{z} = q^2 \bar{z}z \,. \tag{2}$$

We consider formal power series in these variables $z, \bar{z}$ as functions on this space,

$$\mathbb{R}^2_q := \mathbb{R}\langle\langle z, \bar{z}\rangle\rangle / (z\bar{z} - q^2\bar{z}z) \,. \tag{3}$$

Notice that the simple commutation relation (2) are inconsistent with the usual formulas for differentiation and integration. We should therefore first discuss the appropriate differential calculus and invariant integration. Finally, to get physical predictions i.e. real numbers from the abstract algebra, we also need either a representation of the noncommutative algebra, or a realization of the algebra using a star product.

2.1 Covariant Differential Calculus

It is natural to require that there exist deformed spaces of k-forms Ω_q^k which are covariant with respect to $E_q(2)$, and that the exterior differential $d : \Omega_q^k \to \Omega_q^{k+1}$ satisfies the usual Leibniz-rule as well as $d^2 = 0$. One can show that there exists a unique covariant differential calculus with these properties [4]:

$$\begin{aligned} z\mathrm{d}z &= q^{-2}\mathrm{d}z z, & \overline{z}\mathrm{d}z &= q^{-2}\mathrm{d}z\overline{z} \\ z\mathrm{d}\overline{z} &= q^{2}\mathrm{d}\overline{z}z, & \overline{z}\mathrm{d}\overline{z} &= q^{2}\mathrm{d}\overline{z}\,\overline{z} \end{aligned} \tag{4}$$

The following result is particularly useful for the construction of gauge field theories on $\mathbb{R}_q^2$:

Lemma 1. *Consider the one-forms*

$$\theta \equiv \theta^z := z^{-1}\overline{z}\mathrm{d}z \quad , \quad \overline{\theta} \equiv \theta^{\overline{z}} := \mathrm{d}\overline{z}z\overline{z}^{-1} \tag{5}$$

and define

$$\Theta := \theta^i \lambda_i, \quad \textit{where} \quad \lambda_z := \frac{1}{1-q^{-2}}\overline{z}^{-1}, \; \lambda_{\overline{z}} := -\frac{1}{1-q^{-2}}z^{-1}. \tag{6}$$

Then for all functions $f \in \mathbb{R}_q^2$ and one-forms α, the following holds:

$$[\theta, f] = [\overline{\theta}, f] = 0, \tag{7}$$

$$\mathrm{d}f = [\Theta, f] = [\lambda_i, f]\theta^i \tag{8}$$

$$\mathrm{d}\alpha = \{\Theta, \alpha\} \tag{9}$$

denoting with $\{\cdot,\cdot\}$ the anti-commutator.

2.2 Invariant Integral

In order to define an invariant action, we need an integral on $\mathbb{R}_q^2$ which is invariant under quantum group transformations. This means that

$$\int^q f(z,\overline{z}) \triangleleft X = \varepsilon(X) \int^q f(z,\overline{z}) \tag{10}$$

for all $f \in \mathbb{R}_q^2$ and $X \in U_q(e(2))$. Here $U_q(e(2))$ is the q-deformed universal enveloping Lie algebra of the two-dimensional Euclidean group, $\triangleleft$ denotes the right action of $U_q(e(2))$ on $\mathbb{R}_q^2$ and $\varepsilon(X)$ is the counit. Now any function in $\mathbb{R}_q^2$ can be decomposed as $f(z,\overline{z}) = \sum_{m\in\mathbb{Z}} z^m f_m(z\overline{z})$. It can be shown that (10) is satisfied for the following discrete quantum traces [5]

$$\int^{q,(r_0)} f(z,\overline{z}) := r_0^2(q^2-1) \sum_{k=-\infty}^{\infty} q^{2k} f_0(q^{2k} r_0^2), \tag{11}$$

where $r_0 \in \mathbb{R}$ labels the irreducible representations of $\mathbb{R}^2_q$. The most general invariant integral is given by superpositions of these integrals,

$$\int^q f(z,\bar{z}) = \int_1^q dr_0 \mu(r_0) \int^{q,(r_0)} f_0(z\bar{z}) \tag{12}$$

with arbitrary "weight" function $\mu(r) > 0$. It is quite remarkable and useful that for the special choice $\mu(r_0) = \frac{1}{r_0(q^2-1)}$, one recovers the usual Riemannian integral, which is therefore also invariant under $U_q(e(2))$ [3].

3 Star Product Approach

3.1 The Star Product

The noncommutative algebra $\mathbb{R}^2_q$ can be realized on the algebra of commutative functions on $\mathbb{R}^2$ using a new, noncommutative product, called star product. Let us denote the commutative variables on $\mathbb{R}^2$ by greek letters $\zeta, \bar{\zeta}$ to distinguish them from the generators $z, \bar{z}$, and let $q =: \mathrm{e}^h$. Then a hermitian star product for $\mathbb{R}^2_q$ is given by

$$f \star g := \mu \circ \mathrm{e}^{h(\zeta\partial_\zeta \otimes \bar{\zeta}\partial_{\bar{\zeta}} - \bar{\zeta}\partial_{\bar{\zeta}} \otimes \zeta\partial_\zeta)}(f \otimes g) = fg + h\zeta\bar{\zeta}(\partial_\zeta f \partial_{\bar{\zeta}} g - \partial_{\bar{\zeta}} f \partial_\zeta g) + \mathcal{O}(h^2) \,. \tag{13}$$

3.2 Noncommutative Gauge Transformations

The formalism of covariant coordinates was established in [6] for an arbitrary Poisson structure. This leads to problems in the semi-classical limit[1]. Therefore we propose the following approach, taking advantage of the frame $\theta, \bar{\theta}$ which commutes with all functions and the generator Θ of the exterior differential. We define infinitesimal gauge transformations of a matter field as

$$\delta\psi = \mathrm{i}\Lambda \star \psi, \qquad \delta\zeta^i = 0 \,. \tag{14}$$

Let us introduce the "covariant derivative" (or covariant one-form) as

$$D := \Theta - \mathrm{i}A \,. \tag{15}$$

Then requiring that $D\psi$ transforms covariantly, i.e.

$$\delta D \star \psi \stackrel{!}{=} \mathrm{i}\Lambda \star D \star \psi \tag{16}$$

leads to the following gauge transformation property for the gauge field A:

$$\delta A = [\Theta \stackrel{\star}{,} \Lambda] + \mathrm{i}[\Lambda \stackrel{\star}{,} A] = \mathrm{d}\Lambda + \mathrm{i}[\Lambda \stackrel{\star}{,} A] \,. \tag{17}$$

We define the field strength as the two-form

[1] To obtain in the classical limit the classical gauge field a_i we have to invert θ^{ij}. This is only well-defined if θ is invertible, and even then it spoils the covariant transformation property whenever θ is not constant. To maintain covariance one has to "invert θ covariantly" as done in [7], leading to complicated expressions.

$$F := D \wedge_q D \tag{18}$$

where $\wedge_q$ the star-wedge [3]. Then

$$\delta F = \mathrm{i}[\Lambda \stackrel{\star}{,} F]. \tag{19}$$

As a two-form, the field strength can be written as $F = f\theta \wedge_q \bar{\theta}$. Since the frame $\theta, \bar{\theta}$ commutes with functions, f transforms covariantly as well:

$$\delta f = \mathrm{i}[\Lambda \stackrel{\star}{,} f] \,. \tag{20}$$

We note that all transformations have the correct classical limit as $q \to 1$.

3.3 Seiberg-Witten Map

The Seiberg-Witten map [8] allows to express the noncommutative gauge fields in terms of the commutative ones. Hence the noncommutative theory can be interpreted as a deformation of the commutative theory. Its physical predictions can be explicitly obtained by expanding in the deformation parameter h, and the commutative theory is reproduced in the limit $h \to 0$. The Seiberg-Witten map is based on the following requirement:

- The *consistency condition:*

$$\begin{aligned}(\delta_\alpha\delta_\beta - \delta_\beta\delta_\alpha)\Psi &= \delta_{-i[\alpha,\beta]}\Psi \\ \Leftrightarrow \mathrm{i}\delta_\alpha\Lambda_\beta - \mathrm{i}\delta_\beta\Lambda_\alpha + [\Lambda_\alpha \stackrel{\star}{,} \Lambda_\beta] &= \mathrm{i}\Lambda_{-i[\alpha,\beta]},\end{aligned} \tag{21}$$

- Noncommutative gauge transformations are related to commutative ones:

$$\begin{aligned}A_i[a_i] + \delta_\Lambda A_i[a_i] &= A_i[a_i + \delta_\alpha a_i] \\ \Psi[\psi, a_i] + \delta_\Lambda \Psi[\psi, a_i] &= \Psi[\psi + \delta_\alpha\psi, a_i + \delta_\alpha a_i] \,.\end{aligned} \tag{22}$$

Solving these conditions in our case, we obtain the following expression for the field strength expanded in powers of h:

$$\begin{aligned}f = F^0_{12} + h\{&F^0_{12} + \theta^{12}(F^0_{12}F^0_{12} - a_\zeta\partial_{\bar\zeta}F^0_{12} + a_{\bar\zeta}\partial_\zeta F^0_{12}) + \partial_\zeta\theta^{12}(a_\zeta\partial_{\bar\zeta}a_{\bar\zeta} \\ &+ a_{\bar\zeta}\partial_{\bar\zeta}a_\zeta + 2a_{\bar\zeta}\partial_\zeta a_{\bar\zeta}) + \partial_{\bar\zeta}\theta^{12}(a_\zeta\partial_\zeta a_{\bar\zeta} + a_{\bar\zeta}\partial_\zeta a_\zeta + 2a_\zeta\partial_{\bar\zeta}a_\zeta)\} + \mathcal{O}(h^2)\,,\end{aligned} \tag{23}$$

where a_{ζ_i} is the classical gauge field and $F^0_{ij} = \partial_{\zeta_i}a_j - \partial_{\zeta_j}a_i$ is the classical field strength.

3.4 The Action

To define a gauge-invariant action, we need an integral which is cyclic with respect to the star product, since the field strength transforms in the adjoint. The invariant integrals (11) do not have this property, which can be restored by introducing a measure function μ. It can be shown that for the measure function $\mu := \frac{1}{\zeta\bar\zeta}$, we can even drop the star under the integral:

$$\int \mathrm{d}\zeta\mathrm{d}\bar\zeta\mu f \star g = \int \mathrm{d}\zeta\mathrm{d}\bar\zeta\mu f g = \int \mathrm{d}\zeta\mathrm{d}\bar\zeta\mu g \star f \tag{24}$$

Putting all this together, we can now define a gauge invariant action:

$$S := \frac{1}{2} \int \mathrm{d}\zeta \mathrm{d}\bar{\zeta} \mu f \star f \,. \tag{25}$$

Gauge invariance is guaranteed because of (20). Moreover, the classical action for abelian gauge field theory is reproduced in the classical limit $h \to 0$ because of (23).

Another possibility is to replace the measure function μ by a scalar "Higgs" field φ, which transforms such that it restores the gauge invariance of the action. One can in fact find a suitable potential for ϕ which admits a solution $\langle \phi \rangle = \mu = \frac{1}{\zeta\bar{\zeta}}$, leading to a $E_q(2)$-invariant action through spontaneous symmetry breaking.

References

1. W. Pauli: *Scientific Correspondence*, Vol. II (Springer 1985) p 15.
2. M. R. Douglas and N. A. Nekrasov: Rev. Mod. Phys. **73**, 977 (2001).
3. F. Meyer and H. Steinacker: Int. J. Mod. Phys. A, arXiv: hep-th/0309053.
4. M. Chaichian and A. P. Demichev: Phys. Lett. **B320**, 273 (1994).
5. H. T. Koelink: Duke Math. J. **76**, 483 (1994).
6. J. Madore, S. Schraml, P. Schupp and J. Wess: Eur. Phys. J. **C16**, 161 (2000).
7. F. Meyer: *Models of Gauge Field Theory on Noncommutative Spaces*, Diploma-Thesis, Univ. of Munich, Chair Prof. J. Wess (2003), arXiv: hep-th/0308186.
8. N. Seiberg and E. Witten: JHEP **09**, 032 (1999).

A Multispecies Calogero Model

Marijan Mileković[1], Stjepan Meljanac[2], and Andjelo Samsarov[3]

[1] University of Zagreb, Faculty of Science
marijan@phy.hr
[2] meljanac@irb.hr
[3] asamsarov@irb.hr
Rudjer Bošković Institute, Zagreb

It is well known that the Calogero model for N identical particles on the line [1] is connected to Haldane's exclusion statistics [2]. The role of statistical parameter is played by the strenght of the two-body inverse-square interaction ν. In Haldane's formulation, however, there is the possibility of having particles of different species with a mutual statistical coupling parameter depending on the species coupled. This suggest the possible generalization of the Calogero system to the system of N distinguishable particles, distinguishabillity being introduced through the replacements $\nu \to \nu_{ij} = \nu_{ji}$ and $m \to m_i$, m being a mass of the Calogero particles.

A one-dimensional multispecies Calogero model is defined by the following Hamiltonian [3] ($\hbar = \omega = 1; i,j = 1,2 \ldots N$):

$$H = -\frac{1}{2}\sum_i \frac{1}{m_i}\frac{\partial_i^2}{\partial x_i^2} + \frac{1}{2}\sum_i m_i x_i^2 + \frac{1}{4}\sum_{i\neq j}\frac{\nu_{ij}(\nu_{ij}-1)}{(x_i-x_j)^2}\left(\frac{1}{m_i}+\frac{1}{m_j}\right)$$

$$+\frac{1}{2}\sum_{i,j,k\neq}\frac{\nu_{ij}\nu_{jk}}{m_j(x_j-x_i)(x_j-x_k)} . \tag{1}$$

The ground state of H is

$$\Psi_0(x_1,x_2,\cdots x_N) = \Delta e^{-\frac{1}{2}\sum_i m_i x_i^2} = \prod_{i<j}(x_i-x_j)^{\nu_{ij}} e^{-\frac{1}{2}\sum_i m_i x_i^2} , \tag{2}$$

$$E_0 = \left(\frac{N}{2} + \frac{1}{2}\sum_{i\neq j}\nu_{ij}\right) .$$

Notice the appearance of the three-body term in (1). It can be shown that in the limit $\nu_{ij} \to \nu$ and $m_i \to m$ the last term in (1) identically vanish and the Hamiltonian (1) smoothly goes to the Calogero Hamiltonian [1]. (Another nontrivial condition for disappearance of the three-body interaction is $\nu_{ij} = \alpha m_i m_j$ [3]). Performing the non-unitary transformation on Ψ_0, $\tilde{\Psi}_0 = \Delta^{-1}\Psi_0$, one generates Hamiltonian $\tilde{H} = \Delta^{-1} H \Delta$ for which the three-body term apparently disappears:

$$\tilde{H} = -\frac{1}{2}\sum_i \frac{1}{m_i}\frac{\partial^2}{\partial x_i^2} + \frac{1}{2}\sum_i m_i x_i^2 - \frac{1}{2}\sum_{i\neq j}\frac{\nu_{ij}}{(x_i-x_j)}\left(\frac{1}{m_i}\frac{\partial}{\partial x_i} - \frac{1}{m_j}\frac{\partial}{\partial x_j}\right) . \tag{3}$$

For later convenience we define operators $\{T_+, T_-, T_0\}$ as

$$T_+ = \frac{1}{2}\sum_i m_i x_i^2, \qquad T_0 = \frac{1}{2}\left(\sum_i x_i \frac{\partial}{\partial x_i} + E_0\right)$$

$$T_- = \frac{1}{2}\sum_i \frac{1}{m_i}\frac{\partial^2}{\partial x_i^2} + \frac{1}{2}\sum_{i\neq j}\frac{\nu_{ij}}{(x_i - x_j)}\left(\frac{1}{m_i}\frac{\partial}{\partial x_i} - \frac{1}{m_j}\frac{\partial}{\partial x_j}\right) . \tag{4}$$

Operators (4) generate $SU(1,1)$ algebra. In terms of these operators, the Hamiltonian (3) reads $\tilde{H} = T_+ - T_-$. Next, we separate center-of-mass motion (CM) and relative motion (R) of particles by defining variables X and ξ_i:

$$X = \frac{\sum_i m_i x_i}{\sum_i m_i} \equiv \frac{1}{M}\sum_i m_i x_i, \qquad \xi_i = x_i - X .$$

Hamiltonian $\tilde{H}$ then separates as $\tilde{H} = \tilde{H}_{CM} + \tilde{H}_R$ and wave function $\tilde{\Psi}_0$ separates as $\tilde{\Psi}_0(x_1, x_2, \cdots x_N) = \tilde{\Psi}_0(X)\tilde{\Psi}_0(\xi_1, \xi_2 \cdots \xi_N)$.

Now we introduce creation and annihilation operators $\{A_1^+, A_1^-\}$, $\{A_2^+, A_2^-\}$ and $\{B_2^+, B_2^-\}$:

$$A_1^\pm = \frac{1}{\sqrt{2}}\left(\sqrt{M}X \mp \frac{1}{\sqrt{M}}\frac{\partial}{\partial X}\right), \qquad A_2^\pm = \frac{1}{2}(T_+ + T_-) \mp T_0 , \tag{5}$$

$$B_2^\pm = A_2{}^\pm - \frac{1}{2}A_1^{\pm 2} .$$

Introduction of the operators $\{B_2^+, B_2^-\}$ help us to decouple inessential CM-motion, described by the operators $A_1^\pm$, i.e.

$$\tilde{H}_{CM} = \frac{1}{2}\{A_1^-, A_1^+\}_+, \qquad \tilde{H}_R = [B_2^-, B_2^+] ,$$

$$[\tilde{H}_{CM}, A_1^\pm] = \pm A_1^\pm, \qquad [\tilde{H}_R, B_2^\pm] = \pm 2B_2^\pm , \tag{6}$$

$$\tilde{H}_{CM}|\tilde{0}\rangle_{CM} = \frac{1}{2}|\tilde{0}\rangle_{CM}, \qquad \tilde{H}_R|\tilde{0}\rangle_R = \left(\frac{N-1}{2} + \frac{1}{2}\sum_{i\neq j}\nu_{ij}\right)|\tilde{0}\rangle_R .$$

The Fock space now splits into the CM-Fock space, spanned by $A_1^{+n_1}|\tilde{0}\rangle_{CM}$, and the R-Fock space, spanned by $B_2^{+n_2}|\tilde{0}\rangle_R$, with $|\tilde{0}\rangle_{CM} \propto \tilde{\Psi}_0(X)$ and $|\tilde{0}\rangle_R \propto \tilde{\Psi}_0(\xi_1, \xi_2 \cdots \xi_N)$.

The operators just introduced act on the Fock vacuum $|\tilde{0}\rangle \propto \tilde{\Psi}_0(x_1, x_2, \cdots x_N)$ as:

$$A_1^-|\tilde{0}\rangle = A_2^-|\tilde{0}\rangle = B_2^-|\tilde{0}\rangle = 0, \qquad \langle\tilde{0}|\tilde{0}\rangle = 1 .$$

The excited states are built as

$$|n_1, n_2\rangle \propto A_1^{+n_1} B_2^{+n_2}|\tilde{0}\rangle, \qquad \forall n_1, n_2 = 0, 1, \cdots \tag{7}$$

The states $|n_1, n_2\rangle$ are eigenstates of $\tilde{H}$, (3), with the eigenvalues

$$E_{n_1,n_2} = (n_1 + 2n_2) + E_0 \equiv n + E_0 . \tag{8}$$

Thus, the energy spectrum is linear in quantum numbers n_1, n_2 and degenerate. This result is universal, i.e. it holds for all parameters m_i and ν_{ij}. It is evident that for $n = even$, the degeneracy is $(\frac{n}{2} + 1)$ and for $n = odd$, the degeneracy is $(\frac{n+1}{2})$. The dynamical symmetry of degenerate energy levels is $SU(2)$ algebra [4], generated by $A_1^{+2}B_2^-$, $B_2^+A_1^{-2}$ and $A_1^+A_1^-$. This is the minimal symmetry that remains in the generic case.

Closer inspection of the R-Fock space reveals the existence of the universal critical point at which the system exhibits singular behaviour [3,5]. The critical point is defined by the null-vector

$$_R\langle\tilde{0}|B_2^- B_2^+|\tilde{0}\rangle_R = \frac{N-1}{2} + \frac{1}{2}\sum_{i\neq j}\nu_{ij} = 0\ . \tag{9}$$

At the critical point the system described by $\tilde{H}_R$ collapses completely, i.e. the relative coordinates, the relative momenta and the relative energy are all zero. There survives only one oscillator, describing the motion of the centre-of-mass.

The open problem that still remains is the question of integrability and full solvability of the model(s). It would be also interesting to incorporate supersymmetry into the model(s). Finally, the most important question is whether there is a realistic system in physics or in other cross-disciplinary areas where such a model(s) play a role.

References

1. F. Calogero, J. Math. Phys. **10**, 2191 (1969).
2. F. D. M. Haldane, Phys. Rev. Lett. **67**, 937 (1991); M. V. N. Murthy and R. Shankar, Phys. Rev. Lett. **73**, 3331 (1991).
3. S. Meljanac, M. Mileković and A. Samsarov, Phys. Lett. **B 573**, 202 (2003).
4. L. Jonke and S. Meljanac, Phys. Lett. **B 511**, 276 (2001).
5. V. Bardek, L. Jonke, S. Meljanac and M. Mileković, Phys. Lett. **B 531**, 311 (2001).

Divergencies in Noncommutative $SU(2)$ Yang-Mills Theory

Voja Radovanović[1] and Maja Burić[2]

[1] rvoja@ff.bg.ac.yu
[2] majab@ff.bg.ac.yu
Faculty of Physics, P.O.Box 368, 11001 Belgrade, Serbia and Montenegro

1 Introduction

Although discussed for quite some time, the question of renormalizability of noncommutative field theories on $\mathbf{R}^4$ has not been settled in a satisfactory way yet. Intuitively, noncommutativity, that is, a relation of the type

$$[\hat{x}^\mu, \hat{x}^\nu] = \mathrm{i}\theta^{\mu\nu} \tag{1}$$

puts the lower bound on the measurement of the coordinates, and one expects that it regularizes the theory in the UV sector.

In discussions of quantization the noncommutative field $\hat{\Phi}(\hat{x})$ is usually represented by a function $\hat{\Phi}(x)$ on commutative $\mathbf{R}^4$, whereas for the constant noncommutativity parameter $\theta^{\mu\nu}$ the multiplication is given by the Moyal, or $\star$-product:

$$\phi(x) \star \chi(x) = \mathrm{e}^{\frac{i}{2}\theta^{\mu\nu}\frac{\partial}{\partial x^\mu}\frac{\partial}{\partial y^\nu}}\phi(x)\chi(y)|_{y\to x} \ . \tag{2}$$

The lagrangian of the noncommutative theory is obtained from the commutative one by the systematic replacement of multiplication by the $\star$-product. One can quantize theory in the usual perturbative manner; however, as discussed in [1], the UV/IR mixing spoils the renormalizability.

As shown by Seiberg and Witten [2], noncommutative and commutative gauge theories are equivalent. This equivalence is realized by a mapping (SW map) relating the representation of NC gauge symmetry to the fields carrying the representation of its commutative counterpart. SW map is given as a series in powers of $\theta^{\mu\nu}$. Classical action is also expanded in θ: in the zero-th order it reduces to the action of ordinary gauge theory; additional terms can be treated as couplings.

We study the renormalizability of NC $SU(2)$ in the θ-expanded approach [3, 4]. For NC $SU(2)$ Yang-Mills theory the classical action is given by

$$S = \int \mathrm{d}^4x\, \hat{\bar{\psi}} \star (i\gamma^\mu \hat{D}_\mu - m)\hat{\psi} - \frac{1}{4}\int \mathrm{d}^4x\, (\hat{F}_{\mu\nu} \star \hat{F}^{\mu\nu}) \tag{3}$$

where the noncommutative field strength $\hat{F}_{\mu\nu}$ is

$$\hat{F}_{\mu\nu} = \partial_\mu \hat{A}_\nu - \partial_\nu \hat{A}_\mu - \mathrm{i}(\hat{A}_\mu \star \hat{A}_\nu - \hat{A}_\nu \star \hat{A}_\mu) \ , \tag{4}$$

and the covariant derivative

$$\hat{D}_\mu \hat{\psi} = \partial_\mu \hat{\psi} - \mathrm{i}\hat{A}_\mu \star \hat{\psi} \ . \tag{5}$$

Commutative fields $\psi(x)$, $A_\rho(x)$ are related to noncommutative matter and gauge fields $\hat{\psi}(x)$, $\hat{A}_\rho(x)$ by the Seiberg-Witten expansion

$$\hat{A}_\rho(x) = A_\rho(x) - \frac{1}{4}\,\theta^{\mu\nu}\left\{A_\mu(x), \partial_\nu A_\rho(x) + F_{\nu\rho}(x)\right\} + \dots \tag{6}$$

$$\hat{\psi}(x) = \psi(x) - \frac{1}{2}\,\theta^{\mu\nu} A_\mu(x)\partial_\nu\psi(x) + \frac{i}{4}\,\theta^{\mu\nu} A_\mu(x)A_\nu(x)\psi(x) + \dots$$

$\psi(x)$ is in the fundamental representation of $SU(2)$, so the group generators are $T^a = \frac{\sigma_a}{2}$.

Applying the SW map, the classical action in the θ-linear order becomes: $S = S_0 + S_{1,A} + S_{1,\psi}$, with

$$\begin{aligned} S_0 &= \int \mathrm{d}^4x \left(\bar{\psi}(i\gamma^\mu D_\mu - m)\psi - \frac{1}{4}F^{\mu\nu a}F^a_{\mu\nu}\right) , \\ S_{1,A} &= 0 \ , \\ S_{1,\psi} &= \frac{1}{2}\,\theta^{\rho\sigma}\int \mathrm{d}^4x \left(-i\bar{\psi}\gamma^\mu F_{\mu\rho}D_\sigma\psi + \frac{1}{2}\,\bar{\psi}F_{\rho\sigma}(-i\gamma^\mu D_\mu + m)\psi\right) \\ &= \int \mathrm{d}^4x \left(-\frac{1}{8}\,\theta^{\rho\sigma}\Delta^{\mu\nu\alpha}_{\sigma\rho\beta}\bar{\psi}F_{\mu\nu}\gamma^\beta(i\partial_\alpha + A_\alpha)\psi + \frac{m}{4}\,\theta^{\mu\nu}\bar{\psi}F_{\mu\nu}\psi\right) . \end{aligned} \tag{7}$$

To simplify the notation we introduced the symbol $\Delta^{\alpha\beta\gamma}_{\sigma\rho\mu}$:

$$\Delta^{\alpha\beta\gamma}_{\sigma\rho\mu} = \delta^\alpha_\sigma\delta^\beta_\rho\delta^\gamma_\mu - \delta^\alpha_\rho\delta^\beta_\sigma\delta^\gamma_\mu + (\text{cyclic } \alpha\beta\gamma) = -\epsilon^{\alpha\beta\gamma\lambda}\epsilon_{\sigma\rho\mu\lambda} \ . \tag{8}$$

The bosonic θ-linear term $S_{1,A}$ vanishes because it is proportional to the symmetric coefficients $d^{abc} = \mathrm{Tr}(\{\mathrm{T^a}, \mathrm{T^b}\}\mathrm{T^c})$, and for $SU(2)$ these coefficients are zero for all irreducible representations.

2 Divergences and Discussion

In order to find the divergent part of the one-loop effective action we used the background field method. In the θ^0-order we obtain the standard result for the 2-point functions

$$\Gamma_2 = \frac{1}{(4\pi)^2\epsilon}\Big(- 6A^a_\mu\partial^2 A^{\mu a} - 6(\partial_\mu A^{\mu a})^2 + \frac{3i}{2}\bar{\psi}\gamma_\mu\partial^\mu\psi - 6m\bar{\psi}\psi\Big) \ . \tag{9}$$

The contribution of the ghost action is to be added

$$\Gamma_{2,gh} = \frac{1}{(4\pi)^2\epsilon}\,\frac{1}{3}\,F^a_{\mu\nu}F^{\mu\nu a} \ . \tag{10}$$

In the linear order the divergent part of the 2-point function reads [5]:

$$\begin{aligned} \Gamma'_{2\psi} &= \frac{1}{(4\pi)^2\epsilon}\,\theta^{\mu\nu}\Big(\frac{i}{4}\,\epsilon_{\mu\nu\rho\sigma}\bar{\psi}\gamma_5\gamma^\sigma\partial^2\partial^\rho\psi + \frac{1}{2}\,m\bar{\psi}\sigma_{\nu\alpha}\partial_\mu\partial^\alpha\psi - \frac{1}{4}\,m\bar{\psi}\sigma_{\mu\nu}\partial^2\psi\Big) \\ &+ \frac{1}{(4\pi)^2\epsilon}\,\frac{3}{8}\,\theta^{\mu\nu}\Big(- im^2\epsilon_{\mu\nu\alpha\beta}\bar{\psi}\gamma_5\gamma^\beta\partial^\alpha\psi + m^3\bar{\psi}\sigma_{\nu\mu}\psi\Big) \ . \end{aligned} \tag{11}$$

This is a nice result: comparing $\Gamma'_{2\psi}$ with the θ-linear correction for the 2-point functions in NC QED [6], we see that in the $SU(2)$ case only the fermionic propagator gets a correction; the correction is, up to a factor, the same as the QED correction. This has further consequences. In NC QED we argued that the massive terms obstruct renormalization, as only in the case $m = 0$ one can redefine fields in such a way that the divergent terms disappear. The analysis can be repeated for $SU(2)$ without change. Hence, we come to the conclusion: NC gauge theories with massive fermions are not renormalizable.

The divergent part of the 3-point function for the massless theory has the following covariant form:

$$\begin{aligned}\Gamma'_3 = &-\frac{1}{(4\pi)^2\epsilon}\,\frac{1}{2}\,\theta^{\mu\nu}\Big(-\frac{111i}{6}\,\bar\psi\gamma^\alpha F_{\nu\alpha}D_\mu\psi \\ &-\frac{43i}{4}\,\bar\psi\gamma^\alpha F_{\mu\nu}D_\alpha\psi-\frac{3i}{4}\,\bar\psi\gamma^\alpha(D_\alpha F_{\mu\nu})\psi \\ &+2i\bar\psi\gamma_\mu F_{\nu\alpha}D^\alpha\psi+i\bar\psi\gamma_\mu(D^\alpha F_{\nu\alpha})\psi \\ &+\frac{5}{8}\,\epsilon_{\nu\alpha\beta\rho}\bar\psi\gamma_5\gamma^\rho(D_\mu F^{\alpha\beta})\psi-\frac{1}{16}\,\epsilon_{\mu\nu\alpha\beta}\bar\psi\gamma_5\gamma^\sigma(D_\sigma F^{\alpha\beta})\psi \\ &+\frac{1}{8}\,\epsilon_{\mu\nu\alpha\beta}\big(2\bar\psi\gamma_5\gamma^\beta F^{\rho\alpha}D_\rho\psi+\bar\psi\gamma_5\gamma^\beta(D_\rho F^{\rho\alpha})\psi\big)\Big)\,.\end{aligned} \tag{12}$$

4-fermionic vertex has a very important role in the discussion of renormalizability. The divergent part of the 4-point fermionic function is

$$\Gamma'_{4\psi} = \frac{1}{(4\pi)^2\epsilon}\,\frac{9}{32}\,\theta^{\mu\nu}\epsilon_{\mu\nu\rho\sigma}\bar\psi\gamma_5\gamma^\sigma\psi\,\bar\psi\gamma^\rho\psi\,. \tag{13}$$

As pointed out in [7, 8], SW map is not unique and it can be used to subtract the divergent terms in the effective action and regularize the theory. Nonuniqueness allows one to redefine fields $A_\mu^{(n)}$, $\psi^{(n)}$ in the n-th order by adding any gauge covariant expression (of appropriate dimension) with exactly n factors of θ. One can think of such a procedure as a sort of "dressing" of the "bare" fields. Analyzing the expressions $\Gamma'_{2\psi}(m=0)$ and Γ'_3 we can see that they can be cancelled by the field redefinitions.

However, there is a divergent term which spoils the generalized renormalizability: the 4-point function $\Gamma'_{4\psi}$. It predicts the current-current interaction, and there is no simple way to circumvent this noncommutativity-induced coupling. From the power counting we see that, in the linear order, 4ψ is the only potentially problematic vertex. In the quadratic order, possible vertices which do not fit into the field redefinition scheme are, e.g., θ^2F^4, $\theta^2(\bar\psi\gamma\psi)^2F$, etc.

The 4ψ term $\theta^{\mu\nu}\epsilon_{\mu\nu\rho\sigma}\bar\psi\gamma_5\gamma^\sigma\psi\,\bar\psi\gamma^\rho\psi$ is present in both $U(1)$ and $SU(2)$ theories coupled to fermions. An interesting fact, however, is that in both theories it has the same form, whereas the other combinations allowed by covariance, as, e.g. $\theta^{\mu\nu}\epsilon_{\mu\nu\rho\sigma}\bar\psi\gamma_5\gamma^\sigma\frac{\sigma_a}{2}\psi\,\bar\psi\gamma^\rho\frac{\sigma_a}{2}\psi$ or $\theta^{\mu\nu}\bar\psi\gamma_5\gamma_\mu\psi\,\bar\psi\gamma_\nu\psi$, never show up. This opens the possibility that this divergence might cancel in a gauge theory based on the product of gauge groups. However, we are not in favor of a theory which needs too much fine tuning. Thus we are inclined to interpret our results as an indication that NC gauge theories coupled to fermions are not renormalizable, even in the massless case. However, in order to draw a definite conclusion, one should certainly consider the contributions of the mentioned θ^2-order $4A$ or $A4\psi$ vertices.

References

1. M. Hayakawa, "Perturbative analysis on infrared and ultraviolet aspects on noncommutative QED on R^4", preprint hep-th/9912167; Phys. Lett. B **478**, 394 (2000); C. P. Martin and D. Sánchez-Ruiz, Phys. Rev. Lett. **83**,476 (1999); I. F. Riad and M. M. Sheikh-Jabbari, JHEP **0008**, 045 (2000)
2. N. Seiberg and E. Witten, JHEP **9909**, 032 (1999)
3. J. Madore, S. Schraml, P. Schupp and J. Wess, Eur. Phys. Jour. C **16**, 161 (2000)
4. B. Jurčo, L. Möller, S. Schraml, P. Schupp and J. Wess, Eur. Phys. J. C **21**,383 (2001)
5. M. Burić and V. Radovanović, JHEP **0402**, 040 (2004)
6. M. Burić and V. Radovanović, JHEP **0210**, 074 (2002)
7. A. A. Bichl, J. M. Grimstrup, H. Grosse, L. Popp, M. Schweeda and R. Wulkenhaar, JHEP **0106**, 013 (2001)
8. R. Wulkenhaar, JHEP **0203**, 024 (2002)

Gauge Theory on the Fuzzy Sphere and Random Matrices

Harold Steinacker[1]

Institut für theoretische Physik
Ludwig–Maximilians–Universität München
Theresienstr. 37, 80333 München, Germany
harold.steinacker@physik.uni-muenchen.de

1 Introduction

Gauge theories provide the best known description of the fundamental forces in nature. At very short distances however, physics is not known, and it seems unlikely that spacetime is a perfect continuum down to arbitrarily small scales. Indeed, physicists have started to learn in recent years how to formulate field theory on quantized, or noncommutative spaces.

However, most attempts to (second-) quantize these field theories using modifications of the conventional, perturbative methods have failed up to now. It seems therefore worthwhile to try to develop new techniques for their quantization, trying to take advantage of the peculiarities of the non-commutative case. There is indeed one striking feature of some non-commutative gauge theories: they can be formulated as *(multi-) matrix models.*

I will explain here the main ideas of [1], where such a matrix formulation of (pure) $U(n)$ gauge theory on the fuzzy sphere has been used to calculate its partition function in the commutative limit. This is done using matrix techniques which *cannot* be applied in the commutative case.

2 The Model

Consider the matrix model for 3 hermitian $N \times N$ matrices B_i, with action

$$S(B) = \frac{2}{g^2 N} \mathrm{Tr} \left(\left(B_i B^i - \frac{N^2-1}{4} \right)^2 + (B_i + \mathrm{i}\varepsilon_{ijk} B^j B^k)(B^i + \mathrm{i}\varepsilon^{irs} B_r B_s) \right) \quad (1)$$

where g is the coupling constant and N is a (large) integer. This action describes pure gauge theory on the fuzzy sphere S^2_N, cp. [3, 4]. It is invariant under the $U(N)$ gauge symmetry acting as

$$B_i \to U^{-1} B_i U \ .$$

To see that this corresponds to the usual $U(1)$ local gauge symmetry in the classical limit, we first note that the absolute minimum (the "vacuum") of the action is given by

$$B_i = \lambda_i = \pi_N(J_i)$$

up to gauge transformation, where π_N is the N-dimensional representation of $su(2)$ with generators J_i. Upon rescaling $\lambda_i = x_i \sqrt{\frac{N^2-1}{4}}$, on finds the generators x_i of the fuzzy sphere [3] which satisfy

$$\sum x_i x_i = 1, \quad [x_i, x_j] = \mathrm{i}\sqrt{\frac{4}{N^2-1}}\, \epsilon_{ijk} x_k \, .$$

This means that the vacuum of this matrix model is the fuzzy sphere. We can now write any field ("covariant coordinate") as

$$B_i = \lambda_i + \mathcal{A}_i \, . \tag{2}$$

Then

$$B^i + \mathrm{i}\varepsilon^{ikl} B_k B_l = \frac{1}{2}\varepsilon^{ikl} F_{kl}, \quad F_{kl} := \mathrm{i}[\lambda_k, \mathcal{A}_l] - \mathrm{i}[\lambda_l, \mathcal{A}_k] + \mathrm{i}[\mathcal{A}_k, \mathcal{A}_l] + \varepsilon_{klm}\mathcal{A}^m \, .$$

Notice that the kinetic terms in the field strength F_{kl} arise automatically due to the shift (2). The $U(N)$ gauge symmetry acts on $\mathcal{A}_i$ as

$$\mathcal{A}_i \to U^{-1}\mathcal{A}_i U + U^{-1}[\lambda_i, U] \tag{3}$$

which for $U = \exp(ih(x))$ and $N \to \infty$ becomes the usual (abelian!) gauge transformation for a gauge field. One can furthermore show that the "radial" field

$$\varphi := \lambda^i \mathcal{A}_i$$

decouples in the large N limit, and $\frac{1}{N}\mathrm{Tr} \to \int$. Hence the model reduces to the usual $U(1)$ Yang-Mills action

$$S = \frac{1}{g^2}\int F_{mn}F^{mn} \tag{4}$$

in the commutative (= large N) limit. This is readily extended to the nonabelian case by using matrices of size nN, i.e. $B_i = B_{i,\alpha}t^\alpha = \lambda_i\, t^0 + \mathcal{A}_{i,0}\, t^0 + \mathcal{A}_{i,a}\, t^a$ where t^a denote the Gell-Mann matrices of $su(n)$. The action then reduces to the usual $U(n)$ Yang-Mills action

$$S = \frac{1}{g^2}\int (F_{mn,0}F^{mn,0} + F_{mn,a}F^{mn,a}) \tag{5}$$

in the commutative limit.

Furthermore, one can show [1] that all the monopole and instanton sectors are recovered if one admits matrices of arbitrary size $M = nN - m$ for the above action.

3 Quantization

The quantization of $U(n)$ Yang-Mills gauge theory on the usual 2-sphere is well known, see e.g. [2, 5]. In particular, the partition function and correlation functions of Wilson loops have been calculated. Our goal is to calculate the partition function for the YM action (1) on the fuzzy sphere, taking advantage of the formulation as matrix model. This can be achieved by collecting the 3 matrices B_i into a single $2M \times 2M$ matrix

$$C = C_0 + B_i \sigma^i \tag{6}$$

The main observation is that the above action (1) can be rewritten simply as

$$S(B) = \mathrm{Tr} V(C)$$

imposing the constraint $C_0 = \frac{1}{2}$, for the potential

$$V(C) = \frac{1}{g^2 N} \left(C^2 - \left(\frac{N}{2} \right)^2 \right)^2$$

Then we proceed as

$$\begin{aligned} Z &= \int \mathrm{d}B_i \, \exp(-S(B))) \\ &= \int \mathrm{d}C \, \delta\left(C_0 - \frac{1}{2} \right) \, \exp(-\mathrm{Tr}V(C)) \\ &= \int \mathrm{d}\Lambda_i \Delta^2(\Lambda_i) \exp(-\mathrm{Tr}V(\Lambda)) \int \mathrm{d}U \delta\left((U^{-1}\Lambda U)_0 - \frac{1}{2} \right) \end{aligned}$$

where $\mathrm{d}U$ is the integral over $2M \times 2M$ unitary matrices, $C = U^{-1}\Lambda U$, and $\Delta(\Lambda_i)$ is the Vandermonde-determinant of the eigenvalues Λ_i. Here $\delta(C_0 - \frac{1}{2})$ is a product over M^2 delta functions, which can be calculated by introducing $J = \begin{pmatrix} K & 0 \\ 0 & K \end{pmatrix} = K\,\sigma^0$ where K is a $N \times N$ matrix. Then

$$\delta\left((U^{-1}CU)_0 - \frac{1}{2} \right) = \int \mathrm{d}K \exp(i\mathrm{Tr}(U^{-1}\left(C - \frac{1}{2} \right) UJ)) \, .$$

By gauge invariance, the r.h.s. depends only on the eigenvalues Λ_i of C. Hence

$$\begin{aligned} Z &= \int \mathrm{d}K \int \mathrm{d}\Lambda_i \Delta^2(\Lambda_i) \exp(-\mathrm{Tr}V(\Lambda)) \int \mathrm{d}U \exp\left(i\mathrm{Tr}\left(U^{-1}\Lambda UJ - \frac{1}{2}J \right) \right) \\ &= \int \mathrm{d}K \; Z[J] \, \mathrm{e}^{-\frac{i}{2}\mathrm{Tr}\,J} \end{aligned} \tag{7}$$

where

$$Z[J] := \int \mathrm{d}C \; \exp(-\mathrm{Tr}V(C) + \mathrm{i}\mathrm{Tr}(CJ)) \tag{8}$$

depends only on the eigenvalues J_i of J. Diagonalizing $K = V^{-1}\mathrm{k}\mathrm{V}$, we get

$$\begin{aligned} Z = \int \mathrm{d}k_i \Delta^2(k) \int \mathrm{d}\Lambda_i \Delta^2(\Lambda_i) \exp(-\mathrm{Tr}V(\Lambda)) \\ \times \int \mathrm{d}U \exp\left(i\mathrm{Tr}(U^{-1}\left(\Lambda - \frac{1}{2} \right) UJ) \right) \end{aligned}$$

where $\int \mathrm{d}V$ was absorbed in $\int \mathrm{d}U$. The integral over $\int \mathrm{d}U$ can now be done using the Itzykson-Zuber-Harish-Chandra formula [6]

$$\int dU \exp(i\mathrm{Tr}(U^{-1}CUJ)) = \mathrm{const}\frac{\det(e^{i\Lambda_i J_j})}{\Delta(\Lambda_i)\Delta(J_j)} ,$$

which also depends only on the eigenvalues of J and C.

In this step the number of integrals is reduced from N^2 to $2N$. This basically means that the integral over fields on S^2_N is reduced to the integral over functions in one variable. This is a huge step, just like in the usual matrix models. The constraint however forces us to evaluate in addition the integral over k_i, which is quite complicated due to the rapid oscillations in $\det(e^{i\Lambda_i J_j})$; note that $\Lambda_i \approx \pm\frac{N}{2}$. Nevertheless, the integrals can be evaluated for large N [1], with the result

$$Z_m = \sum_{m_1+\ldots+m_n=m} \int_{-\infty}^{\infty} d\kappa_1 \ldots d\kappa_n \, \Delta^2(\kappa)\, e^{i\kappa_i m_i} \exp\left(-\frac{g^2}{2}\sum \kappa_i^2\right) .$$

Here we consider matrices of size $M = nN-m$, which corresponds to the monopole sectors with total $U(1)$ charge $m = m_1 + \ldots + m_n$. This can be rewritten in the "localized" form as a weighted sum of saddle-point contributions, as advocated by Witten [5]:

$$Z_m = \sum_{m_1+\ldots+m_n=m} P(m_i, g) \exp\left(-\frac{1}{2g^2}\sum m_i^2\right)$$

where $P(m_i, g)$ is a totally symmetric polynomial in the m_i which can be given explicitly. In order to include all monopole configurations, we should simply sum over matrices of different sizes $M = nN - m$, for the same action given by $V(C)$. One can indeed find corresponding saddle-points of the action (1) which have the form [1]

$$A_i = \begin{pmatrix} m_1 A_i & 0 & \ldots & 0 \\ 0 & m_2 A_i & \ldots & 0 \\ \vdots & \vdots & \ddots & \vdots \\ 0 & 0 & \ldots & m_n A_i \end{pmatrix}$$

where

$$\boldsymbol{A}_i = \boldsymbol{r} \times \boldsymbol{\mathcal{A}} \approx \frac{m}{2}\frac{1}{1+x_3}\begin{pmatrix} x_2 \\ -x_1 \\ 0 \end{pmatrix}$$

becomes the usual monopole field for large N, and action becomes

$$S(C^{(m_1,\ldots,m_n)}) = \frac{1}{2g^2}\sum_i m_i^2$$

for large N. This is a standard result on the classical sphere.

Hence the full partition function is obtained by summing[1] over all Z_m,

$$Z = \sum_m Z_m = \sum_{m_1,\ldots,m_n=-\infty}^{\infty} \int d\kappa_i \, \Delta^2(\kappa)\, e^{i\kappa_i m_i} \exp\left(-\frac{g^2}{2}\sum \kappa_i^2\right) .$$

Using a Poisson resummation, this can be rewritten in the form

[1] The relative weights of Z_m for different m is not determined here. However, it could be calculated in principle

$$Z = \sum_{p_1,\ldots,p_n \in Z} \Delta^2(p) \exp\left(-2\pi^2 g^2 \sum_i p_i^2\right)$$

or equivalently

$$Z = \sum_R (d_R)^2 \exp(-4\pi^2 g^2 C_{2R}) \; .$$

Here the sum is over all representations of $U(n)$, d_R is the dimension of the representation and C_{2R} the quadratic casimir. This form was found in [2] for the partition function of a $U(n)$ Yang-Mills theory on the ordinary 2-sphere.

We see that the limit $N \to \infty$ of the partition function for $U(n)$ YM on the fuzzy sphere is well-defined, and reproduces the result for YM on the classical sphere. This strongly suggests that the same holds for the full YM theory on the fuzzy sphere, and that there is nothing like UV/IR mixing for pure gauge theory on S^2_N. This is unlike the case of a scalar field, which exhibits a "non-commutative anomaly" [7] related to UV/IR mixing.

References

1. H. Steinacker, *Quantized Gauge Theory on the Fuzzy Sphere as Random Matrix Model*, Nucl. Phys. **B679**, vol 1-2 (2004) 66-98
2. A. A. Migdal, *Recursion equations in gauge theories*, Sov. Phys. JETP **42** (1976) 413; B. E. Rusakov, *Loop averages and partition functions in $U(N)$ gauge theory on two-dimensional manifolds*, Mod. Phys. Lett. **A5** (1990) 693
3. J. Madore, *The Fuzzy Sphere*, Class. Quant. Grav. **9**, 69 (1992)
4. U. Carow-Watamura, S. Watamura, *Noncommutative Geometry and Gauge Theory on Fuzzy Sphere*, Commun. Math. Phys. **212** (2000) 395
5. E. Witten, *Two-dimensional gauge theories revisited*, J. Geom. Phys. **9** (1992) 303
6. C. Itzykson, J. B. Zuber *The planar approximation. 2*, Journ. Math. Phys. **21** (1980), 411
7. C. S. Chu, J. Madore and H. Steinacker, *Scaling limits of the fuzzy sphere at one loop*, JHEP **0108** (2001) 038

Part III

Standard Model – Theory and Experiment

Waiting for Clear Signals of New Physics in B and K Decays

Andrzej J. Buras

Technical University Munich, Physics Department 85748 Garching, Germany

1 Introduction

The quark flavour dynamics of the Standard Model (SM) is consistent with the existing data on K and B meson decays within experimental and theoretical uncertainties. In spite of this, most of us expect that when the precision of experiments and also the theoretical tools improve, some clear signals of new physics (NP) at very short distance scales will be seen. As the search for new phenomena through K and B decays is necessarily an indirect one, it will not be easy to find out what precisely this NP is. This will be even the case in the presence of clear deviations from the SM expectations.

Yet, remembering that decays like $K_L \to \mu^+\mu^-$, $K_L \to \pi\pi$ and the $K^0 - \bar{K}^0$ mixing played an important role in the construction of the SM [1] and led to the GIM mechanism [2] and the CKM matrix [3], we are confident that these indirect signals of NP will give us definitive hints where to go and where not to go. This in turn will allow us to select few competing theories which will be tested directly through high energy collider experiments. Moreover, these signals will hopefully give us additional hints for the fundamental flavour dynamics at ultra short distance scales that cannot be tested directly in a foreseeable future. It is obvious that for this project to become successful, it is essential to

- make the SM predictions for K and B decay observables as accurate as possible, in order to be sure that the observed deviations from the SM expectations originate in NP contributions and are not due to our insufficient understanding of hadron dynamics and/or truncation of the perturbative series.
- consider simultaneously as many processes as possible. Only in this manner the paramaters of a given theory can be fully determined and having them at hand predictions for other observables can be made. In this enterprise correlations between various observables play an important rôle, as they may exclude or pinpoint a given extension of the SM even without a detailed knowledge of the parameters specific to this theory.
- consider observables which while being sensitive to the short distance structure of the theory are only marginally sensitive to the long distance dynamics that is not yet fully under control at present.

These notes review selected aspects of this program. Due to severe space limitations I will follow the strategy of many of my colleagues and will concentrate to a large extent on my own work and the work done in my group in Munich. In Sect. 2, after giving a master formula for weak decay amplitudes, I will group various extentions of the SM in five classes. In subsequent three sections, I will discuss the first three classes that being more predictive than the remaining two, should be easiest to test. A shopping list in Sect. 6, with the hope to be able to distinguish between various NP scenarios, ends this presentation. Recent more detailed reviews can be found in [4].

2 Theoretical Framework

2.1 Master Formula for Weak Decays

The present framework for weak decays is based on the operator product expansion that allows to separate short (μ_{SD}) and long (μ_{LD}) distance contributions to weak amplitudes and on the renormalization group (RG) methods that allow to sum large logarithms $\log \mu_{SD}/\mu_{LD}$ to all orders in perturbation theory. The full exposition of these methods can be found in [5, 6].

An amplitude for a decay of a meson $M = K, B, \ldots$ into a final state $F = \pi\nu\bar{\nu},\ \pi\pi,\ \pi K$ is then simply given by

$$A(M \to F) = \frac{G_F}{\sqrt{2}} \sum_i V^i_{CKM} C_i(\mu) \langle F|Q_i(\mu)|M\rangle \,. \tag{1}$$

Here G_F is the Fermi constant and Q_i are the relevant local operators which govern the decays in question. They are built out of quark and lepton fields. The Cabibbo-Kobayashi-Maskawa factors $V^i_{\rm CKM}$ [3], the matrix elements $\langle F|Q_i(\mu)|M\rangle$ and the Wilson coefficients $C_i(\mu)$, evaluated at the renormalization scale μ, describe the strength with which a given operator enters the amplitude.

Formula (1) can be cast into a master formula for weak decay amplitudes that goes beyond the SM [7]. It reads (we suppress $G_{\rm F}$):

$$\begin{aligned} \text{A(Decay)} &= \sum_i B_i \eta^i_{\rm QCD} V^i_{\rm CKM} [F^i_{\rm SM} + F^i_{\rm New}] \\ &\quad + \sum_k B^{\rm New}_k [\eta^k_{\rm QCD}]^{\rm New} V^k_{\rm New} [G^k_{\rm New}] \,. \end{aligned} \tag{2}$$

The non-perturbative parameters B_i represent the matrix elements of local operators present in the SM. For instance in the case of $K^0 - \bar{K}^0$ mixing, the matrix element of the operator $\bar{s}\gamma_\mu(1-\gamma_5)d \otimes \bar{s}\gamma^\mu(1-\gamma_5)d$ is represented by the parameter $\hat{B}_K$. There are other non-perturbative parameters in the SM that represent matrix elements of operators Q_i with different colour and Dirac structures. The objects $\eta^i_{\rm QCD}$ are the QCD factors resulting from RG-analysis of the corresponding operators and $F^i_{\rm SM}$ stand for the so-called Inami-Lim functions [8] that result from the calculations of various box and penguin diagrams in the SM. They depend on the top-quark mass.

New physics can contribute to our master formula in two ways. It can modify the importance of a given operator, present already in the SM, through the new short distance functions $F^i_{\rm New}$ that depend on the new parameters in the extensions of the SM like the masses of charginos, squarks, charged Higgs particles and $\tan\beta = v_2/v_1$ in the MSSM. These new particles enter the new box and penguin diagrams. In more complicated extensions of the SM new operators (Dirac structures) that are either absent or very strongly suppressed in the SM, can become important. Their contributions are described by the second sum in (2) with $B_k^{\rm New}$, $[\eta_{\rm QCD}^k]^{\rm New}$, $V_{\rm New}^k$, $G_{\rm New}^k$ being analogs of the corresponding objects in the first sum of the master formula. The $V_{\rm New}^k$ show explicitly that the second sum describes generally new sources of flavour and CP violation beyond the CKM matrix. This sum may, however, also include contributions governed by the CKM matrix that are strongly suppressed in the SM but become important in some extensions of the SM. In this case $V_{\rm New}^k = V_{\rm CKM}^k$. Clearly the new functions $F_{\rm New}^i$ and $G_{\rm New}^k$ as well as the factors $V_{\rm New}^k$ may depend on new CP violating phases complicating considerably phenomenological analyses.

2.2 Classification of New Physics

Classification of new physics contributions can be done in various ways. Having the formula (2) at hand let us classify these contributions from the point of view of the operator structure of the effective weak Hamiltonian, the complex phases present in the Wilson coefficients of the relevant operators and the distinction whether the flavour changing transitions are governed by the CKM matrix or by new sources of flavour violation [9]. For the first four classes below we assume that there are only three generations of quarks and leptons. The last class allows for more generations.

Class A

This is the simplest class to which also the SM belongs to. In this class there are no new complex phases and flavour changing transitions are governed by the CKM matrix. Moreover, the only relevant operators are those that are relevant in the SM. Consequently NP enters only through the Wilson coefficients of the SM operators that can receive new contributions through diagrams involving new internal particles.

The models with these properties will be called Minimal Flavour Violation (MFV) models, as defined in [10]. Other definitions can be found in [11, 12]. In this case our master formula simplifies to

$$\mathrm{A(Decay)} = \sum_i B_i \eta_{\rm QCD}^i V_{\rm CKM}^i F_i(v) \qquad F_i = F_{\rm SM}^i + F_{\rm New}^i \quad \text{(real)}\,, \tag{3}$$

where $F_i(v)$ are the *master functions* of MFV models [13]

$$S(v),\ X(v),\ Y(v),\ Z(v),\ E(v),\ D'(v),\ E'(v) \tag{4}$$

with v denoting collectively the parameters of a given MFV model. A very detailed account of the MFV can be found in [13]. In Sect. 3 some of its main features will be recalled. Examples of models in this class are the Two Higgs Doublet Model II and the constrained MSSM if $\tan\beta$ is not too large. Also models with one extra universal dimension are of MFV type.

branching ratios [13]. The ratio $\hat{B}_s/\hat{B}_d$ is known from lattice calculations with a respectable precision [17]:

$$\frac{\hat{B}_s}{\hat{B}_d} = 1.00 \pm 0.03, \qquad \hat{B}_d = 1.34 \pm 0.12, \qquad \hat{B}_s = 1.34 \pm 0.12\,. \tag{10}$$

With a future precise measurement of ΔM_s, the second formula in (9) will give a precise prediction for the ratio of the branching ratios $Br(B_q \to \mu\bar{\mu})$.

3. It is possible to derive an accurate formula for $\sin 2\beta$ that depends only on the $K \to \pi\nu\bar{\nu}$ branching ratios and a calculable $\bar{P}_c(X) = 0.38 \pm 0.06$ [22, 23, 24]:

$$\sin 2(\beta - \beta_s) = \frac{2r_s}{1 + r_s^2}, \qquad r_s = \frac{\varepsilon_1\sqrt{B_1 - B_2} - \bar{P}_c(X)}{\varepsilon_2\sqrt{B_2}}\,, \tag{11}$$

where $\beta_s \approx -1^\circ$ enters $V_{ts} = -|V_{ts}|\exp(-i\beta_s)$, $\varepsilon_i = \pm 1$ and

$$B_1 = \frac{Br(K^+ \to \pi^+\nu\bar{\nu})}{4.78 \cdot 10^{-11}}\,, \qquad B_2 = \frac{Br(K_{\rm L} \to \pi^0\nu\bar{\nu})}{2.09 \cdot 10^{-10}}\,. \tag{12}$$

In the MFV models $\varepsilon_1 = \varepsilon_2 = {\rm sgn}(X)$ [25], where X is the relevant master function.

4. With no weak phases beyond the CKM phase, we also expect

$$(\sin 2\beta)_{\pi\nu\bar{\nu}} = (\sin 2\beta)_{\psi K_S}\,, \qquad (\sin 2\beta)_{\phi K_S} \approx (\sin 2\beta)_{\psi K_S} \tag{13}$$

with the accuracy of the last relation at the level of a few percent [26]. The confirmation of these two relations would be a very important test of the MFV idea. Indeed, in $K \to \pi\nu\bar{\nu}$ the phase β originates in the Z^0 penguin diagram, whereas in the case of $a_{\psi K_S}$ in the $B_d^0 - \bar{B}_d^0$ box diagrams. In the case of the asymmetry $a_{\phi K_S}$ it originates also in $B_d^0 - \bar{B}_d^0$ box diagrams but the second relation in (13) could be spoiled by NP contributions in the decay amplitude for $B \to \phi K_S$ that is non-vanishing only at the one loop level. Interestingly the present data from Belle may indicate the violation of the second relation in (13), although the experimental situation is very unclear at present [27, 28]:

$$(\sin 2\beta)_{\phi K_S} = \begin{cases} +0.45 \pm 0.43 \pm 0.07 & \text{(BaBar)} \\ -0.96 \pm 0.50^{+0.11}_{-0.09} & \text{(Belle)}, \end{cases} \tag{14}$$

A subset of theoretical papers addressing this issue is listed in [29].

5. An important consequence of (11)–(13) is the following one. For a given $\sin 2\beta$ extracted from $a_{\psi K_S}$ and $Br(K^+ \to \pi^+\nu\bar{\nu})$ only two values of $Br(K_{\rm L} \to \pi^0\nu\bar{\nu})$, corresponding to two signs of the master function $X(v)$, are possible in the full class of MFV models, independent of any new parameters present in these models [25]. Consequently, measuring $Br(K_{\rm L} \to \pi^0\nu\bar{\nu})$ will either select one of these two possible values or rule out all MFV models.

6. As pointed out in [30] in most MFV models there exists a correlation between the zero $\hat{s}_0$ in the forward-backward asymmetry $A_{\rm FB}$ in $B \to X_s\mu^+\mu^-$ and $Br(B \to X_s\gamma)$. We show this correlation in Fig. 2.

7. Other correlations between various decays can be found in [25, 31, 32, 33, 34]. For instance there exists in addition to an obvious correlation between $K \to \pi\nu\bar{\nu}$ and $B \to X_q\nu\bar{\nu}$ also a correlation between ε'/ε and rare semileptonic B and K decays. A discussion of correlations between $B \to \pi K$ decays and rare decays within MFV models with enhanced Z^0 penguins can be found in [35].

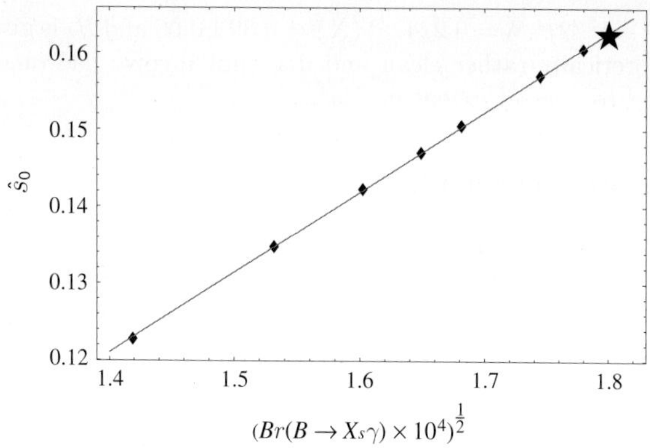

Fig. 2. Correlation between $\sqrt{Br(B \to X_s\gamma)}$ and $\hat{s}_0$ [30]. The dots are the results in the ACD model (see below) with the compactification scale $200, 250, 300, 350, 400, 600$ and $1000\,\mathrm{GeV}$ and the star denotes the SM value

3.2 Model Dependent Relations

$B_q \to \mu\bar{\mu}$ and ΔM_q

The relations [21]

$$Br(B_q \to \mu\bar{\mu}) = 4.36 \cdot 10^{-10} \frac{\tau(B_q)}{\hat{B}_q} \frac{Y^2(v)}{S(v)} \Delta M_q, \qquad (q = s, d) \,. \tag{15}$$

allow to predict $Br(B_{s,d} \to \mu\bar{\mu})$ in a given MFV model, characterized by $Y(v)$ and $S(v)$, with substantially smaller hadronic uncertainties than found by using directly the formulae for these branching ratios that suffer from large uncertainties due to F_{B_q}. In particular in the SM model we find [21]

$$Br(B_s \to \mu\bar{\mu}) = (3.4 \pm 0.5) \cdot 10^{-9}\,, \quad Br(B_d \to \mu\bar{\mu}) = (1.00 \pm 0.14) \cdot 10^{-10}\,, \tag{16}$$

where $\overline{m}_t(m_t) = (167 \pm 5)\,\mathrm{GeV}$, the lifetimes from [17], $\hat{B}_q$ in (10), $\Delta M_d = (0.503 \pm 0.006)/\mathrm{ps}$ and as an example $\Delta M_s = (18.0 \pm 0.5)/\mathrm{ps}$, have been used. These results are substantially more accurate than the ones found in the literature in the past but by orders of magnitude below the experimental upper bounds from CDF(D0) and Belle [36, 37].

$Br(K^+ \to \pi^+\nu\bar{\nu})$, $\Delta M_d/\Delta M_s$ and β

In [38] an upper bound on $Br(K^+ \to \pi^+\nu\bar{\nu})$ in terms of $\Delta M_d/\Delta M_s$ has been derived within the SM. It has been subsequently cast into a useful relation between $Br(K^+ \to \pi^+\nu\bar{\nu})$, $\Delta M_d/\Delta M_s$ and β in [39]. In any MFV model this relation reads

$$Br(K^+ \to \pi^+\nu\bar{\nu}) = 7.54 \cdot 10^{-6} |V_{cb}|^4 X^2_{\mathrm{eff}}(v) \tag{17}$$

$$X^2_{\mathrm{eff}}(v) = X^2(v) \left[\sigma R_t^2 \sin^2\beta + \frac{1}{\sigma} \left(R_t \cos\beta + \frac{\lambda^4 P_c(X)}{|V_{cb}|^2 X(v)} \right)^2 \right], \tag{18}$$

where $\sigma = 1/(1-\lambda^2/2)^2$, $\lambda = 0.224$, $P_c(X) = 0.39\pm0.06$ and R_t is given in (7). This formula is theoretically rather clean and does not involve hadronic uncertainties except for ξ and to a lesser extent in $|V_{cb}|$.

3.3 Maximal Enhancements

How large could various branching ratios in the MFV models be? A detailed numerical analysis of this question is beyond the scope of this presentation but assuming that the dominant NP effects in rare K and B decays come from enhanced Z^0 penguins hidden in the master functions $X(v)$, $Y(v)$ and $Z(v)$, bounding this enhancement by the Belle and BaBar data on $B \to X_s l^+ l^-$ [40] and setting all other parameters at their central values, we find the results in column MFV of Table 1 where also the SM results are shown. While somewhat higher values of branching ratios can still be obtained when the input parameters are varied, this exercise shows that enhancements of branching ratios in the MFV models by more than factors of six relative to the SM should not be expected. A similar analysis in a different spirit and a different set of input parameters can be found in [11].

Table 1. Example of branching ratios for rare decays in the MFV and the SM

Branching Ratios	MFV	SM
$Br(K^+ \to \pi^+\nu\bar{\nu}) \times 10^{11}$	19.1	8.0
$Br(K_{\rm L} \to \pi^0\nu\bar{\nu}) \times 10^{11}$	9.9	3.2
$Br(K_{\rm L} \to \mu^+\mu^-)_{\rm SD} \times 10^9$	3.5	0.9
$Br(K_{\rm L} \to \pi^0 e^+ e^-)_{\rm CPV} \times 10^{11}$	4.9	3.2
$Br(B \to X_s\nu\bar{\nu}) \times 10^5$	11.1	3.6
$Br(B \to X_d\nu\bar{\nu}) \times 10^6$	4.9	1.6
$Br(B_s \to \mu^+\mu^-) \times 10^9$	19.4	3.9
$Br(B_d \to \mu^+\mu^-) \times 10^{10}$	6.1	1.2

3.4 MFV and Universal Extra Dimensions

A detailed analysis of all rare and radiative K and B decays in a particular model with one universal extra dimension (ACD) [41] has been presented in [30]. The nice feature of this extension of the SM is the presence of only one additional parameter, the compactification scale. This feature allows a unique pattern of various enhancements and suppressions relative to the SM expectations. Our analysis shows that all the present data on FCNC processes are consistent with the compactification scale as low as 250 GeV, implying that the Kaluza–Klein particles could in principle be found already at the Tevatron. Possibly, the most interesting results of our analysis is the sizable downward shift of the zero ($\hat{s}_0$) in the $A_{\rm FB}$ asymmetry in $B \to X_s\mu^+\mu^-$ and the suppression of $Br(B \to X_s\gamma)$ (also found in [42]) that are correlated as shown in Fig. 2. Note that a measurement of $\hat{s}_0$ that is higher than the SM estimate would automatically exclude this model as there is no compactification scale for which this could be satisfied.

4 Class B: MSSM at Large $\tan\beta$

An example of a model in this class is the MSSM with large $\tan\beta$ but without any relevant contributions from new flavour violating interactions coming from the non-diagonal elements in squark mass matrices when these are given in the quark mass eigenstate basis. In these models the dominant new effects in K and B decays come from very strongly enhanced flavour changing neutral (FCNC) Higgs couplings to the down quarks. These couplings are generated only at one loop level but being proportional to $(\tan\beta)^2$ become very important for $\tan\beta \geq 30$.

The presence of the enhanced FCNC Higgs couplings implies in turn important contributions of new operators in the effective theory that are strongly suppressed in models of class A. In particular the operators

$$O_S = m_b(\overline{b_R s_L})(\bar{\mu}\mu), \qquad O_P = m_b(\overline{b_R s_L})(\bar{\mu}\gamma_5\mu) \tag{19}$$

fully dominate the branching ratios for $B_{s,d} \to \mu^+\mu^-$ [43] when $\tan\beta \geq 30$. An approximate formula for $Br(B_s \to \mu^+\mu^-)$ is given then by [44, 45]

$$Br(B_s \to \mu^+\mu^-) \approx 3.5 \cdot 10^{-5} \left[\frac{\tan\beta}{50}\right]^6 \left[\frac{m_t}{M_A}\right]^4 F(\varepsilon_i, \tan\beta)\,, \tag{20}$$

$$F(\varepsilon_i, \tan\beta) = \left[\frac{16\pi^2\varepsilon_Y}{(1+\tilde{\varepsilon}_3\tan\beta)(1+\tilde{\varepsilon}_0\tan\beta)}\right]^2\,, \tag{21}$$

where ε_i, depending on the SUSY parameters, are at most $\mathcal{O}(10^{-2})$ and we have set $\tau(B_s)$, F_{B_s} and $|V_{ts}|$ at their central values. M_A is the mass of the pseudoscalar A. The expression for $Br(B_d \to \mu^+\mu^-)$ is obtained by using

$$\frac{Br(B_d \to \mu^+\mu^-)}{Br(B_s \to \mu^+\mu^-)} = \left[\frac{\tau(B_d)}{\tau(B_s)}\right] \left[\frac{F_{B_d}}{F_{B_d}}\right]^2 \left|\frac{V_{td}}{V_{ts}}\right|^2 \left[\frac{M_{B_d}}{M_{B_d}}\right]^5 \tag{22}$$

which differs slightly from the usual MFV formula in that the last factor has the power five instead of two. On the other hand the branching ratios themselves can still be enhanced by almost a factor of 500. If this is indeed the case, $B_{s,d} \to \mu^+\mu^-$ should be observed already at Tevatron and B factories, respectively.

The enhanced neutral Higgs couplings and more generally large $\tan\beta$ effects can play also a significant role in ΔM_s [45] inducing in particular four–fermion operators

$$Q_1^{LR} = (\overline{b_L}\gamma_\mu s_L)(\overline{b_R}\gamma^\mu s_R), \qquad Q_2^{LR} = (\overline{b_R}s_L)(\overline{b_L}s_R), \tag{23}$$

whose Wilson coefficients are negligible in the SM. One finds then

$$\Delta M_s = (\Delta M_s)_{\rm SM}(1+f_s) \approx (\Delta M_s)_{\rm SM} - |\Delta M_s|^{\rm DP} \tag{24}$$

where the new contributions come dominantly from double Higgs penguins (DP), as indicated above. Being proportional to $m_q m_b \tan^4\beta$ for ΔM_q, these contributions can be neglected in ΔM_d that for large $\tan\beta$ is very close to the SM estimate. It turns out that ΔM_s is suppressed for any choice of supersymmetric parameters ($f_s < 0$) with the size of suppression dependent strongly on the stop mixing, M_A and $\tan\beta$.

As a consequence of the mismatch between Higgs contributions to ΔM_d and ΔM_s, the MFV formula for R_t in (7) is modified to

$$R_t = 0.90 \left[\frac{\xi}{1.24}\right] \sqrt{\frac{18.4/ps}{\Delta M_s}} \sqrt{\frac{\Delta M_d}{0.50/ps}} \sqrt{1+f_s}. \tag{25}$$

However, most interesting is the correlation of the enhanced neutral Higgs effects in $B_{s,d} \to \mu^+\mu^-$ and ΔM_s, that is independent of $F(\varepsilon_i, \tan\beta)$ in (21):

$$Br(B_s \to \mu^+\mu^-) \approx 10^{-6} \left[\frac{\tan\beta}{50}\right]^2 \left[\frac{200\,\mathrm{GeV}}{M_A}\right]^2 \left[\frac{|\Delta M_s|^{\mathrm{DP}}}{2.1/\mathrm{ps}}\right]. \tag{26}$$

Consequently a strong enhancement of $B_{s,d} \to \mu^+\mu^-$ implies a significant suppression of ΔM_s. This means that in this scenario, an observation of $B_{s,d} \to \mu^+\mu^-$ at the level of $\mathcal{O}(10^{-7})$ and $\mathcal{O}(10^{-8})$, respectively should be accompanied by ΔM_s below the SM estimates. On the other hand if $(\Delta M_s)_{\mathrm{exp}} > (\Delta M_s)_{\mathrm{SM}}$ this scenario is excluded and the observation of $B_{s,d} \to \mu^+\mu^-$ at this level would point toward other flavour violating sources, coming for instance from non-diagonal elements in the squark mass matrices [46].

The difficult task in testing this scenario will be to demonstrate whether the measured value $(\Delta M_s)_{\mathrm{exp}}$ is indeed smaller or larger than $(\Delta M_s)_{\mathrm{SM}}$. To this end a significant reduction of the uncertainties in the non-perturbative parameters is required. On the other hand an enhancement of $B_{s,d} \to \mu^+\mu^-$ by one or two orders of magnitude with respect to the SM estimates in (16) would be truely spectacular independently of the situation with ΔM_s.

Other interesting correlations are the ones between the ratios $Br(B \to H\mu^+\mu^-)/Br(B \to He^+e^-)$ $(H = K^{(*)}, X_s)$ and $Br(B_s \to \mu^+\mu^-)$ [47].

5 Class C: New Weak Phases

5.1 Preliminaries

In this class of models the dominant operators are as in class A but the master functions become now complex quantities. If the new weak phases are large, the deviations from the SM can be spectacular as we will see below.

5.2 Weak Phases in $\Delta F = 2$ Transitions

In the MFV scenario of Sect. 3 the NP effects enter universally in $K^0-\bar K^0$, $B_d^0-\bar B_d^0$ and $B_s^0-\bar B_s^0$ through the single real function $S(v)$, implying strong correlations between new physics effect in the $\Delta F = 2$ observables of K and B decays. When new complex weak phases are present, the situation could be more involved with $S(v)$ replaced by

$$S_K(v) = |S_K(v)|\mathrm{e}^{\mathrm{i}\theta_K}, \quad S_d(v) = |S_K(v)|\mathrm{e}^{\mathrm{i}\theta_d}, \quad S_s(v) = |S_K(v)|\mathrm{e}^{\mathrm{i}\theta_s}, \tag{27}$$

for $K^0-\bar K^0$, $B_d^0-\bar B_d^0$ and $B_s^0-\bar B_s^0$ mixing, respectively. If these three functions are different from each other, some universal properties found in class A are lost. In addition the mixing induced CP asymmetries in B decays will not measure the angles of the UT but only sums of these angles and of θ_i. Yet, within each class of K, B_d and B_s decays, the NP effects of this sort will be universal. Scenarios of this type have been considered for instance in [48].

5.3 Weak Phases in $\Delta F = 1$ Transitions

New weak phases could enter also decay amplitudes. As now these effects enter in principle differently in each decay, the situation can be very involved with many free parameters, no universal effects and little predictive power.

Here I will only discuss one scenario, discussed first in [31, 32, 49, 50] and recently in the context of a simultaneous analysis of prominent non-leptonic B decays like $B \to \pi\pi$, $B \to \pi K$, $B \to \psi K_S$ and $B \to \phi K_S$ and equally prominent rare decays like $K \to \pi\nu\bar{\nu}$, $K_L \to \pi^0 e^+ e^-$, $B_{s,d} \to \mu^+\mu^-$, $B \to X_{s,d} e^+ e^-$ and ε'/ε in [23, 24]. It is the scenario of enhanced Z^0 penguins with a large complex weak phase in which the only modification with respect to class A is the replacement in the Z^0 penguin function $C(v) \to |C(v)|e^{i\theta_C}$ that makes the master functions $X(v)$, $Y(v)$ and $Z(v)$ complex quantities:

$$X(v) = |X(v)|e^{i\theta_X}, \qquad Y(v) = |Y(v)|e^{i\theta_Y}, \qquad Z(v) = |Z(v)|e^{i\theta_Z}. \tag{28}$$

The magnitudes and phases of these three functions are correlated with each other as they depend only on $|C(v)|e^{i\theta_C}$ and other smaller contributions that can be set to their SM values. This new analysis has been motivated by the following experimental situation in $B \to \pi\pi$ and $B \to \pi K$ decays.

5.4 The $B \to \pi\pi$ Puzzle

The BaBar and Belle collaborations have very recently reported the observation of $B_d \to \pi^0\pi^0$ decays with CP-averaged branching ratios of $(2.1 \pm 0.6 \pm 0.3) \times 10^{-6}$ and $(1.7 \pm 0.6 \pm 0.2) \times 10^{-6}$, respectively [51, 52]. These measurements represent quite some challenge for theory. For example, in a recent state-of-the-art calculation [53] within QCD factorization (QCDF) [54], a branching ratio that is about six times smaller is favoured, whereas the calculation of $B_d \to \pi^+\pi^-$ points towards a branching ratio about two times larger than the current experimental average. On the other hand, the calculation of $B^+ \to \pi^+\pi^0$ reproduces the data rather well. This "$B \to \pi\pi$ puzzle" is reflected by the following quantities:

$$R_{+-}^{\pi\pi} \equiv 2\left[\frac{Br(B^\pm \to \pi^\pm\pi^0)}{Br(B_d \to \pi^+\pi^-)}\right]\frac{\tau_{B_d^0}}{\tau_{B^+}} = 2.12 \pm 0.37 \tag{29}$$

$$R_{00}^{\pi\pi} \equiv 2\left[\frac{Br(B_d \to \pi^0\pi^0)}{Br(B_d \to \pi^+\pi^-)}\right] = 0.83 \pm 0.23. \tag{30}$$

The central values calculated in QCDF [53] are $R_{+-}^{\pi\pi} = 1.24$ and $R_{00}^{\pi\pi} = 0.07$.

5.5 The $B \to \pi K$ Puzzle

In the $B \to \pi K$ system, the CLEO, BaBar and Belle collaborations have measured the following ratios of CP-averaged branching ratios [55]:

$$R_c \equiv 2\left[\frac{Br(B^\pm \to \pi^0 K^\pm)}{Br(B^\pm \to \pi^\pm K^0)}\right] = 1.17 \pm 0.12 \tag{31}$$

$$R_n \equiv \frac{1}{2}\left[\frac{Br(B_d \to \pi^\mp K^\pm)}{Br(B_d \to \pi^0 K)}\right] = 0.76 \pm 0.10, \tag{32}$$

with numerical values following from [6]. As noted in [57], the pattern of $R_c > 1$ and $R_n < 1$, which is now consistently favoured by the separtate BaBar, Belle and CLEO data, is actually puzzling in the framework of QCD factorization that gives typically $R_c \approx R_n \approx 1.15$. On the other hand,

$$R \equiv \left[\frac{Br(B_d \to \pi^{\mp}K^{\pm})}{Br(B^{\pm} \to \pi^{\pm}K)}\right] \frac{\tau_{B^+}}{\tau_{B_d^0}} = 0.91 \pm 0.07 \tag{33}$$

does not show any anomalous behaviour. Since R_c and R_n are affected significantly by colour-allowed EW penguins, whereas this is not the case of R, this "$B \to \pi K$ puzzle" may be a manifestation of NP in the EW penguin sector [23, 24, 35, 57], offering an attractive avenue for physics beyond the SM to enter the $B \to \pi K$ system [58, 59].

5.6 The Analysis of [23, 24]

In view of significant experimental uncertainties, none of these exciting results is conclusive at the moment, but it is legitimate and interesting to take them seriously and to search for possible origins of these "signals" for deviations from the SM expectations. As we are dealing here with non-leptonic decays, the natural question arises whether these signals originate in the NP contributions or/and result from our insufficient understanding of hadron dynamics. The purpose of [23, 24] was to develop a strategy which would allow us to address the $B \to \pi\pi$ and $B \to \pi K$ puzzles in a systematic manner once the experimental data on these decays improve. In order to illustrate this strategy in explicit terms, we considered a simple extension of the SM in which NP enters dominantly through enhanced CP-violating Z^0 penguins. As we will see below, this choice is dictated by the pattern of the data on the $B \to \pi K$ observables and the great predictivity of this scenario. It was first considered in [31, 32, 49] to study correlations between rare K decays and the ratio ε'/ε, and was generalized to rare B decays in [50]. Extending these considerations to non-leptonic B-meson decays, allowed us to confront this NP scenario with many more experimental results. Our strategy consists of three interrelated steps, and has the following logical structure:

Step 1

Since $B \to \pi\pi$ decays and the usual analysis of the UT are only insignificantly affected by EW penguins, the $B \to \pi\pi$ system can be described as in the SM. Employing the $SU(2)$ isospin flavour symmetry of strong interactions and the information on γ from the UT fits [17], we could extract the relevant hadronic $B \to \pi\pi$ parameters, and find large non-factorizable contributions, in variance with the QCD factorization approach. Having these parameters at hand, we could then also predict the direct and mixing-induced CP asymmetries of the $B_d \to \pi^0\pi^0$ channel. A future measurement of one of these observables allows a determination of γ.

Step 2

Using the $SU(3)$ flavour symmetry and plausible dynamical assumptions, we could subsequently determine the hadronic $B \to \pi K$ parameters through the $B \to \pi\pi$

analysis, and calculate the $B \to \pi K$ observables in the SM. Interestingly, we found agreement with the pattern of the B-factory data for those observables where EW penguins play only a minor rôle. On the other hand, the observables receiving significant EW penguin contributions did *not* agree with the experimental picture, thereby suggesting NP in the EW penguin sector. Indeed, a detailed analysis shows that one can describe all the currently available $B \to \pi K$ data through sizeably enhanced EW penguins with a large CP-violating NP phase around $-90°$. A crucial future test of this scenario will be provided by the CP-violating $B_d \to \pi^0 K_S$ observables, which we could predict. Moreover, we could obtain valuable insights into $SU(3)$-breaking effects, which support our working assumptions, and could also determine the UT angle γ, that is in agreement with the UT fits.

Step 3

In turn, the sizeably enhanced EW penguins with their large CP-violating NP phase have important implications for rare K and B decays as well as ε'/ε. Interestingly, several predictions differ significantly from the SM expectations and should easily be identified once the data improve.

The most interesting results of this study, presented in [23, 24] are:

a) For the very clean $K \to \pi\nu\bar{\nu}$ decays, we find

$$\begin{aligned} Br(K^+ \to \pi^+\nu\bar{\nu}) &= (7.5 \pm 2.1) \cdot 10^{-11}, \\ Br(K_L \to \pi^0\nu\bar{\nu}) &= (3.1 \pm 1.0) \cdot 10^{-10}, \end{aligned} \tag{34}$$

to be compared with the SM estimates $(7.7 \pm 1.1) \times 10^{-11}$ and $(2.6 \pm 0.5) \times 10^{-11}$ [60], respectively, and the AGS E787 result $\text{BR}(K^+ \to \pi^+\nu\bar{\nu}) = (15.7^{+17.5}_{-8.2}) \times 10^{-11}$ [61]. The enhancement of $\text{BR}(K_L \to \pi^0\nu\bar{\nu})$ by one order of magnitude and the pattern in (34) are dominantly the consequences of $\beta_X = \beta - \theta_X \approx 110°$, as

$$\frac{Br(K_L \to \pi^0\nu\bar{\nu})}{Br(K_L \to \pi^0\nu\bar{\nu})_{SM}} = \left|\frac{X}{X_{SM}}\right|^2 \left[\frac{\sin\beta_X}{\sin\beta}\right]^2 \tag{35}$$

$$\frac{Br(K_L \to \pi^0\nu\bar{\nu})}{Br(K^+ \to \pi^+\nu\bar{\nu})} \approx 4.4 \times (\sin\beta_X)^2 \approx (4.2 \pm 0.2)\,. \tag{36}$$

Interestingly, the above ratio turns out to be very close to its absolute upper bound in [62]. A spectacular implication of these findings is a strong violation of $(\sin 2\beta)_{\pi\nu\bar{\nu}} = (\sin 2\beta)_{\psi K_S}$ [22], which is valid in the SM and any model with minimal flavour violation as discussed in Sect. 3. Indeed, we find

$$(\sin 2\beta)_{\pi\nu\bar{\nu}} = \sin 2\beta_X = -(0.69^{+0.23}_{-0.41})\,, \tag{37}$$

in striking disagreement with $(\sin 2\beta)_{\psi K_S} = 0.74 \pm 0.05$ in (8).

b) Another implication is the large branching ratio

$$Br(K_L \to \pi^0 e^+ e^-) = (7.8 \pm 1.6) \times 10^{-11}\,, \tag{38}$$

which is governed by direct CP violation in this scenario. On the other hand, the SM result $(3.2^{+1.2}_{-0.8}) \times 10^{-11}$ [63] is dominated by the indirect CP violation. Next, the integrated forward–backward CP asymmetry for $B_d \to K^* \mu^+ \mu^-$ [50] can be very large in view of $\theta_Y \approx -100^\circ$ as it is given by

$$A_{\rm FB}^{\rm CP} = (0.03 \pm 0.01) \times \tan \theta_Y . \quad (39)$$

c) Next, $Br(B \to X_{s,d} \nu \bar{\nu})$ and $Br(B_{s,d} \to \mu^+ \mu^-)$ are enhanced by factors of 2 and 5, respectively. The impact on $K_{\rm L} \to \mu^+ \mu^-$ is rather moderate.

d) As emphasized in [31], enhanced Z^0 penguins may play an important rôle in ε'/ε. The enhanced value of C and its large negative phase suggested by the $B \to \pi K$ analysis require a significant enhancement of the relevant hadronic matrix element of the QCD penguin operator Q_6, with respect to the one of the EW penguin operator Q_8, to be consistent with the ε'/ε data.

e) We have also explored the implications for the decay $B_d \to \phi K_{\rm S}$ [23]. Large hadronic uncertainties preclude a precise prediction, but assuming that the sign of the cosine of a strong phase agrees with factorization, we find that $(\sin 2\beta)_{\phi K_{\rm S}} > (\sin 2\beta)_{\psi K_{\rm S}}$. This pattern is qualitatively different from the present B-factory data [27], which are, however, not yet conclusive as seen in (14). On the other hand, a future confirmation of this pattern could be another signal of enhanced CP-violating Z^0 penguins at work.

6 Shopping List

We have seen that each of the NP scenarios discussed above had some specific features not shared by other scenarios and with a sufficient number of measurements it should be possible to distinguish between them, eventually selecting one of them or demonstrating the necessity for going to scenarios in classes D and E.

There is a number of questions which I hope will be answered in the coming years:

- Probably the most important at present is the clarification of the discrepancy between Belle and BaBar in the measurement of $(\sin 2\beta)_{\phi K_S}$. The confirmation of the significant departure of $(\sin 2\beta)_{\phi K_S}$, with $(\sin 2\beta)_{\phi K_S} < 0$, from $(\sin 2\beta)_{\psi K_S}$, would be a clear signal of new physics that very likely cannot be accommodated within classes A–C.
- Also very important are the measurements of $Br(B_{d,s} \to \mu^+ \mu^-)$ and ΔM_s. The possible enhancements of $Br(B_{d,s} \to \mu^+ \mu^-)$ by factors as high as 500 are the largest enhancements in the field of K and B decays, that are still consistent with all available data. The measurement of ΔM_s is, on the other hand, very important as it may have a considerable impact on the determination of the unitarity triangle. Finding ΔM_s below $(\Delta M_s)_{\rm SM}$ and $Br(B_{d,s} \to \mu^+ \mu^-)$ well above the SM expectations would be a nice confirmation of a SUSY scenario with a large $\tan \beta$ (Sect. 4).
- The improved measurements of several $B \to \pi\pi$ and $B \to \pi K$ observables are very important in order to see whether the theoretical approaches like QCDF [54], PQCD [64] and SCET [65] in addition to their nice theoretical structures are also phenomenologically useful. On the other hand, independently of the outcome of these measurements, the pure phenomenological strategy [23, 24] presented in Section 5, will be useful in correlating the experimental results for $B \to \pi\pi$ and $B \to \pi K$ with those for rare K and B decays, $B_d \to \phi K_S$ and ε'/ε.

- Assuming that the future more accurate data on $B \to \pi\pi$ and $B \to \pi K$ will not modify significantly the presently observed pattern in these decays, the scenario of enhanced Z^0 penguins with a new large complex weak phase will remain to be an atractive possibility. While the enhancement of $Br(K_{\rm L} \to \pi^0 \nu\bar{\nu})$ by one order of magnitude would be very welcome to our experimental colleagues and $(\sin 2\beta)_{\pi\nu\bar{\nu}} < 0$ would be a very spectacular signal of NP, even more moderate departures of this sort from the SM and the MFV expectations could be easily identified in the very clean $K \to \pi\nu\bar{\nu}$ decays as clear signals of NP.
- The improved measurements of $Br(B \to X_s l^+ l^-)$ and $Br(K^+ \to \pi^+ \nu\bar{\nu})$ in the coming years will efficiently bound the possible enhancements of Z^0 penguins, at least within the scenarios A–C discussed here.
- Also very important is an improved measurement of $Br(B \to X_s\gamma)$ as well as the removal of its sensitivity to μ_c in $m_c(\mu_c)$ through a NNLO calculation. This would increase the precision on the MFV correlation between $Br(B \to X_s\gamma)$ and the zero $\hat{s}_0$ in the forward–backward asymmetry $A_{\rm FB}(\hat{s})$ in $B \to X_s l^+ l^-$. A 20% suppression of $Br(B \to X_s\gamma)$ with respect to the SM accompanied by a downward shift of $\hat{s}_0$ would be an interesting confirmation of the correlation in question and consistent with the effects of universal extra dimensions with a low compactification scale of order few hundred GeV. On the other hand finding no zero in $A_{\rm FB}(\hat{s})$ would likely point towards flavour violation beyond the MFV.
- Finally, improved bounds and/or measurements of processes not existing or very strongly suppressed in the SM, like various electric dipol moments and FCNC transitions in the charm sector will be very important in the search for new physics. The same applies to $\mu \to e\gamma$ and generally lepton flavour violation.

We could continue this list for ever, in particular in view of the expected progress at Belle and Babar, charm physics at Cornell, experimental program at LHCb, BeTeV and the planned rare K physics experiments. But the upper bound on the maximal number of pages for my contribution has been already significantly violated which is a clear signal that I should conclude here. The conclusion is not unexpected: in this decade, it will be very exciting to follow the development in this field and to monitor the values of various observables provided by our experimental colleagues by using the strategies presented here and other strategies found in the rich literature.

Acknowledgements

I would like to thank the organizers for inviting me to such a wonderful meeting and enjoyable atmosphere. The work presented here has been supported in part by the German Bundesministerium für Bildung und Forschung under the contract 05HT1WOA3 and the DFG Project Bu. 706/1-2.

References

1. S. Weinberg, arXiv:hep-ph/0401010.
2. S.L. Glashow, J. Iliopoulos and L. Maiani Phys. Rev. **D2**, 1285 (1970).

3. N. Cabibbo, Phys. Rev. Lett. **10**, 531 (1963); M. Kobayashi and T. Maskawa, Prog. Theor. Phys. **49** (1973) 652.
4. R. Fleischer, Phys. Rep. **370** (2002) 537. Y. Nir, hep-ph/0109090, G. Buchalla, hep-ph/0302145; A. J. Buras, arXiv:hep-ph/0307203. A. Ali, arXiv:hep-ph/0312303; T. Hurth, Rev. Mod. Phys. **75**, 1159 (2003).
5. A. J. Buras, arXiv:hep-ph/9806471.
6. G. Buchalla et al., Rev. Mod. Phys **68**, 1125 (1996).
7. A. J. Buras, arXiv:hep-ph/0109197.
8. T. Inami and C.S. Lim, Progr. Theor. Phys. **65**, 297 (1981).
9. A. J. Buras, arXiv:hep-ph/0101336.
10. A. J. Buras, P. Gambino, M. Gorbahn, S. Jager and L. Silvestrini, Phys. Lett. B **500**, 161 (2001).
11. G. D'Ambrosio et al., Nucl. Phys. B **645**, 155 (2002).
12. C. Bobeth, T. Ewerth, F. Kruger and J. Urban, Phys. Rev. D **66**, 074021 (2002).
13. A.J. Buras, [hep-ph/0310208].
14. M. Misiak, S. Pokorski and J. Rosiek, hep-ph/9703442,
15. F. Gabbiani et al., Nucl. Phys. **B477** 321 (1996). M. Ciuchini et al. Phys. Rev. D **67**, 075016 (2003) [Erratum-ibid. D **68**, 079901 (2003)].
16. H. Abele et al., arXiv:hep-ph/0312150.
17. M. Battaglia, A.J. Buras, P. Gambino, A. Stocchi et al., hep-ph/0304132.
18. B. Aubert et al., Phys. Rev. Lett. **89**, 201802 (2002).
19. K. Abe et al., Phys. Rev. **D66**, 071102 (2002).
20. A. J. Buras, F. Parodi and A. Stocchi, JHEP **0301**, 029 (2003).
21. A. J. Buras, Phys. Lett. B **566**, 115 (2003).
22. G. Buchalla, A. J. Buras, Phys. Lett. **B333**, 221 (1994). Phys. Rev. D **54**, 6782 (1996).
23. A.J. Buras, R. Fleischer, S. Recksiegel, F. Schwab, in preparation.
24. A. J. Buras, R. Fleischer, S. Recksiegel and F. Schwab, arXiv:hep-ph/0312259.
25. A. J. Buras and R. Fleischer, Phys. Rev. D **64**, 115010 (2001).
26. Y. Grossman, G. Isidori and M.P. Worah, Phys. Rev. **D58**, 057504 (1998).
27. T.E. Browder, hep-ex/0312024.
28. K. Abe et al. [Belle Collaboration], hep-ex/0308035.
29. R. Fleischer and T. Mannel, Phys. Lett. **B511**, 240 (2001), G. Hiller, Phys. Rev. **D66**, 071502 (2002), A. Datta, Phys. Rev. **D66**, 071702 (2002), M. Ciuchini and L. Silvestrini, Phys. Rev. Lett. **89**, 231802 (2002), M. Raidal, Phys. Rev. Lett. **89**, 231803 (2002), Y. Grossman, Z. Ligeti, Y. Nir and H. Quinn, hep-ph/0303171, S. Khalil and E. Kou, hep-ph/0307024.
30. A. J. Buras, M. Spranger and A. Weiler, Nucl. Phys. B **660**, 225 (2003); A. J. Buras, A. Poschenrieder, M. Spranger and A. Weiler, Nucl. Phys. B **678**, 455 (2004).
31. A. J. Buras and L. Silvestrini, Nucl. Phys. B **546**, 299 (1999).
32. A. J. Buras, G. Colangelo, G. Isidori, A. Romanino and L. Silvestrini, Nucl. Phys. B **566**, 3 (2000).
33. S. Bergmann and G. Perez, Phys. Rev. D **64**, 115009 (2001), JHEP **0008**, 034 (2000).
34. S. Laplace, Z. Ligeti, Y. Nir and G. Perez, Phys. Rev. **D65**, 094040 (2002); G. Buchalla and A. S. Safir, arXiv:hep-ph/0310218.
35. A. J. Buras, R. Fleischer, S. Recksiegel and F. Schwab, Eur. Phys. J. C **32**, 45 (2003).

36. M. Nakao, arXiv:hep-ex/0312041.
37. M.-C. Chang et al., Belle Collaboration, hep-ex/0309069.
38. G. Buchalla and A. J. Buras, Nucl. Phys. B **548**, 309 (1999).
39. G. D'Ambrosio and G. Isidori, Phys. Lett. **B530**, 108 (2002).
40. J. Kaneko et al. [Belle Collaboration], Phys. Rev. Lett. **90**, 021801 (2003), B. Aubert et al. [BABAR Collaboration], arXiv:hep-ex/0308016.
41. T. Appelquist et al., Phys. Rev. D **64**, 035002 (2001).
42. K. Agashe, N. G. Deshpande and G. H. Wu, Phys. Lett. B **514**, 309 (2001).
43. C.-S. Huang and Q.-S. Yan, Phys. Lett. B **442** 209 (1998); C.-S. Huang et al., Phys. Rev. D **59** 011701 (1999); K. S. Babu and C. Kolda, Phys. Rev. Lett. **84** 228 (2000); P. H. Chankowski and L. Slawianowska, Phys. Rev. D **63** 054012 (2001); C. S. Huang et al., Phys. Rev. D **63** 114021 (2001); Phys. Rev. **64** 059902 (2001) (E); C. Bobeth et al., Phys. Rev. D **64** 074014 (2001).
44. G. Isidori and A. Retico, JHEP **0111**, 001 (2001).
45. A. J. Buras, P. H. Chankowski, J. Rosiek and L. Slawianowska, Nucl. Phys. B **619**, 434 (2001); Phys. Lett. B **546**, 96 (2002); Nucl. Phys. B **659**, 3 (2003).
46. P. H. Chankowski and J. Rosiek, Acta Phys. Polon. B **33**, 2329 (2002). G. Isidori and A. Retico, JHEP **0209**, 063 (2002). A. Dedes, arXiv:hep-ph/0309233.
47. G. Hiller and F. Kruger, arXiv:hep-ph/0310219.
48. S. Bertolini, F. Borzumati and A. Masiero, Phys. Lett. B **194**, 545 (1987) [Erratum-ibid. B **198**, 590 (1987)]. Y. Nir and D. J. Silverman, Nucl. Phys. B **345**, 301 (1990); Phys. Rev. D **42**, 1477 (1990); R. Fleischer, G. Isidori and J. Matias, JHEP **0305**, 053 (2003).
49. A.J. Buras, A. Romanino and L. Silvestrini, Nucl. Phys. **B520**, 3 (1998).
50. G. Buchalla, G. Hiller and G. Isidori, Phys. Rev. D **63**, 014015 (2001), D. Atwood and G. Hiller, [hep-ph/0307251].
51. B. Aubert et al. [BaBar Collaboration], [hep-ex/0308012].
52. K. Abe et al. [Belle Collaboration], [hep-ex/0308040].
53. M. Beneke and M. Neubert, Nucl. Phys. **B675** 333 (2003).
54. M. Beneke et al., Phys. Rev. Lett. **83** 1914 (1999).
55. A.J. Buras and R. Fleischer, Eur. Phys. J. **C11** 93 (1999).
56. Heavy Flavour Averaging Group, `http://www.slac.stanford.edu/xorg/hfag/`.
57. A.J. Buras and R. Fleischer, Eur. Phys. J. **C16** 97 (2000).
58. R. Fleischer and T. Mannel, [hep-ph/9706261].
59. Y. Grossman, M. Neubert and A. Kagan, JHEP **9910**, 029 (1999); Phys. Rev. **D60**, 034021 (1999). T. Yoshikawa, Phys. Rev. **D68**, 054023 (2003). M. Gronau and J. L. Rosner, arXiv:hep-ph/0311280.
60. G. Isidori, hep-ph/0307014 and references therein.
61. S. Adler et al., Phys. Rev. Lett. **88**, 041803 (2002).
62. Y. Grossman and Y. Nir, Phys. Lett. **B398**, 163 (1997).
63. G. Buchalla, G. D'Ambrosio and G. Isidori, arXiv:hep-ph/0308008.
64. Y.-Y. Keum et al. Phys. Rev. **D63**, 054008 (2001) and references therein.
65. I.W. Stewart, hep-ph/0208034 and references therein.

Electron-Positron Linear Collider

Klaus Desch[1]

Institut für Experimentalphysik, University of Hamburg, Luruper Chaussee 149, 22761 Hamburg, Germany
Klaus.Desch@desy.de

1 Introduction

Particle physics in the 20th century witnessed the big success of the Standard Model (SM) as the first self-consistent fundamental description of microscopic matter and its interactions by means of a minimal set of fundamental matter particles and a set of local gauge theories to describe their electro-weak and strong interactions. This theory has been extensively tested experimentally. Not only all matter constituents have been discovered but also the structure of the interactions has been tested at the quantum level to per-mil accuracy. As a result, for example the mass of the top quark could be predicted by precision measurements at LEP and SLC before its direct discovery at the Tevatron. With the top quark being discovered and with the help of more precise knowledge of the mass of the W boson, the mass of the last missing SM ingredient, the Higgs boson, could be constrained to be less than approximately 200 GeV within the SM framework.

Due to its theoretical and experimental success the SM provides a solid basis from which deeper questions can be approached. They emerge from both the theoretical incompleteness and recent experimental observations. A clear limitation of the SM is the fact that it does not incorporate gravity as a quantum theory. While it is justified to postpone this deficit to physics at the Planck scale, the assumption that the SM is valid for energies up to this scale is not satisfactory from a theoretical point of view. The reason lies in a severe fine tuning of the bare Higgs boson mass relative to quadratically growing radiative corrections from gauge boson and fermion loops. Also, many important questions such as the origin of flavour, why there are separate forces acting on leptons and quarks, why gravity is so weak compared to the other forces are not addressed by the SM. Finally and most importantly there is recent experimental evidence that physics beyond the SM (and most probably relevant far below Planck scale energies) is present: neutrino oscillations and the total energy budget of the universe with an apparently dominant contribution from dark matter and dark energy.

A multitude of experimental approaches to tackle these problems are being undertaken, ranging from astrophysical experiments, neutrino experiments, high precision experimentes at low energy to high energy colliders. The latter play a central role because new particles can be directly produced under controlled experimental conditions.

Many of the possible solutions to the above puzzles and in particular the still unresolved question of how the electro-weak symmetry is broken point to new phenomena to be expected at TeV scale energies. In this energy regime the Large

Hadron Collider LHC is currently under construction. It provides an excellent opportunity to discover new particles in that regime. The ability of the LHC to unravel the underlying theory associated with these new particles is however limited due to the harsh environment of hadron collisions. It is the cleanliness of the events, the well-defined initial state and rather low backgrounds from SM processes which make an electron-positron collider with up to about 1 TeV energy an ideal tool to complement the LHC in order to explore TeV scale physics and to open up a window to even much higher scales. An electron-positron linear collider is agreed world-wide to be the next large-scale facility in high energy physics.

The physics case for the LC has been studied in great detail over recent years in regional studies in North America [1], Asia [2] and Europe [3] as well as in world-wide workshops. It is summarised below.

2 Physics

2.1 Higgs Bosons

After the discovery and first measurements of a Higgs boson, it is the task of the LC to unambiguously establish the Higgs mechanism as being responsible for electro-weak symmetry breaking (EWSB). This experimental program comprises a complete survey of the Higgs boson properties. All major studies of the feasibility of this program are described in the TESLA TDR [3]. Recent progress achieved in the context of the ECFA/DESY study on physics and detector for a linear collider is summarised in [4]. The most important aspects are the following.

At the LC, any Higgs boson coupling to the Z^0 can be observed independent of its decay properties in the process $e^+e^- \to H^0Z^0 \to H^0\lambda^+\lambda^-$ by the observation of a peak in the recoil mass spectrum of $Z^0 \to \lambda^+\lambda^-$ pairs (see Fig. 1, left). This method allows for a model-independent determination of the Higgs coupling to the Z^0 with precision a of approximately 2–3% for $m_H = 120\,\mathrm{GeV}$ and $500\,\mathrm{fb}^{-1}$ of data at $\sqrt{s} = 350\,\mathrm{GeV}$ [5]. Once the existence of the Higgs boson has been established, various final states can be exploited to measure its mass. The largest sensitivity for the mass measurement comes from the hadronic final state $H^0Z^0 \to b\bar{b}q\bar{q}$, yielding a mass precision of 50 MeV for a light Higgs boson.

Various techniques offer the possibility for determining the Higgs boson spin. It can be most unambiguously determined from a scan of the H^0Z^0 cross section close to the production threshold (Fig. 1, right) [6]. The CP properties of the Higgs boson can be studied both in the angular distributions of the H^0 and Z^0 bosons as well as through the study of transverse spin correlations in the decay $H^0 \to \tau^+\tau^-$ [7].

The measurement of Higgs boson branching ratios is an unambiguous way to investigate if the coupling strength of Higgs bosons to fermions and gauge bosons is indeed proportional to their mass. The measurement takes advantage of the unprecedented flavour tagging capabilities of the LC detector. For a light Higgs boson, all decay modes with branching ratios in excess of approximately 10^{-3} can be measured with per-cent level precision. In the SM, this comprises the decays into $b\bar{b}$, $c\bar{c}$, gg, $\tau^+\tau^-$, W^+W^-, Z^0Z^0, $\gamma\gamma$ (see Fig. 2, left) [8]. These measurements will be precise enough to gain sensitivity to the structure of the Higgs sector itself.

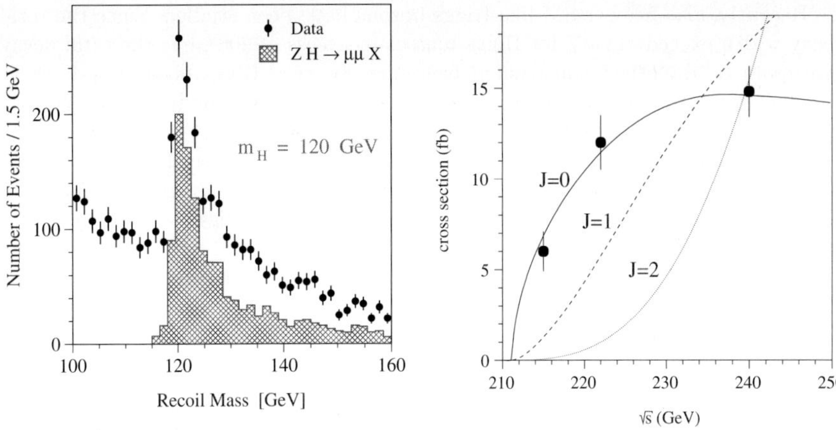

Fig. 1. *Left*: Recoil mass of events with two isolated muons consistent with a Z^0. The shaded histogram represents the contribution from $e^+e^- \to H^0Z^0$ events ($\sqrt{s} = 350\,\text{GeV}$, $500\,\text{fb}^{-1}$, $m_H = 120\,\text{GeV}$. *Right*: Cross section for $e^+e^- \to H^0Z^0$ for different hypotheses of the Higgs spin. The dots with error bars represent simulated measurements corresponding to $10\,\text{fb}^{-1}$ per point

For example, small deviations of the branching ratios from their SM values may be interpreted within extensions of the SM. In the MSSM, the Yukawa couplings differ from their SM values depending on the mass of the heavy pseudo-scalar Higgs boson A^0 (see Fig. 2, right).

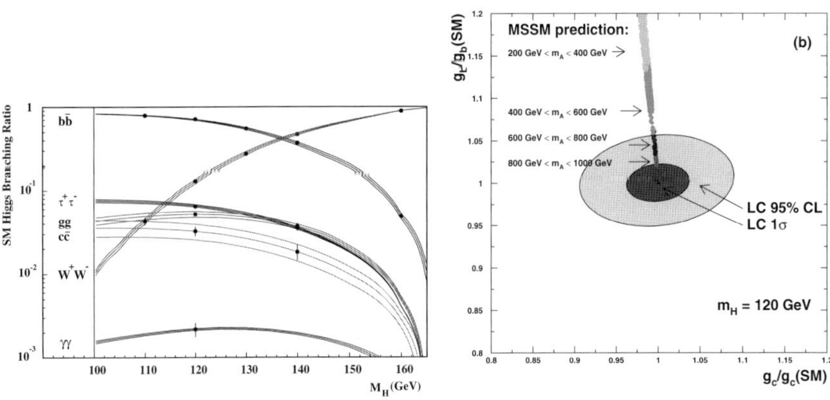

Fig. 2. *Left*: Branching ratios of a light SM Higgs boson. The bands denote the current theoretical uncertainties on the prediction, the dots with error bars show the expected LC errors for $\sqrt{s} = 350\,\text{GeV}$, $500\,\text{fb}^{-1}$. *Right*: Higgs boson couplings (relative to the SM couplings) from a global fit to Higgs boson observables at the LC. The error ellipses are for $\sqrt{s} = 350\,\text{GeV}$, $500\,\text{fb}^{-1}$, the dots represent predictions for the lightest MSSM Higgs boson for various values of the mass of the CP-odd Higgs boson

Recently, also heavier SM-like Higgs bosons have been studied. Since the total decay width exceeds 1 GeV for Higgs masses in excess of 200 GeV, the total decay width can be directly reconstructed from the observed Higgs boson lineshape at the 20% level. At a second phase LC with $\sqrt{s} \sim 1\,\mathrm{TeV}$, also the top quark Yukawa coupling becomes accessible in the mass range below 200 GeV to a precision of 7–15% [4]. This represents an important test of the mechanism for fermion mass generation since the heaviest quark may be most sensitive to new physics. A crucial test of the Higgs mechanism is the measurement of the Higgs self-coupling.It is sensitive to the shape of the Higgs field potential. At the LC the Higgs self-coupling is accessible through the rare $e^+e^- \to H^0H^0Z^0$ process. Given the large luminosity and very good jet energy resolution expected at the LC, the Higgs self-coupling can be measured to a precision of approximately 20% for $m_H = 120\,\mathrm{GeV}$ with 1 ab^{-1} at $\sqrt{s} = 500\,\mathrm{GeV}$ [10].

Beyond the measurements involving SM-like Higgs bosons, Higgs bosons of extended models are of high interest. The heavy CP-even and CP-odd Higgs bosons H^0 and A^0 of the MSSM have been studied recently in the context of the LC. In the process $e^+e^- \to H^0A^0$ both the $b\bar{b}b\bar{b}$ and the $b\bar{b}\tau^+\tau^-$ final states can be observed for Higgs masses up to a few tens of GeV below the kinematical threshold. In particular, a in large part of the parameter space at intermediate values of $\tan\beta$ in which the LHC would only observe the lightest supersymmetric Higgs boson[11], the heavy Higgs could be discovered at the LC. Also in extended models like general two Higgs doublet models, the next-to-minimal supersymmetric model (NMSSM) involving an additional Higgs singlet and extra-dimensional models involving radions, good prospects exist for the LC to distinguish the Higgs sector from the SM and constrain the model parameters.

2.2 Supersymmetry

Low energy Supersymmetry (SUSY) represents one of the most promising extensions of the SM since it addresses many of the SM deficits such as the hierarchy problem, it provides a natural path to grand unification and also provides an excellent candidate for cold dark matter in the universe. However, since SUSY must be broken, already its extension with minimal particle content (the MSSM) introduces more than 100 free parameters most of which are connected to SUSY breaking. After a possible discovery of SUSY at the LHC, it is the most important to learn about SUSY breaking in a model-independent way. Therefore it is the task of the LC to measure with high precision the fundamental SUSY parameters at the EW scale and reconstruct the underlying SUSY breaking mechanism in a bottom-up approach.

In typical examples of SUSY breaking scenarios, in particular those which assume GUT relations between mass parameters at high scale, the colourless SUSY partners tend to be light enough to be copiously produced at an LC with $\sqrt{s} = 500\,\mathrm{GeV}$. Squarks and Gluinos on the other hand will be the dominantly produced SUSY particles at the LHC. Therefore the main emphasis in LC SUSY studies lies on sleptons, charginos and neutralinos. If R-Parity is conserved they are pair produced leading to well distinguishable and cleanly measurable final states. Precise mass determinations can be achieved both in the continuum from the measurement of the endpoints of the energy spectra of the decay products, and from

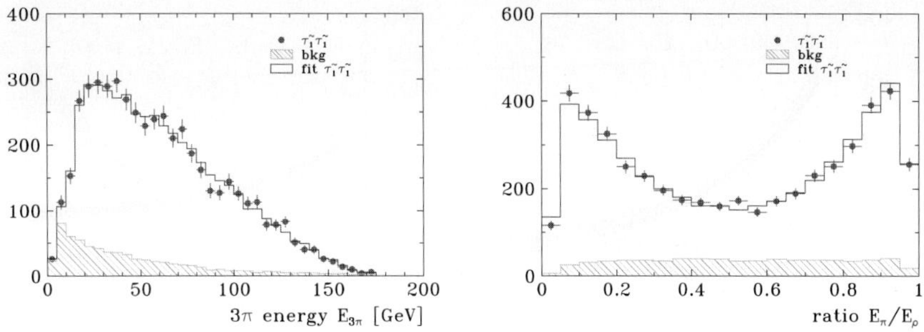

Fig. 3. *Left*: Reconstructed energy of visible τ decay products in $\tau^\pm \to \pi^\pm\pi^\mp\pi^\pm\nu_\tau$ decays from $e^+e^- \to \tilde{\tau}\tilde{\tau}$ production together with the expected SM and MSSM background ($\sqrt{s} = 400\,\text{GeV}$, SUSY point SPS1a). *Right*: Ratio of the energy spectra for $\tau \to \pi\nu_\tau$ and $\tau \to \rho\nu_\tau$ decays. The left distribution can be used to extract the $\tilde{\tau}$ mass, the right distribution yields sensitivity to the average τ polarisation

dedicated scans of the production thresholds taking advantage of the tunable centre-of-mass energy at the LC. Mass precisions in the region of several hundred MeV can be achieved, in some cases even better than 100 MeV [3, 12].

Recently, more attention has been given to scenarios in which the $\tilde{\tau}$ is the next-to-lightest SUSY particle. In this case many SUSY decay chains end in final states involving τ leptons. In a study of the process $e^+e^- \to \tilde{\tau}\tilde{\tau} \to \tau^+\tau^-\chi_1^0\chi_1^0$ it was shown that the endpoint method for mass determination is still applicable in spite of the missing neutrinos of the τ decays [13], see Fig. 3. Furthermore, different hadronic τ decays can be distinguished allowing for a measurement of the average τ polarisation. In conjunction with the possibility of polarised initial electrons and positron, this information can be used to obtain further sensitivity to SUSY parameters such as the $\tilde{\tau}$ mixing angle and $\tan\beta$.

As a consequence of the precise measurements, the fundamental EW scale SUSY parameters can be uniquely determined. For example, from the measurement of the masses of $\chi_1^\pm, \chi_1^0$, and χ_2^0 and the (polarised) production cross sections for $e^+e^- \to \chi_1^+\chi_1^-$ and $\chi_1^0\chi_2^0$, the tree level parameters of the chargino-neutralino-system, $M_1, M_2, \mu, \tan\beta$ can be determined [14]. These parameters along with those of the sfermion sector and the gluino mass parameter M_3 can be extrapolated to high energy scales using renormalisation group equations and be tested against various unification and SUSY breaking patterns [15] (see Fig. 4).

2.3 Top Quark

At $\sqrt{s} \sim 350\,\text{GeV}$ the channel $e^+e^- \to t\bar{t}$ opens up. From a threshold scan, the top quark mass can be determined to a precision of 100 MeV [3]. Recently improved NLO QCD calculations have shown [16] that this experimental precision can indeed be matched by a well suited theoretical definition of the top mass. The top mass is an important SM parameter and is therefore interesting as such. However, equally important, it enters the predictions of beyond-SM models. In particular SUSY predictions, often show a strong top mass dependence. Therefore, in order to match

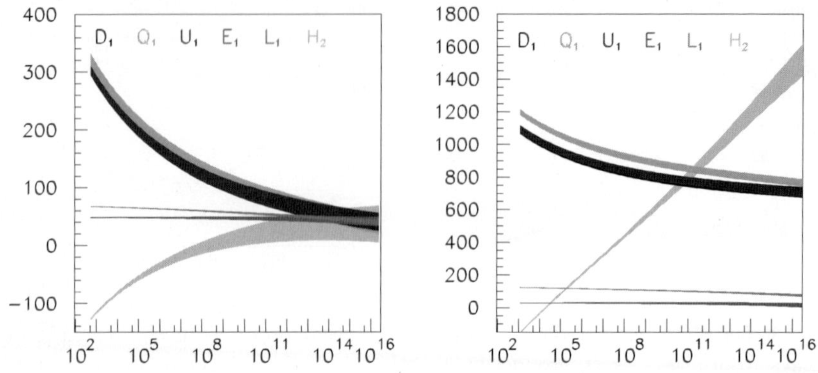

Fig. 4. Evolution of first generation sfermion mass parameters from low to high energy scales. Starting points on the left are for a mSugra SSB scenario, on the right for a GMSB scenario. The width of the error bands corresponds to LC/LHC measurement errors at one standard deviation

the expected precision e.g. of supersymmetric Higgs observables a high top mass precision in mandatory.

2.4 Telescopic Sensitivity

While in the previous examples (Higgs bosons, SUSY particles, top quark) the role of the LC lies in a direct "microscopic" observation and analysis of real new particles a second strength lies in its ability to gain "telescopic" sensitivity to phenomena whose characteristic energy scale (or the mass of the associated particles if any) lies far above the actual centre-of-mass energy of the LC. This strength originates from the ability of precise measurements of the properties of SM processes, such that virtual corrections from the new physics become visible.

No Higgs Boson

If no Higgs boson with mass below ~1 TeV is present, the scattering amplitude for longitudinal W bosons, $W_LW_L \to W_LW_L$ diverges and violates unitarity at approximately 1.3 TeV. Therefore, in the absence of the Higgs mechanism a new strong interaction must set in at this energy scale. The longitudinal degrees of freedom of the massive gauge bosons are then the Goldstone bosons which can be viewed as fermion-condensates bound by the new interaction. Experimental consequences for centre-of-mass energies below the threshold of new vector-like resonances are best describable in terms of effective anomalous triple and quartic gauge couplings. Triple gauge couplings can be measured at the LC in the $e^+e^- \to W^+W^-$, $e^+e^- \to e^+\nu_e W^-$ and $e^+e^- \to \nu_e\bar{\nu_e}\gamma$ processes. At 800 GeV centre-of-mass energy, anomalous couplings can be constrained to less than 10^{-3}–10^{-4}, corresponding to an effective energy scale of approximately 8 TeV for the new physics [17]. Quartic gauge couplings are accessible in the $e^+e^- \to W^+(Z)W^-(Z)\nu_e(e^-)\bar{\nu_e}(e^+)$ processes. Here, an effective energy scale of up to 3 TeV can be probed. Thus the complete threshold region of a new interaction can be probed [18].

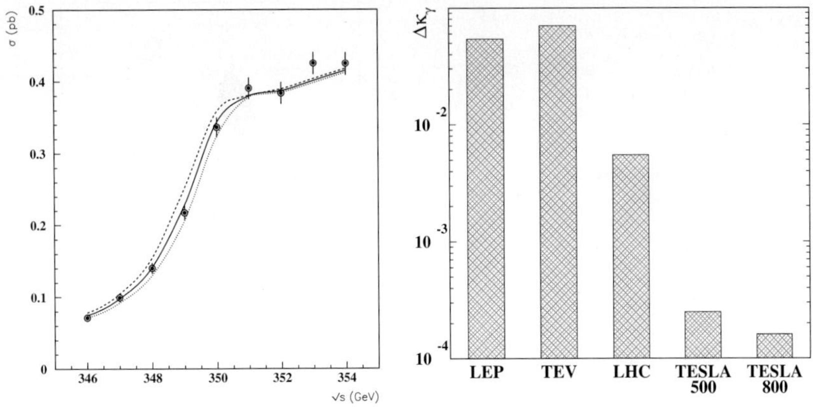

Fig. 5. *Left*: Excitation curve of $t\bar{t}$ quarks including initial-state radiation and beamstrahlung. The errors of the data points correspond to an integrated luminosity of $100\,\mathrm{fb}^{-1}$. The *dotted curves* indicate shifts of the top mass by $\pm 100\,\mathrm{MeV}$. *Right*: Comparison of the precision to measure the anomalous trilinear gauge coupling $\Delta\kappa_\gamma$ at different machines. For LHC and TESLA three years of running are assumed (LHC: $300\,\mathrm{fb}^{-1}$, TESLA $\sqrt{s} = 500\,\mathrm{GeV}$: $900\,\mathrm{fb}^{-1}$, TESLA $\sqrt{s} = 800\,\mathrm{GeV}$: $1500\,\mathrm{fb}^{-1}$)

New Gauge Bosons

New Gauge bosons at TeV scale energies often appear when explicit assumptions of GUT symmetry breaking mechanisms are made. Clearly, if a new Z' boson would be within the LC kinematic reach it can be studied in great detail. But also for Z' masses above $\sqrt{s}$ the modification of fermion couplings from mixing of the Z' propagator with the Z^0 and γ propagators leads to clearly visible effects. Depending on the assumed Z' properties, Z' masses between 4 and 14 TeV can be probed at $\sqrt{s} = 1$ TeV and 1 ab^{-1} of data using both electron and positron beam polarisation. If a Z' should be discovered at the LHC in the 1–3 TeV region, the LC can precisely measure its coupling structure (see Fig. 6) [19].

Large Extra Dimensions

Models with large or warped extra space dimensions provide an attractive alternative for a solution of the hierarchy problem. For the LC the ADD model [20] of large extra dimensions and the RS model [21] of warped extra dimensions have been studied in detail. In ADD, effects from both real graviton emission and virtual graviton exchange can be observed. From the measurement of single photon production at two different centre-of-mass energies the number of extra dimensions can be extracted. Contributions to $e^+e^- \to f\bar{f}$ from graviton exchange can be used to prove that the exchanged particle has indeed spin 2. In the RS scenario, KK excitations of the graviton can be observed if they are within kinematic reach.

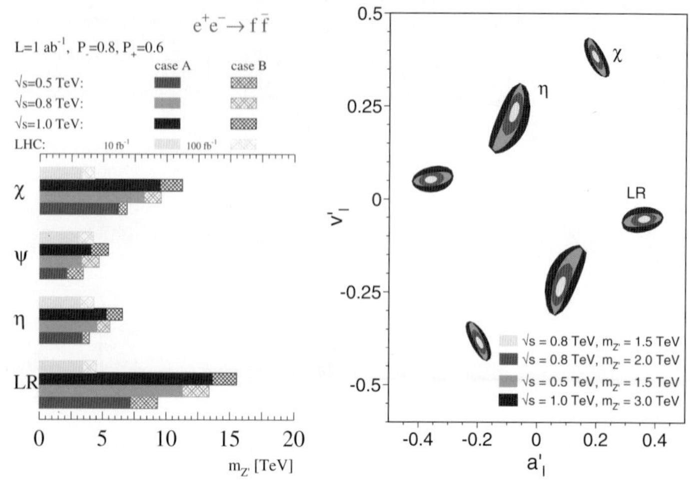

Fig. 6. *Left*: Sensitivity to lower bounds on the Z′, Z_R masses (95% C.L.) in E_6 (χ, ψ and η realization) and left–right symmetric models. The integrated luminosity is $1000\,\mathrm{fb}^{-1}$ for conservative (A) and optimistic (B) assumptions on systematic errors. *Right*: Resolution power (95% CL) for different $m_{Z'}$ based on measurements of leptonic observables at $\sqrt{s} = 500\,\mathrm{GeV}$, 800 GeV, 1 TeV with $1000\,\mathrm{fb}^{-1}$

3 The TESLA Project

The technical realisation of the Linear Collider is a great challenge. In order to keep the total cost reasonable, high accelerating gradients have to be achieved. In order to provide the very high luminosity, mainly dictated by the 1/s decrease of cross sections and by the goal to provide precision measurements, low emittance and high charge density beams have to be produced, transported through the accelerator and finally squeezed to unprecedented small size at the interaction point. In recent years two different designs for a TeV-class LC have been developed. The North American/Asian NLC/GLC design uses normal conducting resonators at 11.4 GHz to accelerate the beams [24]. The superconducting TESLA [23] design developed by the TESLA collaboration led by DESY uses 1.3 GHz superconducting niobium resonators. The TESLA technology offers a number of technical advantages: superconducting cavities provide an efficient transfer of the RF power into the accelerated beam with small losses in the resonators themselves. Also, due to the lower frequency, the structures are larger in diameter allowing for larger alignment tolerances due to smaller wakefields.

The big challenge was to develop cavities which deliver much higher gradients than ever achieved before. The TESLA collaboration is an international collaboration involving 49 institutes from 12 countries. Since its foundation in 1991 the achievable gradient in SC cavities could be raised by a factor 4–5 while reducing the cost approximately by a factor 4. The original goal of the collaboration to reach 23 MV/m, needed for a 500 GeV LC design, could be routinely achieved. The main technological advances came from an improved pre-selection of the niobium sheets, improved welding techniques and cavity production and treatment in an ultra-clean

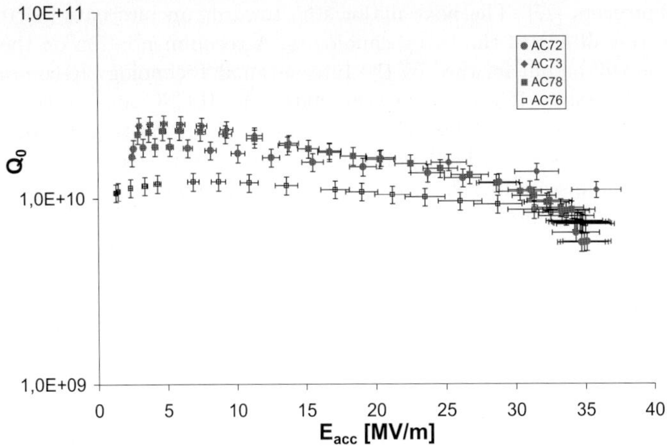

Fig. 7. Excitation curves of the 4 best electro-polished nine-cell cavities after the electro-polishing at Nomura Plating. Plotted is the quality factor Q_0 as a function of the accelerating field. The tests have been performed at 2 K

dust-free environment. The whole accelerator technology was proven to function in a test accelerator constructed at DESY, the TESLA Test Facility, with more than 10000 hours of operation.

Recently further progress was achieved by an improved surface treatment (electro polishing). Electro-polished nine-cell cavities reached more than 35 MV/m in a test setup with full pulsed RF power as required in LC operation and full cavity infrastructure attached. Long-term stability was demonstrated in more than 1000 hours of operation [25].

4 International Realization

The world-wide consensus that a TeV-class linear collider should be the next large accelerator for particle physics was endorsed by the major regional particle physics organisations, ACFA, ECFA and HEPAP. After the publication of the TESLA TDR the German science council requested that the federal government gives its binding consent to a German participation in the LC. In February 2003 the ministry for education and research decided not to put forward a German site at this time. However it encouraged DESY to continue to work on TESLA to facilitate German participation in a future global project. Since then the international community made big efforts to define a single global linear collider project. On the initiative of ICFA the International Linear Collider Steering Committee (ILCSC) was formed [26]. In order to work out technical and sociological aspects of a truly global accelerator being operated from control rooms located at various places around the world, several workshops on the idea of a Global Accelerator Network (GAN) took place. Meanwhile the Office of Science of the US Department of Energy has published a roadmap plan for large scale science facilities in which a LC ranks first of the mid-term projects [27]. The next major step towards an international realisation is the selection of one of the two technologies. A recommendation on the technology decision will be put forward by the International Technology Recommendation Panel (ITRP) [28] to ICFA in the year 2004. The ILCSC also agrees that after the technology decision a global LC design based on the existing design should be developed. In parallel high-level political talks have started in the context of the OECD [29] at ministerial level. After the completion of an international design and the definition of organisational matters the community aims at an approval of the LC around 2006/07. This would allow for a start of operation in the middle of the next decade.

Acknowledgements

The author would like to thank the organizers of the 9th Adriatic Meeting on Particle Physics and the Universe for their warm hospitality and for a very inspiring workshop.

References

1. T. Abe et al. [American Linear Collider Working Group], *Linear collider physics resource book for Snowmass 2001,* hep-ex/0106055 (part 1), hep-ex/0106056 (part 2), hep-ex/0106057 (part 3), and hep-ex/0106058 (2001).
2. K. Abe et al. [ACFA Linear Collider Working Group], *Particle physics experiments at JLC,* hep-ph/0109166.
3. ECFA/DESY LC Physics Working Groups, J. A. Aguilar-Saavedra et al., *TESLA Technical Design Report Part III: Physics at an e+e- Linear Collider,* hep-ph/0106315, DESY-2001-011.
4. K. Desch [Higgs Working Group of the Extended ECFA/DESY Study], hep-ph/0311092.
5. P. Garcia-Abia and W. Lohmann, Eur. Phys. J. direct C **2** (2000) 2.
6. S. Y. Choi, D. J. Miller, M. M. Mühlleitner and P. M. Zerwas, *Identifying the Higgs Spin and Parity in Decays to Z Pairs,* Phys. Lett. B **553** (2003) 61, hep-ph/0210077, LC-TH-2003-036 (2003).
7. K. Desch, Z. Was and M. Worek, *Measuring the Higgs boson parity at a linear collider using the tau impact parameter and tau → rho nu decay,* Eur. Phys. J. C **29** (2003) 491, hep-ph/0302046, LC-PHSM-2003-003 (2003).
8. M. Battaglia, hep-ph/9910271.
9. N. Meyer, *Measuring Resonance Parameters of Heavy Higgs Bosons at TESLA,* LC-PHSM-2003-066, hep-ph/0308142 (2003).
10. C. Castanier, P. Gay, P. Lutz and J. Orloff, *Higgs self coupling measurement in e+e- collisions at center-of-mass energy of 500* GeV, LC-PHSM-2000-061 (2000).
11. ATLAS Collaboration. ATLAS Detector and Physics Performance Technical Design Report. CERN-LHCC 99-14.
12. J. Kalinowski, hep-ph/0309235.

13. E. Boos, H. U. Martyn, G. Moortgat-Pick, M. Sachwitz, A. Sherstnev and P. M. Zerwas, Eur. Phys. J. C **30** (2003) 395, hep-ph/0303110.
14. K. Desch, J. Kalinowski, G. Moortgat-Pick, M. M. Nojiri and G. Polesello, hep-ph/0312069.
15. G. A. Blair, W. Porod and P. M. Zerwas, Eur. Phys. J. C **27** (2003) 263, hep-ph/0210058.
16. A. Brandenburg,
17. W. Menges, DESY-THESIS-2003-043.
18. R. Chierici, S. Rosati and M. Kobel, LC-PHSM-2001-038 *Prepared for 5th International Linear Collider Workshop (LCWS 2000), Fermilab, Batavia, Illinois, 24-28 Oct 2000*
19. S. Riemann, LC-TH-2001-007
20. N. Arkani-Hamed, S. Dimopoulos, and G. R. Dvali, Phys. Lett. **B429** (1998), 263.
21. L. Randall and R. Sundrum, *Phys. Rev. Lett.* **83** (1999) 3370, *Phys. Rev. Lett.* **83** (1999) 4690.
22. G. F. Giudice, R. Rattazzi, and J. D. Wells, *Nucl. Phys.* **B595** (2001) 250. C. Csaki, M. L. Graesser, and G. D. Kribs, *Phys. Rev.* **D63** (2001) 065002.
23. R. Brinkmann, K. Flöttmann, J. Rossbach, P. Schmüser, N. Walker and H. Weise, *TESLA: The superconducting electron positron linear collider with an integrated X-ray laser laboratory. Technical design report. Pt. 2: The accelerator*, DESY-01-011.
24. The Next Linear Collider, http://www-project.slac.stanford.edu/lc/NLC-tech.html
25. L. Lilje et al., *Achievement of 35-MV/m in the superconducting nine-cell cavities for TESLA*, arXiv:physics/0401141.
26. International Linear Collider Steering Committee (ILCSC) http://www.fnal.gov/directorate/icfa/International_ILCSC.html
27. US Department of Energy, Office of Science, http://www.sc.doe.gov/Sub/Facilities_for_future/facilities_future.htm
28. International Technology Recommendation Panel (ITRP), http://www.ligo.caltech.edu/ donna/ITRP_Meeting_One.htm
29. OECD Global Science Forum, http://www.oecd.org/document/18/0,2340,en_2649_34319_1949714_1_1_1_1,00.html

New Source of CP Violation in B Physics?

Nilendra G. Deshpande and Dilip Kumar Ghosh

Institute of Theoretical Science
University of Oregon, Eugene, OR 97403
desh@oregen.uoregon.edu

1 Introduction

Time dependent asymmetries measured in the decay $B \to \phi K_S$ both by BaBar and Belle collaborations [1, 2, 3, 4] show significant deviation from the standard model and this has generated much theoretical speculation regarding physics beyond the standard model [5-20]. In the standard model, the process $B \to \phi K_S$ is purely penguin dominated and the leading contribution has no weak phase. The coefficient of $\sin(\Delta m_B t)$ in the asymmetry therefore should measure $\sin 2\beta$, the same quantity that is involved in $B \to \psi K_S$ in the standard model. The most recent measured average values of asymmetries are [4, 21]

$$\begin{aligned} S_{\psi K_S} &= 0.734 \pm 0.055 \\ S_{\phi K_S} &= -0.15 \pm 0.33 \end{aligned} \tag{1}$$

The value for $S_{\psi K_S}$ agrees with theoretical expectation from the CKM matrix of $S_{\psi K_S} = \sin 2\beta = 0.715^{+0.05}_{-0.045}$ [22]. This leads to the conclusion that CP phase in $B - \bar{B}$ mixing is consistent with the standard model. The deviation in the ϕK_S is intriguing because a penguin process being a loop induced process is particularly sensitive to new physics which can manifest itself in a loop diagram through exchange of heavy particles. In this talk [23] we consider effects arising from non universal squark mixing in the second and third generation of the down type squarks in supersymmetric theory as the origin of additional contributions to the amplitude within the mass insertion approximation scheme. In particular, the exchanges of gluinos ($\tilde{g}$) and squark ($\tilde{q}$) with left-right mixing can enhance the Wilson coefficient of the gluonic dipole penguin operator $\mathcal{O}_{8g}$ by a factor of $m_{\tilde{g}}/m_b$ compared with the standard model prediction and we take into account its effect on the process $B \to \phi K_S$. In our analysis we take the $B - \bar{B}$ mixing phase the same as in the standard model as required by ψK_S data, and permitted in SUSY by requiring that the first and third generation squark mixing to be small. We study $B \to \phi K$ in QCD improved factorization scheme (BBNS approach) [24]. This method incorporates elements of naive factorization approach (as its leading term) and perturbative QCD corrections (as sub-leading contributions) and allows one to compute systematic radiative corrections to the naive factorization for the hadronic B decays.

In supersymmetry, assuming masses of squarks ($\tilde{q}$) and gluinos ($\tilde{g}$), the new source of CP violation can be parameterized by the complex quantity $\delta^{bs}_{LR(RL)}$ written in the form $\rho e^{i\psi}$ We identify the region in $\rho - \psi$ plane allowed by the experimental data on $B \to \phi K$ time dependent asymmetries $S_{\phi K_S}$ and $C_{\phi K_S}$ and the branching ratio. This allowed region is dependent of the QCD scale μ, therefore we illustrate the region for two values of $\mu = m_b$ and $m_b/2$. The same contribution should also be present in other penguin mediated process. We study the effect of $LR(RL)$ mass insertion to the $B \to \phi K^*$ decay mode which is also a pure penguin process using QCD improved factorization method. We then estimate the branching ratio $\mathcal{B}(B \to \phi K^*)$ and the CP asymmetry $\mathcal{A}_{CP}$ in the parameter space of $\delta^{bs}_{LR(RL)}$ allowed by $B \to \phi K$ data. In this vector vector final state, one can also construct more CP violating observables [25]. We compute these observables in the same range of parameter space as that allowed by $B \to \phi K$.

2 CP Asymmetry of $B \to \phi K$

The time dependent CP asymmetry of $B \to \phi K_S$ is described by :

$$\mathcal{A}_{\phi K_S}(t) = \frac{\Gamma(\overline{B^0}(t) \to \phi K_S) - \Gamma(B^0(t) \to \phi K_S)}{\Gamma(\overline{B^0}(t) \to \phi K_S) + \Gamma(B^0(t) \to \phi K_S)} \tag{2}$$

$$= -C_{\phi K_S} \cos(\Delta m_B t) + S_{\phi K} \sin(\Delta m_B t) \tag{3}$$

where $S_{\phi K}$ and $C_{\phi K_S}$ are given by

$$S_{\phi K} = \frac{2Im\ \lambda_{\phi K_S}}{1+ \mid \lambda_{\phi K_S} \mid^2}, \qquad C_{\phi K_S} = \frac{1- \mid \lambda_{\phi K_S} \mid^2}{1+ \mid \lambda_{\phi K_S} \mid^2} \tag{4}$$

and $\lambda_{\phi K_S}$ can be expressed in terms of decay amplitudes:

$$\lambda_{\phi K_S} = -\mathrm{e}^{-2i\beta} \frac{\overline{\mathcal{M}}(\overline{B^0} \to \phi K_S)}{\mathcal{M}(B^0 \to \phi K_S)} \tag{5}$$

The branching ratio and the direct CP asymmetries of both the charged and neutral modes of $B \to \phi K$ have been measured [1, 2, 3, 4, 21, 26]:

$$\mathcal{B}(B^0 \to \phi K_S) = (8.0 \pm 1.3) \times 10^{-6} \tag{6}$$

$$\mathcal{B}(B^+ \to \phi K^+) = (9.4 \pm 0.9) \times 10^{-6}, \tag{7}$$

$$S_{\phi K_S} = +0.45 \pm 0.43 \pm 0.07 \ \text{(BaBar)}; \tag{8}$$

$$= -0.96 \pm 0.50^{+0.09}_{-0.11} \ \text{(Belle)}; \tag{9}$$

$$C_{\phi K_S} = -0.19 \pm 0.30 \tag{10}$$

$$\mathcal{A}_{CP}(B^+ \to \phi K^+) = (3.9 \pm 8.8 \pm 1.1)\% \tag{11}$$

3 The Exclusive $B \to \phi K$ Decay

In the standard model, the effective Hamiltonian for charmless $B \to \phi K(\phi K^*)$ decay is given by [24]

$$\mathcal{H}_{eff} = -\frac{G_F}{\sqrt{2}} V_{tb} V_{ts}^* \Bigg\{ C_1(\mu) \mathcal{O}_1(\mu) + C_2(\mu) \mathcal{O}_2(\mu) + \sum_{i=3}^{10} C_i(\mu) \mathcal{O}_i(\mu) + C_{7\gamma} \mathcal{O}_{7\gamma} + C_{8g} \mathcal{O}_{8g} \Bigg\} \tag{12}$$

where the Wilson coefficients $C_i(\mu)$ are obtained from the weak scale down to scale μ by running the renormalization group equations. The definitions of the operators and different Wilson coefficients can be found in [24].

4 $B \to \phi K$ in the QCDF Approach

In the QCD improved factorization scheme, the $B \to \phi K$ decay amplitude due to a particular operator can be represented in following form:

$$\langle \phi K \mid \mathcal{O} \mid B \rangle = \langle \phi K \mid \mathcal{O} \mid B \rangle_{fact} \left[1 + \sum r_n \alpha_s^n + O(\Lambda_{QCD}/m_b) \right] \tag{13}$$

where $\langle \phi K \mid \mathcal{O} \mid B \rangle_{fact}$ denotes the naive factorization result. The second and third term in the bracket represent higher order α_s and Λ_{QCD}/m_b correction to the hadronic transition amplitude. Following the scheme and notations presented in [27, 28], we write down the $B \to \phi K$ amplitude in the heavy quark limit.

$$\begin{aligned} \mathcal{M}(B^+ \to \phi K^+) &= \\ \mathcal{M}(B^0 \to \phi K^0) &= \frac{G_F}{\sqrt{2}} m_B^2 f_\phi F_1^{B\to K}(m_\phi^2) V_{pb} V_{ps}^* \left[a_3^p + a_4^p + a_5^p \right. \\ &\quad \left. - \frac{(a_7^p + a_9^p + a_{10}^p)}{2} + a_{10a}^p \right] \end{aligned} \tag{14}$$

where p is summed over u and c. The coefficients a_i^p are given by

$$\begin{aligned} a_3^u &= a_3^c = C_3 + \frac{C_4}{N_c} \left[1 + \frac{C_F \alpha_s}{4\pi} (V_\phi + H_\phi) \right] , \\ a_4^p &= C_4 + \frac{C_3}{N_c} \left[1 + \frac{C_F \alpha_s}{4\pi} (V_\phi + H_\phi) \right] + \frac{C_F \alpha_s}{4\pi N_c} P_\phi^p , \\ a_5^u &= a_5^c = C_5 + \frac{C_6}{N_c} \left[1 + \frac{C_F \alpha_s}{4\pi} (-12 - V_\phi) \right] , \\ a_7^u &= a_7^c = C_7 + \frac{C_8}{N_c} \left[1 + \frac{C_F \alpha_s}{4\pi} (-12 - V_\phi - H_\phi) \right] , \\ a_9^u &= a_9^c = C_9 + \frac{C_{10}}{N_c} \left[1 + \frac{C_F \alpha_s}{4\pi} (V_\phi + H_\phi) \right] , \\ a_{10}^u &= a_{10}^c = \left[1 + \frac{C_F \alpha_s}{4\pi} (V_\phi + H_\phi) \right] , \\ a_{10a}^u &= a_{10a}^c = \frac{C_F \alpha_s}{4\pi N_c} Q_\phi \end{aligned} \tag{15}$$

with $C_F = (N_c^2 - 1)/2N_c$ and $N_c = 3$. The quantities V_ϕ, H_ϕ, P_ϕ^p and Q_ϕ^p are hadronic parameters that contain all nonperturbative dynamics.

$$V_\phi = -12\ln\frac{\mu}{m_b} - 18 + f^I_\phi ,$$

$$f^I_\phi = \int_0^1 \mathrm{d}x g(x)\Phi_\phi(x); g(x) = 3\frac{1-2x}{1-x}\ln x - 3i\pi ,$$

$$H_\phi = \frac{4\pi^2}{N_c}\frac{f_B f_K}{F_1^{B\to K}(0)m_B^2}\int_0^1 \mathrm{d}z\frac{\Phi_B(z)}{z}\int_0^1 \mathrm{d}x\frac{\Phi_K(x)}{x}\int_0^1 \mathrm{d}y\frac{\Phi_\phi(y)}{y} ,$$

$$P^p_\phi = C_3\left[G_\phi(s_s) + G_\phi(s_b)\right] + C_2 G_\phi(s_p) + (C_4 + C_6)\sum_{f=u}^{b}\tilde{G}_\phi(s_f) + C^{eff}_{8g}G_{\phi g}$$

$$Q_\phi = (C_8 + C_{10})\frac{3}{2}\sum_{f=u}^{b} e_f G_\phi(s_f) + C_9\frac{3}{2}\left[e_s G_\phi(s_s) + e_b G_\phi(s_b)\right]$$

$$G_\phi(s) = \frac{2}{3} - \frac{4}{3}\ln\frac{\mu}{m_b} + 4\int_0^1 \mathrm{d}x\ \Phi_\phi(x)\int_0^1 \mathrm{d}u\ u\ (1-u)\ln\left[s - u(1-u)(1-x)\right]$$

$$\tilde{G}_\phi(s) = G_\phi(s) - (2/3)$$

$$G_{\phi g} = -\int_0^1 \mathrm{d}x\frac{2}{(1-x)}\Phi_\phi(x) \tag{16}$$

where, $s_i = m_i^2/m_b^2$. Here, V_ϕ represent contributions from the vertex correction and H_ϕ correspond to hard gluon-exchange interactions with spectator quarks. P^p_ϕ and Q^p_ϕ represent QCD penguin contributions. We neglect order α_{em} EW penguin corrections to a_i. f_B, f_K are the B and K meson decay constants and $F_1^{B\to K}$ denotes the form factor for $B \to K$ transitions. $\Phi_B(z), \Phi_K(x)$, and $\Phi_\phi(y)$ are the B, K, and ϕ meson wave functions respectively. In this analysis we take following forms for them [27]

$$\Phi_B(x) = N_B\ x^2\ (1-x)^2 \exp\left[-\frac{m_B^2 x^2}{2\omega_B^2}\right] ,$$

$$\Phi_{K,\phi}(x) = 6\ x\ (1-x) \tag{17}$$

where, N_B is a normalization factor satisfying $\int_0^1 \mathrm{d}x\ \Phi_B(x) = 1$, and $\omega_B = 0.4$.

For the sake of completeness, we give the branching ratio for $B \to \phi K$ decay channel in the rest frame of the B meson.

$$\mathcal{BR}(B \to \phi K) = \frac{\tau_B}{8\pi}\frac{\mid P_{cm}\mid}{m_B^2}\mid \mathcal{M}(B \to \phi K)\mid^2 \tag{18}$$

where, τ_B represents the B meson lifetime and the kinematical factor $\mid P_{cm} \mid$ is written as

$$\mid P_{cm} \mid = \frac{1}{2m_B}\sqrt{[m_B^2 - (m_K + m_\phi)^2]\,[m_B^2 - (m_K - m_\phi)^2]} \tag{19}$$

5 SUSY Gluino Contributions to $B \to \phi K$

In order to study the new physics contribution to the CP violating phase of amplitude $\mathcal{M}(B \to \phi K)$, we compute the effect of flavor changing contribution to $B \to \phi K$ arising from $q - \tilde{q} - \tilde{g}$ interactions in supersymmetric theory under the mass insertion approximation scheme [29, 30]. In this approximation, the flavor changing contribution is parameterized in terms of $\delta^{ij}_{AB} = \Delta^{ij}_{AB}/\tilde{m}^2$, where, Δ represents the off-diagonal entries of the squark mass matrices, $\tilde{m}$ is an average squark mass, $A, B = L, R$ and i, j are the generation indices. The $LR(RL)$ mass insertion can enhance the Wilson coefficients $C_{7\gamma}$ and C_{8g} by a factor of $m_{\tilde{g}}/m_b$ compared to the standard model contribution. This leads to a strong limit of order $O(10^{-2})$ on the $LR(RL)$ insertions $\mid \delta^{bs}_{LR(RL)} \mid$ from the $\mathcal{B}(B \to X_s\gamma)$ [30, 31] while the limit on the LL and RR ones is rather mild [30, 31]. Thus, although larger values for LL and RR mixings are allowed, when one considers $B \to \phi K$, the effect of their mixings are only significant in the parameter space where the squark and gluino masses are at the edge of their experimental constraints [17]. Motivated by this fact, we only concentrate on $LR(RL)$ down type squark mixing in hereafter. Thus, the new physics effect is very sensitive to $\delta^{bs}_{LR(RL)}$.

In general, these contributions $LR(RL)$ can generate gluonic dipole interactions with the same as well as opposite chiral structure as the standard model. In our analysis we will consider each of them separately. Furthermore we will only consider the gluonic dipole moment operator, which is the dominant operator for this process.

The effective Wilson coefficient for C^{SUSY}_{8g} obtained in the mass insertion approximation is given by for the same chiral structure as the standard model [32, 33]

$$C^{SUSY}_{8g}(m_{\tilde{q}}) = -\frac{\sqrt{2}\pi\alpha_s}{G_F(V_{ub}V^*_{us} + V_{cb}V^*_{cs})m^2_{\tilde{g}}}\,\delta^{bs}_{LR(RL)}\,\frac{m_{\tilde{g}}}{m_b}\,G(x)\;, \tag{20}$$

with

$$G(x) = \frac{x}{3(1-x)^4}\left[22 - 20x - 2x^2 + 16x\ln(x) - x^2\ln(x) + 9\ln(x)\right]\;, \tag{21}$$

where $x = m^2_{\tilde{g}}/m^2_{\tilde{q}}$ is the ratio of the gluino and squark mass.

Using the renormalization group equation one can evolve the coefficient C^{SUSY}_{8g} from the high scale $m_{\tilde{q}}$ to the scale m_b relevant for $B \to \phi K$ decay [32]

$$C^{SUSY}_{8g}(m_b) = \eta C^{SUSY}_{8g}(m_{\tilde{q}})\;, \tag{22}$$

with

$$\eta = (\alpha_s(m_{\tilde{q}})/\alpha_s(m_t))^{2/21}\,(\alpha_s(m_t)/\alpha_s(m_b))^{2/23} \tag{23}$$

One can obtain C^{SUSY}_{8g} for opposite chirality, by adding one more operator similar to $\mathcal{O}_{8g}$ with $(1+\gamma_5) \to (1-\gamma_5)$ and $\delta^{bs}_{LR} \to \delta^{bs}_{RL}$. However, in $B \to \phi K$ process, both LR and RL contribute with the same sign because B and K parity are both 0^-, and the process is parity conserving.

The effective Wilson coefficient C^{eff}_{8g} is defined as $C^{eff}_{8g} = C_{8g} + C^{SUSY}_{8g}$. This effective C^{eff}_{8g} will contribute to the amplitude $\mathcal{M}(B \to \phi K)$ through the function P^p_ϕ of Equation 16. C^{eff}_{8g} depends on the magnitude and phase of the $(\delta^{bs}_{LR(RL)})$, value of squark mass $(m_{\tilde{q}})$ and the ratio x $(= m^2_{\tilde{g}}/m^2_{\tilde{q}})$. The variation of C^{eff}_{8g} with x is determined by the function $\mid G(x) \mid$ as shown in Fig. 1. From this figure, it is clear that SUSY gluino contribution to $B \to \phi K$ first increases with increase in x, and then after some value of $x = 0.5$, it starts decreasing asymptotically with further increase in x.

The different input parameters and their values used in numerical calculation of branching ratio and CP asymmetries are given in [23].

5.1 $LR(RL)$ Mixing

In this section we study the effect of $LR(RL)$ mixing in $B \to \phi K$ process. This $LR(RL)$ mixing of the down type squark sector can also affect the $B \to \gamma X_s$ process and $B_s - \bar{B}_s$ mixing. Hence we need to take into account the limit on $LR(RL)$ mixing parameter $\delta^{bs}_{LR(RL)}$ from the above two experimental data in the present analysis. In the first case, it has been shown in [30] that from the measurement of $\mathcal{B}(B \to \gamma X_s)$ one gets $| \delta^{bs}_{LR(RL)} | < 1.0 \times 10^{-2}$ and 3.0×10^{-2} for $x = 0.3$ and 4 respectively, with $m_{\tilde{q}} = 500\,\mathrm{GeV}$. It is interesting to note that the lower the x value stronger the limit on $| \delta^{bs}_{LR(RL)} |$, which can be explained by the x dependent behavior of the $C^{SUSY}_{7\gamma}$.

The current experimental data on $B_s - \bar{B}_s$ mixing is $\Delta M_s > 14.4\ \mathrm{ps}^{-1}$ (at 95% C.L.) [35]. We have found that the $LR(RL)$ mixing does not change the value of ΔM_s significantly from the standard model prediction in the allowed range of $| \delta^{bs}_{LR(RL)} |$.

In our analysis we consider $m_{\tilde{q}} = 500\,\mathrm{GeV}$ and take two values of $x = 0.3$ and 4.0, which will determine the gluino masses. In Fig. 1 we show the 1σ allowed region in $\rho - \psi$ plane from $B \to \phi K$ data on $S_{\phi K}, C_{\phi K}$ and $\mathcal{B}$. The gray band indicate the parameter space which is allowed by $S_{\phi K}$. The area outside the two dotted contours is allowed by $C_{\phi K}$, while the area enclosed by the solid curves is allowed by the $\mathcal{B}(B \to \phi K_S)$ measurement. The region (marked by Z) in gray band enclosed by the solid curves is the only parameter space left in $\rho - \psi$ plane which is allowed by the experimentally measured $S_{\phi K}, C_{\phi K}$ and $\mathcal{B}$ within 1σ.

The Figs. 1(a) and (b), correspond to contour plots for $x = 0.3$ and 4.0 respectively at the scale $\mu = m_b$. For $x = 0.3$, we get two allowed regions each at positive and negative values of the new phase ψ. On the other hand for $x = 4.0$, we get only one allowed region which lies at the negative value of ψ and at much higher value of $\rho > 2.2 \times 10^{-2}$. We have noticed before that the constraint on $LR(RL)$ mixing parameter from the $\mathcal{B}(B \to X_s\gamma)$ is stronger at $x = 0.3$ compared to the limit at $x = 4.0$. This behavior is also reflected in the $B \to \phi K$ process, where we find that, for $x = 4.0$, the 1σ constraint from $S_{\phi K}, C_{\phi K}$ and $\mathcal{B}$ is much weaker compared to the constraint shown in Fig. 1(a) correspond to $x = 0.3$.

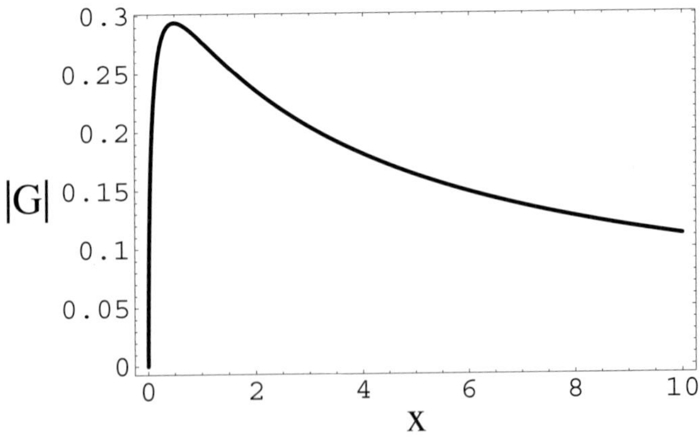

Fig. 1. Variation of $| G(x) |$ with $x(= m^2_{\tilde{g}}/m^2_{\tilde{q}})$

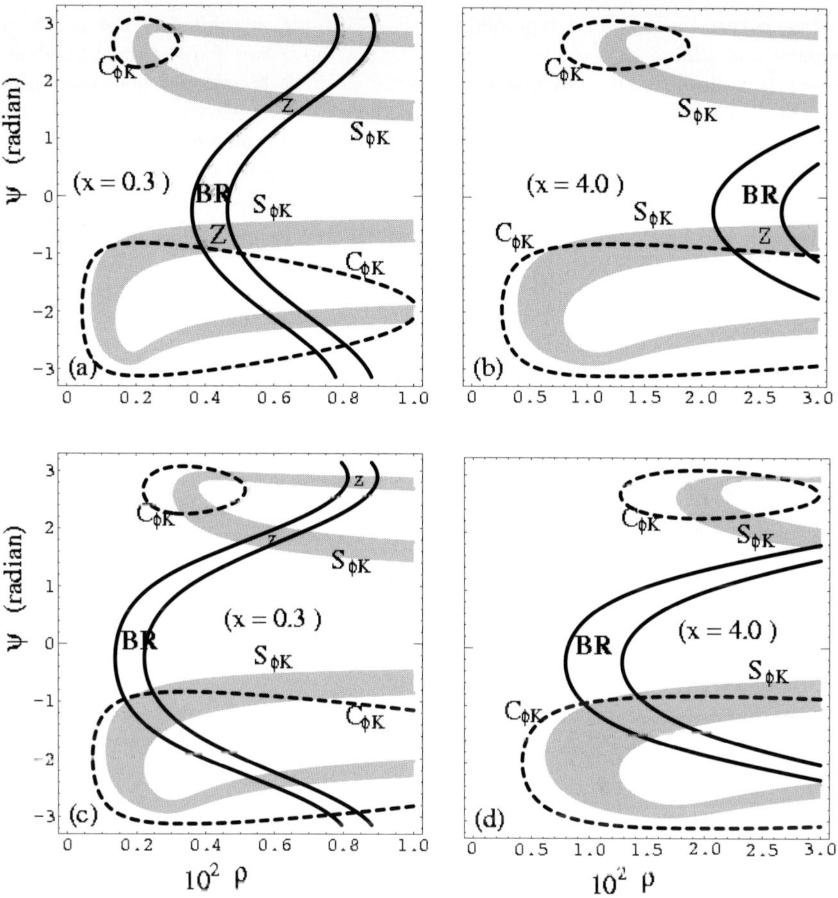

Fig. 2. Contour plots of $S_{\phi K}, C_{\phi K}$ and $\mathcal{B}(B \to \phi K_S)$ in $\rho - \psi$ plane for two values of $x = 0.3$ (a, c) and 4.0 (b, d) for $LR(RL)$ mixing with $m_{\tilde{q}} = 500\,\mathrm{GeV}$. The scale $\mu = m_b$ for Figs. (a) and (b), while it is $m_b/2$ for Figs. (c) and (d). The 1σ allowed regions of $S_{\phi K}$, $\mathcal{B}$ and $C_{\phi K}$ are two gray bands, area within the solid curves and area outside the two dotted contours respectively

Similar allowed regions are shown in Fig. 1(c) for a different choice of the QCD scale $\mu = m_b/2$. One can see that the allowed parameter space does depend on μ. In this case, both the allowed regions are confined at the positive value of ψ. For $x = 4.0$ (Fig. 1(d)), there are no allowed regions. From the $S_{\phi K}$ and branching ratio contour one can see that the allowed region from $B \to \phi K$ require some higher value of ρ which lies beyond $B \to X_s\gamma$ limit.

Before we conclude this section, we would like to compare our predictions with some of the existing literatures on $B \to \phi K$ process [10, 11, 17].

We agree qualitatively with the results of [11, 17] in places where we overlap. Similar to our approach, both of these analyses were based upon the QCD improved factorization scheme. However, there are some quantitative differences between these papers and our analysis. For example, we differ in the choice of squark and gluino masses, the authors of the above two papers considered degenerate squark and gluino masses, whereas we have considered non-degenerate squark-gluino masses. We have fixed the squark mass at 500 GeV and considered two values of the gluino masses, determined by the parameter x defined earlier. Secondly, we have performed our analysis for two values the QCD scale, $\mu = m_b/2$ and m_b. Our results depend strongly on the choice of the ratio x and also on the scale μ. However, in a broad sense, we do agree that to satisfy $B \to \phi K$ data, one requires $| \delta^{bs}_{LR(RL)} | \sim 10^{-3}$–$10^{-2}$.

In [10], authors made a detailed investigation of a scenario in which the LR and RR operators co-exist. Moreover, because of the large mixing, the calculation was done in the mass eigenbasis with more model dependence than ours. It has been shown in this analysis that RR insertion (which arises due to a large mixing between $\tilde{s}_R$ and $\tilde{b}_R$) could show sizable effect on $S_{\phi K}$, but only for very light gluino mass, near the experimental bound. Such a large RR mixing also modify ΔM_s significantly which can be observed at the Tevatron Run II. In their second case, they have the combination of both large right-right and left-right squark mixing ($LR + RR$). In this case the squark and gluinos could be sufficiently heavy to have no significant enhancement of the ΔM_s.

From our analysis we observe that SUSY leads to a comprehensive understanding of $B \to \phi K_S$ data though in a very limited parameter space of $\delta^{bs}_{LR(RL)}$. In rest of the paper we now explore the consequence of such $LR(RL)$ mixing of squarks in the $B \to \phi K^*$ process.

6 $B \to \phi K^*$ Decay

In this section we will study the effect of $LR(RL)$ mixing of down type squarks to $B \to \phi K^*$ process through the gluonic dipole moment operator C_{8g}. We will study the $B \to \phi K^*$ process by using the QCD improved factorization. Using this method one can compute nonfactorizable corrections to the above process in the heavy quark limit. Recently the $B \to VV$ process has been computed using QCD improved factorization method [36]. In rest of analysis we will follow [36].

The most general Lorentz invariant decay amplitude for the process $B \to VV$ can be expressed as

$$\mathcal{M}(B(p_B) \to V_1(\epsilon_1, p_1) V_2(\epsilon_2, p_2)) \propto \epsilon_1^{*\mu} \epsilon_2^{*\nu} \left[a g_{\mu\nu} + b p_{B\mu} p_{B\nu} + i c \epsilon_{\mu\nu\alpha\beta} p_1^\alpha p_2^\beta \right] \tag{24}$$

where the coefficients c correspond to the p-wave amplitude, and a, b to the mixture of s and d wave amplitudes. Using these a, b and c coefficients one can construct the three helicity amplitudes:

$$\begin{aligned} H_{00} &= \frac{1}{2 m_{V_1} m_{V_2}} \left[\left(m_B^2 - m_{V_1}^2 - m_{V_2}^2 \right) a + 2 m_B^2 p_{cm}^2 b \right] \\ H_{\pm\pm} &= a \mp m_B p_{cm} c \end{aligned} \tag{25}$$

where p_{cm} is the center of mass momentum of the vector meson in the B rest frame and $m_{V_1} (m_{V_2})$ is the mass of the vector meson $V_1 (V_2)$. These helicity amplitudes H_{00} and $H_{\pm\pm}$ can be related to the spin amplitudes in the transverse basis

$(A_0, A_{||}, A_\perp)$ defined in terms of linear polarization of the vector mesons:

$$\begin{aligned} A_0 &= H_{00} \\ A_{||} &= \frac{1}{\sqrt{2}}\,(H_{++} + H_{--}) \\ A_\perp &= \frac{1}{\sqrt{2}}\,(H_{++} - H_{--}) \end{aligned} \tag{26}$$

The decay rate can be written as

$$\Gamma(B \to V_1 V_2) = \frac{p_{cm}}{8\pi m_B^2}\left[|\,H_{00}\,|^2 + |\,H_{++}\,|^2 + |\,H_{--}\,|^2\right] \tag{27}$$

Neglecting the annihilation contributions (which are expected to be small) to $B \to \phi K^*$, H_{00} and $H_{\pm\pm}$ are given by:

$$\begin{aligned} H_{00} &= \frac{G_F}{\sqrt{2}}\frac{a^n(\phi K^*) f_\phi}{2m_{K^*}}\Bigg\{ \left(m_B^2 - m_{K^*}^2 - m_\phi^2\right)\left(m_B + m_{K^*}\right) A_1^{BK^*}\left(m_\phi^2\right) \\ &\quad - \frac{4m_B^2 p_c^2}{m_B + m_{K^*}} A_2^{BK^*}\left(m_\phi^2\right)\Bigg\} \\ H_{\pm\pm} &= \frac{G_F}{\sqrt{2}} a^n(\phi K^*) m_\phi f_\phi \Bigg\{ \left(m_B + m_{K^*}\right) A_1^{BK^*}\left(m_\phi^2\right) \\ &\quad \mp \frac{2m_B p_c}{m_B + m_{K^*}} V^{BK^*}\left(m_\phi^2\right)\Bigg\} \end{aligned} \tag{28}$$

where, $a^n(\phi K^*) = a_3^n + a_4^n + a_5^n - (a_7^n + a_9^n + a_{10}^n)/2$. The effective parameters a_i appearing in the helicity amplitudes H_{00} and $H_{\pm\pm}$ given in [36, 23] and other input parameters are given in [23].

In Table 1, we display the experimentally (BaBar, CLEO and Belle) measured branching ratios and the weighted averaged values for the $B^+ \to \phi K^{*+}$ and $\bar{B}^0 \to \phi \bar{K}^{*0}$. The theoretical predictions in the SM for two different form factor models, the LCSR and BSW models are given in [36].

Table 1. Experimental data of $B \to \phi K^*$ decays from BaBar [39], CLEO [40] and Belle [41] and their weighted average

Branching ratio		Data	Weighted average
$B^+ \to \phi K^{*+}$	BaBar	$\left(12.1^{+2.1}_{-1.9} \pm 1.5\right) \times 10^{-6}$	
	CLEO	$\left(10.6^{+6.4+1.8}_{-4.9-1.6}\right) \times 10^{-6}$	$(9.9 \pm 1.23) \times 10^{-6}$
	Belle	$(9.4 \pm 1.1 \pm 0.7) \times 10^{-6}$	
$\bar{B}^0 \to \phi \bar{K}^{*0}$	BaBar	$\left(11.1^{+1.3}_{-1.2} \pm 1.1\right) \times 10^{-6}$	
	CLEO	$\left(11.5^{+4.5+1.8}_{-3.7-1.7}\right) \times 10^{-6}$	$(10.6 \pm 1.3) \times 10^{-6}$
	Belle	$\left(10^{+1.6+0.7}_{-1.5-0.8}\right) \times 10^{-6}$	

6.1 $LR(RL)$ Mixing Contributions to $B \to \phi K^*$

In this section we will study the effect of $LR(RL)$ mixing in $B \to \phi K^*$ process. This $LR(RL)$ mass insertion can enhance the Wilson coefficient C_{8g} by a factor of $m_{\tilde{g}}/m_b$ compared to the standard model contribution in the same way as shown in Sect. 5 for $B \to \phi K$ process. Hence, one need to impose the constrain on $LR(RL)$ mixing from experimentally measured $\mathcal{B}(B \to X_s\gamma)$ and also from $S_{\phi K}, C_{\phi K}$ and $\mathcal{B}(B \to \phi K)$ as obtained Sect. 5.

In this scenario, the new weak phase ψ, (the phase of the $LR(RL)$ mixing) will contribute to direct CP-violating asymmetry $\mathcal{A}_{CP}$ defined as :

$$\mathcal{A}_{CP} = \frac{\Gamma(B^+ \to \phi K^{*+}) - \Gamma(B^- \to \phi K^{*-})}{\Gamma(B^+ \to \phi K^{*+}) + \Gamma(B^- \to \phi K^{*-})} \tag{29}$$

in terms of partial widths. Recently BaBar and Belle Collaboration has presented their measurement of CP violating asymmetries for $B^0 \to \phi K^{*0}$ and $B^\pm \to \phi K^{*\pm}$ [39, 41]

$$\mathcal{A}_{CP}(\bar{B}^0 \to \phi \bar{K}^{*0}) = 0.04 \pm 0.12 \pm 0.02, \quad 0.07 \pm 0.15^{+0.05}_{-0.03} \tag{30}$$

$$\mathcal{A}_{CP}(B^\pm \to \phi K^{*\pm}) = +0.16 \pm 0.17 \pm 0.04, \quad -0.13 \pm 0.29^{+0.08}_{-0.11} \tag{31}$$

where, in each asymmetry result, the first number correspond to the BaBar data while the second one correspond to Belle measurement. The standard model value for this asymmetry is less than 1%. The new physics (SUSY) contributions from the new penguin operator appeared due to $LR(RL)$ mixing ($\delta^{bs}_{LR(RL)}$) can modify the sign and magnitude of $\mathcal{A}_{CP}(B^\pm \to \phi K^{*\pm})$ within the allowed parameter space of $\delta^{bs}_{LR(RL)}$.

To get the numerical values of $\mathcal{B}(B^+ \to \phi K^{*+})$, and $\mathcal{A}_{CP}(B^+ \to \phi K^{*+})$, we fix $x = 0.3$ and 4.0. Then for a given QCD scale μ, we select some points in the allowed parameter space of $\rho - \psi$ plane (as marked by Z in Fig. 1) for both values of x. In this computation, we include ($\pm 10\%$) theoretical uncertainties.

In Table 2, we present the branching ratio $\mathcal{B}(B^+ \to \phi K^{*+})$ and the CP rate asymmetry $\mathcal{A}(B^+ \to \phi K^{*+})$ for $\mu = m_b$ and selected values of $x = 0.3$ and 4.0 for values of ρ and ψ allowed by $B \to \phi K_S$ data for LR mass insertion (RL is shown

Table 2. $\mathcal{B}(B^+ \to \phi K^{*+})$ and $\mathcal{A}_{CP}(B^+ \to \phi K^{*+})$ at the QCD scale $\mu = m_b$ for LR mass insertion for selected points in the allowed $\rho - \psi$ space. The numbers in the parenthesis correspond to the RL mass insertion. The standard model branching ratio corresponding to this scale is $(6.18^{+1.29}_{-1.15}) \times 10^{-6}$. The errors are due to $\pm 10\%$ theoretical uncertainties in the calculation

x	(ρ, ψ)	$\mathcal{B}^{\rm SUSY}$ (in units of 10^{-6})	$\mathcal{A}_{CP}$ (in %)
	$(0.4 \times 10^{-2}$, -0.5)	$23.37^{+4.88}_{-4.42}$ $(21.76^{+4.56}_{-4.13})$	$-4.7(-4.4)$
0.3	$(0.4 \times 10^{-2}$, -0.7)	$21.50^{+4.49}_{-4.06}$ $(20.17^{+4.22}_{-3.83})$	-7.0 (-6.5)
	$(0.6 \times 10^{-2}$, 1.5)	$27.46^{+5.75}_{-5.2}$ $(26.82^{+5.62}_{-5.08})$	17.7 (15.7)
	$(2.4 \times 10^{-2}, -0.5)$	$24.33^{+5.08}_{-4.6}$ $(22.65^{+4.75}_{-4.3})$	-4.7 (-4.4)
4.0	$(2.6 \times 10^{-2}$, -0.45)	$26.98^{+5.6}_{-5.1}$ $(25.11^{+5.27}_{-4.76})$	-4.2 (-3.9)
	$(2.8 \times 10^{-2}$, -0.8)	$25.31^{+5.3}_{-4.8}$ $(23.87^{+5.01}_{-4.52})$	-8.0 (-7.5)

Table 3. $\mathcal{B}(B^+ \to \phi K^{*+})$ and $\mathcal{A}_{CP}(B^+ \to \phi K^{*+})$ at the QCD scale $\mu = m_b/2$ for LR mass insertion for selected points in the allowed $\rho-\psi$ space. The numbers in the parenthesis correspond to the RL mass insertion. The standard model branching ratio corresponding to this scale is $(14.92^{+3.08}_{-2.78}) \times 10^{-6}$. The errors are due to $\pm 10\%$ theoretical uncertainties in the calculation

x	(ρ, ψ)	$\mathcal{B}^{\rm SUSY}$ (in units of 10^{-6})	$\mathcal{A}_{CP}$ (in %)
	$(0.55 \times 10^{-2}, 1.8)$	$31.83^{+6.6}_{-6.0}$ $(32.45^{+6.7}_{-6.1})$	$19.39^{+0.01}_{-0.02}$ $(16.17^{+0.05}_{-0.07})$
0.3	$(0.82 \times 10^{-2}, 2.8)$	$14.50^{+3.04}_{-2.74}$ $(21.62^{+4.44}_{-4.02})$	20.2 $(10.96^{+0.06}_{-0.08})$
	$(0.82 \times 10^{-2}, 2.9)$	$12.39^{+2.59}_{-2.34}$ $(19.82^{+4.05}_{-3.67})$	15.73 $(8.04^{+0.05}_{-0.06})$

in the parenthesis). The branching ratio with SUSY turn out to be much higher than the standard model value of $6.18^{+1.29}_{-1.15}$, which is lower than the experimental data (Table 1). Even the lower range of theory prediction is much higher than the upper range of experimental data within 1σ. The rate asymmetry has much less error and is consistently within the range $\sim -4\%$ to $\sim 18\%$.

Similarly in Table 3, we show $\mathcal{B}$ and $\mathcal{A}_{CP}$ calculated for QCD scale $\mu = m_b/2$. In this case, there are two allowed regions from the combined $B \to \phi K$ and $B \to X_s\gamma$ constraints corresponding to $x = 0.3$. For $x = 4.0$, there are no allowed regions from $B \to \phi K$ data. The standard model branching ratio is much larger compared to the one computed at $\mu = m_b$. In SUSY, apart from the QCD scale μ, the branching ratio also depend on the values of ρ and ψ. Moreover, the selected points in $\rho - \psi$ plane are different in the two cases. In the case, with $\mu = m_b/2$, and at $\rho = 0.82 \times 10^{-2}$ and $\psi = 2.8, 2.9$ radians, with LR mixing, lower ranges of the theory predictions are consistent with the upper range of experimental data at one sigma. For the other value of ρ and ψ, the theoretical prediction for branching ratio is much higher than the standard model theory as well as experimental data. With RL mixing, the predicted branching ratio is much larger compared to both the standard model prediction and experimental data. The asymmetries for both LR and RL mixing case are always positive with less errors.

We conclude that for some selected points in $\rho - \psi$ plane allowed by $B \to \phi K$ and $B \to X_s\gamma$ at $\mu = m_b/2$ provide a satisfactory understanding of $B \to \phi K^*$ process. We also note that, at $\mu = m_b$ the SUSY contribution to the branching ratio of $B \to \phi K^*$ is too large to be consistent with the experimental data.

We have also studied other CP violating asymmetries that can arise in vector-vector final state. The set of observables are defined in terms of $A_0, A_{||}$ and $A_\perp$ as follows [25].

$$\begin{aligned}
\Lambda_\lambda &= \frac{|A_\lambda|^2 + |\bar{A}_\lambda|^2}{2}, \qquad \Sigma_{\lambda\lambda} = \frac{|A_\lambda|^2 - |\bar{A}_\lambda|^2}{2}, \\
\Lambda_{\perp i} &= -\,\mathrm{Im}\left(A_\perp A_i^* - \bar{A}_\perp \bar{A}_i^*\right), \qquad \Lambda_{||0} = \mathrm{Re}\left(A_{||}A_0^* + \bar{A}_{||}\bar{A}^*{}_0\right), \\
\Sigma_{\perp i} &= -\mathrm{Im}\left(A_\perp A_i^* + \bar{A}_\perp \bar{A}^*{}_i\right), \qquad \Sigma_{||0} = \mathrm{Re}\left(A_{||}A_0^* - \bar{A}_{||}\bar{A}^*{}_0\right), \\
\rho_{\perp i} &= \mathrm{Re}\left(\frac{q}{p}\left[A_\perp^* \bar{A}_i + A_i^* \bar{A}_\perp\right]\right), \qquad \rho_{\perp\perp} = \mathrm{Im}\left(\frac{q}{p} A_\perp^* \bar{A}_\perp\right), \\
\rho_{||0} &= -\mathrm{Im}\left(\frac{q}{p}\left[A_{||}^* \bar{A}_0 + A_0^* \bar{A}_{||}\right]\right), \qquad \rho_{ii} = -\mathrm{Im}\left(\frac{q}{p} A^* \bar{A}_i\right)
\end{aligned} \tag{32}$$

where $\lambda = \{0, \|, \perp\}$ and the observables where $i = \{0, \|\}$, We restrict ourselves to the study of helicity dependent CP asymmetry defined as $\Sigma_{\lambda\lambda}/\Lambda_{\lambda\lambda}$ [25]. For the purpose of illustration we select last two sample points from the Table 3. At these values of ρ and ψ, with LR mass insertion, the lower range of the $\mathcal{B}^{SUSY}$ is consistent with the upper range of the experimental data on $\mathcal{B}(B \to \phi K^*)$ at one sigma. We then compute $\Sigma_{\lambda\lambda}/\Lambda_{\lambda\lambda}$ for each values of λ for these two sets of ρ and ψ and is shown in Table 4. As before, in this case also we include $\pm 10\%$ theoretical uncertainties in our calculation. We only show the helicity dependent asymmetries for LR mass insertion, since with RL mass insertion the SUSY contribution to the branching ratio is too large to be consistent with the data.

Table 4. Helicity dependent CP asymmetry at the QCD scale $\mu = m_b/2$ for LR mass insertion for selected points in the allowed $\rho - \psi$ space. The errors consist of $\pm 10\%$ theoretical uncertainties

x	(ρ, ψ)	Σ_{00}/Λ_{00}	$\Sigma_{\|\|}/\Lambda_{\|\|}$	$\Sigma_{\perp\perp}/\Lambda_{\perp\perp}$
0.3	$(0.82 \times 10^{-2}, 2.8)$	0.19 ± 0.00	$0.61^{+0.009}_{-0.012}$	$0.57^{+.011}_{-0.013}$
	$(0.82 \times 10^{-2}, 2.9)$	0.15 ± 0.00	$0.71^{+0.020}_{-0.026}$	$0.65^{+0.022}_{-0.027}$

7 Conclusions

In this talk, we considered the SUSY contribution to the gluonic dipole moment operator to $B \to \phi K_S$ process. We found that the $LR(RL)$ mass insertion can enhance the gluonic dipole moment operator significantly. We then used the experimentally measured quantities, such as $S_{\phi K}, C_{\phi K}$ and $\mathcal{B}(B \to \phi K_S)$ to constrain the parameter space of $LR(RL)$ mixing. Interestingly, we find that the constraints from $B \to \phi K$ data is consistent with the $B \to X_s\gamma$ limit. It turned out that the same enhancement of gluonic dipole moment operator could also affect other penguin dominated process, such as $B \to \phi K^*$, which is a pure penguin process like $B \to \phi K_S$. In standard model, the predicted $\mathcal{A}_{CP}(B \to \phi K^*)$ is less than 1%. We calculated such asymmetries and also the branching ratio for the set of parameters allowed by $B \to \phi K$ data. At $\mu = m_b$, for both LR and RL mass insertion, we observed that the predicted branching ratio is well above the experimentally measured one. On the other hand, at the QCD scale $\mu = m_b/2$ with LR mass insertion, we found that the theoretically computed branching ratio is consistent with the data with in one sigma error. At this second choice of allowed parameter space of $\delta^{bs}_{LR(RL)}$, we found $\mathcal{A}_{CP}(B^+ \to \phi K^{*+})$ in the range 15% to 20%, which is significantly higher than the standard model prediction but is still consistent with the present data. Finally, we also presented helicity dependent CP asymmetries in the same parameter space of δ^{bs}_{LR}.

Acknowledgements

This work was supported in part by US DOE contract numbers DE-FG03-96ER40969.

References

1. B. Aubert et al. [BaBar Collaboration], arXiv:hep-ex/0207070.
2. K. Abe et al. [Belle Collaboration], arXiv:hep-ex/0207098
3. K. Abe et al. [Belle Collaboration], Phys. Rev. D **67**, 031102 (2003).
4. G. Hamel De Monchenault, arXiv:hep-ex/0305055.
5. R. Barbieri and A. Strumia, Nucl. Phys. **B508**, 3 (1997).
6. A. Kagan, arXiv:hep-ph/9806266; talk at the 2nd International Workshop on B physics and CP Violation, Taipei, 2002; SLAC Summer Institute on Particle Physics, August 2002.
7. G. Hiller, Phys. Rev. D **66**, 071502 (2002).
8. A. Datta, Phys. Rev. D **66**, 071702 (2002).
9. M.Ciuchini and L. Silvestrini, Phys. Rev. Lett. **89**, 231802 (2002);
10. R. Harnik, D. T. Larson, H. Murayama and A. Pierce, arXiv:hep-ph/0212180.
11. M. Ciuchini, E. Franco, A. Masiero and L. Silvestrini, Phys. Rev. D **67**, 075016 (2003) [Erratum-idib. D **68**, 079901 (2003)].
12. B. Dutta, C.S. Kim and S. Oh, Phys. Rev. Lett. **90**, 011801 (2003).
13. S. Khalil and E. Kou, Phys. Rev. D **67**, 055009 (2003).
14. C.W. Chiang and J.L. Rosner, Phys. Rev. D **68**, 014007 (2003).
15. A. Kundu and T. Mitra, Phys. Rev. D **67**, 116005 (2003).
16. K. Agashe and C.D. Carone, Phys. Rev. D **68**, 035017 (2003).
17. G.L. Kane, P. Ko, H. b. Wang, C. Kolda, J.h. Park and L.T. Wang, Phys. Rev. Lett. **90**, 141803 (2003).
18. D. Chakraverty, E. Gabrielli, K. Huitu and S. Khalil, Phys. Rev. D **68**, 095004 (2003).
19. J.F. Cheng, C.S. Huang and X.h. Wu, arXiv: hep-ph/0306086.
20. R. Arnowitt, B. Dutta and B. Hu, Phys. Rev. D **68**, 075008 (2003).
21. K. Abe et al. [Belle Collaboration], arXiv:hep-ex/0308035; T. Browder, *CKM Phases (β/ϕ_1)*, Talk presented at Lepton-Photon 2003.
22. A.J. Buras, arXiv:hep-ph/0210291.
23. C. Dariescu, M.A. Dariescu, N.G. Deshpande and D. K. Ghosh, arXiv: hep-ph/0308305.
24. M. Beneke, G. Buchalla, M. Neubert and C.T. Sachrajda, Phys. Rev. Lett. **83**, 1914 (1999); Nucl. Phys. **B591**, 313 (2000); Nucl. Phys. **B606**, 245 (2001).
25. D. London, N. Sinha, and R. Sinha, Phys. Rev. Lett. **85**, 1807 (2000).
26. R.A. Briere et al. , [CLEO Collaboration], Phys. Rev. Lett. **86**, 3718 (2002).
27. X.-G. He, J.P. Ma and C.-Y. Wu, Phys. Rev. D **63**, 094004 (2001);
28. C.-S. Huang and S. Zhu, Phys. Rev. D **68**, 114020 (2003).
29. L.J. Hall, V.A.Kostelecky and S. Raby, Nucl. Phys. **B267**, 415 (1986).
30. F. Gabbiani, E. Gabrielli, A. Masiero and L. Silvestrini, Nucl. Phys. **B477**, 321 (1996).
31. M. B. Causse and J. Orloff, Eur. Phys. J. **C23**,749 (2002); S. Khalil and E. Kou in Ref.[5].

32. A.J. Buras, G. Colangelo, G. Isidori, A. Romanino, and L. Silvestrini, Nucl. Phys. **B566**, 3 (2000).
33. X.-G. He, J.-Y. Leou and J.-Q. Shi, Phys. Rev. D **64**, 094018 (2001).
34. A. Ali, G. Kramer and Cai-Dian Lu, Phys. Rev. D **58**, 094009 (1998).
35. A. Stocchi, Phys. Proc. Suppl. **117**, 145 (2003).
36. H.-Y. Cheng and K.-C. Yang, Phys. Lett. **B511**, 40 (2001).
37. P. Ball and V.M. Braun, Phys. Rev. D **58**, 094016 (1998); P. Ball, J. High Energy Phys. **09**, 005 (1998).
38. M. Wirbel, B. Stech and M. Bauer, Z. Phys. **C29**, 637 (1985); M. Bauer, B. Stech and M. Wirbel, Z. Phys. **C34**, 103 (1987).
39. B. Aubert et al., [BaBar Collaboration], arXiv: hep-ex/0303020.
40. R.A. Briere et al., [CLEO Collaboration], Phys. Rev. Lett. **86**, 3718 (2001).
41. K.-F. Chen et al., [Belle Collaboration], Phys. Rev. Lett. **91**, 201801 (2003).

LHC Physics

Fabiola Gianotti

CERN, PH Division, 1211 Genève 23, Switzerland
fabiola.gianotti@cern.ch

1 Introduction

The Large Hadron Collider (LHC, [1]) is a machine presently under construction at CERN, which will provide pp collisions at the unprecedented centre-of-mass energy of $\sqrt{s} = 14\,\text{TeV}$ and luminosity of $L \sim 10^{34}\,\text{cm}^{-2}\,\text{s}^{-1}$. It will also provide heavy ion collisions, for instance lead-lead collisions, at a centre-of-mass energy of about 1000 TeV. Data taking should start in April 2007.

The machine, which will be installed in the 27 km ring previously used for the LEP e^+e^- collider, is technologically very challenging. Achieving a beam energy of 7 TeV in a 27 km ring requires the bending power of about 1200 superconducting dipole magnets providing a field of 8.3 T. The construction is progressing well. At the time of writing this paper, about 180 dipoles had been delivered at CERN.

Four main experiments will take data at the LHC: two general-purpose detectors, ATLAS [2] and CMS [3], which have a very broad physics programme; an experiment, LHCb [4], dedicated to the study of B-hadrons and CP violation; an experiment, ALICE [5], which will study ion-ion and p-ion physics. In addition, the TOTEM detector [6], integrated with the CMS experiment, will measure the total pp cross-section. Here only some physics topics addressed by ATLAS, CMS and TOTEM are discussed.

2 Motivations for the LHC

It is well known that the Standard Model (SM), the theory which describes the elementary particles and their interactions, has been verified with spectacular accuracy, i.e. to 0.1% or better in most cases [7], by experiments performed at various machines (e.g. at the CERN LEP collider).

It is also known, however, that one essential piece is missing, the Higgs boson, which in the SM is needed to account for the particles masses and for which there is no experimental evidence so far.

In addition, particle physicists have good reasons to believe that new (and presently unknown) physics exists beyond the Standard Model. Figure 1 shows the energy scale Λ up to which the SM is valid, as a function of the Higgs mass. It can be seen that if the Higgs mass is around 115 GeV, i.e. just above the present experimental lower limit from direct searches at LEP, the SM works only up to an energy scale of $\sim 10^6$ GeV, which means that new physics should appear at this scale or below. If, on the other hand, the Higgs mass is in the range $\sim$130–180 GeV,

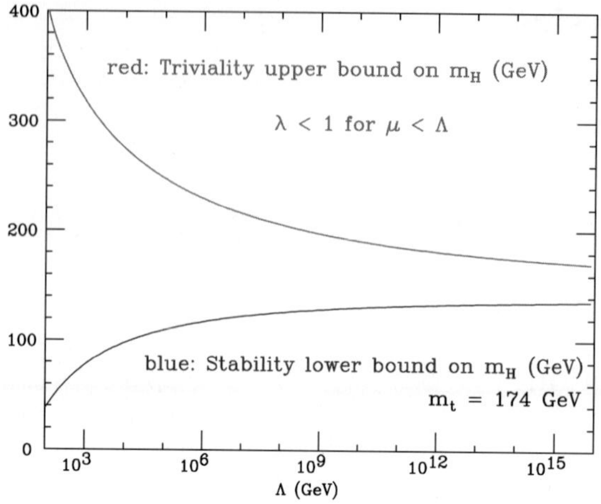

Fig. 1. Theoretical bounds on the Higgs boson mass (in GeV) as a function of the energy scale (Λ) up to which the SM works. The region above the top curve and the region below the bottom curve are theoretically forbidden

then the Standard Model technically works up to the Planck scale. This scenario, not particularly attractive from the experimental point of view, introduces two problems: the so-called "hierarchy problem", i.e. the huge difference between the electroweak scale (~100 GeV) and the Planck scale (~10^{19} GeV); and the so-called "naturalness problem", i.e. the fact that in the SM the Higgs mass increases as Λ^2, and some tricks are needed to keep it down at the electroweak scale.

In addition, there are many other questions of fundamental importance, such as the origin of three fermion families with apparently random mass patterns, the size of the cosmological constant (e.g. the predicted contribution of the Higgs field is ~55 orders of magnitude too large), etc., to which the SM is unable to give fully satisfactory answers.

The above arguments indicate that the Standard Model is most likely not the ultimate theory of particle interactions, but rather an approximation, valid at the energies tested so far, of a more fundamental theory. There are today several candidate scenarios for physics beyond the Standard Model, among which Supersymmetry, Technicolour and theories with Extra-dimensions. All of them predict that something new, new particles or new phenomena, should manifest at the TeV scale. This calls for a machine able to explore directly and in detail the TeV energy scale. This machine is the LHC.

3 The ATLAS and CMS Experiments

Since we don't know how new physics will manifest, the two multi-purpose experiments ATLAS and CMS must detect and measure all known particles (charged leptons, neutrinos, jets coming from the fragmentation of quarks and gluons, photons, etc.) over a broad energy range (from a few GeV to a few TeV). In addition, they must provide excellent performance in terms of energy resolution and particle identification capabilities, in order to extract possible signals of new physics (which may be very tiny) from the huge background of known processes.

ATLAS and CMS have the typical structure of any collider experiment (see the left picture in Fig. 2). From the interaction point outward, they consist of the following sub-detectors:

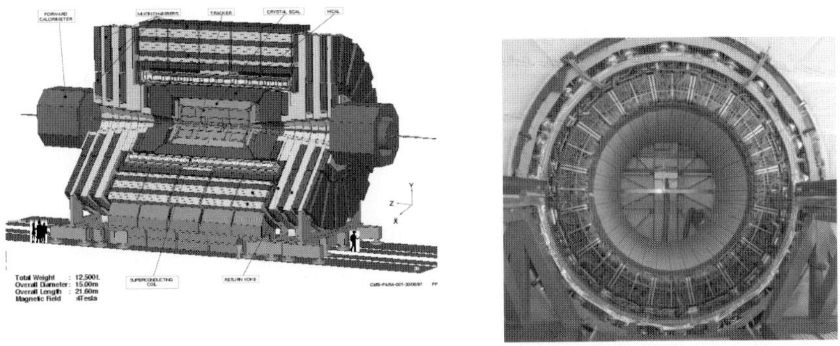

Fig. 2. *Left*: Schematic view of the CMS detector. *Right*: Half of the ATLAS barrel liquid-argon electromagnetic calorimeter installed inside its cryostat

- An inner tracker immersed in a magnetic field (B = 2 T in ATLAS and B = 4 T in CMS), used to detect charged particles and to measure their momenta and charge, as well as to tag secondary vertices (e.g. from b-quark decays). Layers of Silicon pixels and strips are used in both ATLAS and CMS. ATLAS has in addition a Transition Radiation Detector for particle identification (e/π separation).
- An electromagnetic calorimeter, which measures the energy and position of electrons and photons and contributes to their identification. It also participates to the measurement of hadronic jets and of the event missing transverse energy (see below). CMS uses a homogeneous crystal calorimeter, while ATLAS has a lead-liquid argon sampling calorimeter.
- Hadron calorimeters, devoted to the measurement of the energy and position of hadrons and jets, as well as to the measurement of the total event missing transverse energy (used to detect indirectly the presence of neutrinos in the final state). In ATLAS, a Fe-scintillator calorimeter has been chosen for the central part and Cu-liquid argon and W-liquid argon calorimeters for the forward part. CMS uses a brass-scintillator sampling calorimeter over most of the coverage, complemented by a Fe-quartz detector in the very forward regions.
- An external muon spectrometer, which identifies muons, provides the muon trigger, and measures the muon momentum (together with the inner detector). The ATLAS muon chambers are immersed in a toroidal magnetic field in air, whereas the CMS muon chambers are embedded in the iron of the return joke of the solenoid which produces the 4 T field in the inner cavity.

ATLAS has a length of about 45 m and a diameter of about 25 m. CMS, more compact, has a length of about 20 m and a diameter of about 15 m.

The drawback of operating at a luminosity of $10^{34}\,\mathrm{cm}^{-2}\,\mathrm{s}^{-1}$, which is almost a factor 500 higher than that achieved so far at the Tevatron collider, and which is needed to observe rare physics processes, is the harsh experimental environment (pp inelastic interactions are expected to occur at a rate of $\sim 10^9$ per second). As a consequence, compared to previous collider experiments, the ATLAS and CMS detectors must have a fast response time, typically $\lesssim 50$ ns, and a fine readout granularity (which implies a large number of electronic channels, $\sim 10^8$), and must be very radiation hard.

The construction of both detectors is progressing well and several components are already completed (an example is shown in the right picture of Fig. 2). In addition, many beam tests of sub-detector prototypes or production modules have been performed and have shown performances at the level required by the specifications.

4 Examples of the LHC Physics Potential

Thanks to its high energy and luminosity, the LHC will be a factory of all particles with masses at the TeV scale or below. Table 1 shows, for some representative processes both from known and new physics, the expected event rates in the initial phase of the LHC operation, when the machine luminosity will be a factor of ten lower than the design luminosity. Even in these less favourable conditions, millions of events should be collected in many physics channels in only one year of data taking, a statistics much larger than the total event samples accumulated by previous colliders over their whole life.

As a consequence, the mass reach for the discovery of new heavy particles extends up to masses of ~ 5 TeV. Some examples, namely the discovery potential for the SM Higgs boson, supersymmetric particles and extra-dimensions, are presented in Sects. 4.1–4.3 (see [2, 3] for a more complete discussion of the LHC physics programme). The possible impact of these studies on astroparticle physics and on our knowledge of the universe is mentioned, and an additional relevant example is discussed in Sect. 4.4.

Table 1. For the physics channels listed in the first column, the expected numbers of events at production in each experiment (ATLAS or CMS) per second (second column) and in one year of data taking (third column), for a luminosity $L = 10^{33}\,\mathrm{cm}^{-2}\mathrm{s}^{-1}$. The total event samples collected (or expected to be collected) at colliders operating before the LHC start-up are given in the last column

Process	Evts/s	Evts/year	Total samples from previous colliders
$\mathrm{W} \to e\nu$	15	10^8	10^4 events LEP, 10^7 events Tevatron
$\mathrm{Z} \to \mathrm{ee}$	1.5	10^7	10^7 events LEP
$t\bar{t}$	1	10^7	10^4 events Tevatron
$b\bar{b}$	10^6	$10^{12} - 10^{13}$	10^9 events BaBar, Belle
$\tilde{\mathrm{g}}\tilde{\mathrm{g}}$ $m = 1$ TeV	0.001	10^4	–
Higgs $m_H = 130$ GeV	0.02	10^5	–

4.1 The Standard Model Higgs boson

Our present knowledge about this elusive particle can be summarised as follows. The value of the Higgs mass m_H is not specified by the theory, which gives only an upper bound of ~1 TeV. Direct searches performed at LEP have set a mass lower limit of 114.4 GeV [8]. In addition, in the year 2000, the last year of LEP operation, a few events with features compatible with those expected from the production of a Higgs boson of mass ~115 GeV have been observed, but the size of the effect [8] is not large enough to claim a discovery. Finally, a fit of the Standard Model to the ensemble of data collected by various machines (e.g. the LEP, Tevatron and SLC colliders), with the Higgs mass as a free parameter, gives an upper bound on m_H of about 200 GeV [7]. Although this is an indirect limit, it nevertheless indicates that the present data favour a light Higgs boson.

On the other hand, the Higgs decay modes are known, because the SM predicts that the Higgs particle couples to fermions and bosons with strength proportional to their masses. Therefore, for $m_H < 120$ GeV the Higgs boson is expected to decay mainly into $b\bar{b}$ pairs, since b-quarks are the most massive particles which are accessible in this m_H range. For larger Higgs masses, decays into heavier particles, like W pairs and Z pairs, open and dominate.

At the LHC a SM Higgs boson can be detected in a large variety of channels. Among the best ones are $H \to \gamma\gamma$ decays, which are observable for masses below ~150 GeV, and $H \to ZZ \to 4\lambda$ (where $\lambda = e, \mu$) decays, which are relevant in the mass range 130–800 GeV.

The left panel in Fig. 3 summarises the ATLAS discovery potential (CMS has a similar reach) in terms of signal significance. The latter is defined as the number of expected signal events divided by the square root of the number of background events. To claim a discovery, a signal significance of at least 5σ is required. It can be seen that a SM Higgs boson can be discovered over the full allowed mass range up to 1 TeV with only 10 fb^{-1} of integrated luminosity, which corresponds to (in principle) only one year of LHC operation (in practice, a lot of time will be needed at the beginning to understand and calibrate complex detectors like ATLAS and CMS, and extracting a convincing Higgs signal will most likely take more than one year).

The main conclusion is that the LHC should be able to say the final word about the SM Higgs mechanism, i.e. if nothing is found this mechanism is wrong. If, on the other hand, the Higgs boson will be observed, then ATLAS and CMS should also be able to perform first precise measurements of the Higgs properties, e.g. measure m_H to the remarkable ultimate precision of 0.1% (see the right panel in Fig. 3). This should provide further insight to the origin of particle masses and allow powerful tests of the SM internal consistency.

4.2 Supersymmetry

Supersymmetry (SUSY, [9]) is an appealing symmetry relating fermions and bosons, i.e. matter fields and force fields. Only a brief and simplified phenomenological introduction is given here.

Supersymmetry states that for each SM particle p there exists a supersymmetric partner $\tilde{p}$, with identical couplings and quantum numbers except the spin which

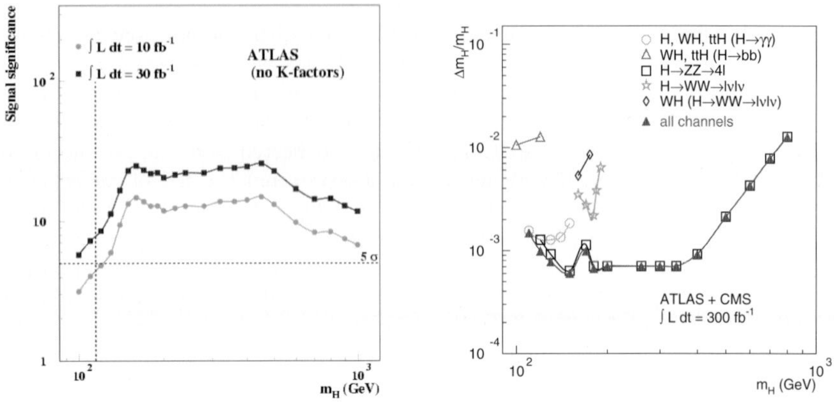

Fig. 3. *Left*: The expected signal significance for a SM Higgs boson in ATLAS as a function of mass, for integrated luminosities of 10 fb^{-1} (*dots*) and 30 fb^{-1} (*squares*). The *vertical line* shows the lower limit from LEP. The *horizontal line* indicates the minimum significance (5σ) needed for discovery. *Right*: The expected experimental precision on the measurement of the SM Higgs mass at the LHC, as a function of mass. The different symbols indicate the various detection channels. An integrated luminosity of 300 fb^{-1} per experiment has been assumed

differs by half a unit. The SUSY particles (sparticles) predicted by minimal SUSY models, such as the MSSM [9], are listed in Table 2. In addition, there are five Higgs bosons, h, H, A, $H^{\pm}$.

Table 2. Standard Model particles and their supersymmetric partners in the MSSM

SM particles	SUSY partners	Spin of SUSY partners
quarks	squarks $\tilde{q}$	0
leptons	sleptons $\tilde{\lambda}$	0
gluon	gluino $\tilde{g}$	1/2
$W^{\pm}, H^{\pm}$	charginos $\chi^{\pm}_{1,2}$	1/2
Z, γ, H	neutralinos $\chi^{0}_{1,2,3,4}$	1/2

Important phenomenological consequences come from the fact that the theory contains a multiplicative quantum number, called R-parity, with values +1 for ordinary particles and −1 for SUSY particles. If R-parity is conserved, as it is assumed here, then sparticles are produced in pairs and the Lightest Superymmetric Particle (LSP), to which all sparticles eventually decay, must be stable. In most models the LSP is the lightest neutralino χ^0_1, which is a stable, massive and weakly-interacting particle, and therefore an excellent candidate for the universe cold dark matter.

Despite the numerous motivations in favour of Supersymmetry, no experimental evidence for this theory has been found as of today. Direct searches for SUSY particles performed at LEP and Tevatron have been unsuccessful, and have been able to set lower bounds on the sparticle masses in the range 90–300 GeV.

In addition, the LEP experiments have obtained an indirect lower limit on the mass of the lightest neutralino χ_1^0, which has cosmological implications if this particle is the cold dark matter constituent. The LEP limit is $m(\chi_1^0) > 46$ GeV at the 95% C.L. [10]. It can be compared to the results of direct searches for cold dark matter performed by experiments looking for neutralinos, or more generally for Weakly-Interacting Massive Particles (WIMP's), coming from the galactic halo and interacting with the detectors via WIMP-nuclei scattering. Figure 4 shows the region of the plane of the WIMP mass versus WIMP cross-section favoured by the DAMA experiment at Gran Sasso if the observed annual modulation in the rate of nuclear recoils [11] is attributed to galactic neutralinos. Also shown are the regions excluded by other experiments of similar scope. It can be seen that the LEP limit $m(\chi_1^0) > 46$ GeV is complementary to these results because it does not depend on the neutralino-nucleon cross-section, and therefore is able to exclude the small cross-section region of the plane where direct searches have no sensitivity.

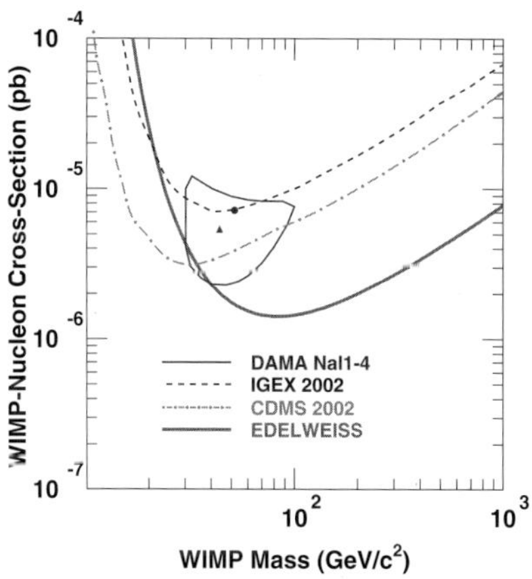

Fig. 4. The regions of the plane of the WIMP mass versus WIMP-nucleon spin-independent cross-section favoured at the 3σ level by the DAMA experiment (closed contour, [11]), or excluded at the 90% C.L. by the EDELWEISS experiment (*solid line*, [12]), by the CDMS experiment (*dash-dotted line*, [13]), and by the IGEX experiment (*dashed line*, [14]). From [12]

Because of the present experimental limits, SUSY particles, if they exist, must be relatively heavy. They cannot however be too heavy, because Supersymmetry can solve the naturalness problem mentioned in Sect. 2 only if the sparticle masses are at the TeV scale or below. Hence Supersymmetry should not escape discovery at the LHC.

At this machine, the dominant SUSY process is expected to be the production of pairs of squarks or gluinos, e.g. $q\bar{q} \to \tilde{g}\tilde{g}$, because these are strongly-interacting particles and therefore have huge cross-sections. For instance, a sample of about 10^4

$\tilde{q}\tilde{q}$, $\tilde{g}\tilde{g}$ and $\tilde{q}\tilde{g}$ events should be produced in only one year of data taking at the initial luminosity of $10^{33}\,\mathrm{cm}^{-2}\,\mathrm{s}^{-1}$ if squarks and gluinos have masses of $\sim$1 TeV. These sparticles are expected to decay through long chains with several intermediate steps, and hence to give rise to crowded final states with many jets, leptons and missing energy (the latter is due to the fact that the lightest neutralino is neutral and weakly-interacting, and therefore escapes experimental detection). Such spectacular signatures can be easily recognised from SM processes. As an example, Fig. 5 shows, for events of the type $pp \to \tilde{q}_L\tilde{g}$ with $\tilde{q}_L \to q\chi_2^0 \to q\lambda^+\lambda^-\chi_1^0$, the reconstructed invariant mass of the two leptons in the final state. In this case the squark and gluino masses are about 400 GeV. A striking SUSY signal should be visible, on top of a negligible background from SM processes, after only a few months of data taking.

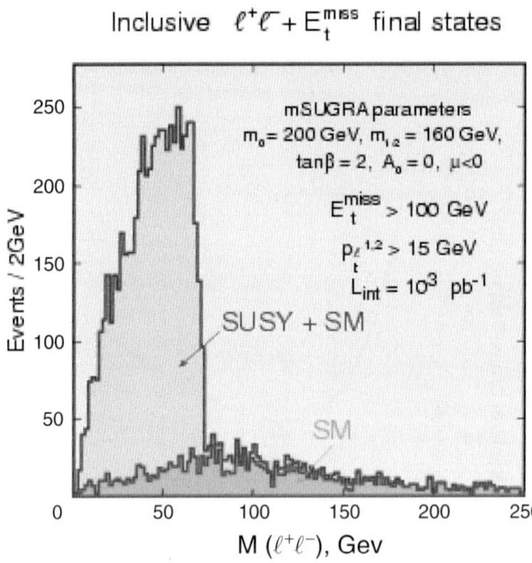

Fig. 5. The reconstructed invariant mass of pairs of opposite-sign same-flavour leptons (e^+e^-, $\mu^+\mu^-$) for a sample of simulated SUSY events of the type described in the text, as expected in the CMS experiment [15] for an integrated luminosity of $10^3\,\mathrm{pb}^{-1}$

Thus, SUSY discovery at the LHC should be relatively fast and easy. The ultimate mass reach of this machine is up to $\sim$3 TeV for squarks and gluinos. Therefore if nothing is found, TeV-scale Supersymmetry will most likely be ruled out, for the reasons mentioned above.

On the other hand, if SUSY is there, ATLAS and CMS should be able to perform several precise measurements of the sparticle masses, and thus determine the fundamental parameters of the theory (at least for minimal models like mSUGRA [9]) with a precision of $\sim$10% or better in many cases.

An important question is whether it will be possible to measure the mass of the lightest neutralino, the dark matter candidate. This particle is not directly observable. However its mass can be constrained indirectly by performing several

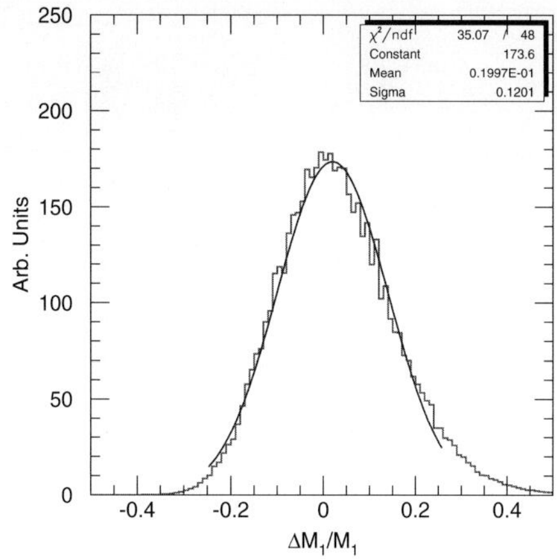

Fig. 6. The distribution of the fractional difference between the reconstructed and the true χ_1^0 mass (here called M1) at one point of the SUSY parameter space (see text), from an ATLAS simulation [2]

measurements of mass spectra and kinematic distributions for the visible sparticles in the final state (like the one shown in Fig. 5). An example is presented in Fig. 6 for a point in the mSUGRA parameter space (the so-called "LHC Point 5" [2]) where $m(\tilde{q}) \simeq 700\,\mathrm{GeV}$, $m(\tilde{g}) \simeq 800\,\mathrm{GeV}$ and $m(\chi_1^0) \simeq 120\,\mathrm{GeV}$. In this case, the LSP mass can be determined indirectly with a precision of $\sim 10\%$ from various kinematic measurements of the two main decay chains of left-handed squarks: $\tilde{q}_L \to q\chi_2^0 \to qh\chi_1^0$ and $\tilde{q}_L \to q\chi_2^0 \to q\lambda^+\lambda^-\chi_1^0$. Furthermore, a fit of the mSUGRA model to the ensemble of experimental measurements should provide a measurement of the density of the universe cold dark matter (assuming it is composed only of relic neutralinos). The expected precision is ~2% at Point 5. Then, comparisons of these and other results to astroparticle measurements (like the recent WMAP [16] results) and cosmological predictions should tell whether the features of the neutralinos measured at the LHC are compatible with this particle being the cold dark matter constituent.

A rich phenomenology is also expected from the SUSY Higgs sector which consists, as already mentioned, of five particles, three neutral (h, H, A) and two charged ($H^{\pm}$) bosons. The mass of the lightest one, h, is predicted to be below 135 GeV, whereas the others are expected to be heavier and essentially mass-degenerate in most models.

The left panel in Fig. 7 shows the parameter space of the minimal SUSY Higgs sector, which is usually described in terms of the mass of one of the Higgs bosons (m_A) and of the parameter tanβ (the ratio of the vacuum expectation values of the two Higgs doublets). The regions where the various Higgs bosons can be discovered at the LHC (at the 5σ level or more) through their decays into SM particles are indicated. It can be seen that an integrated luminosity of only $10\,\mathrm{fb}^{-1}$, collected with well-understood detectors, should allow exploration of most of the plane. Furthermore, as the right panel in Fig. 7 shows, with more luminosity the LHC experiments should be able to detect two or more SUSY Higgs bosons over a large part of the

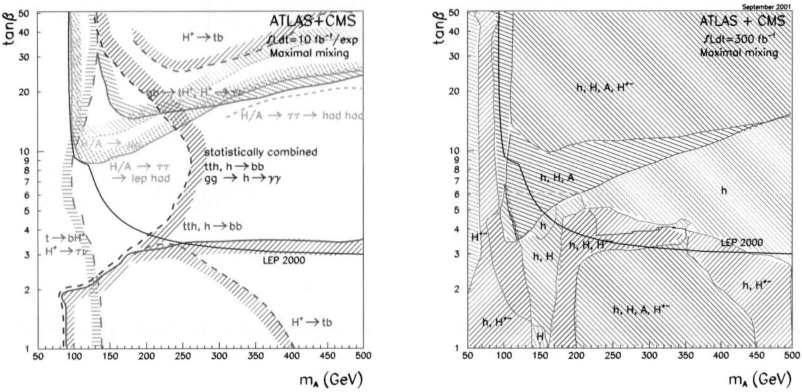

Fig. 7. The regions of the minimal SUSY plane m_A-tanβ where the various SUSY Higgs bosons can be discovered at $\geq 5\sigma$ at the LHC through their decays into SM particles. The left panel (which shows some of the decay modes) is for an integrated luminosity of $10\,\text{fb}^{-1}$ per experiment, the right panel for $300\,\text{fb}^{-1}$ per experiment

parameter space. The exception is the region at large m_A and moderate tanβ, where only h can be discovered at the LHC, unless the heavier Higgs bosons have observable decays into SUSY particles (e.g. $A/H \to \chi_2^0\chi_2^0 \to 4\lambda$). As a consequence the LHC, although very powerful, may miss part of the SUSY Higgs spectrum (observation of the complete spectrum may require a multi-TeV Linear Collider).

ATLAS and CMS should also be able to measure several features of the observed Higgs bosons, and hence constrain the underlying parameters of the theory. As an example, tanβ can be determined with an ultimate precision of a few percent by measuring the production rates of the $A/H \to \tau^+\tau^-, \mu^+\mu^-$ channels, which have a strong tanβ-dependence.

4.3 Theories with Large Extra-Dimensions

One of the aims of these theories [17] is to solve the hierarchy problem mentioned in Sect. 2 by lowering the fundamental scale of gravity from the Planck scale to ~1 TeV. In this way the electroweak and gravity scales become similar. This is possible without contradicting Newton's laws and experimental observations if there exist n additional dimensions, and gravity propagates in the $4+n$-dimensional world whereas the Standard Model is confined to a 4-dimensional wall.

One of the expected signatures of Extra-dimensions at hadron colliders is the associated production of a quark or gluon with gravitons, leading to final states consisting of a high-p_T jet and large missing transverse energy (gravitons couple very weakly with matter and therefore escape detection). Using this signature, the LHC experiments should be able to explore up to 4 additional dimensions and gravity scales up to 9 TeV.

Another manifestation may be the production of mini black holes. The theory and the LHC potential for this physics have not been worked out in detail yet.

However, since this is an intriguing topic, and may have important implications for our understanding of the universe, it is briefly mentioned here.

Mini black holes can be produced according to theories with large Extra-dimensions for the following reason. Because the gravity scale is of order 1 TeV, the black hole Schwarzschild radius, that is the radius from within which nothing escapes attraction (not even light), is much larger than in the standard case with the gravity scale at the Planck energies. Therefore mini black holes with masses of only a few TeV can exist and be produced. This can happen if two colliding partons (e.g. two quarks) interact with a centre-of-mass energy of the order of the black hole mass and at a distance smaller than the Schwarzschild radius. The cross-section is expected to be huge, and the LHC may become a black hole factory. These mini black holes decay immediately (fortunately) by Hawking radiation, which is a democratic evaporation into all elementary particles. The expected experimental signature is quite spectacular: high-multiplicity final states, with a lot of jet activity and little missing energy.

Mini black holes may also be produced by energetic cosmic neutrinos and detected by the Auger observatory [18]. In parallel the LHC should be able to produce these objects in the lab and measure their mass (from the invariant mass of the final-state products) and the Hawking temperature (from the energy spectra of leptons and photons). This should allow extraction of the fundamental parameters of the theory, namely the gravity scale and the number of extra-dimensions.

4.4 Measurement of the *pp* Cross-Section

As a further example of the links with astroparticle physics, LHC physics studies which can have an impact on the understanding of high-energy cosmic rays are briefly discussed in this section.

If the LHC is converted into a fixed-target accelerator, its centre-of-mass energy is equivalent to an incident proton beam energy of ~100 PeV. This is almost two orders of magnitude larger than the equivalent beam energy at today's highest-energy machine, the Fermilab Tevatron $p\bar{p}$ collider, where $\sqrt{s} = 2\,\mathrm{TeV}$. It can be seen that the LHC will be the first machine able to probe the energy range beyond the "knee" structure of the cosmic ray spectrum.

Obviously, particularly relevant to high-energy cosmic rays will be the studies of the most energetic particles produced in the *pp* collisions (and comparisons with Monte Carlo hadronic interaction models) and the measurements of the *pp*, *pA* and *AA* cross-sections. This requires particle detection in the forward regions. The two general-purpose experiments ATLAS and CMS cover the central rapidity region $|\eta| \leq 5$, because they have been designed for the study of high-p_T processes. However, in the vast majority of inelastic *pp* collisions, which are low-p_T processes, most of the event energy (more than 90%, amounting to several TeV) is taken by only a few particles emitted in the very forward regions. This forward part of the phase space, which is most relevant to high-energy cosmic rays, will be explored by the TOTEM experiment and possibly also by extensions to the ATLAS and CMS detectors (presently under discussion).

The main goal of the TOTEM experiment is the measurement of the total *pp* cross-section. At the highest energies, above 2 TeV, only cosmic ray measurements exist, which are affected by large normalization uncertainties (>10%).

The TOTEM experiment [6], should provide a precise measurement at $\sqrt{s} = 14\,\text{TeV}$ with an expected error of only ∼1%. TOTEM will likely consist of four stations of Roman pots, located at 120–400 m from the interaction centre, which will be used to detect protons from elastic *pp* scattering down to angles of only 20 μrad.

5 Conclusions

The LHC has very compelling and ambitious physics goals. It will explore in detail the strongly-motivated TeV scale, with a direct discovery potential for new particles up to masses of ∼5 TeV. It should therefore say the final word about the SM Higgs mechanism, Supersymmetry, and other TeV-scale predictions. In addition, the LHC experiments will be able to perform many precise measurements (e.g. of the W and top masses, of CP-violation) with unprecedented accuracy, and perform several studies in an energy range overlapping with the high-energy part of the cosmic rays spectrum. There is also a very interesting heavy-ion programme, addressing in particular the study of the quark-gluon plasma.

The LHC should therefore add many crucial pieces to our knowledge of fundamental physics, and consequently have also a big impact on our understanding of the universe (e.g. the dark matter composition, the matter-antimatter asymmetry). These goals are possible thanks to a machine and experiments of unprecedented performance and complexity.

References

1. The LHC Study Group: *The Large Hadron Collider conceptual design*, CERN/AC/95-05, 1995.
2. ATLAS Collaboration: *Detector and physics performance Technical Design Report*, CERN/LHCC/99-15, 1999.
3. CMS Collaboration: *The Compact Muon Solenoid, Technical Proposal*, CERN/LHCC/94-38, 1994.
4. LHCb Collaboration: *A Large Hadron Collider Beauty experiment for precision measurements of CP-violation and rare decays Technical Proposal*, CERN/LHCC/98-4, 1998.
5. ALICE Collaboration: *A Large Ion Collider Experiment, Technical Proposal*, CERN/LHCC/95-71, 1995.
6. TOTEM Collaboration: *Total cross-section, elastic scattering and diffraction dissociation at the LHC, Technical Proposal*, CERN/LHCC 99-7, 1999.
7. LEP Collaborations: *A combination of preliminary electroweak measurements and constraints on the Standard Model*, hep-ex/0312023, http://lepewwg.web.cern.ch/LEPEWWG/.
8. LEP Collaborations: *Search for the Standard Model Higgs Boson at LEP*, CERN-EP/2003-011, http://lephiggs.web.cern.ch/LEPHIGG/papers/index.html.
9. For a phenomenological review see for instance P. Fayet and S. Ferrara: Phys. Rep. C **32** 249 (1997); H. P. Nilles: Phys. Rep. C **110** 1 (1984); H. E. Haber and G. L. Kane: Phys. Rep. C **117** 75 (1985).

10. LEP SUSY Working Group, http://lepsusy.web.cern.ch/lepsusy/.
11. R. Bernabei et al.: Phys. Lett. B **480** 23 (2000).
12. A. Benoit et al.: *Improved exclusion limits from the EDELWEISS WIMP Search*, astro-ph/0206271.
13. D. Abrams et al.: *Exclusion limits on the WIMP-nucleon cross-section from the Cryogenic Dark Matter Search*, astro-ph/0203500.
14. A. Morales et al.: Phys. Lett. B **532** 8 (2002).
15. http://cmsdoc.cern.ch/cms/outreach/html/CMSdocuments/CMSdocuments.html.
16. C. L. Bennet et al.: Astrophys. J. Suppl. **148** 1 (2003); D. N. Spergel et al.: Astrophys. J. Suppl. **148** 175 (2003); H. V. Peiris et al.: Astrophys. J. Suppl. **148** 213 (2003).
17. See for instance N. Arkani-Hamed, S. Dimopoulos and C. Dvali, Phys. Lett. B **429** 263 (1998).
18. J. Feng and A. Shapere: Phys. Rev. Lett. **88** 021303 (2002).
19. V. Avati, K. Eggert and C. Taylor: *FELIX: a full acceptance detector at LHC*, hep-ex/9801021.
20. TOTEM Collaboration: Letter of Intent, CERN/LHCC 97-49, 1997, http://totem.web.cern.ch/Totem/.

Precision Calculations in the MSSM

Wolfgang Hollik

Max-Planck-Institut für Physik (Werner-Heisenberg-Institut)
Föhringer Ring 6, 80805 Munich, Germany
hollik@mppmu.mpg.de

1 Introduction

High-precision experiments at electron-positron and hadron colliders together with the highly accurate measurements of the muon lifetime and gyromagnetic factor impose stringent tests on the standard model and possible extensions. The experimental accuracy in the electroweak observables has reached the level of the quantum effects, and requires the highest standards on the theoretical side as well. A sizeable amount of work has continuously contributed over the last two decades to a steadily rising improvement of the standard model predictions, pinning down the theoretical uncertainties to the level required for the proper interpretation of the precision data. Also for the minimal supersymmetric standard model (MSSM), remarkable progress has to be reported in predicting the precision observables with similar accuracy as in the standard model. Table 1 summarizes the present experimental precision for those high-energy parameters where essential improvements are expected from future collider experiments at the Tevatron (Run II), the LHC, and a e^+e^- Linear Collider with an additional high-luminosity GigaZ option. Moreover, the Z-boson mass and the Fermi constant with their tiny uncertainties [1], $\delta M_Z = 2.1\,\mathrm{MeV}$ and $\delta G_F/G_F = 1 \cdot 10^{-5}$, will also be at our disposal. The availability of both highly accurate measurements and theoretical predictions, at the level of 0.1% precision and better, provides unique tests of the quantum structure of the models and yields indirect informations on yet unexplored heavy sectors.

Table 1. Present experimental accuracies and expectations for future colliders (see [2] and references therein)

Error for	now	Tev/LHC	LC	GigaZ
M_W [MeV]	33	15	15	6 – 7
$\sin^2\theta_{\mathrm{eff}}$	0.00017	0.00021		0.000013
m_{top} [GeV]	5.1	2	0.2	0.13
M_{Higgs} [GeV]	–	0.1	0.05	0.05

2 Electroweak Precision Observables

The possibility of performing precision tests is based on the formulation of the standard model and the MSSM as renormalizable quantum field theories preserving their predictive power beyond tree-level calculations.

2.1 Muon Decay and the Vector-Boson Mass Correlation

The basic physical quantity for the M_W–M_Z correlation is the muon lifetime τ_μ, which defines the Fermi constant G_F according to

$$\frac{1}{\tau_\mu} = \frac{G_F^2\, m_\mu^5}{192\pi^3}\, F\left(\frac{m_e^2}{m_\mu^2}\right)\left(1+\frac{3}{5}\frac{m_\mu^2}{M_W^2}\right)(1+\Delta_{\rm QED})\ , \tag{1}$$

with $F(x) = 1 - 8x - 12x^2 \ln x + 8x^3 - x^4$. By convention, the QED corrections within the Fermi Model, $\Delta_{\rm QED}$, are included in this defining equation for G_F. The one-loop result for $\Delta_{\rm QED}$ [3] has already been known for several decades; it has recently been supplemented by the two-loop correction [4], yielding

$$\Delta_{\rm QED} = 1 - 1.81\,\frac{\alpha(m_\mu)}{\pi} + 6.7\left(\frac{\alpha(m_\mu)}{\pi}\right)^2\ , \quad \text{with} \quad \alpha(m_\mu) \simeq \frac{1}{135.90}\ . \tag{2}$$

The tree-level W-propagator effect giving rise to the (numerically insignificant) term $3m_\mu^2/(5M_W^2)$ in (1), is conventionally also included in the definition of G_F, although not part of the Fermi Model prediction. From the precisely measured muon-decay width the value [1] $G_F = (1.16637 \pm 0.00001)\,10^{-5}\,\mathrm{GeV}^{-2}$ for the Fermi constant is derived.

Calculating the muon lifetime within the standard model or the MSSM and comparing the result with (1) yields the relation

$$M_W^2\left(1-\frac{M_W^2}{M_Z^2}\right) = \frac{\pi\alpha}{\sqrt{2}G_F}\,(1+\Delta r)\ , \tag{3}$$

where the radiative corrections are summarized in the quantity Δr. Thereby, a set of infrared-divergent QED-correction graphs has to be removed, which reproduce the QED-correction factor of the Fermi-model result in (1). They have no influence on the relation between G_F and the model parameters.

In the standard model the quantum correction Δr has received a lot of attention. The one-loop result [5] has been improved over the last two decades by numerically important QCD and electroweak higher-order terms, establishing thus a powerful relation that can be used to predict M_W within the SM (or possible extensions), to be confronted with the experimental result for M_W. The quantity Δr depends on the entire set of input parameters. It contains, among others, the on-shell mass counterterms and the photon vacuum polarization from charge renormalization in the classical limit.

The photon vacuum polarization is a basic entry in the predictions for electroweak precision observables. The difference

$$\mathrm{Re}\,\hat{\Pi}^\gamma(M_Z^2) = \mathrm{Re}\,\Pi^\gamma(M_Z^2) - \Pi^\gamma(0) \tag{4}$$

is a finite quantity. The purely fermionic part corresponds to standard QED and does not depend on the details of the electroweak theory. It can be split into a leptonic and a hadronic contribution, yielding the quantity

$$\Delta\alpha = \Delta\alpha_{\text{lept}} + \Delta\alpha_{\text{had}} = -\operatorname{Re}\hat{\Pi}^{\gamma}_{\text{lept}}(M_Z^2) - \operatorname{Re}\hat{\Pi}^{\gamma}_{\text{had}}\left(M_Z^2\right) , \tag{5}$$

which represents a QED-induced shift in the electromagnetic fine structure constant

$$\alpha \to \alpha(1+\Delta\alpha) . \tag{6}$$

It can be resummed [6] according to the renormalization group, accommodating all the leading logarithms of the type $\alpha^n \log^n(M_Z/m_f)$, to give an effective fine-structure constant at the Z mass scale,

$$\alpha\left(M_Z^2\right) = \frac{\alpha}{1-\Delta\alpha} . \tag{7}$$

The leptonic content of $\Delta\alpha$ can be directly evaluated in terms of the known lepton masses, yielding at three-loop order [7]

$$\Delta\alpha_{\text{lept}} = 314.97687 \cdot 10^{-4} . \tag{8}$$

For the light hadronic part, perturbative QCD is not applicable and quark masses are no reasonable input parameters. Instead, the 5-flavour contribution to $\hat{\Pi}^{\gamma}_{\text{had}}$ can be derived from experimental data with the help of a dispersion relation

$$\Delta\alpha_{\text{had}} = -\frac{\alpha}{3\pi} M_Z^2 \operatorname{Re}\int_{4m_\pi^2}^{\infty} \mathrm{d}s' \frac{R^{\gamma}(s')}{s'(s'-M_Z^2-i\varepsilon)} \tag{9}$$

where

$$R^{\gamma}(s) = \frac{\sigma(e^+e^- \to \gamma^* \to \text{hadrons})}{\sigma(e^+e^- \to \gamma^* \to \mu^+\mu^-)}$$

is an experimental input quantity for the low energy range. Recent updates including new data points from BES [8] and CMD [9] yield the value $\Delta\alpha = 0.02769 \pm 0.00035$ [10], respectively $\Delta\alpha = 0.02755 \pm 0.00023$ [11].

The heavy quark doublet (t, b) contributes predominantly via the ρ parameter [12],

$$\Delta\rho = \frac{3G_F m_t^2}{8\pi^2\sqrt{2}} \tag{10}$$

to Δr, which thus has a simple form in the light- and heavy-fermion terms at one-loop,

$$\Delta r = \Delta\alpha - \frac{c_W^2}{s_W^2}\Delta\rho + \Delta r_{\text{rem}} . \tag{11}$$

$\Delta\alpha$ contains the large logarithmic corrections from the light fermions and $\Delta\rho$ the quadratic correction from m_t. All other terms are collected in the remainder Δr_{rem}, which has a typical size of the order $\sim$0.01.

Beyond the one-loop order, higher-order 1-particle reducible and irreducible 2- and 3-loop contributions to the ρ parameter have been obtained with electroweak and QCD terms [13]. QCD corrections to Δr beyond the contribution via $\Delta\rho$ are known at $\mathcal{O}(\alpha\alpha_s)$ [14] and $\mathcal{O}(\alpha\alpha_s^2)$ [15]. First approximative electroweak two-loop calculations were performed based on expansions for asymptotically large values of M_H [16] and m_t [17].

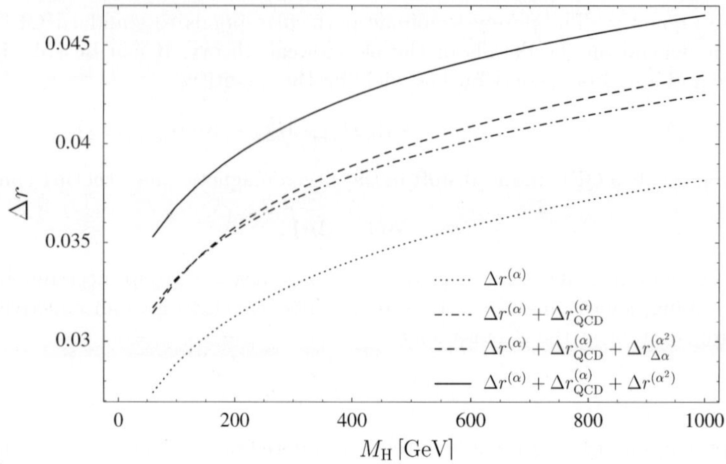

Fig. 1. Various stages of Δr, as a function of M_H. The one-loop contribution, $\Delta r^{(\alpha)}$, is supplemented by the two-loop and three-loop QCD corrections, $\Delta r^{(\alpha)}_{\mathrm{QCD}} \equiv \Delta r^{(\alpha\alpha_s)} + \Delta r^{(\alpha\alpha_s^2)}$, and the fermionic electroweak two-loop contributions, $\Delta r^{(\alpha^2)} \equiv \Delta r^{(N_{\mathrm{f}}\alpha^2)} + \Delta r^{(N_{\mathrm{f}}^2\alpha^2)}$. For comparison, the effect of the two-loop corrections induced by a resummation of $\Delta\alpha$, $\Delta r^{(\alpha^2)}_{\Delta\alpha}$, is shown separately

In the meantime, the complete electroweak two-loop result in the standard model has become available: the fermionic two-loop terms [18] with all two-loop diagrams for the muon-decay amplitude containing at least one closed fermion loop, and the residual class of the two-loop purely bosonic diagrams [19, 20]. Their influence is displayed in Fig. 1 for the fermionic and in Fig. 2 for the bosonic contributions, in terms of Δr and M_W.

2.2 *Z* Boson Observables

With M_Z used as a precise input parameter, together with α and G_F, the predictions for the width, partial widths and asymmetries can conveniently be calculated in terms of effective neutral current coupling constants for the various fermions (see e.g. [21]):

$$\begin{aligned} J_\nu^{\mathrm{NC}} &= \left(\sqrt{2}G_F M_Z^2\right)^{1/2} \left(g_V^f \gamma_\nu - g_A^f \gamma_\nu\gamma_5\right) \\ &= \left(\sqrt{2}G_F M_Z^2 \rho_f\right)^{1/2} \left((I_3^f - 2Q_f s_f^2)\gamma_\nu - I_3^f \gamma_\nu\gamma_5\right) . \end{aligned} \tag{12}$$

The subleading 2-loop corrections $\sim G_F^2 m_t^2 M_Z^2$ for the leptonic mixing angle [22] s_λ^2 have also been obtained in the meantime, as well as for ρ_λ [23].

The effective mixing angles are of particular interest, since they determine the on-resonance asymmetries via the combinations

$$A_f = \frac{2 g_V^f g_A^f}{\left(g_V^f\right)^2 + \left(g_A^f\right)^2} , \tag{13}$$

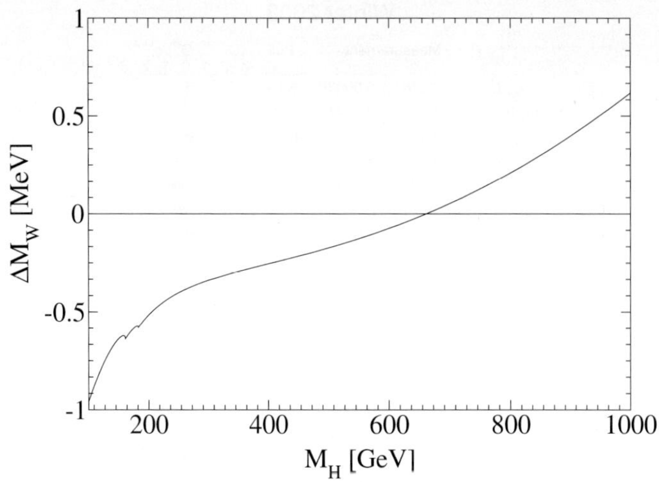

Fig. 2. The shift in M_W from the two-loop bosonic contributions to Δr (from [19])

namely

$$A_{\rm FB} = \frac{3}{4} A_e A_f , \quad A_\tau^{\rm pol} = A_\tau , \quad A_{\rm LR} = A_e \, . \tag{14}$$

Measurements of the asymmetries hence are measurements of the ratios

$$g_V^f / g_A^f = 1 - 2Q_f s_f^2 \tag{15}$$

or the effective mixing angles, respectively.

3 Standard Model and Precision Data

3.1 Global Fit and Higgs-Boson Mass

The Z-boson observables form LEP 1 and SLC together with M_W and the top-quark mass from LEP 2 and the Tevatron, constitute the set of high-energy quantities entering a global precision analysis (see [24] for a recent review). From low-energy experiments, the quantity $s_W^2 = M_W/M_Z$ can indirectly be measured in deep-inelastic neutrino–nucleon scattering. The result of the NuTeV collaboration [25] can be expressed as follows,

$$\begin{aligned} s_W^2 &= 0.2277 \pm 0.0013 \pm 0.0009 \\ &\quad -0.00022 \frac{m_t^2 - (175 \text{ GeV})^2}{(50 \text{ GeV})^2} \\ &\quad +0.00032 \ln(M_H/150 \text{ GeV}) \, . \end{aligned}$$

Global fits within the standard model to the electroweak precision data contain M_H as the only free parameter, yielding an upper limit to the Higgs mass at the 95% C.L. of $M_H < 219\,\text{GeV}$ [24], including the present theoretical uncertainties of

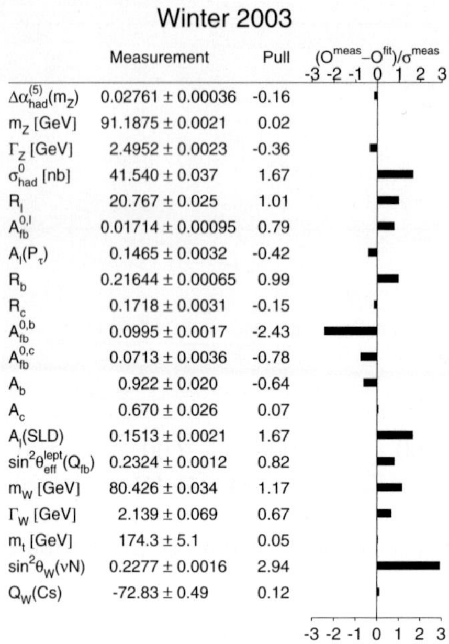

Fig. 3. Experimental results and pulls from a standard model fit [24]. Pull = obs(exp)-obs(SM)/(exp.error)

the standard model predictions. Figure 3, showing the deviation of the individual quantities from the standard model best-fit values, points out the forward-backward asymmetry for b quarks and s_W^2 from deep-inelastic neutrino scattering as the largest deviations from the standard model value for the best fit.

3.2 Muon Anomalous Magnetic Moment

The anomalous magnetic moment of the muon,

$$a_\mu = \frac{g_\mu - 2}{2} \tag{16}$$

provides a precision test at low energies. The new experimental result of E 821 at Brookhaven National Laboratory [26] has reached a substantial improvement in accuracy. It shows a deviation from the standard model prediction by 2.7 [1.4] standard deviations depending on the evaluation of the hadronic vacuum polarization from data based on e^+e^- annihilation [hadronic τ decays together with isospin rotation], as discussed in [27]. Other recent analyses were performed in [10, 11, 35].

4 The MSSM and Precision Data

Among the extensions of the standard model, the minimal supersymmetric standard model (MSSM) is the theoretically favoured scenario as the most predictive framework beyond the standard model. A definite prediction of the MSSM is the existence of a light Higgs boson with mass below ~135 GeV [28]. The detection of a light Higgs boson could be a significant hint for supersymmetry.

The structure of the MSSM as a renormalizable quantum field theory allows a similarly complete calculation of the electroweak precision observables as in the standard model in terms of one Higgs mass (usually taken as the CP-odd 'pseudoscalar' mass M_A) and $\tan\beta = v_2/v_1$, together with the set of SUSY soft-breaking parameters fixing the chargino/neutralino and scalar fermion sectors. The general discussion of renormalization of the MSSM to all orders with implications on the structure of the counter terms is given in [29]. Complete 1-loop calculations are available for Δr [30] and for the Z boson observables [31].

A possible mass splitting between $\tilde{b}_L$ and $\tilde{t}_L$ yields a contribution to the ρ-parameter of the same sign as the standard top term. As a universal loop contribution, it enters the quantity Δr and the Z boson couplings and is thus significantly constrained by the data The 2-loop α_s-corrections have been computed in [32], and the electroweak 2-loop contribution from the Yukawa couplings in [33].

As an example, Fig. 4 displays the range of predictions for M_W in the minimal model and in the MSSM, together with the present experimental errors and the expectations for the future collliders LHC and LC. As can be seen, the MSSM prediction is in better agreement with the present data, although not conclusive as yet. Future increase in the experimental accuracy, however, will become decisive for the separation between the models.

Especially for the muonic $g-2$, the MSSM can significantly improve the agreement between theory and experiment: relatively light scalar muons, muon-sneutrinos and charginos/neutralinos, together with a large value of $\tan\beta$ can

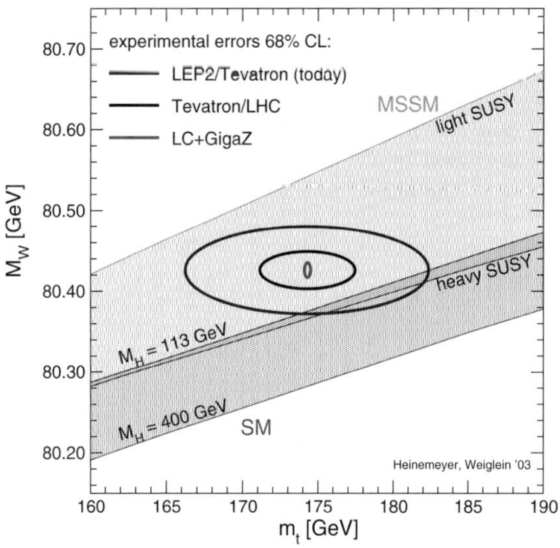

Fig. 4. The W mass range in the standard model (*lower band*) and in the MSSM (*upper band*). Bounds are from the non-observation of Higgs bosons and SUSY particles

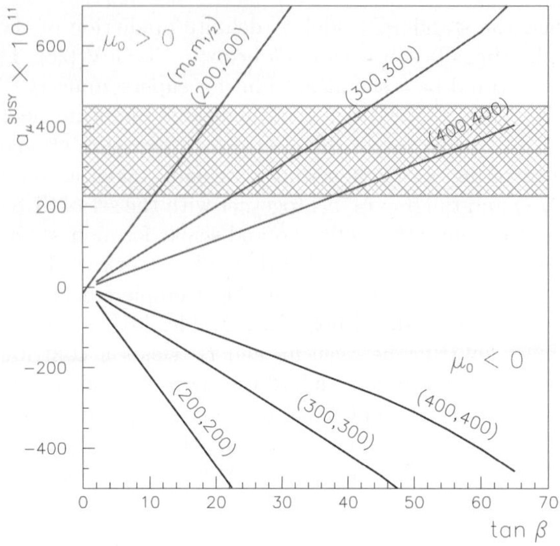

Fig. 5. Supersymmetric contribution to a_μ [37]. The deviation of the measured value from the standard model prediction is indicated by the horizontal band

provide a positive contribution Δa_μ which can entirely explain the difference $a_\mu^{\rm exp} - a_\mu^{\rm SM}$ [34]. Figure 5 illustrates the MSSM contribution for universal SUSY scalar mass parameters m_0 and spin-1/2 mass parameters $m_{1/2}$.

The MSSM yields a comprehensive description of the precision data, in a similar way as the standard model does. Global fits, varying the MSSM parameters, are available [36] to all electroweak precision data. They have been updated [37], showing that the description within the MSSM is slightly better than in the standard model. This is mainly due to the improved agreement for a_μ (see Fig. 6). The situation for $A_{\rm FB}^{\rm b}$, however, remains unaltered.

As far as the deviation of the NuTeV result (16) from the standard model prediction is concerned, the MSSM fails to improve the situation [38].

5 The Light Higgs Boson of the MSSM

The existence of a light Higgs boson, in the mass range below 135 GeV, is a definite prediction of the MSSM. In contrast to the standard model, its mass m_h is not a free parameter but depends on the other parameters of the model. The prediction for m_h, therefore, is a crucial theoretical tool to probe the MSSM parameter space. From the experimental side, the Higgs mass can be measured with high accuracy [2]: 100 MeV at the LHC, and 50 MeV at a Linear e^+e^- Collider. m_h is, hence, another precision observable in the MSSM. The precise theoretical value is very sensitive to higher-order effects (see [28] for recent results and references given therein). An illustrative example is shown in Fig. 7.

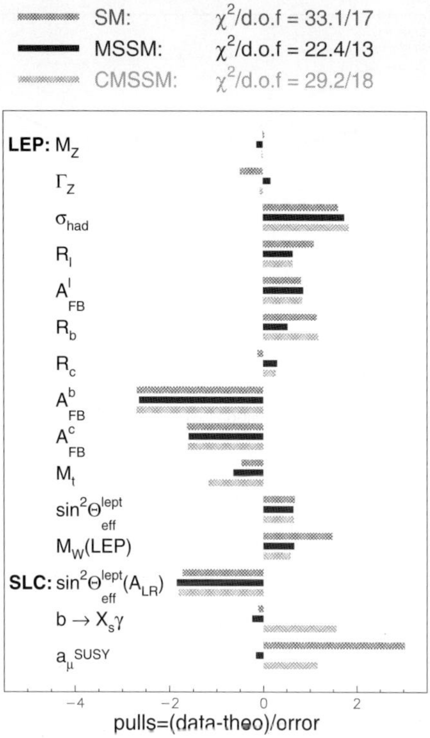

Fig. 6. Best fits in the SM and in the MSSM, normalized to the data [37]. Error bars are those from data

6 SUSY Particles

Experiments at future high-energy colliders will be able to discover supersymmetric particles and to investigate their properties. Provided their masses are not too high, a linear electron-positron collider will be the best environment for precision studies of supersymmetric models [40], especially of the MSSM. From precise measurements of masses, cross sections and asymmetries in chargino and neutralino production, the fundamental parameters can be reconstructed [41], to shed light on the mechanism of SUSY breaking.

In view of the experimental prospects it is inevitable to include higher-order terms in the calculation of the measureable quantities in order to achieve theoretical predictions matching the experimental accuracy. Studies on chargino-pair production [42, 43, 44], neutralino-pair production [45], scalar-quark decays [46], and sfermion-pair production [47] have demonstrated that Born-level predictions can be influenced significantly by one-loop radiative corrections. In [44, 46], with complete one-loop calculations performed in on-shell renormalization schemes, it was shown explicitly that besides the fermion- and sfermion-loop contributions also the virtual contributions from the supersymmetric gauge and Higgs sector are not negligible.

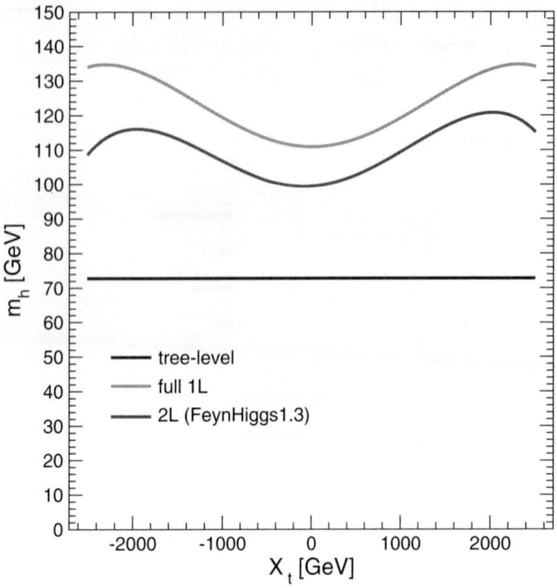

Fig. 7. The lightest Higgs-boson mass in the MSSM, in various orders of perturbation theory [39]. SUSY parameters: $\tan\beta = 3$, $M_{\tilde{Q}} = M_2 = \mu = M_A = 1$ TeV, $m_{\tilde{g}} = 800\,\text{GeV}$. X_t is the non-diagonal entry in the top-squark mass matrix

Since the masses of charginos and neutralinos are among the precision observables with lots of information on the SUSY-breaking structure, the relations between the particle masses and the SUSY parameters as well as the relations between the masses themselves are important theoretical objects for precision calculations. Previous studies were done in the $\overline{\text{MS}}$ renormalization scheme [48, 49] with running parameters. In [50, 51] on-shell scheme calculations were performed yielding the one-loop corrected mass eigenvalues. In a similar way, the sfermion-mass spectrum, including complete one-loop contributions, was obtained [52].

On-shell renormalization is specified in particular by treating all particle masses as pole masses, and, optionally, with field renormalization implemented in a way that allows to formulate the renormalized 2-point vertex functions as UV-finite matrices which become diagonal for external momenta on-shell. The masses of the two charginos and of one neutralino are used as input to fix the MSSM parameters μ, M_1, M_2. Since only the gaugino-mass parameters M_1, M_2 and the Higgsino-mass parameter μ can be renormalized independently in terms of three pole masses, with all other renormalization constants fixed in the gauge and Higgs sector, the residual eigenvalues of the tree-level mass matrices are no longer the pole positions of the corresponding dressed propagators; the pole masses hence receive a shift versus the tree-level masses, which is calculable in terms of the renormalized self-energies. At one-loop order, the following relation

$$m_{\tilde{\chi}_i^0} = m_i \left[1 - \text{Re}\,\hat{\Sigma}_{ii}^L\left(m_i^2\right)\right] - \text{Re}\,\hat{\Sigma}_{ii}^{SL}\left(m_i^2\right) \tag{17}$$

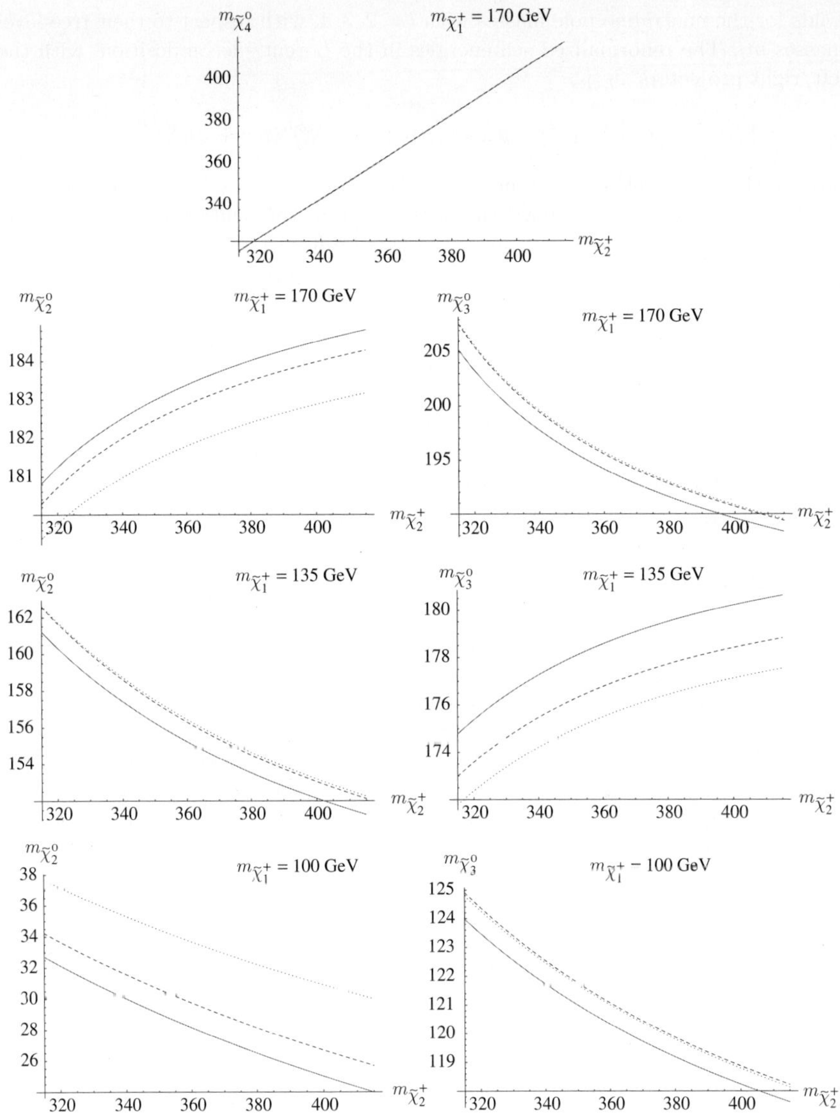

Fig. 8. Dependence of the derived neutralino masses (in GeV) on the chargino masses $m_{\tilde{\chi}_1^+}$ and $m_{\tilde{\chi}_2^+}$. Born approximation (*dotted, black*), including loop corrections with (s)fermions only (*dashed, blue*), and with the complete one-loop contributions (*solid, red*). The input neutralino mass is chosen as $m_{\tilde{\chi}_1^0} = 110\,\mathrm{GeV}$ throughout all diagrams. The plots for $m_{\tilde{\chi}_4^0}$ are practically indistinguishable for all three different values of $m_{\tilde{\chi}_1^+}$

holds for the neutralino pole masses with $i = 2, 3, 4$, with respect to their tree-level masses m_i. The renormalized self-energies in the Lorentz-decomposition, with the left/right projectors $\omega_{L/R}$,

$$\hat{\Sigma}_{ij}(p) = \not{p}\,\omega_L \hat{\Sigma}^L_{ij}(p^2) + \not{p}\,\omega_R \hat{\Sigma}^R_{ij}(p^2) + \omega_L \hat{\Sigma}^{SL}_{ij}(p^2) + \omega_R \hat{\Sigma}^{SR}_{ij}(p^2)\,,$$

provide the scalar self-energy functions in (17).

As an example, Fig. 8 shows the dependence of the neutralino masses on the input mass $m_{\tilde{\chi}_2^+}$ of the heavy chargino at three different values for the mass $m_{\tilde{\chi}_1^+}$ of the light chargino. Depicted are: the tree-level approximation of the calculated neutralino masses, their values after including the corrections from the (s)fermionic loops only, and finally the complete one-loop corrected masses with all MSSM particles in the virtual states. For the heaviest neutralino mass $m_{\tilde{\chi}_4^0}$ the shift is small, not more than 175 MeV throughout the whole scanned parameter space, and nearly invisible in the graphical illustration. Accordingly, the relative corrections are less than 0.05% everywhere, and hence we do not give more than one graph in the figure. More details can be found in [50].

7 Conclusions

The experimental data for tests of the standard model have achieved an impressive accuracy. In the meantime, many theoretical contributions have become available to improve and stabilize the standard model predictions and to reach a theoretical accuracy clearly better than 0.1%.

The MSSM, mainly theoretically advocated, is competitive to the standard model in describing the data with improvements in specific observables, although not conclusive. Since the MSSM predicts the existence of a light Higgs boson, the detection of a Higgs particle could be an indication of supersymmetry. It is therefore highly important to study the different features of such a Higgs boson in the various models at a level of high precision. Moreover, precision studies of supersymmetric particles will become necessary for revealing the mechanism of SUSY breaking and will require a proper inclusion of higher-order effects as well.

References

1. Particle Data Group, K. Hagiwara et al., *Phys. Rev.* D **66**, 010001 (2002)
2. J. Erler, S. Heinemeyer, W. Hollik, G. Weiglein, P.M. Zerwas, *Phys. Lett.* B **486**, 125 (2000); U. Baur et al., hep-ph/0111314
3. R.E. Behrends, R.J. Finkelstein, A. Sirlin, Phys. Rev **101**, 866 (1956); S.M. Berman, Phys. Rev. **112**, 267 (1958); T. Kinoshita, A. Sirlin, Phys. Rev. **113**, 1652 (1959)
4. T. van Ritbergen, R.G. Stuart, *Phys. Rev. Lett.* **82**, 488 (1999); *Phys. Lett.* B **564**, 343 (2000);
   M. Steinhauser, T. Seidensticker, *Phys. Lett.* B **467**, 271 (1999)
5. A. Sirlin, *Phys. Rev.* D **22**, 971 (1980); W.J. Marciano, A. Sirlin, *Phys. Rev.* D **22**, 2695 (1980)

6. W.J. Marciano, *Phys. Rev.* D **20**, 274 (1979)
7. G. Källén, A. Sabry, *K. Dan. Vidensk. Selsk. Mat.-Fys. Medd.* **29** (1955) No. 17; M. Steinhauser, *Phys. Lett.* B **429**, 158 (1998)
8. BES Collaboration, J.Z. Bai et al., *Phys. Rev. Lett.* **84**, 594 (2000); **88**, 101802 (2002)
9. CMD-2 Collaboration, R.R. Akhmetshin et al., *Phys. Lett.* B **527**, 161 (2002)
10. F. Jegerlehner, *J. Phys.* G **29**, 101 (2003); hep-ph/0312372
11. K. Hagiwara, A.D. Martin, D. Nomura, T. Teubner, hep-ph/0312250
12. D. Ross, M. Veltman, *Nucl. Phys.* B **95**, 135 (1975); M. Veltman, *Nucl. Phys.* B **123**, 89 (1977); M.S. Chanowitz, M.A. Furman, I. Hinchliffe, *Phys. Lett.* B **78**, 285 (1978)
13. M. Consoli, W. Hollik, F. Jegerlehner, *Phys. Lett.* B **227**, 167 (1989); J. van der Bij, F. Hoogeveen, *Nucl. Phys.* B **283**, 477 (1987; R. Barbieri, M. Beccaria, P. Ciafaloni, G. Curci, A. Vicere, *Phys. Lett.* B **288**, 95 (1992); *Nucl. Phys.* B **409**, 105 (1993); J. Fleischer, O.V. Tarasov, F. Jegerlehner, *Phys. Lett.* B **319**, 249 (1993) and *Phys. Rev.* D **51**, 3820 (1995); A. Djouadi, C. Verzegnassi, *Phys. Lett.* B **195**, 265 (1987); L. Avdeev, J. Fleischer, S.M. Mikhailov, O. Tarasov, *Phys. Lett.* B **336**, 560 (1994); E: **349**, 597 (1995); K. Chetyrkin, J. Kühn, M. Steinhauser, *Phys. Lett.* B **351**, 331 (1995); J. van der Bij, K. Chetyrkin, M. Faisst, G. Jikia, T. Seidensticker, *Phys. Lett.* B **498**, 156 (2001); M. Faisst, J. Kühn, T. Seidensticker, O. Veretin, *Nucl. Phys.* B **665**, 649 (2003)
14. B.A. Kniehl, *Nucl. Phys.* B **347**, 89 (1990); F. Halzen, B.A. Kniehl, *Nucl. Phys.* B **353**, 567 (1991); B.A. Kniehl, A. Sirlin, *Nucl. Phys.* B **371**, 141 (1992), *Phys. Rev.* D **47**, 883 (1993); A. Djouadi, P. Gambino, *Phys. Rev.* D **49**, 3499 (1994);
15. K. Chetyrkin, J. Kühn, M. Steinhauser, *Phys. Rev. Lett.* **75**, 3394 (1995)
16. J. van der Bij, M. Veltman, *Nucl. Phys.* B **231**, 205 (1984)
17. G. Degrassi, P. Gambino, A. Vicini, *Phys. Lett.* B **383**, 219 (1996); G. Degrassi, P. Gambino, M. Passera, A. Sirlin, *Phys. Lett.* B **418**, 209 (1998)
18. A. Freitas, W. Hollik, W. Walter, G. Weiglein, *Phys. Lett.* B **495**, 338 (2000); *Nucl. Phys.* B **632**, 189 (2002); M. Awramik, M. Czakon, *Phys. Lett.* B **568**, 48 (2003)
19. M. Awramik, M. Czakon, *Phys. Rev. Lett.* **89**, 241801 (2002)
20. A. Onishchenko, O. Veretin, *Phys. Lett.* B **551**, 111 (2003); M. Awramik, M. Czakon, A. Onishchenko, O. Veretin, *Phys. Rev.* D **68**, 053004 (2003)
21. D. Bardin et al., *Reports of the Working Group on Precision Calculations for the Z Resonance*, p. 7, CERN 95-03 (1995), eds. D. Bardin, W. Hollik, G. Passarino; hep-ph/9709229
22. G. Degrassi, P. Gambino, A. Sirlin, *Phys. Lett.* B **394**, 188 (1997)
23. G. Degrassi, P. Gambino, *Nucl. Phys.* B **567**, 3 (2000)
24. M. Grünewald, hep-ex/0304023; The LEP Collaborations, the LEP Electroweak Working Group, the SLD Electroweak and Heavy Flavour Working Groups, hep-ex/0312023
25. NuTeV Collaboration, G. Zeller et al., *Phys. Rev. Lett.* **88**, 091802
26. H.N. Brown et al., *Phys. Rev. Lett.* **86**, 2227 (2001); G.W. Bennett et al., *Phys. Rev. Lett.* **89**, 101804 (2002) and hep-ex/0401008
27. M. Davier, S. Eidelman, A. Höcker, Z. Zhang, *Eur. Phys. J.* C **27**, 497 (2003); S. Ghozzi, F. Jegerlehner, *Phys. Lett.* B **583**, 222 (2004)
28. G. Degrassi, S. Heinemeyer, W. Hollik, P. Slavich, G. Weiglein, *Eur. Phys. J.* C **28**, 133 (2003)

29. W. Hollik, E. Kraus, M. Roth, C. Rupp, K. Sibold, D. Stöckinger, *Nucl. Phys.* B **639**, 3 (2002)
30. P. Chankowski, A. Dabelstein, W. Hollik, W. Mösle, S. Pokorski, J. Rosiek, *Nucl. Phys.* B **417**, 101 (1994); D. Garcia, J. Solà, *Mod. Phys. Lett.* A **9**, 211 (1994)
31. D. Garcia, R. Jiménez, J. Solà, *Phys. Lett.* B **347**, 309 and 321 (1995); D. Garcia, J. Solà, *Phys. Lett.* B **357**, 349 (1995); A. Dabelstein, W. Hollik, W. Mösle, in *Perspectives for Electroweak Interactions in e^+e^- Collisions*, Ringberg Castle 1995, ed. B.A. Kniehl, World Scientific 1995 (p. 345); P. Chankowski, S. Pokorski, *Nucl. Phys.* B **475**, 3 (1996); J. Bagger, K. Matchev, D. Pierce, R. Zhang, *Nucl. Phys.* B **491**, 3 (1997)
32. A. Djouadi, P. Gambino, S. Heinemeyer, W. Hollik, C. Jünger, G. Weiglein, *Phys. Rev. Lett.* **78**, 3626 (1997) and *Phys. Rev.* D **57**, 4179 (1998)
33. S. Heinemeyer, G. Weiglein, *JHEP* **0210**, 072 (2002) and hep-ph/0301062
34. A. Czarnecki, W. Marciano, *Phys. Rev.* D **64**, 013014 (2001)
35. J.F. de Trocóniz, F.J. Ynduráin, hep-ph/0402285
36. W. de Boer, A. Dabelstein, W. Hollik, W. Mösle, U. Schwickerath, *Z. Phys.* C **75**, 627 (1997)
37. W. de Boer, C. Sander, Proceedings of *SUSY 02*, eds. P. Nath and P.M. Zerwas, Hamburg 2002
38. S. Davidson, S. Forte, P. Gambino, N. Rius, A. Strumia, *JHEP* **0202**, 037 (2002); A. Kurylov, M.J. Ramsey-Musolf, S. Su, *Nucl. Phys.* B **667**, 321 (2003)
39. S. Heinemeyer, W. Hollik, G. Weiglein, *Comp. Phys. Comm.* **124**, 76 (2000). The updated version of *FeynHiggs* is available at `http://www.feynhiggs.de`
40. TESLA Technical Design Report, DESY 2001-011, Part III, R. Heuer, D. Miller, F. Richard, P. Zerwas (eds.), hep-ph/0106315
41. S.Y. Choi, A. Djouadi, M. Guchait, J. Kalinowski, H.S. Song, P.M. Zerwas, *Eur. Phys. J.* C **14**, 535 (2000) G. Moortgat-Pick, A. Bartl, H. Fraas, W. Majerotto, hep-ph/0004181; S.Y. Choi, J. Kalinowski, G. Moortgat-Pick, P.M. Zerwas, hep-ph/0108117 and hep-ph/00202139; J. Kalinowski, G. Moortgat-Pick, hep-ph/0202083; hep-ph/0202084
42. M. A. Díaz, S. F. King, D. A. Ross, *Nucl. Phys.* B **529**, 23 (1998)
43. S. Kiyoura, M. M. Nojiri, D. M. Pierce, Y. Yamada, *Phys. Rev.* D **58**, 075002 (1998)
44. T. Blank, W. Hollik, hep-ph/0011092
45. W. Öller, H. Eberl, W. Majerotto, hep-ph/0402134
46. J. Guasch, W. Hollik, J. Solà, *Phys. Lett.* B **510**, 211 (2001) and *JHEP* **10**, 040 (2002)
47. A. Arhrib, W. Hollik, hep-ph/0311149; K. Kovarik, C. Weber, H. Eberl, W. Majerotto, hep-ph/0401092
48. D. Pierce, A. Papadopoulos, *Phys. Rev.* D **50**, 565 (1994) and *Nucl. Phys.* B **430**, 278 (1994)
49. A. B. Lahanas, K. Tamvakis, N. D. Tracas, *Phys. Lett.* B 324, 387 1994
50. T. Fritzsche, W. Hollik, *Eur. Phys. J.* C **24**, 619 (2002)
51. H. Eberl, M. Kincel, W. Majerotto, Y. Yamada, *Phys. Rev.* D **64**, 115013 (2001);
W. Öller, H. Eberl, W. Majerotto, C. Weber, *Eur. Phys. J.* C **29**, 563 (2002)
52. W. Hollik, H. Rzehak, *Eur. Phys. J.* C **32**, 127 (2003)

Theoretical Aspects of Heavy Flavour Physics

Thomas Mannel

Theoretische Physik 1, Siegen University, 57068 Siegen, Germany
mannel@hep.physik.uni-siegen.de

1 Introduction

Favour physics and in particular heavy flavour physics is currently one of the most active fields in particle physics. Large experimental efforts are made to investigate the flavour sector of the standard model, which have to be supported by theoretical progress in order to perform a precise test of our picture of flavour mixing and CP violation.

The heart of the standard-model flavour sector is the CKM matrix

$$V_{CKM} = \begin{pmatrix} V_{ud} & V_{us} & V_{ub} \\ V_{cd} & V_{cs} & V_{cb} \\ V_{td} & V_{ts} & V_{tb} \end{pmatrix} . \tag{1}$$

V_{CKM} is a unitary matrix carrying a single irreducible phase which is the origin of CP violation in the standard model. Unitarity and the phase are usually depicted in the form of the unitarity triangle which is shown in Fig. 1. This triangle represents the unitarity relation obtained from multiplying the first column with the conjugate complex of the last column.

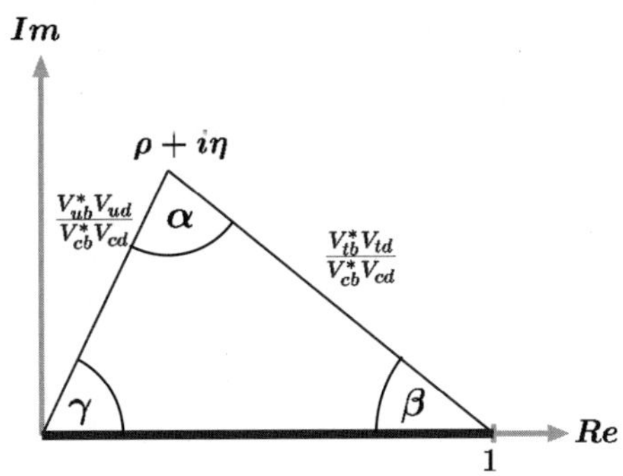

Fig. 1. The standard unitarity triangle with the definitions of ρ, η and the CKM angles α, β and γ

The area of the triangle is a measure of CP violation; hence any trivial values of the CKM angles α, β and γ (0° or 180°) would mean vanishing CP violation. It was a major breakthrough, when the B factories established a non-trivial value of γ by measuring the time dependent CP asymmetry in $B \to J/\Psi K_s$ in the year 2001 [1].

It is the goal of the current experiments in flavour physics to overconstrain this triangle to test CKM unitarity as stringently as possible. Any significant inconsistency would indicate new physics in the flavour sector. However, this also requires further progress in the theoretical description of (heavy) meson weak decays to reduce the uncertainties stemming from hadronic matrix elements.

Our current understanding of flavour is in fact very unsatisfactory:

- A large number of parameters originates from the Yukawa couplings which are the source of quark-flavour mixing in the standard model: Out of the 27 parameters of the standard model (including lepton-flavour mixing assuming Majorana neutrinos) 22 are related to the Yukawa couplings.
- The hierarchical structure of the CKM matrix elements remains a mystery.
- The quark masses (with the exception of the top) have very small values relative to the electroweak vacuum-expectation value, i.e. the Yukawa couplings are unnaturally small.
- There is no theoretical ansatz why the number of families is three.
- CP violation exhibits a specific pattern, e.g. CP violation in flavour diagonal processes (such as electric dipole moments) has not been observed; this is in accordance with the standard model.
- The amount of CP violation in the standard model is too small to explain the observed matter-antimatter asymmetry.

I this lecture I give a brief overview over the theoretical tools that have been developed in order to extract the information on the unitarity triangle from experiment. In particular for heavy flavours it is the heavy mass expansion which allows us to perform systematic expansion in powers of Λ_{QCD}/m_Q where m_Q is the mass of the heavy quark.

2 The Heavy Mass Expansion

The main obstacle in expressing the observed flavour mixing and CP violation in terms of the fundamental parameters of the standard model is the presence of strong interactions. Hence the primary challenge for theory is to develop model-independent methods to deal with strong interactions.

As far as decays of heavy quarks are concerned, the main tool to deal with strong interactions is the heavy quark expansion (HQE) [2] and / or heavy quark effective theory (HQET) [3]. Here one makes use of the fact that the quark mass is large compared to the scale parameter of QCD Λ_{QCD}. In reality we certainly have $m_b \gg \Lambda_{\text{QCD}}$, while the relation $m_c \gg \Lambda_{\text{QCD}}$ can be debated.

In both HQE and HQET one performs an expansion in the quantities

$$\frac{\Lambda_{\text{QCD}}}{m_Q} \quad \text{and} \quad \alpha_s(m_Q) \tag{2}$$

which can be done using methods of effective field theory and operator product expansion.

The heavy mass limit has additional symmetries beyond the ones of QCD which allow us to relate matrix elements of static heavy mesons moving with velocities v and v'. The master example is the heavy-quark-symmetry relation [3]

$$\left.\begin{array}{l}\langle B(v)|\bar{b}\Gamma c|D(v')\rangle \\ \langle B(v)|\bar{b}\Gamma c|D^*(v')\rangle\end{array}\right\} = C_\Gamma(v\cdot v')\xi(v\cdot v') + \mathcal{O}(1/m_Q, \alpha_s(m_c)) \tag{3}$$

which defines the Isgur-Wise function ξ containing all the non-perturbative information in the heavy-mass limit. The coefficient $C_\Gamma(v\cdot v')$ is calculable from heavy quark symmetries and depends on the Dirac matrix Γ in the current and on the velocities of the initial and the final state.

In the heavy mass limit the Isgur-Wise function is normalized to unity at $v = v'$; corrections to this normalization can be discussed in HQET. In particular, for currents related to the generators of the heavy-quark symmetries Lukes theorem [4] ensures that first order ($\mathcal{O}(1/m)$) corrections to the normalization at $vv' = 1$ are absent.

2.1 Determination of V_{cb} from Exclusive Decays

The relations given above are used to perform a determination of V_{cb} from exclusive decays. One exploits the relation

$$\lim_{v\to v'} \frac{1}{\sqrt{(vv')^2-1}}\frac{\mathrm{d}\Gamma}{\mathrm{d}(vv')} = \frac{G_F^2}{4\pi^3}|V_{cb}|^2(m_B - m_{D^*})^2 m_{D^*}^3 |\xi_{A1}(vv'=1)|^2 \tag{4}$$

where $\xi_{A1}(vv')$ is one of the form factors of the axial-vector current, which becomes the Isgur-Wise function in the heavy mass limit for both the charm and the bottom quark. According to Lukes theorem $\xi_{A1}(1)$ is protected against $1/m_c$ corrections and, including $\alpha_s(m_c)^2$ [5] and an estimate of the $1/m_c^2$ corrections, the currently best estimate for this quantity is [6]

$$\xi_{A1}(vv'=1) = 0.91^{+0.03}_{-0.04} \tag{5}$$

Estimates of the $1/m_c^2$ terms are model dependent, the results from different estimates agree and are shown in the left plot of Fig. 2. The right plot shows a probability distribution suggested in [6], the broader one represents the theoretical uncertainty today, the more narrow curve is a projection into the future assuming further progress e.g. from the lattice.

The extrapolation usually uses a linear fit in which also the slope ρ^2 defined by

$$\xi_{A1}(vv') = \xi_{A1}(1)\left(1 - \rho^2[vv'-1] + \cdots\right) \tag{6}$$

is extracted. From the theoretical side the slope is restricted from unitarity and analyticity [7]. The current results from the various experiments have been collected and averaged by the heavy-flavour-averaging group [6] and are shown in Fig. 3. From this the value

$$V_{cb}^{\text{excl}} = (40.2 \pm 0.9_{\text{exp}} \pm 1.8_{\text{theo}}) \times 10^{-3} \tag{7}$$

has been extracted [6].

The uncertainty quoted in (4) is mainly from the unknown contributions of order $1/m_c^2$ and higher and constitutes a limitation of this method unless lattice determinations will improve our knowledge of the higher order terms. First progress has been made on this, see [9].

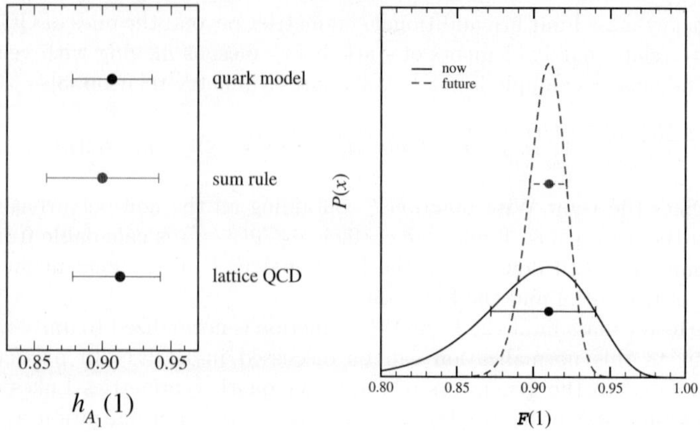

Fig. 2. Current status of the form factor $\xi_{A1} = h_{A1} = F$ in the transiton $B \to D^*$ at $v = v'$. The different entries in the left plot correspond to different methods to estimate the $1/m_c^2$ corrections. The right plot shows probability distributions for the theoretical uncertainty; the narrow curve is a projection into the future. The plot is taken from the CERN Yellow Book [6]

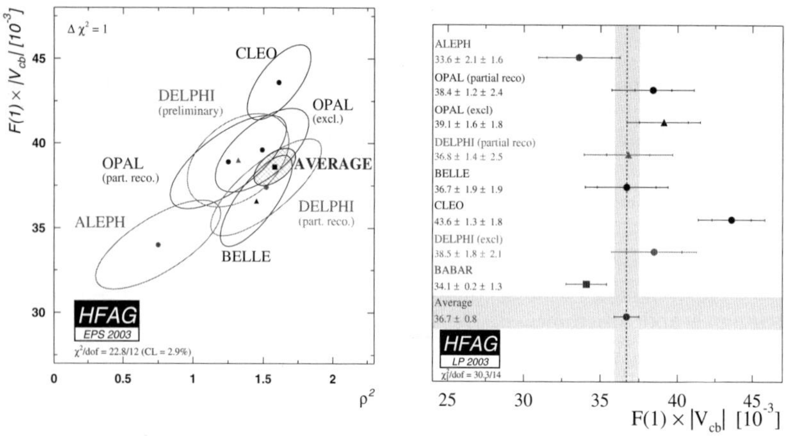

Fig. 3. Current situation of V_{cb} from different experiments. *Left plot*: Correlation ellipses in the ρ^2-$\xi_{A1}(1)V_{cb}$ plane. *Right plot*: Results for $\xi_{A1}(1)V_{cb}$. The plots are taken from [6]

3 Heavy Mass Expansion for Inclusive Decays

One may also obtain a heavy mass expansion for inclusive processes by combining the heavy mass expansion considered above with the method of operator-product expansion (OPE). In this case one can write an inclusive (differential) rate as

$$d\Gamma = d\Gamma_0 + \frac{1}{m_Q} d\Gamma_1 + \frac{1}{m_Q^2} d\Gamma_2 + \frac{1}{m_Q^3} d\Gamma_3 + \cdots \tag{8}$$

Here $d\Gamma_0$ is the partonic rate, i.e. this terms does not depend on an unknown hadronic matrix element. Due to heavy quark symmetries $d\Gamma_1$ always vanishes, while $d\Gamma_2$ can be expressed in terms of two parameters λ_1 and λ_2

$$2M_H\lambda_1 = \langle H(v)|\bar{Q}_v(iD)^2 Q_v|H(v)\rangle \quad (9)$$

$$6M_H\lambda_2 = \langle H(v)|\bar{Q}_v\sigma_{\mu\nu}[iD^\mu, iD^\nu]Q_v|H(v)\rangle \quad (10)$$

The contribution $d\Gamma_3$ is currently under investigation [10], the main contribution comes from the "Darwin term"

$$2M_H\rho_1 = \langle H(v)|\bar{Q}_v(iD)_\mu(ivD)(iD)^\mu Q_v|H(v)\rangle \quad (11)$$

The expressions for inclusive rates thus have the following inputs:

- **The heavy quark mass**, which enters superficially at fifth power, needs to be defined in a suitable scheme. It has been argued that the pole mass is not the best choice, since it suffers from a renormalon ambiguity [12, 13]. A better choice is a short distance mass such as the $\overline{MS}$ mass or the kinetic mass. We will not go into any details concerning this issue; we quote only the uncertainties obtained in recent determinations [14] which is of the order of 50 MeV.
- **Perturbative contributions** from electroweak and QCD radiative corrections are known at $\mathcal{O}(\alpha_s)$ and for the limit $m_c \to 0$ also at $\mathcal{O}(\alpha_s^2)$. The size of the QCD radiative corrections is strongly correlated with the definition of the mass of the b quark. In particular, a short distance mass like the kinetic mass seems to result in a good convergence of the perturbative series. Recently also the terms of order $\alpha_s^{n+1}\beta_0^n$ have been included [11].
- **The ratio m_c/m_b** enters through the phase space functions z_0 and z_2. This ratio is usually determined from the spin averaged meson masses using

$$\overline{M}_B - \overline{M}_D = m_b - m_c - \lambda_1\left(\frac{1}{2m_c} - \frac{1}{2m_b}\right) + \mathcal{O}(1/m_c^2) \quad (12)$$

  which, however, introduces corrections as a series in inverse powers of the *charm* mass. With more data this could be avoided by fitting also the charm mass using the moments of the lepton-energy spectrum.
- **The heavy quark parameters** λ_1 and λ_2 (or μ_π^2 and μ_G^2) at order $1/m_b^2$ and ρ_1 and ρ_2 (or ρ_D^3 and ρ_{LS}^3) at order $1/m_b^3$ are determined from moments of either the lepton-energy spectrum, the hadronic mass spectrum (see below) or from the photon-energy spectrum of the radiative decay $B \to X_s\gamma$. Comparison of the different methods provides a consistency check of the HQE.

3.1 Determination of V_{cb} from Inclusive Decays

Applying this method to the inclusive semi-leptonic decays one obtains a relation of the form

$$\Gamma = |V_{cb}|^2\hat{\Gamma}_0 m_b^5(\mu)(1 + A_{ew})A^{\text{pert}}(r,\mu)\left[z_0(r) + z_2(r)\left(\frac{\lambda_1}{m_b^2}, \frac{\lambda_2}{m_b^2}\right) + \dots\right] \quad (13)$$

where we only quote the structure of the formula, the details are in [11].

Estimating the uncertainties of the inputs discussed above one can expect a theoretical uncertainty in this kind of determination as good as $\Delta V_{cb}/V_{cb} \sim 2\%$. The current value from the inclusive determination is [16]

$$V_{cb}^{\mathrm{incl}} = (40.8 \pm 0.9) \times 10^{-3} \tag{14}$$

It is satisfactory to note that the two results for V_{cb} from the exclusive and the inclusive determination are consistent and can be averaged. Recent concerns, that there may be still uncertainties stemming from parton-hadron duality making the naive average difficult, are probably not valid. This issue is discussed in the next subsection.

3.2 Parton-Hadron Duality: A Potential Problem?

In the context of HQE a concern that has been around over the last few years is the question, if parton-hadron duality is valid in inclusive decays. This is indeed a relevant question once one is aiming at precision determinations of CKM matrix elements.

In order to discuss this issue one first has to give the notion of "duality" a precise meaning. I will follow the arguments given in [17, 18] and argue that very likely violations of duality will be still too small to be relevant.

The common folklore is that "sufficiently" inclusive quantities can be calculated in terms of quarks and gluons [19]. The basis for this is the OPE, which is the $1/m$ expansion for the case at hand. Thus a precise definition of "duality violations" is that these are terms that are not described by the OPE or, likewise, render the OPE non-convergent in the same sense as the perturbative expansion yields at best an asymptotic series.

However, to make this statement more quantitative, one would need an exact solution of QCD. Typically the OPE is performed in the euclidean region and it is known to miss exponentially small terms (such as e.g. an instanton contribution) behaving like $\exp(\sqrt{-q^2})$. However, the analytic continuation to minkowskian q^2 as needed for HQE (where $q^2 = m^2$) turns exponentially small terms into oscillatory ones, which could have a significant effect.

In [17] various model dependent estimates have been given, which typically yield

$$\text{Duality Violations} \sim \frac{\sin(m\rho)}{m^k} \tag{15}$$

where ρ is a hadronic mass scale and k is a power which turns out to be large $k \sim 5$ in the models considered.

Thus no reliable calculation is possible on purely theoretical grounds. However, one may use data to check the validity of the OPE and of duality. One obvious possibility is to extract the HQE parameters from different sources and to check consistency; duality violations would manifest themselves in unnaturally large higher order corrections. The most recent comparisons have been given in [20] and are shown in Fig. 4.

Another observable which can be calculated reliably is the dependence of the first hadronic moments on the lower cut on the lepton energy, at least not for too high values of the cut. Recent data from CLEO [16] show a consistent picture for this observable. The data from CLEO and BaBar are shown in Fig. 5 Thus in conclusion, concerning violations of duality there is no significant hint to a problem.

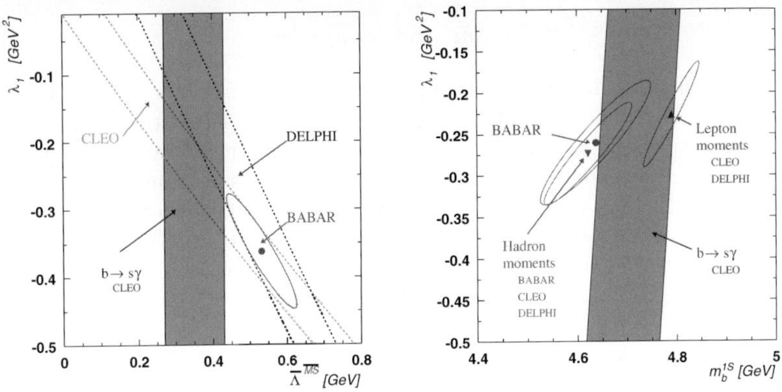

Fig. 4. Extraction of $\bar{\Lambda}$ and λ_1 from different observables. The plots are from [20]

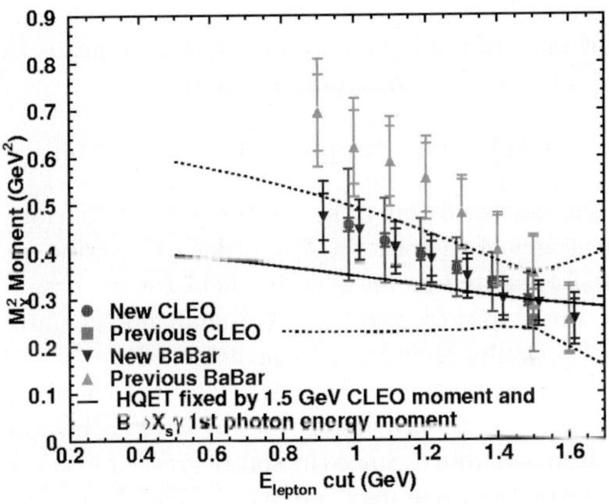

Fig. 5. Dependence of the first hadronic moment on the lower cut on the lepton energy, data versus theory. The band between the two dashed lines indicates the range of the theoretical prediction. Plot is taken from [16]

4 Soft Collinear Effective Theory and QCD Factorization

HQET and HQE are well suited for applications in which the light degrees of freedom are soft, i.e. with momenta of the order Λ_{QCD}. However, in heavy hadron decays we can also have the kinematic situation where the light degrees of freedom carry a large energy but have a small invariant mass. In order to discuss the point, I consider inclusive decays, but – with a few modifications – similar arguments apply to exclusive decays.

The endpoint region of inclusive heavy-to-light decays is defined by the situation where the invariant mass p^2 of the outgoing hadrons is small of the order of $\Lambda_{\mathrm{QCD}} m$, but the hadronic energy (in the rest frame of the decaying B meson) is still of the order m. Thus one has to consider

$$p^2 \sim \lambda m^2 \text{ and } (v \cdot p) \sim m \tag{16}$$

where p is the momentum of the outgoing light degrees of freedom, v is the velocity of the decaying heavy meson, and $\lambda = \Lambda_{\text{QCD}}/m$. This kinematic situation is realized in inclusive decays such as $B \to X_s\gamma$ in the region where the photon energy is close to maximal, or in the endpoint region of the lepton energy spectrum of $B \to X_u\lambda\bar{\nu}_\lambda$.

This particular limit can be formulated an an effective field theory, the so called soft-collinear effective field theory (SCET) [21]. Similar to HQET one splits off a "large" part of the momentum of the heavy quark, which is identified by using the light-cone vectors n_+ and n_-, defined by $v = (n_+ + n_-)/2$ and by a second light-like direction which is e.g. the photon momentum in $B \to X_s\gamma$ or the lepton momentum in $B \to X_u\lambda\bar{\nu}_\lambda$.

The momentum of the light degrees of freedom is written as

$$p = \frac{1}{2}(n_-p)n_+ + \frac{1}{2}(n_+p)n_- + p_\perp \tag{17}$$

and small invariant mass of the light degrees of freedom means that $p^2 \sim \lambda m^2$, and from $(vp) \sim m$ we infer the following power counting

$$(n_-p) \sim m_b \quad (n_+p) \sim \lambda m_b \quad p_\perp \sim \sqrt{\lambda} m_b \tag{18}$$

This shows that one needs not only the soft degrees of freedom scaling as λ, but one has to include scales of the order $\sqrt{\lambda}$ in order to describe inclusive processes.

SCET may be formulated as an effective field theory based on a Lagrangian, implementing the above power counting. We shall not go into any details here, rather we quote the results that have been achieved so far. The main result of SCET is the proof of factorization in $B \to D\pi$ [22]. It is based on the fact, that the light degrees of freedom in a decaying B meson have to be soft according to the counting scheme discussed above. Since the soft degrees of freedom can be decoupled by a unitary transformation, one may remove the coupling to the other degrees of freedom by field redefinitions. Based on this one can show that the exclusive non-leptonic decay $B \to D\pi$ factorizes at leading order in the large mass expansion into the $B \to D$ form factor and the pion-decay constant. This statement is proven to all order in α_s using SCET. However, a poof of a similar factorization in $B \to \pi\pi$ on the basis of SCET has not yet been constructed.

A parallel development is the so called QCD factorization [23], which uses the same kinematical limit but investigates the corresponding Feynman diagrams in full QCD. In this approach also the decays $B \to \pi\pi$ and $B \to K\pi$ have been investigated which are important for the determination of the CKM angle γ [24]. With this method factorization in the same sense as in $B \to D\pi$ has been shown, but up to now only on the basis of the one-loop QCD results.

The typical result of factorization to leading order of the $1/m$ expansion in exclusive non-leptonic heavy-to-light transitions may be diagrammatically described as in Fig. 6. The corresponding expression of the amplitude is

$$\begin{aligned} \langle \pi K | Q_i | B \rangle = {} & F_0^{B\to\pi} \, T^{\text{I}}_{K,i} * f_K \Phi_K + F_0^{B\to K} \, T^{\text{I}}_{\pi,i} * f_\pi \Phi_\pi \\ & + T^{\text{II}}_i * f_B \Phi_B * f_K \Phi_K * f_\pi \Phi_\pi \end{aligned} \tag{19}$$

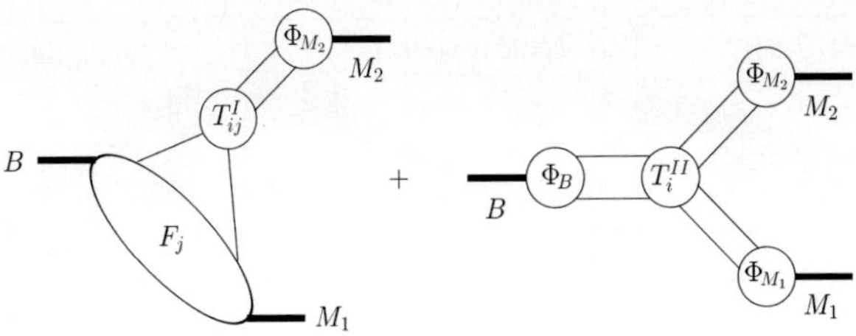

Fig. 6. Graphical interpretation of the QCD factorization theorem in charmless B decays. Figure is taken from [24]

where the $T^{\rm I/II}_{M,i}$ are the hard scattering kernes which are perturbatively calculable, the $F_0^{B\to\pi/K}$ are soft non-perturbative contributions to the form factors and Φ_M are the light-cone distribution functions of the meson M, which are also non-perturbative inputs. Furthermore, the $*$ denotes a convolution, involving a light-cone variable.

Without going into any details, an important general feature of these new approaches is apparent. To leading order in the $1/m$ expansion, any imaginary part of an amplitude must originate from the hard scattering kernels $T^{\rm I/II}_{M,i}$ and thus is perturbatively calculable. From this one concludes that the strong phases entering the analysis of CP violation has to be small, namely it is either of the order $\alpha_s(m_b)$ or it is suppressed by powers of $1/m_b$. Hence the rescattering, which has been discussed intensively in connection with bounds on the CKM angle γ [25], has to be small.

QCD factorization has been applied to charmless non-leptonic B decays [24] and may now be confronted with the data, which have become quite precise. The rates of $B \to K\pi$ and $B \to \pi\pi$ decays depend on the CKM angle γ through the interference of tree and penguin contributions. Evaluating the hadronic matrix elements in QCD factorization allows us to predict the rates of these processes as a function of γ. Furthermore, one may also take ratios of rates, in which some of the theoretical uncertainties cancel.

Figure 7 shows the theoretical predictions for the ratio of branching ratios in comparison with the data. The light-shaded region is the data of last year while the pink (dark) shaded bar is the data published since last year. The new data show that the leading order terms of QCD factorization do not reproduce the data very well. As an example, the ratio of the $B \to K\pi$ decays of the neutral B mesons hint at a small value of the CKM angle γ, while the ratio of the neutral to charged $B \to \pi\pi$ decays hint at a large value of this CKM angle.

Furthermore, the very recently published data on $B^0 \to \pi^0\pi^0$ [26, 27] also is not compatible with the predictions of QCD factorization. The right plot in the second row involves this newly measured branching ratio, and the central value (averaged over the measurements of BaBar and Belle) is even outside of the range of the plot.

It is worthwhile to point out that QCD factorization and SCET are systematic approaches based on an expansion of QCD. In contrast to models, where the dependence on the specific model has to be treated like a systematic uncertainty, a method based on an expansion of QCD allows us at least an estimate of the remaining uncertainties. Thus it is necessary and possible to investigate the failure

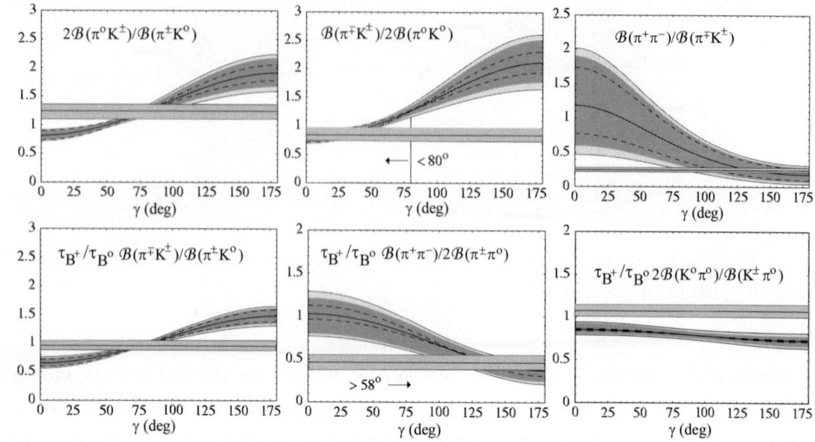

Fig. 7. Predictions of QCD factorization [24] for ratios of rates versus the CKM angle γ in comparison with data. The width of the bands indicate the theoretical and experimental uncertainties [16]

of the leading order predictions of QCD factorization by studying the subleading terms and identifying the sources of possibly large corrections.

Acknowledgements

I want to thank the organizers of the conference for the opportunity to lecture and to discuss physics in such a beautiful and thus stimulating environment.

References

1. B. Aubert et al. [BABAR Collaboration], Phys. Rev. Lett. **87**, 091801 (2001) [arXiv:hep-ex/0107013], A. Abashian et al. [BELLE Collaboration], Phys. Rev. Lett. **86**, 2509 (2001) [arXiv:hep-ex/0102018].
2. I. Bigi, N. Uraltsev and A. Vainshtein, Phys. Lett. **B293** (1992) 430, I. Bigi et al., Phys. Rev. Lett. **71** (1993) 496.
3. N. Isgur and M. Wise, Phys. Lett. **B232** (1989) 113 and **B237** (1990) 527 M. Voloshin and M. Shifman, Sov. J. Nucl. Phys. **45** (1987) 292 and **47** (1988) 511, E. Eichten and B. Hill, Phys. Lett. **B234** (1990) 511, B. Grinstein, Nucl. Phys. **B339** (1990) 253, H. Georgi, Phys. Lett. **B240** (1990) 447, A. Falk, H. Georgi, B. Grinstein and M. Wise, Nucl. Phys. **B343** (1990) 1, A textbook presentation is given in A. V. Manohar and M. B. Wise, Cambridge Monogr. Part. Phys. Nucl. Phys. Cosmol. **10**, 1 (2000), including a more complete set of references.
4. M. E. Luke, Phys. Lett. B **252**, 447 (1990).
5. A. Czarnecki and K. Melnikov, Phys. Rev. Lett. **78**, 3630 (1997) [arXiv:hep-ph/9703291].

6. M. Battaglia et al., Proceedings of the workshop "The CKM matrix and the unitarity triangle," CERN-2003-002, arXiv:hep-ph/0304132.
7. I. Caprini, L. Lellouch and M. Neubert, Nucl. Phys. B **530**, 153 (1998) [arXiv:hep-ph/9712417].
8. The homepage of the HFAG is http://www.slac.stanford.edu/xorg/hfag/
9. S. Hashimoto, A. S. Kronfeld, P. B. Mackenzie, S. M. Ryan and J. N. Simone, Phys. Rev. D **66**, 014503 (2002) [arXiv:hep-ph/0110253].
10. N. Uraltsev, Phys. Lett. B **545**, 337 (2002) [arXiv:hep-ph/0111166].
11. D. Benson, I. I. Bigi, T. Mannel and N. Uraltsev, Nucl. Phys. B **665**, 367 (2003) [arXiv:hep-ph/0302262].
12. I. I. Bigi, M. A. Shifman, N. G. Uraltsev and A. I. Vainshtein, Phys. Rev. D **50**, 2234 (1994) [arXiv:hep-ph/9402360].
13. M. Beneke and V. M. Braun, Nucl. Phys. B **426**, 301 (1994) [arXiv:hep-ph/9402364].
14. J. H. Kuhn and M. Steinhauser, Nucl. Phys. B **619**, 588 (2001) [Erratum-ibid. B **640**, 415 (2002)] [arXiv:hep-ph/0109084].
15. P. Ball, M. Beneke and V. M. Braun, Phys. Rev. D **52**, 3929 (1995) [arXiv:hep-ph/9503492].
16. S. Stone, contribution to the EPS2003, Aachen, Germany, 17-23 Jul 2003.
17. M. A. Shifman, contribution to the Boris Ioffe Festschrift 'At the Frontier of Particle Physics / Handbook of QCD', ed. M. Shifman (World Scientific, Singapore, 2001), arXiv:hep-ph/0009131.
18. I. I. Bigi and N. Uraltsev, Int. J. Mod. Phys. A **16**, 5201 (2001) [arXiv:hep-ph/0106346], I. I. Bigi and T. Mannel, Contribution to the Workshop on CKM Unitarity Triangle (CERN 2002-2003), Geneva, Switzerland, 13-16 Feb 2002, arXiv:hep-ph/0212021.
19. E. C. Poggio, H. R. Quinn and S. Weinberg, Phys. Rev. D **13**, 1958 (1976).
20. U. Langecker, contribution to the EPS2003, Aachen, Germany, 17-23 Jul 2003.
21. C. W. Bauer, S. Fleming, D. Pirjol and I. W. Stewart, Phys. Rev. D **63**, 114020 (2001) [arXiv:hep-ph/0011336], C. W. Bauer, D. Pirjol and I. W. Stewart, Phys. Rev. D **65**, 054022 (2002) [arXiv:hep-ph/0109045], M. Beneke, A. P. Chapovsky, M. Diehl and T. Feldmann, Nucl. Phys. B **643**, 431 (2002) [arXiv:hep-ph/0206152].
22. C. W. Bauer, D. Pirjol and I. W. Stewart, Phys. Rev. Lett. **87**, 201806 (2001) [arXiv:hep-ph/0107002].
23. M. Beneke, G. Buchalla, M. Neubert and C. T. Sachrajda, Phys. Rev. Lett. **83**, 1914 (1999) [arXiv:hep-ph/9905312], M. Beneke, G. Buchalla, M. Neubert and C. T. Sachrajda, Nucl. Phys. B **591**, 313 (2000) [arXiv:hep-ph/0006124].
24. M. Beneke, G. Buchalla, M. Neubert and C. T. Sachrajda, Nucl. Phys. B **606**, 245 (2001) [arXiv:hep-ph/0104110].
25. R. Fleischer and T. Mannel, Phys. Rev. D **57**, 2752 (1998) [arXiv:hep-ph/9704423], A. J. Buras, R. Fleischer and T. Mannel, Nucl. Phys. B **533**, 3 (1998) [arXiv:hep-ph/9711262], M. Neubert and J. L. Rosner, Phys. Lett. B **441**, 403 (1998) [arXiv:hep-ph/9808493], M. Neubert and J. L. Rosner, Phys. Rev. Lett. **81**, 5076 (1998) [arXiv:hep-ph/9809311].
26. B. Aubert et al. [BABAR Collaboration], arXiv:hep-ex/0308012.
27. K. Abe et al. [Belle Collaboration], arXiv:hep-ex/0308040.

Hard Exclusive Processes and Higher-Order QCD Corrections

Kornelija Passek-Kumerički

Theoretical Physics Division, Rudjer Bošković Institute,
P.O. Box 180, 10002 Zagreb, Croatia
passek@thphys.irb.hr

1 Introduction

Quantum Chromodynamics (QCD) offers the description of hadrons in terms of quarks and gluons. There are two basic ingredients of that picture: bound state dynamics of hadrons and fundamental interactions of quarks and gluons. While the former is still rather elusive to existing theoretical tools, the latter is rather well understood. The description of the hadronic processes at large momentum transfer is realized by making use of the factorization of high and low energy (short and long distance) dynamics. The existence of asymptotic freedom makes then the high energy part tractable to the perturbative calculation, i.e., the perturbative QCD (PQCD).

Exclusive processes are defined as the scattering reactions in which the kinematics of all initial and final state particles are specified, like, for example, the processes defining the hadron form factors ($\gamma^* \gamma^{(*)} \to \pi$, $\gamma^* \pi \to \pi$, $\gamma^* \to \pi\pi$, $cp \to cp, \cdots$), the two-photon annihilation processes ($\gamma\gamma \to \pi\pi$, $\gamma\gamma \to \overline{p}p, \cdots$), the hadron scatterings ($\pi p \to \pi p$, $pp \to pp, \cdots$), the decays of heavy hadrons ($J/\psi \to \pi\pi\pi$, $B \to \pi\pi, \cdots$) etc. The hard exclusive reactions, i.e., the exclusive reactions at large momentum transfer (or wide-angle), can be described by the so-called hard-scattering picture [1, 2]. The basis of this picture is the factorization of short and long distance dynamics, i.e, the factorization of the hard-scattering amplitude into the elementary hard-scattering amplitude and hadron distribution amplitudes one for each hadron involved in the process. Usually the following standard approximations are made. Hadron is replaced by the valent Fock state, collinear approximation, in which hadron constituents are constrained to be collinear, is adopted, and the masses are neglected. For example, in the case of the pion this leads to replacing the pion by $|\pi\rangle \longrightarrow |q\overline{q}\rangle$ (correct flavour structure has to be taken into account), adopting $p_q = x\,p$, $p_{\overline{q}} = (1-x)p$, where p, p_q and $p_{\overline{q}}$ are pion, quark and antiquark momenta, respectively, while $0 < x < 1$ is the longitudinal momentum fraction, and taking $m_q = m_{\overline{q}} = 0\,, m_\pi = 0$.

Generally, the hard-scattering amplitude is then represented by the following convolution formula:

$$\mathcal{M}(Q^2) = \int_0^1 [\mathrm{d}x] \quad T_H\left(x_j,\, Q^2,\, \mu_F^2\right) \quad \prod_{h_i} \Phi_{h_i}\left(x_j,\, \mu_F^2\right) \;, \tag{1}$$

$$[\mathrm{d}x] = \prod_{j=1}^{n_{h_i}} \mathrm{d}x_j \, \delta\left(1 - \sum_{k=1}^{n_{h_i}} x_k\right) \;,$$

where T_H is the process-dependent elementary hard-scattering amplitude, Φ_{h_i} is the process-independent distribution amplitude (DA) of the hadron h_i, Q^2 denotes the large momentum transfer while μ_F^2 is the factorization scale at which the separation between short and long distance dynamics takes place.

Within this framework leading-order (LO) predictions have been obtained for many exclusive processes. It is well known, however, that, unlike in QED, the LO predictions in PQCD do not have much predictive power, and that higher-order corrections are essential for many reasons. In general, they have a stabilizing effect reducing the dependence of the predictions on the schemes and scales. Therefore, to achieve a complete confrontation between theoretical predictions and experimental data, it is very important to know the size of radiative corrections to the LO predictions.

The list of exclusive processes at large momentum transfer analyzed at next-to-leading order (NLO) is very short and includes only three processes: the meson electromagnetic form factor [3, 4, 5, 6, 7, 8, 9], the meson transition form factor [10, 11, 7, 12], and the process $\gamma\gamma \to M\overline{M}$ ($M = \pi$, K) [13][1].

We note here that the meson transition form factor belongs to the same class of processes as the deeply virtual Compton scattering (DVCS) [16] ($\gamma^* p \to \gamma^* p$), which recently has been extensively studied in the context of general parton distributions (GPDs) [17]. Regarding the elementary hard-scattering amplitude, these two processes, or correspondingly subprocesses $\gamma^*\gamma^* \to (q\bar{q})$ and $\gamma^* q \to \gamma^* q$, differ only in kinematic region and are related by crossing. Still, the NLO correction to DVCS has been calculated independently [18, 19, 20]. Analogous connection exists between the pion electromagnetic form factor and the deeply virtual electroproduction of mesons (DVEM) [21] ($\gamma^* p \to Mp$, the momentum transfer t between the initial and the final proton is negligible, while the virtuality of the photon is large). In the context of subprocesses, there is a connection between $\gamma^*(q\bar{q}) \to (q\bar{q})$ and $\gamma^* q \to (q\bar{q})q$. In [22] the use has been made of the NLO results for the pion electromagnetic form factor to obtain the NLO prediction for the specific case of DVEM (electroproduction of the pseudoscalar flavour non-singlet mesons). In this work we mostly discuss the meson form factor calculations.

At the next-to-next-to-leading order (NNLO) only the β_0-proportional terms for the deeply virtual Compton scattering and pion transition form factor have been explicitly calculated [23, 24]. The use of conformal constraints made it possible to circumvent the explicit calculation and to obtain the full NNLO results for the pion transition form factor [25].

Finally, let us note that apart from the deeply virtual region, also the wide angle region has been investigated in the literature in the context of the Compton scattering (WACS), as well as, electroproduction of mesons (WAEM) [26]. The NLO corrections were calculated only for the WACS [27, 28].

[1] In contrast to the above introduced standard hard-scattering approach (sHSA), in the so-called modified hard-scattering approach (mHSA) the Sudakov suppression and the transverse momenta of the constituents are taken into account. The LO predictions have again been obtained for number of processes while at NLO order only the pion transition form factor [14] has been calculated. In order to estimate the NLO correction in the mHSA, in [15] the use has been made of the NLO results for the pion electromagnetic and transition form factors obtained using the sHSA.

In this paper the introduction to hard-scattering picture for exclusive processes is given in Sect. 2. The characteristic properties of the PQCD predictions regarding the importance of higher-order corrections and the renormalization scale ambiguities are explained in Sect. 3. Section. 4 is devoted to the short review of exclusive processes calculated to the next-to-leading order (NLO) in the strong coupling constant α_S: meson electromagnetic form factor ($\gamma^* M \to M$), photon-to-meson transition form factor ($\gamma^*\gamma^{(*)} \to M$), meson pair production: $\gamma\gamma \to M\bar{M}$. In Sect. 5 the next-to-next-to-leading order (NNLO) prediction for the photon-to-pion transition form factor obtained using conformal symmetry constraints is explained. Finally, in Sect. 6 the summary and conclusions are given.

2 Introduction to the Hard-Scattering Picture

Let us explain the basic ingredients of the standard hard-scattering picture by taking as an example the simplest exclusive quantity, i.e., the photon-to-pion transition form factor $F_{\pi\gamma^{(*)}}$ appearing in the amplitude of the process $\gamma^*(q_1,\mu)\gamma^{(*)}(q_2,\nu) \to \pi(p)$. At least one photon virtuality has to be large and we take here the simple case : $-q_1^2 = Q^2 \gg$ and $q_2^2 = 0$. The full amplitude is of the form

$$\Gamma^\mu = \mathrm{i}\, e^2\, F_{\pi\gamma}(Q^2)\, \varepsilon^{\mu\nu\alpha\beta}\, q_{1\alpha} q_{2\beta}\, \epsilon_\nu(q_2)\ , \tag{2}$$

and the transition form factor can be represented by a convolution

$$F_{\pi\gamma}(Q^2) = T_H\left(x, Q^2, \mu_F^2\right) \otimes \Phi\left(x, \mu_F^2\right) \tag{3}$$

Here $A(x) \otimes B(x) \equiv \int_0^1 \mathrm{d}x A(x) B(x)$ and μ_F^2 is a factorization scale.

The elementary hard-scattering amplitude T_H obtained from $\gamma^*\,\gamma \to q\bar{q}$ is calculated using the PQCD. By definition, T_H is free of collinear singularities and has a well-defined expansion in $\alpha_S(\mu_R^2)$, with μ_R^2 being the renormalization (or coupling constant) scale of the hard-scattering amplitude. Thus, one can write

$$\begin{aligned} T_H(x, Q^2) &= T_H^{(0)}(x, Q^2) + \frac{\alpha_S(\mu_R^2)}{4\pi} T_H^{(1)}\left(x, Q^2, \mu_F^2\right) \\ &\quad + \frac{\alpha_S^2(\mu_R^2)}{(4\pi)^2} T_H^{(2)}\left(x, Q^2, \mu_F^2, \mu_R^2\right) + \cdots\ . \end{aligned} \tag{4}$$

The diagrams contributing to LO and representative diagrams contributing to NLO order are displayed in Figs. 1 and 2, respectively. When evaluating the NLO amplitude one encounters the UV and collinear singularities. The former are removed by coupling constant (α_S) renormalization introducing the scale μ_R^2, while the latter are factorized into the DA at the scale μ_F^2.

The pion distribution amplitude is defined in terms of the matrix elements of composite operators: $\langle 0|\Psi(-z)\gamma^+\gamma_5 \Omega \Psi(z)|\pi\rangle$. While the DA form is taken as an (nonperturbative) input at some lower scale μ_0^2 ($\Phi(x,\mu_0^2)$), its evolution to the factorization scale μ_F^2 ($\Phi(x,\mu_F^2)$) is governed by PQCD. The DA can hence be written in a form

$$\Phi\left(x, \mu_F^2\right) = \phi_V\left(x, y, \mu_F^2, \mu_0^2\right) \otimes \Phi\left(y, \mu_0^2\right)\ , \tag{5}$$

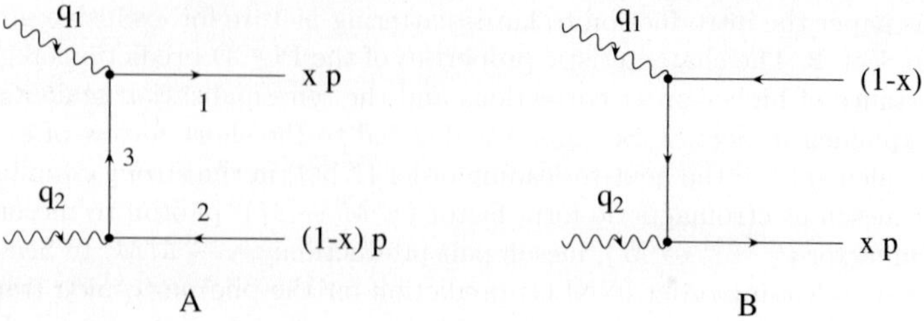

Fig. 1. Lowest-order Feynman diagrams contributing to the $\gamma^*\gamma \to q\bar{q}$ amplitude

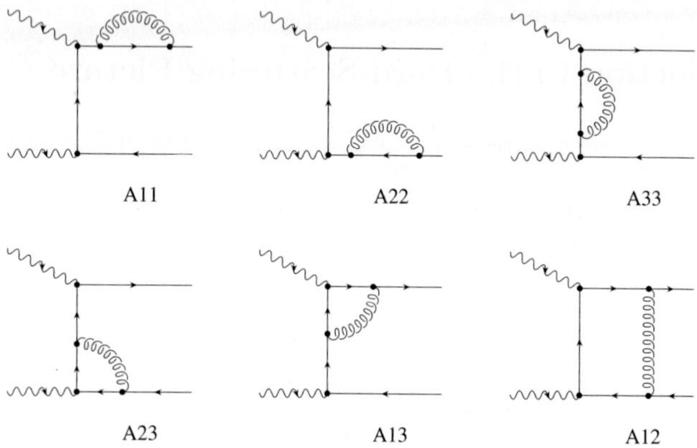

Fig. 2. Distinct one-loop Feynman diagrams contributing to $\gamma^*\gamma \to q\bar{q}$

where ϕ_V denotes the evolution part of the DA. In latter the resummation of $(\alpha_S \ln(\mu_F^2/\mu_0^2))^n$ terms is usually included, and ϕ_V is obtained by solving the evolution equation

$$\mu_F^2 \frac{\partial}{\partial \mu_F^2} \phi_V = V \otimes \phi_V \ , \tag{6}$$

where

$$V = \frac{\alpha_S(\mu_F^2)}{4\pi} V_1 + \frac{\alpha_S^2(\mu_F^2)}{(4\pi)} V_2 + \cdots \tag{7}$$

represents the perturbatively calculable evolution kernel.

One often introduces the distribution amplitude ϕ normalized to unity $\int_0^1 \mathrm{d}x \ \ \phi(x,\mu_F^2) = 1$, and related to $\Phi(x,\mu_F^2)$ by $\Phi = f_\pi/(2\sqrt{2N_c})\,\phi$ where $f_\pi = 0.131\,\mathrm{GeV}$ is the pion decay constant and N_c is the number of colours. The solutions of the evolution equation (6) combined with the nonperturbative input can then be written in a form of an expansion over Gegenbauer polynomials $C_n^{3/2}$ which represent the eigenfunctions of the LO evolution equation:

$$\phi\left(x,\mu_F^2\right) = 6x(1-x)\left[1 + {\sum_{n=2}^{\infty}}' B_n\left(\mu_F^2\right)\, C_n^{3/2}(2x-1)\right] \ . \tag{8}$$

Here $\sum'$ denotes the sum over even indices. The nonperturbative input $B_n(\mu_0^2)$ as well as the evolution is now contained in B_n coefficients. They have a well defined expansion in α_S:

$$B_n(\mu_F^2) = B_n^{LO}(\mu_F^2) + \frac{\alpha_S(\mu_F^2)}{4\pi} B_n^{NLO}(\mu_F^2) + \cdots , \tag{9}$$

where

$$B_n^{LO}(\mu_F^2) = f(\mu_F^2, \mu_0^2, B_n(\mu_0^2)) , \qquad B_n^{NLO}(\mu_F^2) = g(\mu_F^2, \mu_0^2, B_{k(k\leq n)}(\mu_0^2)) \tag{10}$$

represent the LO and NLO [29] parts whose exact form in $\overline{\text{MS}}$ factorization scheme is given in, for example [9, 24].

As the DA input one often takes the asymptotic function

$$\phi_{as} \equiv \phi(x, \infty) = 6x(1-x) \tag{11}$$

being the solution of the DA evolution equation for $\mu_F^2 \to \infty$ and the simplest possibility. We list here two more choices from the literature

$$\begin{aligned} \phi_{CZ}[30] &: B_2(0.25\,\text{GeV}^2) = 2/3 , \\ \phi_{BMS}[31] &: B_2(1\,\text{GeV}^2) = 0.188 \qquad B_4(1\,\text{GeV}^2) = -0.13 . \end{aligned} \tag{12}$$

The CZ distribution amplitude is nowadays mostly ruled out (see [32] and references therein), and it is believed that even at lower energies the pion DA is close to asymptotic form but probably end-point suppressed like the BMS DA.

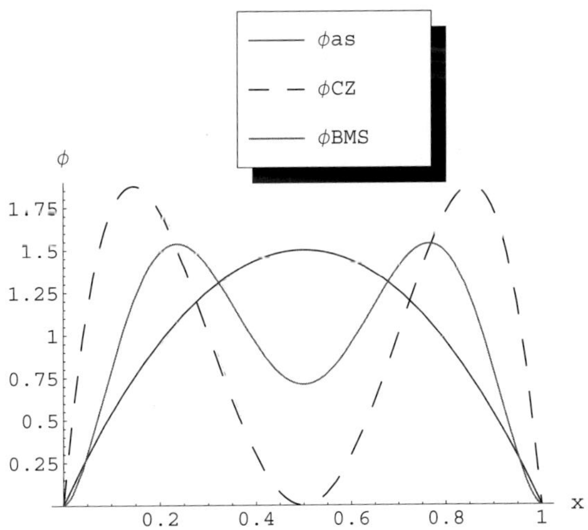

Fig. 3. DA candidates

3 PQCD Prediction

In this section we would like to discuss some properties inherent to all PQCD predictions.

Let us first briefly discuss the expansion parameter, i.e., the QCD coupling constant. The QCD β function given by

$$\beta(\alpha_S(\mu^2)) = \mu^2 \frac{\partial}{\partial \mu^2} \alpha_S(\mu^2) = -\frac{\alpha_S^2(\mu^2)}{4\pi} \beta_0 - \cdots \tag{13}$$

is negative ($\beta_0 = 1/3(11N_c - 2n_f)$, n_f is the number of active flavours) and theory is asymptotically free. The usual one-loop solution of the renormalization group equation (13) is given by

$$\alpha_S(\mu^2) = \frac{4\pi}{\beta_0 \ln(\mu^2/\Lambda^2)} \tag{14}$$

and, obviously, $\alpha_S(\infty) = 0$, while, for example, $\alpha_S((1)4\,\mathrm{GeV}^2) \approx (0.43)0.3$, ($\Lambda = 0.2\,\mathrm{GeV}, n_f = 3$). Obviously, low-energy behaviour of such α_S represents a problem due to the existence of the Landau pole: $\alpha_S(\Lambda^2) \to \infty$. In the literature one can encounter several prescriptions to improve α_S in low-energy region. For example, "frozen" coupling constant

$$\alpha_S(\mu^2) = \frac{4\pi}{\beta_0 \ln((\mu^2 + m_g^2)/\Lambda^2)} \tag{15}$$

or analytical coupling [33]

$$\alpha_S(\mu^2) = \frac{4\pi}{\beta_0} \left[\frac{1}{\ln(\mu^2/\Lambda^2)} + \frac{\Lambda^2}{\Lambda^2 - \mu^2} \right] . \tag{16}$$

But even these improved forms of α_S give rather large values at lower energies. Thus, LO QCD predictions do not have much predictive power and higher-order corrections are important.

Generally, the PQCD amplitude can be written in a form

$$\mathcal{M}(Q^2) = \mathcal{M}^{(0)}(Q^2) + \frac{\alpha_S(\mu_R^2)}{4\pi} \mathcal{M}^{(1)}(Q^2) + \frac{\alpha_S^2(\mu_R^2)}{(4\pi)^2} \mathcal{M}^{(2)}(Q^2, \mu_R^2) + \cdots , \tag{17}$$

where Q^2 is some large momentum and as usual μ_R^2 represents the renormalization scale. The truncation of the perturbative series to finite order introduces the residual dependence of the results on the renormalization scale μ_R and scheme (to the order we are calculating these dependences can be represented by one parameter, say, the scale). Inclusion of higher order corrections decreases this dependence. Nevertheless, we are still left with intrinsic theoretical uncertainty of the perturbative results. One can try to estimate this uncertainty (see, for example [9]) or one can try to find the "optimal" renormalization scale μ_R (and scheme) on the basis of some physical arguments . In the latter case, one can assess the size of the higher order corrections and of the expansion parameter. These values can then serve as a sensible criteria for the convergence of the expansion.

The simplest and widely used choice for μ_R is $\mu_R^2 = Q^2$, and the justification is mainly pragmatic. However, physical arguments suggest that the more appropriate scale μ_R is lower. Namely, since each external momentum entering an exclusive reaction is partitioned among many propagators of the underlying hard-scattering amplitude, the physical scales that control these processes are inevitably much softer than the overall momentum transfer. There are number of suggestions in the literature. According to fastest apparent convergence (FAC) procedure [34], the scale μ_R is determined by the requirement that the NLO coefficient in the perturbative expansion of the physical quantity in question vanishes, i.e., one demands $\mathcal{M}^{(2)}(Q^2, \mu_R^2) = 0$. On the other hand, following the principle of minimum sensitivity (PMS) [35] one mimics the independence of the all order expansion on the scale μ_R, and one chooses the renormalization scale μ_R at the stationary point of the truncated perturbative series: $d\mathcal{M}_{\text{finite order}}(Q^2, \mu_R^2)/\mathrm{d}\mu_R^2 = 0$. In the Brodsky-Lepage-Mackenzie (BLM) procedure [36], all vacuum-polarization effects from the QCD β-function are resummed into the running coupling constant. According to BLM procedure, the renormalization scale best suited to a particular process in a given order can, in practice, be determined by setting the scale demanding that β-proportional terms should vanish:

$$\mathcal{M}^{(2)}(Q^2, \mu_R^2) = \beta_0\, \mathcal{M}^{(2,\beta_0)}(Q^2, \mu_R^2) + \mathcal{M}^{(2,\text{rest})}(Q^2) \tag{18}$$

and

$$\mathcal{M}^{(2,\beta_0)}(Q^2, \mu_R^2) = 0\,. \tag{19}$$

As it is known, the relations between physical observables must be independent of renormalization scale and scheme conventions to any fixed order of perturbation theory. In [37] was argued that applying the BLM scale-fixing to perturbative predictions of two observables in, for example, $\overline{\text{MS}}$ scheme and then algebraically eliminating $\alpha_{\overline{MS}}$ one can relate any perturbatively calculable observables without scale and scheme ambiguity, where the choice of BLM scale ensures that the resulting "commensurate scale relation" (CSR) is independent of the choice of the intermediate renormalization scheme. Following this approach, in paper by Brodsky et al.[38] the several exclusive hadronic amplitudes were analyzed in α_V scheme, in which the effective coupling $\alpha_V(\mu^2)$ is defined from the heavy-quark potential $V(\mu^2)$. The α_V scheme is a natural, physically based scheme, which by definition automatically incorporates vacuum polarization effects. The μ_V^2 scale which then appears in the α_V coupling reflects the mean virtuality of the exchanged gluons. Furthermore, since α_V is an effective running coupling defined from the physical observable it must be finite at low momenta, and the appropriate parameterization of the low-energy region should in principle be included. The scale-fixed relation between the $\alpha_{\overline{MS}}$ and α_V couplings is given by [38]

$$\alpha_{\overline{MS}}(\mu_{BLM}^2) = \alpha_V(\mu_V^2)\left(1 + \frac{\alpha_V(\mu_V^2)}{4\pi}\,\frac{8C_A}{3} + \cdots\right), \tag{20}$$

where $\alpha_V(\mu_V^2)$ is defined from the heavy-quark potential $V(\mu_V^2)$ and

$$\mu_V^2 = \mathrm{e}^{5/3}\,\mu_{BLM}^2\,. \tag{21}$$

4 Exclusive Processes at Higher Order: Explicit Calculations

As mentioned in Sect. 1, only the small number of exclusive processes have been analyzed in higher orders. When higher order calculations are explicitly performed, usually the dimensional regularization together with the $\overline{\text{MS}}$ renormalization scheme is applied.

4.1 Photon-to-π (η, η') Transition Form Factor

The photon-to-π transition form factor appearing in the amplitude $\gamma^*\gamma \to \pi^0$ takes the form of an expansion

$$F_{\pi\gamma}(Q^2) = F^{(0)}_{\pi\gamma}(Q^2) + \frac{\alpha_S(\mu_R^2)}{4\pi} F^{(1)}_{\pi\gamma}(Q^2) + \frac{\alpha_S^2(\mu_R^2)}{(4\pi)^2} \left[\beta_0 \, F^{(2,\beta_0)}_{\pi\gamma}(Q^2, \mu_R^2) + \cdots \right] + \cdots \,, \tag{22}$$

where only the parts that can be found in the literature as explicitly calculated from the contributing Feynman diagrams are written.

There are 2 LO diagrams contributing to the subprocess amplitude $\gamma^*\gamma \to (q\bar{q})$ and displayed in Fig. 1. Furthermore, there are 12 one-loop diagrams contributing at NLO order [10, 11, 7]. The representative diagrams are given in Fig. 2. In the case of the photon-to-η (η') transition form factor the two-gluon states also contribute ($\gamma^*\gamma \to (gg)$) giving rise to 6 more diagrams at NLO [12]. In [24] the β_0-proportional NNLO terms were determined from the 12 two-loop Feynman diagrams obtained from the one-loop diagrams by adding the gluon vacuum polarization bubble.

The numerical predictions for $F_{\pi\gamma}(Q^2)$ are displayed in Fig. 4. Obviously, the results obtained using the CZ DA overshoot the experimental data. The BLM scale for the asymptotic DA amounts to $(\mu^2_{BLM})^{as} \approx Q^2/9$, while $\alpha_S \leq 0.5$ for $Q^2 > 4\,\text{GeV}^2$. In the α_V scheme for the coupling constant scale one obtains $(\mu_V^2)^{as} \approx Q^2/2$.

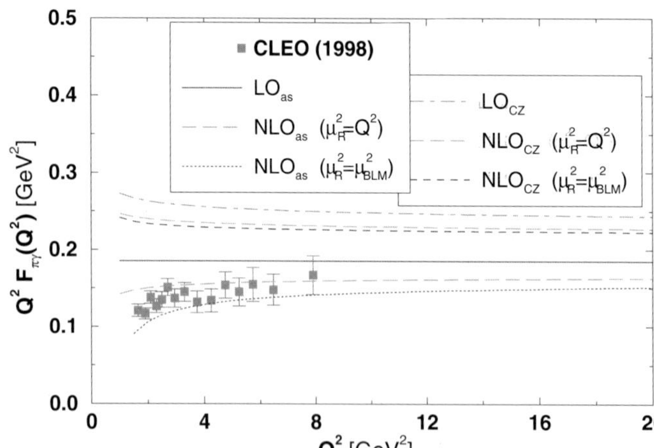

Fig. 4. LO and NLO predictions for the photon-to-pion transition form factor

4.2 Pion Electromagnetic Form Factor

The spacelike[2] pion electromagnetic form factor $F_\pi(Q^2)$ appearing in the amplitude $\gamma^*\pi^{+(-)} \to \pi^{+(-)}$, takes the form of an expansion

$$F_\pi(Q^2) = \frac{\alpha_S(\mu_R^2)}{4\pi} F_\pi^{(1)}(Q^2) + \frac{\alpha_S^2(\mu_R^2)}{(4\pi)^2} F_\pi^{(2)}(Q^2, \mu_R^2) + \cdots . \qquad (23)$$

There are 4 diagrams that contribute to the amplitude $\gamma^*(q_1\bar{q}_2) \to (q_1\bar{q}_2)$ at LO (see Fig. 5), and 62 one-loop diagrams at NLO [3, 4, 5, 6, 8, 7, 9] (see Fig. 6).

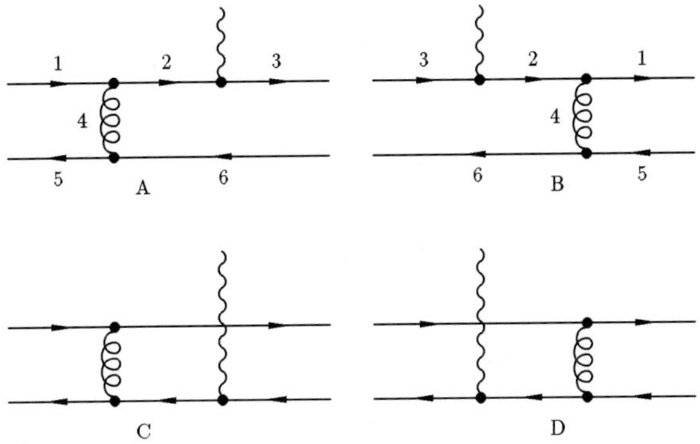

Fig. 5. Lowest-order Feynman diagrams contributing to $\gamma^*(q\bar{q}) \to (q\bar{q})$

Numerical predictions for $F_\pi(Q^2)$ are displayed in Fig. 7. We comment the asymptotic DA results. For $\mu_R^2 = Q^2$ NLO corrections are rather large: the ratio (NLO correction/LO prediction) is $> 30(50)\%$ until $Q^2 > 500(10)\,\mathrm{GeV}^2$ is reached! On the other hand, the BLM scale $\mu_R^2 = (\mu_{BLM}^2)^{as} \approx Q^2/106$ is very small and hence α_S is large. The α_V scheme offers the possible way out. In this scheme the scale amounts to $\mu_R^2 = (\mu_V^2)^{as} \approx Q^2/20$ ($\alpha_S < 0.5$ and NLO corrections <27% for $Q^2 > 20\,\mathrm{GeV}^2$) [40].

4.3 Pion Pair Production

Finally, the amplitude of the process $\gamma\gamma \to \pi^+\pi^-$ takes the form

$$\mathcal{M}(s,t) = \frac{\alpha_S(\mu_R^2)}{4\pi} \mathcal{M}^{(1)}(s,t) + \frac{\alpha_S^2(\mu_R^2)}{(4\pi)^2} \mathcal{M}^{(2)}(s,t,\mu_R^2) + \cdots . \qquad (24)$$

There are 20 diagrams that contribute at LO order $\gamma\gamma \to (q_1\bar{q}_2)(q_2\bar{q}_1)$. At NLO order 454 one-loop diagrams contribute to the NLO prediction. The existing result from the literature [13] covers only the special case of the equal momenta DA, i.e., $\phi(x) = \delta(x - 1/2)$. The numerical result is thus not particularly realistic. New (general) NLO calculation is in preparation and for that purpose convenient general analytical method for evaluation of one-loop Feynman integrals has been developed [41].

[2] For the discussion of the timelike form factor see, for example [39].

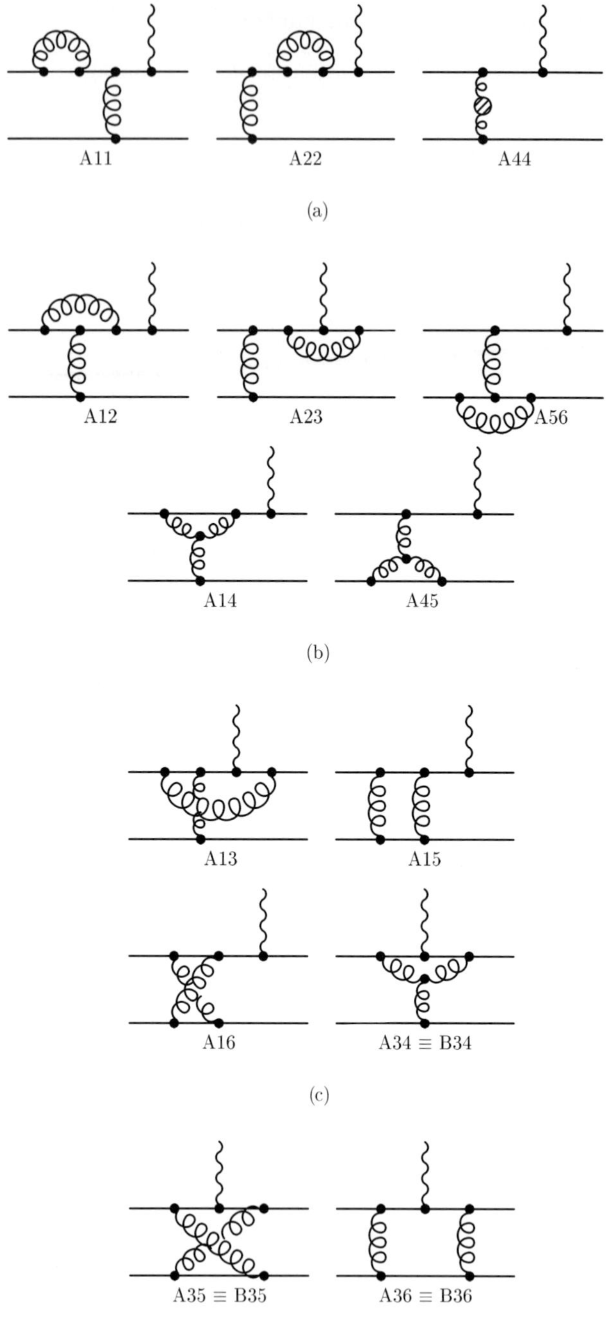

Fig. 6. Distinct one-loop Feynman diagrams contributing to the $\gamma^*(q\bar{q}) \to (q\bar{q})$ amplitude

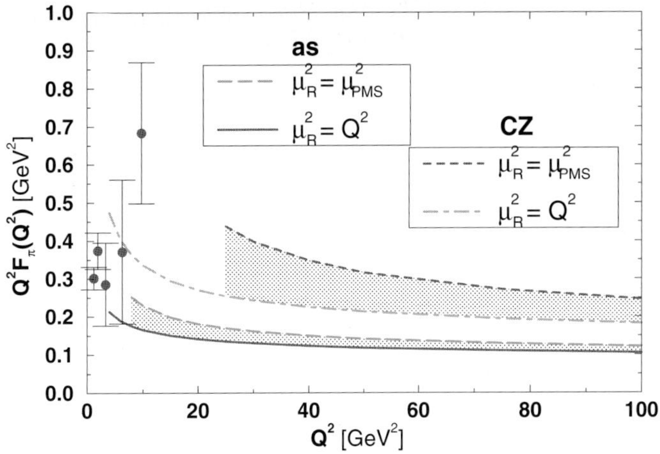

Fig. 7. NLO prediction for $F_\pi(Q^2)$. The shaded area denotes the range of the total NLO prediction and offers the way to asses the theoretical uncertainty

5 NNLO Prediction for the Photon-to-Pion Transition Form Factor Using Conformal Symmetry Constraints

Recently, the conformal symmetry constraints were used to obtain the NNLO prediction for the photon-to-pion transition form factor [25]. The crucial ingredients of this approach lie in the fact that the massless PQCD is invariant under conformal transformations provided that the β function vanishes, and that $F_{\pi\gamma^*}$ belongs to a class of two-photon processes calculable by means of the operator product expansion (OPE). One can then make use of the predictive power of the conformal OPE (COPE), the DIS results for the nonsinglet coefficient function of the polarized structure function g_1 known to NNLO order [42] and the explicitly calculated β-proportional NNLO terms [23, 24].

Let us first introduce the basic ingredients of the formalism. For the general case of the pion transition form factor $\gamma^*(q_1)\gamma^*(q_2) \to \pi(p)$ one expresses the results in terms of

$$\bar{Q}^2 = -\frac{q_1^2 + q_2^2}{2} \qquad \text{and} \qquad \omega = \frac{q_1^2 - q_2^2}{q_1^2 + q_2^2} \, . \tag{25}$$

It is convenient to turn the convolution formula (3) into the sum over conformal moments

$$F_{\pi\gamma^*}(\omega, \bar{Q}^2) = f_\pi \sum_{j=0}^{\infty}{}' \, T_j(\omega, \bar{Q}^2, \mu_F^2) \, \langle \pi | \mathcal{O}_{jj}(\mu_F^2) | 0 \rangle \, . \tag{26}$$

Here

$$T_j(\omega, \bar{Q}^2, \mu_F^2) = \int_0^1 \mathrm{d}x \, T_H(\omega, x, \bar{Q}^2, \mu_F^2) \, \frac{x(1-x)}{2\sqrt{2N_c}N_j} \, C_j^{3/2}(2x-1) \tag{27}$$

is the jth conformal moment of the elementary hard-scattering amplitude, while

$$\phi(x, \mu_F^2) = \sum_{j=0}^{\infty}{}' \frac{x(1-x)}{N_j} C_j^{3/2}(2x-1) \langle \pi | \mathcal{O}_{jj}(\mu_F^2) | 0 \rangle \,, \tag{28}$$

where $\mathcal{O}(\mu_F^2)$ represents composite conformal operator, and $N_j = (j+1)(j+2)/4(2j+3)$.

For the thorough review of the conformal transformations and their applications we refer to [43]. On the quantum level conformal symmetry is broken owing to the regularization and renormalization of UV divergences: coupling constant renormalization resulting in β-proportional terms and the renormalization of composite operators. The latter represents the origin of non-diagonal NLO anomalous dimensions in $\overline{\text{MS}}$ scheme and can be removed by finite renormalization of the hard-scattering and distribution amplitude, i.e., by the specific choice of the factorization scheme.

First, we pose the question weather we can find a factorization scheme in which conformal symmetry holds true up to β-proportional terms? Renormalization group equation for the operators $\mathcal{O}$, equivalent to the DA evolution equation (6), is given by

$$\mu \frac{\mathrm{d}}{\mathrm{d}\mu} \mathcal{O}_{jl} = - \sum_{k=0}^{j} \gamma_{jk} \mathcal{O}_{kl} \,, \tag{29}$$

where the anomalous dimension matrix, corresponding to the evolution kernel (7), is given by

$$\gamma_{jk} = \frac{\alpha_s}{2\pi} \delta_{jk} \gamma_j^{(0)} + \frac{\alpha_s^2}{(2\pi)^2} \gamma_{jk}^{(1)} + \frac{\alpha_s^3}{(2\pi)^3} \gamma_{jk}^{(2)} + O(\alpha_s^4) \tag{30}$$

with $\gamma_j \equiv \gamma_{jj}$. Since the conformal symmetry holds at LO, the anomalous dimensions are diagonal at LO. Similarly non-diagonal terms present in the $\overline{\text{MS}}$ scheme beyond LO, originate in breaking of conformal symmetry due to the renormalization of the composite operators. In contrast, the conformal subtraction (CS) scheme defined by

$$\mathcal{O}^{\text{CS}} = \hat{B}^{-1} \mathcal{O}^{\overline{\text{MS}}}, \qquad B_{jk} = \delta_{jk} + \frac{\alpha_s}{2\pi} B_{jk}^{(1)} + O(\alpha_s^2) \tag{31}$$

and

$$\gamma_{jk}^{\text{CS}} = \delta_{jk} \gamma_j + \theta(j > k) \frac{\beta}{g} \Delta_{jk} \tag{32}$$

preserves the conformal symmetry up to β-proportional terms.

Second, we ask weather and how can we use the predictive power of conformal symmetry? The process of interest belongs to quite a large class of two-photon processes calculable by means of OPE [16]. DVCS, deeply inelastic lepton–hadron scattering (DIS) and production of various hadronic final states by photon–photon fusion belong to this class of processes. Such processes can be described by a general scattering amplitude given by the time-ordered product of two-electromagnetic currents sandwiched between the hadronic states. For a specific process, the generalized Bjorken kinematics at the light-cone can be reduced to the corresponding kinematics, while the particular hadron content of the process reflects itself in the non-perturbative part of the amplitude. Hence, the generalized hard-scattering amplitude enables us to relate predictions of different two-photon processes on partonic level.

Conformal OPE (COPE) for two-photon processes works under the assumption that conformal symmetry holds (CS scheme and $\beta = 0$), and the Wilson coefficients are then, up to normalization, fixed by the ones appearing in DIS structure function g_1 (calculated to NNLO order). Conformal symmetry breaking terms proportional to β function alter COPE result. One can make use of β-proportional NNLO terms explicitly calculated in $\overline{\text{MS}}$ scheme [23, 24]. We note here that there exists freedom in defining β-proportional terms in CS scheme and hence we speak of CS scheme, $\overline{\text{CS}}$ scheme, ... (for detailed explanation see [25]).

Finally we list the numerical results for the special case $\omega = \pm 1$ and $j = 0$, i.e., asymptotic DA. For $\mu_R^2 = 2\bar{Q}^2$

$$F_{\pi\gamma}(\bar{Q}^2) = \frac{\sqrt{2} f_\pi}{2\bar{Q}^2} \left[1 - \frac{\alpha_s(2\bar{Q}^2)}{\pi} - \left\{ \begin{matrix} 7.23 \\ 5.14 \end{matrix} \right\} \frac{\alpha_s^2(2\bar{Q}^2)}{\pi^2} \right. \tag{33}$$

$$\left. + O(\alpha_s^3) \right] \quad \text{in} \quad \left\{ \begin{matrix} \text{CS} \\ \overline{\text{CS}} \end{matrix} \right. \text{-scheme} , \tag{34}$$

while for the BLM prescription $\mu_R^2 = \mu_{\text{BLM}}^2$

$$F_{\pi\gamma}(\bar{Q}^2) = \frac{\sqrt{2} f_\pi}{2\bar{Q}^2} \left[1 - \frac{\alpha_s(\mu_{\text{BLM}}^2)}{\pi} + 0.92 \frac{\alpha_s^2(\mu_{\text{BLM}}^2)}{\pi^2} + O(\alpha_s^3) \right] . \tag{35}$$

Here

$$\mu_{\text{BLM}}^2 = 2Q^2 \left\{ \begin{matrix} 1/37.43 \\ 1/14.78 \end{matrix} \right\} \quad \text{in} \quad \left\{ \begin{matrix} \text{CS} \\ \overline{\text{CS}} \end{matrix} \right. \text{-scheme} . \tag{36}$$

One notices that, similarly to the pion electromagnetic form factor results presented in the preceding section, for μ_R^2 equal to the characteristic scale of the process the QCD corrections are large[3], while for the BLM scale these corrections are smaller but the scale itself is also rather small leading to large expansion parameter α_S. The α_V scheme could as in the case of the pion electromagnetic form factor offer the way out and physically better motivated description of the transition form factor.

We mention, that, as already noticed in [44] and shown in [25], the significance of higher-conformal moments decreases with $|\omega|$ and that with decreasing $|\omega|$ the difference between various schemes also decreases. Hence, small $|\omega|$ region is suitable for a novel test of PQCD.

6 Conclusions

Although the higher-order QCD corrections are important, only few exclusive processes have been explicitly calculated to NLO order. The inclusion of higher-order corrections stabilizes the dependence on renormalization scale. Still, the usual choice μ_R^2 = [characteristic scale of the process] leads to large corrections. Other choices of scales (BLM, α_V scheme) are preferable and more physical. More effort in calculating higher-order corrections are needed and some tools applicable to the kinematic region of interest are underway. Furthermore, for some processes (example: NNLO calculation of photon-to-pion transition form factor), one can make use of the predictive power of conformal symmetry to avoid cumbersome higher-order calculations.

[3] Note that for the case of the meson transition form factor NLO correction represents actually LO QCD correction, while NNLO correction is NLO QCD correction etc.

Acknowledgements

I would like to take the opportunity to thank P. Kroll, B. Melić, D. Müller and B. Nižić for fruitful collaborations in challenging higher order calculations, my collaborators A. P. Bakulev, N. G. Stefanis and W. Schroers for interesting extensions of these investigations, as well as, H. W. Huang, R. Jakob, M. Schürmann and W. Schweiger for collaborating on stimulating and demanding LO calculations. This work was supported by the Ministry of Science and Technology of the Republic of Croatia under Contract No. 0098002.

References

1. G. P. Lepage and S. J. Brodsky: Phys. Lett. B **87**, 359 (1979); G. P. Lepage and S. J. Brodsky: Phys. Rev. Lett. **43**, 545 (1979) [Erratum-ibid. **43**, 1625 (1979)]; A. V. Efremov and A. V. Radyushkin: Theor. Math. Phys. **42**, 97 (1980) [Teor. Mat. Fiz. **42**, 147 (1980)]; A. V. Efremov and A. V. Radyushkin: Phys. Lett. B **94**, 245 (1980); A. Duncan and A. H. Mueller: Phys. Lett. B **90**, 159 (1980); A. Duncan and A. H. Mueller: Phys. Rev. D **21**, 1636 (1980)
2. G. P. Lepage and S. J. Brodsky: Phys. Rev. D **22**, 2157 (1980)
3. R. D. Field, R. Gupta, S. Otto and L. Chang: Nucl. Phys. **B186**, 429 (1981)
4. F. M. Dittes and A. V. Radyushkin: Sov. J. Nucl. Phys. **34**, 293 (1981) [Yad. Fiz. **34**, 529 (1981)]
5. M. H. Sarmadi: Ph. D. thesis, University of Pittsburgh, 1982, UMI 83-18195
6. R. S. Khalmuradov and A. V. Radyushkin: Sov. J. Nucl. Phys. **42**, 289 (1985) [Yad. Fiz. **42**, 458 (1985)]
7. E. P. Kadantseva, S. V. Mikhailov and A. V. Radyushkin: Yad. Fiz. **44**, 507 (1986) [Sov. J. Nucl. Phys. **44**, 326 (1986)]
8. E. Braaten and S. Tse: Phys. Rev. D **35**, 2255 (1987)
9. B. Melić, B. Nižić and K. Passek: Phys. Rev. D **60**, 074004 (1999) [hep-ph/9802204]
10. F. del Aguila and M. K. Chase: Nucl. Phys. **B193**, 517 (1981)
11. E. Braaten: Phys. Rev. D **28**, 524 (1983)
12. P. Kroll and K. Passek-Kumerički: Phys. Rev. D **67** (2003) 054017 [hep-ph/0210045];
13. B. Nižić: Phys. Rev. D **35**, 80 (1987)
14. I. V. Musatov and A. V. Radyushkin: Phys. Rev. D **56**, 2713 (1997) [hep-ph/9702443]
15. N. G. Stefanis, W. Schroers and H. C. Kim: Eur. Phys. J. C **18**, 137 (2000) [hep-ph/0005218]
16. D. Müller, D. Robaschik, B. Geyer, F.-M. Dittes, and J. Hořejši: Fortschr. Phys. 42, 101 (1994) [hep-ph/9812448].
17. X. D. Ji: Phys. Rev. D **55**, 7114 (1997) [hep-ph/9609381]; A. V. Radyushkin: Phys. Rev. D **56**, 5524 (1997) [hep-ph/9704207].
18. L. Mankiewicz, G. Piller, E. Stein, M. Vanttinen and T. Weigl: Phys. Lett. B **425**, 186 (1998) [hep-ph/9712251]
19. X. D. Ji and J. Osborne: Phys. Rev. D **58**, 094018 (1998) [hep-ph/9801260]
20. A. V. Belitsky and D. Müller: Phys. Lett. B **417**, 129 (1998) [hep-ph/9709379]

21. A. V. Radyushkin: Phys. Lett. B **385**, 333 (1996) [hep-ph/9605431]; J. C. Collins, L. Frankfurt and M. Strikman: Phys. Rev. D **56**, 2982 (1997) [hep-ph/9611433]
22. A. V. Belitsky and D. Müller: Phys. Lett. B **513**, 349 (2001) [hep-ph/0105046]
23. A.V. Belitsky and A. Schäfer: Nucl. Phys. **B527**, 235 (1998) [hep-ph/9801252]
24. B. Melić, B. Nižić and K. Passek: Phys. Rev. D **65**, 053020 (2002) [hep-ph/0107295]
25. B. Melić, D. Müller and K. Passek-Kumerički: Phys. Rev. D **68**, 014013 (2003) [hep-ph/0212346]
26. H. W. Huang and P. Kroll, Eur. Phys. J. **C17**, 423 (2000), [hep-ph/0005318].
27. H. W. Huang, P. Kroll and T. Morii: Eur. Phys. J. C **23**, 301 (2002) [Erratum-ibid. C **31**, 279 (2003)] [hep-ph/0110208].
28. H. W. Huang and T. Morii: Phys. Rev. D **68**, 014016 (2003) [hep-ph/0305132].
29. D. Müller: Phys. Rev. D **49**, 2525 (1994); Phys. Rev. D **51**, 3855 (1995) [hep-ph/9411338]
30. V. L. Chernyak and A. R. Zhitnitsky: Phys. Rept. **112**, 173 (1984)
31. A. P. Bakulev, S. V. Mikhailov and N. G. Stefanis: Phys. Lett. B508, 279 (2001) [hep-ph/0103119]
32. B. Melić, B. Nižić and K. Passek: FizikaB **8**, 327 (1999);
33. D. V. Shirkov and I. L. Solovtsov: Phys. Rev. Lett. **79**, 1209 (1997) [hep-ph/9704333]
34. G. Grunberg: Phys. Lett. B **95**, 70 (1980); Phys. Rev. D **29**, 2315 (1984)
35. P. M. Stevenson: Phys. Lett. B **100**, 61 (1981); Phys. Rev. D **23**, 2916 (1981); Nucl. Phys. B **203**, 472 (1982); Nucl. Phys. B **231**, 65 (1984)
36. S. J. Brodsky, G. P. Lepage and P. B. Mackenzie: Phys. Rev. D **28**, 228 (1983)
37. S. J. Brodsky and H. J. Lu, Phys. Rev. D **51**, 3652 (1995) [hep-ph/9405218]
38. S. J. Brodsky, C. Ji, A. Pang and D. G. Robertson, Phys. Rev. D **57**, 245 (1998) [hep-ph/9705221]
39. A. P. Bakulev, A. V. Radyushkin and N. G. Stefanis: Phys. Rev. D **62**, 113001 (2000) [hep-ph/0005085]
40. B. Melić, B. Nižić and K. Passek: in Proc. of Joint INT/JLab Workshop, Newport News 1999, 279-286, hep-ph/9908510
41. G. Duplancic and B. Nizic: Eur. Phys. J. C **20**, 357 (2001) [hep-ph/0006249]; Eur. Phys. J. C **24**, 385 (2002) [hep-ph/0201306]; hep-ph/0303184.
42. E. B. Zijlstra and W. L. van Neerven: Nucl. Phys. **B417**, 61 (1994) [Erratum-ibid. B426:245, (1994)]
43. V. M. Braun, G. P. Korchemsky and D. Muller: Prog. Part. Nucl. Phys. **51**, 311 (2003) [hep-ph/0306057]
44. M. Diehl, P. Kroll, and C. Vogt: Eur. Phys. J. C22,439 (2001) [hep-ph/0108220].

Strings in the Yang-Mills Theory: How They Form, Live and Decay

Adi Armoni[1] and Mikhail Shifman[2]

[1] Theory Division, CERN, 1211 Geneva 23, Switzerland
[2] William I. Fine Theoretical Physics Institute, University of Minnesota, Minneapolis, MN 55455, USA
shifman@umn.edu

1 Introduction

In confining theories, such as the Yang-Mills theory, non-supersymmetric or supersymmetric ($\mathcal{N}=1$ gluodynamics), heavy (probe) color sources in the fundamental representation are connected by color flux tubes, fundamental QCD strings. The fundamental string tension is of the order of Λ^2 where Λ is the dynamical scale parameter of the gauge theory under consideration, and its transverse size is of the order of Λ^{-1}. Both parameters are independent of the number of colors (in what follows the gauge group is assumed to be SU(N)) and, besides Λ, can contain only numerical factors.

Significant effort has been invested recently in studies of the flux tubes induced by color sources in higher representations of SU(N), mostly in connection – but not exclusively – with high-precision lattice calculations, see e.g. [1, 2, 3, 4, 5, 6, 7] and references therein. The composite flux tubes attached to such color sources are also known as k-strings, where k denotes the N-ality of the color representation under consideration. The N-ality of the representation with λ upper and m lower indices (i.e. λ fundamental and m anti-fundamental) is defined as $k=|\lambda-m|$.

Some qualitative features of the k-strings are well understood. The most important feature is that the string tension does not depend on the particular representation of the probe color source, but only on its N-ality. Indeed, the particular Young tableau of the representation plays no role, since all representations with the given N-ality can be converted into each other by emitting an appropriate number of soft gluons. For instance, the adjoint representation has zero N-ality; therefore, the color source in the adjoint can be completely screened by gluons, and the flux tube between the adjoint color sources should not exist. The same is true for any representation with N fundamental or N anti-fundamental indices. Theoretical ideas regarding confinement in lattice gauge theories and the role of N-ality are extensively reviewed in [8].

We view a k-string as a bound state of k elementary fundamental strings (see Fig. 1). This standpoint is natural in the gauge theory/supergravity correspondence [9] where the k-string world sheet (Wilson loop) is described as k coincident elementary world sheets [10, 11]. Accordingly, in field theory one can *define* the k-string Wilson loop as k coincident fundamental Wilson loops. In doing so, one will deal with the probe source which belongs to a *reducible* representation with k upper indices. The string (or the flux tube) which develops between two fundamental sources will be considered as given. Qualitative ideas regarding the emergence of these strings (flux tubes) can be traced back to the classic treatises [12], while

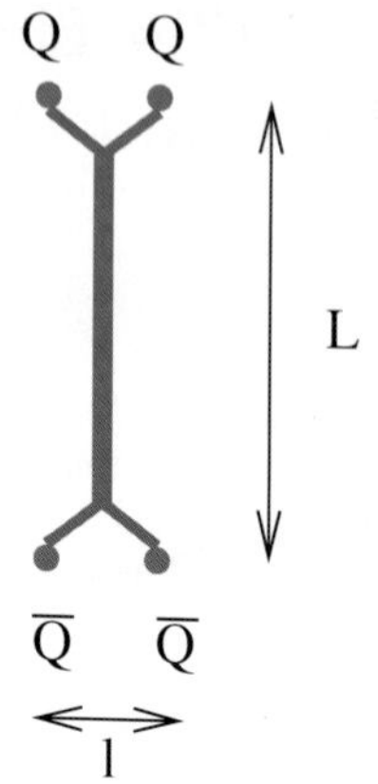

Fig. 1. A flux tube for the 2-string

some new theoretical insights are described e.g. in [13]. Our task is the study of the composite strings – stable, quasistable and unstable. We will analyze mechanisms of the fundamental string attraction at large N (here N is the number of colors), tension defects and what can be guessed regarding cross-sections of the composite strings. A large variety of strings that do not fall into the class of k-strings was detected on the lattices, for instance, the so-called adjoint strings. We will demonstrate that such strings are quasistable – this is the second focus of our presentation. We will show how one can calculate the decay rates of the quasistable strings and explain why they are strongly suppressed at large N. This talk is based mainly on the results obtained in [14, 15].

2 Theoretical Background, Lattice Data

The purpose of this section is to define and review the notion of stable k-strings and review the way their tensions are measured.

2.1 k-Strings: Theoretical Background

Suppose we wish to measure a long-distance quark-antiquark potential in pure SU(N) gauge theory or in its minimal supersymmetric extension, supersymmetric gluodynamics. We assume that the quark-antiquark pair is in a specific representation R. The expected long-distance potential is

$$V = \sigma L \, . \tag{1}$$

The stable string tension σ should not depend on the specific representation R of the quark-antiquark pair, but rather on its N-ality k, since a soft gluon can transform a representation R into a representation R' within the same N-ality. For this reason we can evaluate the string tension by considering a *reducible* representation with a given N-ality k instead of a specific representation R, provided this reducible representation contains R. For instance, one can consider, as a probe source, an ensemble of k heavy quarks (in the fundamental representation) nailed in close proximity from each other. Another source, at distance L from the former, will be composed of k heavy anti-quarks. Since the fully anti-symmetric representation is expected to have the lowest energy, the string attached to a source in a reducible representation will evolve into the anti-symmetric string after a certain time τ. The time τ is definitely $\gtrsim$ than a typical inverse splitting between the energies of the anti-symmetric string and those in other representations. In fact, sometimes it may be much larger.

By the same token, if the probe charges have N-ality zero, even if they are connected by a string at time zero, it will inevitably evolve into a no-string state since such charges can be totally screened. For instance, the adjoint probe heavy quark is screened by a gluon. Another example is as follows: if one has N fundamental quarks separated from N fundamental anti-quarks by a large distance L eventually each of the two N-quark ensembles will develop string junctions (a baryon vertex), and there will be no string connecting the two ensembles.

A graphic illustration is presented in Fig. 2. Suppose that one calculates the expectation value of a Wilson loop in the adjoint representation (i.e. induced by an adjoint probe "quark"). Dynamically, a Wilson "counter-loop" in the adjoint representation will be formed thus screening the non-dynamical probe quark. In physical terms, a gluon lump is produced which combines with the probe adjoint source to form a color singlet.

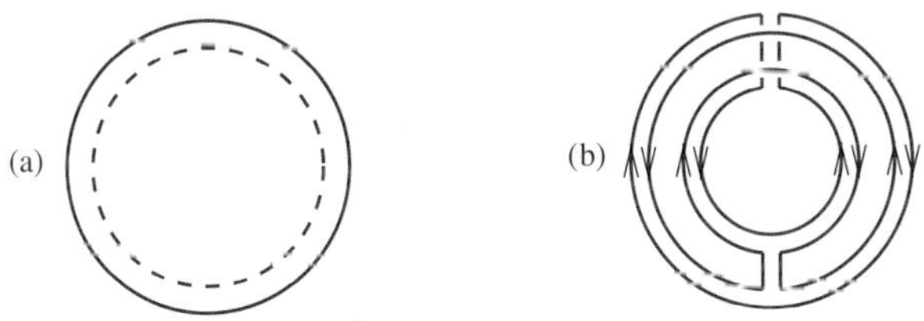

Fig. 2. (**a**). A screening of an adjoint Wilson loop by a dynamical adjoint loop. (**b**). The same process described in 't Hooft double index notation. The process in non-planar

The fundamental string, by definition, is the one that connects a fundamental heavy quark with an anti-quark. The interaction between the fundamental strings is via the glueball exchanges. The process is *non-perturbative* in the coupling g^2N. Although one uses the Feynman graphs in the 't Hooft representation in order to analyze the N dependence, by no means it is implied that the results thus obtained are perturbative in the gauge coupling and "correspond to few-gluon exchange." As is standard in the 't Hooft framework, since the $1/N$ analysis is based only on topology of the graphs (e.g. planarity vs. non-planarity), the results are expected to fully represent the non-perturbative strong coupling gauge dynamics.

Since we deal with the interactions between color-singlet objects and the interaction is non-planar, it will be controlled by $O(1/N^2)$, see [15]. At $N = \infty$ the interaction vanishes, and we have k free fundamental strings with $\sigma_k = k\sigma_{\rm f}$. A sample $1/N^2$ interaction between two fundamental Wilson loops is depicted in Fig. 3a.

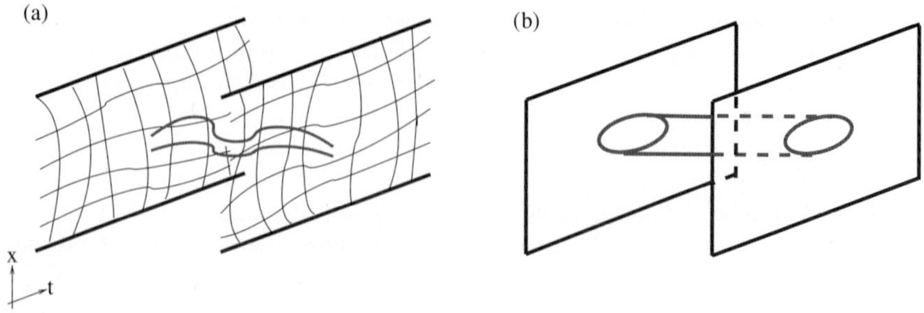

Fig. 3. The interaction of two "fundamental" strings: (**a**) Field-theory picture – (two)gluon exchange; (**b**) String-theory picture – exchange of a closed string between two world sheets

The fact that the corrections to the free strings relation $\sigma_k = k\sigma_1$ run in powers of $1/N^2$ was used [15] to exclude the Casimir scaling hypothesis. The idea is as follows. Expanding the Casimir relation

$$\sigma_k = \Lambda^2 \frac{k(N-k)}{N-1} \tag{2}$$

in powers of $1/N$, yields

$$\sigma_k = \Lambda^2 k \left(1 - \frac{k-1}{N} + \ldots\right) . \tag{3}$$

The leading correction is the first power of $1/N$. This is in contradiction with SU(N) gauge dynamics, as the genus expansion always gives even powers of $1/N$.

On general grounds one can assert that at distances $\gtrsim \Lambda^{-1}$ the fundamental strings attract each other [15], while at short distances there is a repulsion (see the end of this section), so that composite k-strings develop.

This picture is supported by string theory. String theorists prefer to call the Wilson loop a string world sheet. The realization of the k-string will be merely k coincident elementary (fundamental) strings, or more precisely, a bound state of k elementary string world sheets, see Fig. 3b.

Within the AdS/CFT correspondence [9] we can elevate this model to a quantitative level [10, 11]. Indeed, the AdS/CFT correspondence is a natural framework for calculating the k-string tension, since in the dual string picture we work at strong 't Hooft coupling. In addition, $1/N$ is represented by $g_{\rm st}$. In this framework the Wilson loop on the AdS boundary is described by a fundamental world sheet that extends inside the bulk AdS [10]. The value of the Wilson loop is the area of the minimal surface, given simply by the Nambu-Goto action

$$\langle \mathcal{W} \rangle = \exp\left(-S_{\rm NG}\right) . \tag{4}$$

Similarly, the k-string Wilson loop (8) is described by k coincident world sheets. In the supergravity approximation, $g_{\rm st} = 0$, and we obtain $\sigma_k = k\sigma_1$ [11], as expected. In order to calculate $1/N$ corrections, one has to consider string interactions, namely to go beyond the lowest order in $g_{\rm st}$. It is clear, however, that in the closed string theory (and, in particular, in type IIB) the expansion parameter is $g_{\rm st}^2 = 1/N^2$.

The fact that the interaction between the fundamental strings is proportional to $1/N^2$ and, thus, vanishes at $N \to \infty$ is explained in detail in [15]. Moreover, from the same work we know that at distances $\gtrsim \Lambda^{-1}$ the interaction is attractive. An attractive potential between the fundamental strings is also obtained [16] in lattice strong-coupling calculations. The key question is what is the nature of the inter-string interaction at distances of the order of the fundamental string thickness.

Logically there are two options. Either the parallel strings attract at all distances (then k closely situated 1-strings will glue together forming a structureless flux tube carrying k units of N-ality), or the attraction gives place to repulsion at shorter distances (then the k-string will have a substructure in the transverse plane reminiscent of that of a nucleus). The large-N expansion of the k-string tension suggests that it is the latter option that is realized in QCD. Indeed, on general grounds, with no model dependence, one can show that at $N \gg 1$ and $k \gg 1$

$$\left(\frac{\sigma_k}{k\sigma_{\rm f}} - 1\right) \propto \frac{k^2}{N^2}\,.$$

The above k dependence of the binding "energy" has no natural explanation in the picture of a forced "collectivization" inevitable under the assumption that the parallel 1-strings attract at all distances, see Sect. 4.

At the same time, the repulsion at shorts distances, that naturally leads to a nucleus-type structure of the slice of the k-string, *and* explains the above k dependence of the binding "energy," is an immediate consequence of the following consideration.

The trace of the energy-momentum tensor θ^μ_μ in pure Yang-Mills theory or in QCD with massless fermions has the form (the scale anomaly)

$$\theta^\mu_\mu = \frac{-b}{32\pi^2} G^a_{\mu\nu}\, G^{\mu\nu\,,\,a}$$

where $G^a_{\mu\nu}$ is the *operator* of the gluon field strength tensor (b is the first coefficient of the Gell-Mann–Low function). One can use $\theta^\mu_\mu(x)$ as a local probe of the energy density. Indeed, the value of $\theta^\mu_\mu \propto G^2$, in the presence of a Wilson loop $W(C)$, measures the string tension [17],

$$\langle W(C)\rangle^{-1} \int {\rm d}^3x\, \langle \theta^\mu_\mu(x)\,,\, W(C)\rangle_{\rm c,\ eucl} = 2\sigma R\,, \tag{5}$$

where the subscript c stands for the connected part. The chromomagnetic part is presumably negligible for the static flux tubes attached to static color sources. This implies that the energy density is proportional to the expectation value of the operator $\boldsymbol{E}^2(x)$, where $\boldsymbol{E}$ is the chromoelectric field, and the arrow is used to represent both vectorial and color indices. Then the string tension (the energy per unit length in the z direction) is given by the integral

$$\sigma \propto \int {\rm d}^2x\, \langle \boldsymbol{E}^2(x)\rangle\,,$$

where the integral runs in the perpendicular plane.

Let us assume that two one-strings comprising a 2-string significantly overlap, as in Fig. 4. We will see momentarily that this assumption is inconsistent.

Two overlapping fundamental flux tubes are depicted in Fig. 4. Now

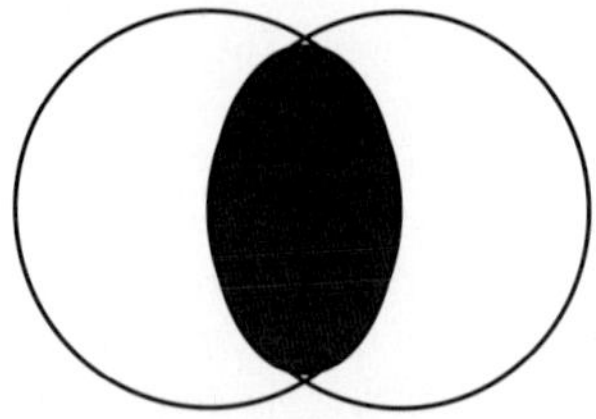

Fig. 4. The spatial slice of two overlapping fundamental strings

$$\sigma_2 \propto \int \mathrm{d}^2x \left\langle (\boldsymbol{E}_{\mathrm{f}\,1}(x) + \boldsymbol{E}_{\mathrm{f}\,2}(x))^2 \right\rangle = 2\,\sigma_{\mathrm{f}} + 2 \int \mathrm{d}^2x \, \langle \boldsymbol{E}_{\mathrm{f}\,1}(x)\, \boldsymbol{E}_{\mathrm{f}\,2}(x) \rangle$$

The fluxes are fixed by the given (static) color sources. This implies that the interference term is positive. It is not difficult to see that it is also suppressed by $1/N^2$, because generically $\boldsymbol{E}_{\mathrm{f}\,1}$ is orthogonal to $\boldsymbol{E}_{\mathrm{f}\,2}$ in the color space; only the Cartan components of the chromoelectric field are important. Thus, if two parallel 1-strings overlap, the energy per unit length is larger than $2\,\sigma_{\mathrm{f}}$ by Λ^2/N^2, i.e. overlapping flux tubes repel each other.

Certainly, the arguments presented above are somewhat quasiclassical. We believe, however, that they are qualitatively correct; they certainly lead to a self-consistent overall picture.

2.2 A Mini-Review of the Lattice Literature on k-Strings

The Euclidean/lattice formulation of the problem is usually given through the expectation value of a rectangular Wilson loop,

$$\langle \mathcal{W}(C) \rangle = \left\langle \operatorname{tr} \exp \left(i \int A^a_\mu T^a_R \, \mathrm{d}x^\mu \right) \right\rangle \,, \tag{6}$$

where T^a_R are the generators of $\mathrm{SU}(N)$ in the given representation, which may or may not be irreducible. Sometimes, for practical purposes, it is more convenient to consider a correlation function of two Polyakov lines. Assume that the (Euclidean) time direction is compactified, and one defines two Polyakov lines in the time direction, separated by a distance L. The Polyakov line is defined through the integral similar to (6), with the generators in the given representation R. If time interval is large enough, the measurement of the correlation function of two Polyakov lines must yield $\exp(-\sigma_R L\, T)$.

The generators of the reducible representation with N-ality k are given by tensor products of the fundamental representation. For example, in the case of N-ality 2 we have

$$\square \otimes \square = \begin{array}{c}\square\\[-4pt]\square\end{array} + \square\!\square \,. \tag{7}$$

Therefore we can evaluate the k-string tension by considering the following Wilson loop

$$\left\langle \mathcal{W}_{k\text{-string}}(C) \right\rangle \equiv \left\langle \left[\mathcal{W}_{\text{fund}}(C) \right]^k \right\rangle \,. \tag{8}$$

Physically, the above definition (8) represents k-fundamental coincident Wilson loops. It is instructive to think of it as of k fundamental probe sources placed at one point. The resulting composite probe source is in the reducible representation of the N-ality k, which includes a mixture of irreducible representations, starting from fully anti-symmetric, up to fully symmetric.

One of the main tasks of this talk is to bridge a gap between theoretical studies of k-strings from the string theory side and large-N field theory side on the one hand, and lattice studies, on the other hand. Therefore, it is natural to give a brief summary of some lattice works devoted to k-strings.

Let us start with analytic studies, namely the strong coupling expansion. Since the seminal work of Wilson [18] it is known that in the Yang-Mills theory with no dynamical matter in the fundamental representation, the expectation value of a *large* Wilson loop induced by a heavy (non-dynamical) fundamental quark will trivially exhibit an area law

$$\left\langle \mathrm{tr}\, \exp\left(i \int A_\mu \mathrm{d}x^\mu \right) \right\rangle \sim \exp\left(-\sigma\, \mathcal{A}\right) \, . \tag{9}$$

Here σ the string tension, $\sigma \sim (\ln g^2) a^{-2}$, where a is the lattice cite and g is the gauge coupling, $g \to \infty$.

The area law for the fundamental Wilson loop is also well established in the *continuum limit*, through numerical simulations. To detect the area law there is an obvious necessary condition on the area of the loop,

$$\mathcal{A} \gg \Lambda^{-2} \, . \tag{10}$$

The slope in front of the area (9) – the fundamental string tension – is measured to a reasonable accuracy. Moreover, a perpendicular slice of the fundamental string was studied too. A typical transverse dimension of the fundamental string is $\sim 0.7\,\mathrm{fm}$.

On the other hand, an area law has been also detected for the adjoint Wilson loop for contours satisfying the condition (10). For instance, typical distances in [6] were ~1.5 to 2 fm. The adjoint string tension was measured. The reported ratio $\sigma_{\mathrm{adj}}/\sigma_{\mathrm{f}}$ is close to the Casimir formula which for SU(3) yields 9/4. (Please, note that this number is *larger* than 2; we will return to this point in Sect. 5.) In fact, the adjoint strings emerge practically in all lattice studies. So far, only one work [5] reports an observation of the adjoint string breaking.[1]

A linear potential between an adjoint probe quark and anti-quark implies the existence of a *quasi-stable string*. It is certain that for asymptotically large areas, the area law *must* give place to the perimeter one; the linear potential *must* flatten off at the level corresponding to the creation of two gluelumps [5]. The effect is not seen, however, in the existing simulations. The question to ask is: "what is the critical size/time needed for detecting the adjoint string breaking?"

Now, the very same dynamical adjoint "counter-loop" as in Fig. 2 (or a few "counter-loops") can transform external probe quarks in a given representation R to a different representation R' with the same N-ality. In physical terms, if we have, say, an anti-symmetric source $Q^{[ij]}$, we can convert it into an object with two symmetric indices $\{ij\}$ by adding a soft gluon. It is clear, then, why one expects the k-string tension to depend on the N-ality, and not on the specific representation R. The genuine stable string for the given N-ality is expected to be attached to the probe source in the anti-symmetric representation. The probe

[1] This work is based on a technique different from all other calculations.

sources in other representations with k (upper) indices give rise to quasi-stable strings whose tension is supposed to depend not only on the N-ality, but on the particular representation. If the probe sources have $k > [N]/2$, we will call such strings *oversaturated*. Representations which have both upper and lower indices can be viewed as bound states of k-strings and some number of adjoint strings. Obviously, they are also quasi-stable. We will refer to such strings as *bicomposites*.

Lattice measurements of the Wilson loops or Polyakov line correlators seem to defy the above argument. They yield string tensions which depend on the particular representation under consideration rather than on the N-ality of the representation. For instance, in [3] which treats the SU(4) and SU(6) cases and measures the anti-symmetric and symmetric Wilson loops, distinct tensions are obtained for the anti-symmetric and symmetric two-index representations. It was argued [3] that the symmetric string is not stable and that it will have to decay into two fundamental strings. In light of the above, again, the most crucial question is what contour sizes are needed to exhibit the decay of the symmetric string into anti-symmetric. It is clear that the answer depends on the decay rate of the symmetric string (Sect. 6).

Returning to the tensions of the stable (anti-symmetric) strings, there is no consensus in the lattice literature on the k-string tension. The Casimir scaling is often reported. In [2] is it argued that in three dimensions the k-string tension is very close to the Casimir scaling. A similar claim is made in [7] for four-dimensional SU(3) theories, where the Casimir behavior is found for symmetric representations. On the other hand, there are dedicated studies [3] which favor the sine formula for the antisymmetric strings.

As we have already mentioned, the Casimir scaling behavior for a stable string [2] is in gross contradiction with analytical considerations [15]: the string tension should be expandable in powers of $1/N^2$ rather than $1/N$. This is just another manifestation of the fact that lattice results should be interpreted with extreme care.

3 k-Strings: Saturation Limit and the Sine Formula

The strings attached to the sources which have k fundamental indices (with no anti-fundamental), or *vice versa* are generically referred to as k-strings. The stable k-strings (or *bona fide* k-strings) are those attached to the fully anti-symmetric sources. Others are quasi-stable k-strings. In this section we will assume that k scales as N^1 at large N.

3.1 Saturation Limit: Generalities

Let us address the issue of the stable k-string tension in a limit which we call *saturation*,

$$N \to \infty\,, \quad k \to \infty\,, \quad \frac{\pi k}{N} \equiv x \quad \text{fixed}\,. \tag{11}$$

We will combine the knowledge obtained from (i) symmetry properties; (ii) the saturation limit; (iii) $N \to \infty$ limit with k fixed; and (iv) $1/N^2$ expansion to prove that the general formula for the k-string tension in the saturation limit is as follows:

$$\frac{\sigma_k}{k} = \sigma_{\rm f}\, f(x)\,, \tag{12}$$

where the function f can be expanded in *even* powers of x,

$$f(x) = 1 + a_1\, x^2 + a_2\, x^4 + \ldots \tag{13}$$

We then comment on how this peculiar k dependence can be physically understood.

The central element of the proof is the Z_N symmetry, i.e. the fact that the *bona fide* string tension σ_k does not change under the replacements

$$k \to N - k\,, \quad \text{and} \quad k \to k + N\,. \tag{14}$$

The function f is dimensionless and, generally speaking, depends on two parameters, k and N. Equation (14) implies that k may enter only in a very special way, namely, through powers of $|\sin x|$. Thus, in general, one can write

$$\sigma_k = \frac{\sigma_{\rm f}}{\pi}\left[c_1(N)\,|\sin x| + c_2(N)\,|\sin x|^2 + c_3(N)\,|\sin x|^3 + \ldots\right]\,, \tag{15}$$

where $c_\lambda(N)$ are coefficients.

Now, we make use of the facts that (a) the saturation limit should be smooth, and (b) for fixed k the expansion parameter is N^{-2} rather than N^{-1}. Then we conclude that the coefficients $c_{2\lambda+1} = O(N^1)$ while $c_{2\lambda+2} = O(N^0)$ (here $\lambda = 0, 1, 2, \ldots$). Moreover, from the condition (iii) above it follows that $c_1(N) = N$.

Omitting terms vanishing in the saturation limit, we get

$$\sigma_k = \frac{\sigma_{\rm f}}{\pi}\, N\,\left\{\,|\sin x| + \; C_3\,|\sin x|^3 + \ldots\right\}\,, \tag{16}$$

which is equivalent to (13) (here C_3 and possibly C_5, C_7 and so on are numerical coefficients). Numerical lattice evidence suggests [19] that C_3 is strongly suppressed. Assuming that only C_3 exists, and fitting the results for the k-strings, Del Debbio, Panagopoulos, and Vicari obtain [19]

$$C_3 = -0.01 \pm 0.02\,.$$

At $N = \infty$ and k fixed (16) yields that $\sigma_k = k\,\sigma_1$. Including the next to leading term in the x-expansion we get

$$\sigma_k = \sigma_{\rm f}\, k\,\left(1 - \text{const}\;\frac{k^2}{N^2}\right)\,. \tag{17}$$

The term $\sigma_{\rm f}\, k^3/N^2$ can be interpreted as the binding energy. The $1/N^2$ factor is well understood, see [15]. A challenging question is to understand why the next-to-leading correction to σ_k scales as k^3. Alternatively, one can say that the binding energy per one fundamental string in the k-string compound scales as k^2. Experience based on numerous discussions with our colleagues tells us that the first guess is, rather, that it is the first power of k that must appear in the binding energy per one fundamental string. We outline a physical picture explaining k^3 and the resulting (17) in Sect. 4.

3.2 How Exact is the Sine Formula (in the Saturation Limit)?

The tension of the k string, σ_k, a crucial parameter of the confinement dynamics, is under intense scrutiny since the mid-1980's. Most frequently were discussed two competing hypotheses: (a) the Casimir scaling and (b) the sine formula originally suggested by Douglas and Shenker [20] (for extensive reviews and representative list of references see e.g. [21, 22]). as was mentioned, the Casimir scaling hypothesis is inconsistent with the large N expansion. In the same paper [15] we argued that the sine formula should hold; in our terms the sine formula can be presented as follows:

$$f(x) = \frac{\sin x}{x} \,. \tag{18}$$

It amounts to keeping only the first term in the general expansion (16). The suppression of higher terms in this expansion cannot be inferred on general grounds alone; it is a dynamical statement following from the model [15].

In the previous section we wrote down the most general expression for the k-string tension (15), that is compatible with the Z_N symmetry of the problem and $1/N^2$ nature of corrections. In the saturation limit the expression simplifies further to (16). It is tempting to speculate that the first term, the sine, controls the dynamics in the saturation limit (11), namely, that the k-string tension is exactly given by

$$\sigma_k = \Lambda^2 N \sin \pi \frac{k}{N} \,. \tag{19}$$

While there is no proof that the relation (19) is exact, there are arguments suggesting that it might be exact or, at least, present a good approximation (although we hasten to add that a parameter controlling this approximation is yet to be discovered). Below we will summarize theoretical evidence in favor of the above assertion.

The QCD string is not obviously a BPS object. Therefore, in principle, it could get higher order corrections in powers of sine, as in (16). The statement that (19) is exact is equivalent to the statement that the QCD-string approaches a "BPS status" in the saturation limit.

The first hint in favor of the sine formula came from the large-N analysis of $\mathcal{N}=2$ SYM theory softly broken to $\mathcal{N}=1$ SYM by the adjoint mass term m, due to Douglas and Shenker [20]. In slightly broken $\mathcal{N}=2$ SYM the QCD string is BPS-saturated to the leading order in m; therefore, no surprise that the sine formula $\sigma_k = m\Lambda N \sin \pi k/N$ was found. Details are as follows. In the softly broken $\mathcal{N}=2$ theory Douglas and Shenker found that

$$\sigma_k \sim m N^2 M_{k,k+1} \,, \tag{20}$$

where $M_{k,j}$ are the masses of the BPS W-bosons,

$$M_{k,j} = |m_k - m_j| \,.$$

Here

$$m_k \sim \Lambda N \left(\sin \frac{\pi k}{N} - \sin \frac{\pi (k-1)}{N} \right) .$$

At large N, to the leading order in $1/N$, (keeping k/N fixed), one then arrives at the sine formula,

$$M_{k,k+1} \sim \frac{\Lambda}{N} \sin \frac{\pi k}{N}, \qquad \sigma_k \sim m\,\Lambda\,N \sin \frac{\pi k}{N}.$$

At order m^2 the strings cease to be BPS, and, accordingly, the m^2 correction to the string tension was found [23] to defy the sine formula.

The second evidence came from MQCD. Here again the sine formula is obtained [24], without an *a priori* reason. Though the MQCD theory is not QCD, it is not clear why the obtained result is the exact sine. A possible reason is that MQCD possesses more symmetries than $\mathcal{N}=1$ SYM, but no such symmetry is known at present.

The third hint is the derivation of the k-string tension via supergravity [25]. Here the result is model-dependent. For the Klebanov-Strassler background the sine formula was found to be an excellent approximation, but not exact. It was found to be exact for the Maldacena-Nuñez background. While both backgrounds are conjectured to be a dual description of pure $\mathcal{N}=1$ gluodynamics in the far infrared, they have different ultraviolet contents. For this reason, presumably, different results were obtained.

The last, and for us the most convincing, argument in favor of the sine formula is the relation between the k-string tension and the k-wall tension in $\mathcal{N}=1$ gluodynamics advocated in [15]. It is known that the BPS domain wall tensions in $\mathcal{N}=1$ gluodynamics are exactly given by the formula

$$T_k = N^2 \Lambda^3 \sin \pi \frac{k}{N}. \tag{21}$$

If one accepts the picture advocated in [15], that domain walls are built from a network of QCD-strings connected by baryon junctions, one immediately arrives at the sine formula. In this picture, the QCD-string effectively becomes a BPS object in the saturation limit. The suggestion of the "walls built of the string network" is admittedly a model, albeit motivated by various arguments. The most crucial question we see at the moment is whether the expansion parameter controlling the accuracy of this model is numerical or related to some well-defined limits, such as large N, saturation limit, and so on. In the absence of the analytic answer to this question one may resort to lattices, see the remark after (16). While the saturation limit is clearly difficult to achieve on the lattice, the effort is worthwhile.

4 Physical Picture Beyond the Sine Formula for Stable k-Strings

The argument of Sect. 3.1 shows that the expansion of the k string tension has the form (17). While the $1/N^2$ factor in the first correction is perfectly clear, it is instructive to get a physical understanding of the k^3 scaling of the $1/N^2$ correction.

First of all let us note that at $N \gg 1$ the 1-string interactions are weak, and a quasiclassical picture is applicable. In order to reconcile the facts that both the flux of the k-string and its tension (in the leading order) grow as k we have to conclude that the total transverse area of the k-string scales as k at $k \gg 1$, see Fig. 5. Thus, the k-string presents an ensemble of loosely bound 1-strings, with k non-intersecting cores of 1-strings. The interaction occurs only at the periphery, in the gaps. This is very similar to the structure of molecules.

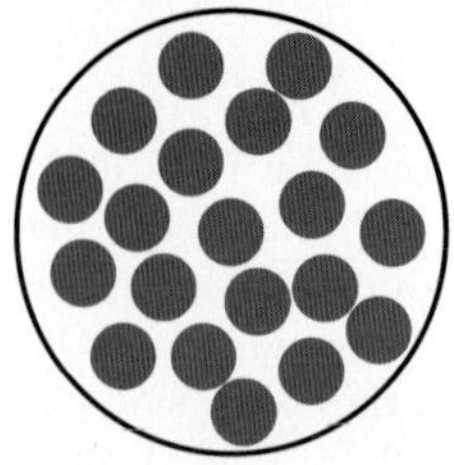

Fig. 5. A slice of a k-string

There are of the order of k gaps altogether, each having the area of the order of Λ^{-2}. Let us compare the chromoelectric fields at the periphery of the 1-string core in two cases: (i) isolated 1-string; (ii) 1-string which is bound inside the k-string. It is clear that the distortion of the field at the periphery in passing from (i) to (ii) is of the order of k/N. The energy per one gap (which is proportional to the chromoelectric field squared) thus scales as $(k/N)^2$. Given that the number of the gaps scales as k, we naturally obtain the k^3/N^2 regime for the first correction in the k-string tension.

Within this picture the expansion parameter $(k/N)^2$, with the overall k factor, occurs naturally, leading to the scaling parameter x introduced in Sect. 3.1.

The linear growth of the string cross section with k at large k is a prediction following from our analysis.

5 More on the Casimir Scaling

In Sect. 2.2 we mentioned that: (a) many lattice measurements detect quasi-stable strings, with no hint on their breaking; and (b) observe the Casimir scaling in the ratio of the tension of the given string to that of the fundamental string. For instance, in [6] ample data are presented regarding the adjoint and a number of other quasi-stable strings in SU(3). No deviations from the Casimir scaling are seen withing the achieved accuracy.

According to the Casimir formula,

$$\sigma_{\text{adj}}/\sigma_{\text{f}} = 9/4\,, \qquad (\text{SU}(3)\ \text{Casimir})\,. \tag{22}$$

The adjoint string is built of two fundamental ones, and at distances $\gtrsim \Lambda^{-1}$ they attract, so that placing these two strings at an appropriate distance one must get the above ratio less than 2. Even if we take the separation distance between the two 1-strings involved to be very large, the ratio will equal 2. How can one get this ratio larger than 2?

Typical string sizes on the lattice are $\lesssim$ 2 fm, while the transverse size of the fundamental string is $\sim$0.7 fm. As we have argued above, the transverse size of the composite string must be even larger. Thus, the length/thickness ratio is in fact not large.

If one could measure the tension of the configuration depicted in Fig. 6a, because of the attraction, one would get less than $2\sigma_{\text{f}}$. The only way to get more than $2\sigma_{\text{f}}$ is to force the fundamental strings to overlap. Then, because of the repulsion of the

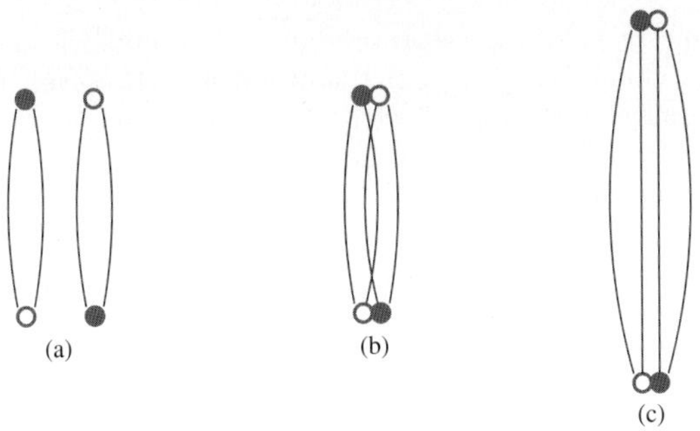

Fig. 6. Various flux tubes. (**a**). Two separated fundamental strings. (**b**). A short adjoint string. (**c**). A long adjoint string

overlapping fluxes, (prior to the string decay), see the end of Sect. 2, one will get $>2\sigma_{\rm f}$ for the energy per unit length.

This is precisely what happens if one considers the probe quark of the form Q_i^j, see Fig. 6b. Because of the insufficient separation between Q_i^j and $\overline{Q_k^l}$ the overlap has to be substantial. For *bona fide* long strings (Fig. 6c) the overlap would be insignificant, and will affect only the part of the potential which does not scale with L as L^1.

Still, the question remains: Why a sausage-like configuration (Fig. 6b), which develops at intermediate distances, is characterized by energy per unit length which follows the Casimir formula?

Here we suggest a tentative "semi-empiric" answer. It is known since the 1970's [26] that basic properties of low-lying quark mesons [2] are to a large extent determined by the gluon condensate. The coefficient in front of the gluon condensate is proportional to the quadratic Casimir operator. Therefore, as long as the higher condensates are not very important numerically, the masses of the lowest mesons will be related to the quadratic Casimir operators. These latter masses, in turn, are determined by "short strings."

If this explanation is correct, approximate Casimir scaling should take place at intermediate distances when the overlap between constituents of the composite string is significant.

6 Quasistable String Decay Rates (Quasiclassical Tunelling)

Not to overburden ourt discussion, we will limit ourselves to $k = 2$. In this case there are two Young tableaux – full symmetrization and full anti-symmetrization, $Q^{\{ij\}}$ and $Q^{[ij]}$. The antisymmetric string is stable, while the symmetric one is unstable. We want to show that the symmetric string does not decay into the anti-symmetric one to any finite order in $1/N^2$. The decay rate is exponential. Indeed, in order to

[2] By low-lying we mean the lowest radial excitations in each channel with the given spin and parity.

convert the symmetric color representation into anti-symmetric one has to produce a pair of gluons. This takes energy of the order of Λ. However, the string is not entirely broken, rather it is restructured, with the tension splitting $\sim\Lambda^2 N^{-2}$. To collect enough energy, the gluon creation should take place not locally, but, rather at the interval of the length $\sim\Lambda^{-1} N^2$. This is then a typical tunneling process.

Under the circumstances it is not surprising that the decay rate can be found quasiclassically. The world sheet of the symmetric string is shown in Fig. 7a. Its decay proceeds via a bubble creation, see Fig. 7b. The world sheet of the symmetric string is two-dimensional "false vacuum," while inside the bubble we have a "true vacuum," i.e. the surface spanned by the anti-symmetric string. The tension difference – in the false vacuum decay problem, the vacuum energy difference – is $\mathcal{E} \sim \Lambda^2 N^{-2}$, while the energy T of the bubble boundary (per unit length) is $T \sim \Lambda$. This means that the thin wall approximation [27, 28] is applicable, and the decay rate Γ (per unit length of the string per unit time) is [27, 28]

$$\Gamma_{\square\square \to \genfrac{}{}{0pt}{}{\square}{\square}} \sim \Lambda^2 \exp\left(-\frac{\pi T^2}{\mathcal{E}}\right) \sim \Lambda^2 \exp\left(-\gamma N^2\right) , \tag{23}$$

where γ is a positive constant of the order of unity. Once the true vacuum domain is created through tunneling, it will expand in the real time pushing the boundaries (i.e. the positions of the gluons responsible for the conversion) toward the string ends.

Other quasi-stable strings can be treated in the same manner [15].

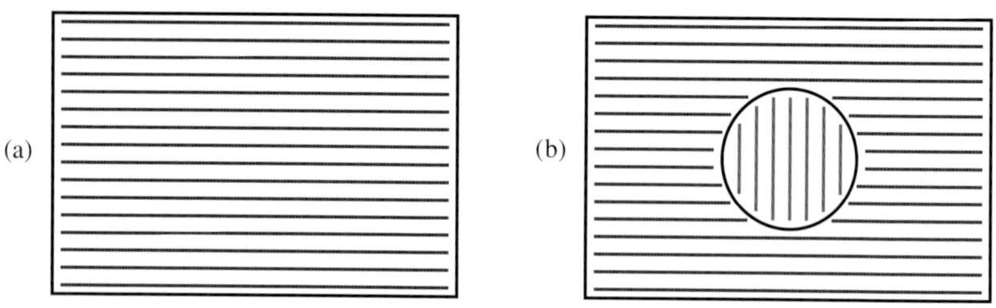

Fig. 7. The world-sheet of a 2-string. (**a**). A purely symmetric string (denoted by *horizontal lines*). (**b**). The decay into an antisymmetric string (*vertical lines*) via an expanding bubble

7 Conclusions

Despite extensive effort invested in the subject of the k-strings during twenty years of its development surprisingly many questions were left unanswered or answered incompletely. Systematic application of the $1/N$ expansion, in conjunction with ideas borrowed from string theory and supersymmetry, allowed to us to advance in a significant manner. We addressed and elucidated long-standing issues such as the Casimir scaling versus the sine formula and observability of the adjoint and other quasistable strings.

Acknowledgements

We would like to thank L. Del Debbio, H. Panagopoulos, and E. Vicari for useful discussions. M.S. is grateful to J. Trampetić for warm hospitality in Dubrovnik. The work of M.S. is supported in part by DOE grant DE-FG02-94ER408.

References

1. L. Del Debbio, M. Faber, J. Greensite and Š. Olejnik, Phys. Rev. D **53**, 5891 (1996); M. Faber, J. Greensite, and Š. Olejnik, Phys. Rev. D **57**, 2603 (1998).
2. B. Lucini and M. Teper, Phys. Rev. D **64**, 105019 (2001).
3. L. Del Debbio, H. Panagopoulos, P. Rossi and E. Vicari, Phys. Rev. D **65**, 021501 (2002); JHEP **0201**, 009 (2002).
4. V. I. Shevchenko and Y. A. Simonov, Phys. Rev. Lett. **85**, 1811 (2000).
5. S. Kratochvila and P. de Forcrand, *String breaking with Wilson loops?*, hep-lat/0209094; S. Kratochvila and P. de Forcrand, *Observing string breaking with Wilson loops*, hep-lat/0306011.
6. S. Deldar, Phys. Rev. D **62**, 034509 (2000); *The String Tension of* SU(3) *Representations,* UMI-99-59930, PhD Thesis, Washington University, St. Louis.
7. G. S. Bali, Phys. Rev. D **62**, 114503 (2000).
8. J. Greensite, Prog. Part. Nucl. Phys. **51**, 1 (2003).
9. J. M. Maldacena, Adv. Theor. Math. Phys. **2**, 231 (1998); [Int. J. Theor. Phys. **38**, 1113 (1999)]; S. S. Gubser, I. R. Klebanov and A. M. Polyakov, Phys. Lett. B **428**, 105 (1998); E. Witten, Adv. Theor. Math. Phys. **2**, 253 (1998).
10. J. M. Maldacena, Phys. Rev. Lett. **80**, 4859 (1998).
11. D. J. Gross and H. Ooguri, Phys. Rev. D **58**, 106002 (1998).
12. S. Mandelstam, *Phys. Reports* **23**, 245, 1976; G. 't Hooft, in *1981 Cargése Summer School Lecture Notes on Fundamental Interactions, NATO Adv. Study Inst. Series B: Phys.*, Vol. 85, ed. M. Lévy et al. (Plenum Press, New York, 1982) [reprinted in G. 't Hooft, *Under the Spell of the Gauge Principle* (World Scientific, Singapore, 1994), page 514]; *Nucl. Phys.* **B190**, 455, 1981.
13. M. Shifman and A. Yung, *Localization of non-Abelian gauge fields on domain walls at weak coupling (D-brane prototypes II)*, hep-th/0312257.
14. A. Armoni and M. Shifman, Nucl. Phys. B **671**, 67 (2003).
15. A. Armoni and M. Shifman, Nucl. Phys. B **664**, 233 (2003).
16. L. Del Debbio, M. Faber and J. Greensite, Nucl. Phys. B **414**, 594 (1994).
17. V. A. Novikov, M. A. Shifman, A. I. Vainshtein and V. I. Zakharov, Nucl. Phys. B **191**, 301 (1981) (see Appendix B; for further discussions see C. Michael, Nucl. Phys. B **280**, 13 (1987); H. G. Dosch, O. Nachtmann and M. Rueter, *String formation in the model of the stochastic vacuum and consistency with low-energy theorems,* hep-ph/9503386; M. Shifman, *ITEP Lectures on Particle Physics and Field Theory*, Chap. 2, Sect. 11.)
18. K. G. Wilson, Phys. Rev. D **10**, 2445 (1974).
19. L. Del Debbio, H. Panagopoulos, and E. Vicari, private communication.
20. M. R. Douglas and S. H. Shenker, Nucl. Phys. B **447**, 271 (1995).
21. M. J. Strassler, Nucl. Phys. Proc. Suppl. **73**, 120 (1999).

22. M. J. Strassler, *Millenial Messages for QCD from the Superworld and from the String*, in *At the Frontier of Particle Physics/Handbook of QCD*, Ed. M. Shifman, (World Scientific, Singapore, 2001), Vol. 3, p. 1859.
23. R. Auzzi and K. Konishi, New J. Phys. **4**, 59 (2002).
24. A. Hanany, M. J. Strassler and A. Zaffaroni, Nucl. Phys. B **513**, 87 (1998).
25. C. P. Herzog and I. R. Klebanov, Phys. Lett. B **526**, 388 (2002).
26. M. A. Shifman, A. I. Vainshtein and V. I. Zakharov, Nucl. Phys. B **147**, 385, 448 (1979).
27. I. Y. Kobzarev, L. B. Okun and M. B. Voloshin, Yad. Fiz. **20**, 1229 (1974) [Sov. J. Nucl. Phys. **20**, 644 (1975)].
28. M. B. Voloshin, *False vacuum decay,* Lecture given at International School of Subnuclear Physics: *Vacuum and Vacua: the Physics of Nothing*, Erice, Italy, July 1995, in *Vacuum and vacua: the physics of nothing*, Ed. A. Zichichi (World Scientific, Singapore, 1996), pp. 88–124.

Constraining New Physics from the Muon Decay

Astrid Bauer

Institut für Physik, Johannes Gutenberg-Universität, Staudingerweg 7,
55099 Mainz, Germany
astrid.bauer@thep.physik.uni-mainz.de

We consider the Standard Model (SM) as an effective field theory extended by higher-dimensional operators consisting of SM fields. Operators contributing to the processes $\mu^- \to e^- + \nu_1 + \nu_2$ (ν_1, ν_2 can be any type of neutrino or antineutrino of electron or muon flavor) are compared with the high precision data of the decay spectrum. Upper bounds for the unknown coupling constants arising with each new operator are obtained. Taking other processes into account, which can also be realized by these operators, improves the constraints partially.

It is a reasonable alternative to think of the SM not as a basic, but as an effective theory, which has to be extended by new interactions at high energies. This is supported by the convincing experimental evidences for solar and atmospheric neutrino oscillations (Homestake [1], (Super-)Kamiokande [2]), which can not be described by the minimal SM, as well as by theoretical reasons. The intention of an effective field theory is the description of a physical system at a low energy scale $E < \Lambda$, where the form of the underlying interaction cannot be directly seen. Λ denotes the energy scale where new degrees of freedom are revealed. In this approach we follow the idea of a power series expansion of the SM Lagrangian in terms of Λ^{-1} [3, 4]

$$\mathscr{L}_{\text{eff}} = \sum_{k=0}^{\infty} \Lambda^{-k} \mathscr{L}_k, \text{ where } \mathscr{L}_0 = \mathscr{L}_{SM} . \tag{1}$$

Each higher-dimensional term in (1) consists of operators $\mathcal{O}_j$ of mass dimension $k+4$ which include a coupling strength α_j.

In order to allow for a feasible analysis we consider only new operators consisting of fields appearing in the SM. It is also reasonable to study new effects caused by only one additional operator. Otherwise there are too many unrestricted constants as to give sensible constraints. In contrast to previous analysis [5, 6], we drop lepton number conservation.

There are no relevant operators of dimension five, which would contribute to the muon decay. The only operator, which can be built from SM fields gives rise to a Majorana mass term. In order to achieve sufficient small Majorana masses, the new energy scale Λ has to be larger than 10^{13} GeV and is therefore much above the region we are interested in.

For a complete discussion of lepton number conserving dim$\,=6$ operators derived from SM fields we refer to [7]. We include into our considerations also lepton number violating operators, which can be obtained by allowing terms consisting of lepton fields with different flavor. Generalizing [7] in this way, we find eight operators, which generate the decay of the muon into electron and two (anti-)neutrinos

Table 1. Operators contributing to the muon lifetime

Dimension six	Flavor	
$\mathcal{O}_{ll(1)} = \alpha_{ll(1)}(\bar{l}\gamma^\mu l)(\bar{l}\gamma_\mu l)$	$(srrs)$, $(eeee)$	$\lvert\mathrm{Re}(\alpha)\rvert <$ 0.08, 0.54
	$(rsee)$, $(eers)$	$\lvert\alpha\rvert < 3.3 \cdot 10^{-5}$
	$(rrss)$, $(rsrs)$, $(rs\mu\mu)$, $(\mu\mu rs)$	$\lvert\alpha\rvert <$ 0.9 – 3.3
$\mathcal{O}_{ll(3)} = \alpha_{ll(3)}(\bar{l}\gamma^\mu\tau^i l)(\bar{l}\gamma_\mu\tau_i l)$	$(rrss)$, $(srrs)$, $(eeee)$	$\lvert\mathrm{Re}(\alpha)\rvert <$ 0.04 – 0.54
	$(rsee)$, $(eers)$	$\lvert\alpha\rvert < 3.3 \cdot 10^{-5}$
	$(rs\mu\mu)$, $(\mu\mu rs)$	$\lvert\alpha\rvert <$ 3.4
$\mathcal{O}_{le} = \alpha_{le}(\bar{l}e)(\bar{e}l)$	$(\mu ee\mu)$, $(eeee)$	$\lvert\mathrm{Re}(\alpha)\rvert <$ 0.54, 23.30
	$(rrss)$, $(rsrs)$, $(\mu\mu e\mu)$, $(\mu e\mu\mu)$	$\lvert\alpha\rvert <$ 6.5 – 11.5
	$(eers)$, $(rsee)$	$\lvert\alpha\rvert < 6.6 \cdot 10^{-5}$
$\mathcal{O}_{lW} = i\alpha_{lW}\bar{l}\tau^i\gamma_\mu D_\nu l W_i^{\mu\nu}$	$(e\mu)$, (μe)	$\lvert\alpha\rvert < 1.5 \cdot 10^{-5}$
$\mathcal{O}_{\varphi l(3)} = i\alpha_{\varphi l(3)}(\varphi^+ D_\mu\tau^i\varphi)(\bar{l}\gamma^\mu\tau_i l)$	(ee), $(\mu\mu)$	$\lvert\mathrm{Re}(\alpha)\rvert <$ 0.2
	(ee), $(\mu\mu)$	$\lvert\alpha\rvert < 9.2 \cdot 10^{-5}$
$\mathcal{O}_{De} = \alpha_{De}\bar{l}(D_\mu e)D^\mu\varphi$	$(e\mu)$, (μe)	$\lvert\alpha\rvert <$ 1.4
$\mathcal{O}_{\bar{D}e} = \alpha_{\bar{D}e}(D_\mu\bar{l})eD^\mu\varphi$	$(e\mu)$, (μe)	$\lvert\alpha\rvert <$ 1.4
$\mathcal{O}_{eW} = \alpha_{eW}(\bar{l}\sigma^{\mu\nu}\tau_i e)\varphi W_{\mu\nu}^i$	$(\mu\mu)$	$\lvert\alpha\rvert <$ 31.4
	$(e\mu)$, (μe)	$\lvert\alpha\rvert < 2.3 \cdot 10^{-9}$
Dimension seven	**Flavor**	
$\mathcal{O}_{lCl(1)} = i\alpha_{lCl(1)}(\tilde{l}Cl)(\bar{e}_R\tilde{\varphi}l)$	$(srsm)$, $(srrm)$	$\lvert\alpha\rvert <$ 20.8
$\mathcal{O}_{lCl(3)} = i\alpha_{lCl(3)}(\tilde{l}C\tau_i l)(\bar{e}_R\tilde{\varphi}\tau^i l)$	$(rrrs)$, $(rsrs)$, $(srrs)$	$\lvert\alpha\rvert <$ 0.5 – 0.7
$\mathcal{O}_{lCl(1)T} = i\alpha_{lCl(1)T}(\tilde{l}C\sigma_{\mu\nu}l)(\bar{e}_R\tilde{\varphi}\sigma^{\mu\nu}l)$	$(rrsm)$	$\lvert\alpha\rvert <$ 0.02
	$(srrm)$, $(rsrm)$	$\lvert\alpha\rvert <$ 0.04
$\mathcal{O}_{lCl(3)T} = i\alpha_{lCl(3)T}(\tilde{l}C\tau_i\sigma_{\mu\nu}l)(\bar{e}_R\tilde{\varphi}\tau^i\sigma^{\mu\nu}l)$	$(rsrr)$, $(srrr)$	$\lvert\alpha\rvert <$ 0.04
	$(rrrs)$, $(rrsr)$, $(rsrs)$, $(srrs)$	$\lvert\alpha\rvert <$ 0.03
Dimension nine	**Flavor**	
$\mathcal{O}_{lCl\varphi} = \alpha_{lCl\varphi}(\bar{l}e\varphi)(lCl\varphi^\dagger\varphi)$	$(srsm)$, $(srrm)$	$\lvert\alpha\rvert <$ 32.6
$\mathcal{O}_{lCl\varphi^\dagger} = \alpha_{lCl\varphi^\dagger}(\bar{e}\varphi^\dagger l)(lCl\varphi^\dagger\varphi)$	$(srsm)$, $(srrm)$	$\lvert\alpha\rvert <$ 32.6

of arbitrary flavor ($f = e, \mu$). They are listed in Table 1 together with dimension seven operators discussed in [8]. l denotes the left-handed $SU(2)$ doublet, e the right-handed lepton singlet, $\varphi = (\varphi^+, \varphi^0)$ the Higgs doublet with vacuum expectation value $(0, v)$. $\tilde{l}$ and $\tilde{\varphi}$ are given by $\tilde{l}_i = \epsilon^{ij} l_j$ with the total antisymmetric tensor ϵ. τ^i and γ^μ are the usual Pauli and Dirac matrices. C is the charge conjugation matrix, $D_\mu = \partial_\mu - ig\frac{1}{2}\tau_i W_\mu^i - ig' Y B_\mu$ the covariant derivative with the hypercharges $Y(l) = -\frac{1}{2}$, $Y(e) = -1$ and $Y(\varphi) = \frac{1}{2}$ and the gauge bosons W_μ^i and B_μ. The $SU(2)$ and $U(1)$ coupling constants are denoted by g and g'.

Note that Table 1 shows only lepton number violating 4-fermion operators with dimension seven. In addition, we include in our list operators of dimension nine, which have been proposed recently [9] as a possible explanation of the controversial LSND result [10]. Each operator given in Table 1 is only a substitute for a group of operators, characterized by the general structure shown there. To obtain explicit results, the fields have to be labeled with flavor indices. For simplicity these indices are suppressed in Table 1.

Most of the experimental data lack information concerning the neutrino flavor. Therefore we include into our considerations also lepton number violating processes. With respect to the decay of the muon one obtains three different kinds of processes characterized by the possible violation of lepton ($\Delta L \neq 0$) or lepton family number ($\Delta L_f \neq 0,\ f = e, \mu$). $\mu^- \to e^- \bar{\nu}_e \nu_\mu$ ($\Delta L = 0,\ \Delta L_e = \Delta L_\mu = 0$) leads to the usual final state expected by the SM. It differs only by the coupling strength or the way the fermions couple to each other. $\mu^- \to e^- \bar{\nu}_e \nu_e$ ($\Delta L = 0,\ \Delta L_e = -\Delta L_\mu = 1$) realize final states different from the SM prediction.

The effective ansatz provides a model independent method to derive constraints to coupling constants of the effective field theory. The results given in Table 1 scale according to their Λ dependence. First, we derive new operator's contributions to the muon lifetime which is measured with high precision $\tau_\mu = (2.19703 \pm 0.00004) \cdot 10^{-6}$ s [11]. The experimental error is negligible. The dominant contribution to the error of the theoretical prediction of the SM comes from the uncertainty of the W-boson mass. Comparing the error of the SM prediction $\Delta\Gamma_{SM}$ with the prediction Γ_j obtained from the new effective operator $\mathcal{O}_j$ finally leads to the relation $|\Gamma_j / \Gamma_{SM}| < 5.9 \cdot 10^{-3}$ and therefore constrains the new couplings α_j. Now effects could also be observed in high-precision measurements of the spectrum, whose shape might deviate from the SM prediction. We divide the energy of the electron into two bins with about the same numbers of events. The ratio $R = \int_{bin1} \mathrm{d}\Gamma / \int_{bin2} \mathrm{d}\Gamma = 1 + \Delta R_{bSM}$ leads to a limit $\Delta R_{bSM} < 2\sqrt{2}\varepsilon$ which results in a constraint for the coupling constant. $\varepsilon = \Delta\Gamma/\Gamma \sim 2 \cdot 10^{-6}$ is expected to be the precision of future experiments [12].

In addition to the constraints gained with respect to the muon decay for the operators with dimension six, we also compute restrictions from crossed processes, processes involving neutrinos in place of charged leptons and exotic decays. The process $\nu_\mu e^- \to \mu^- \nu_e$ has been measured by the CHARM collaboration [13]. Note that the CHARM detector does not determine the flavor of the outgoing neutrino. Therefore we consider $\nu_\mu e^- \to \nu_{e/\mu} \mu^-$ and $\nu_\mu e^- \to \bar{\nu}_{e/\mu} \mu^-$ as well. The scattering cross sections of $\nu_\mu e^- \to \nu_\mu e^-$ and $\bar{\nu}_\mu e^- \to \bar{\nu}_\mu e^-$ are given in [14]. We consider both cases $\nu_i, \bar{\nu}_i$ with $i = e, \mu$. The LAMPF collaboration [15] measured the $\nu_e e^- \to \nu_e e^-$ cross section. We consider the final state consisting of (anti-)neutrinos with electron or muon flavor. We also take exotic decays like $Z \to e^\pm \mu^\mp$ [16], $\mu^- \to e^- \gamma$ [17], $\mu^- \to e^- e^+ e^-$ [18] into account.

The results are shown in Table 1. The numbers are obtained under the assumption $\Lambda = 1\,\mathrm{TeV}$. Constraints, which imply that the new energy scale has to be several orders of magnitude larger than 1 TeV in order that the couplings are smaller than ~10 are not listed in Table 1. The flavor indices are given in the 3rd column ($r, s, m \subset \{e, \mu\},\ r \neq s$). For conciseness, we summarize some flavor combinations in one line. For further details see [19].

References

1. R.J. Davis, D.S. Harmer and K.C. Hoffmann: Phys. Rev. Lett **20**, 1205 (1968), R. Davis: Prog. Part. Nucl. Phys. **32**, 13 (1994)
2. Y. Fukuda et al: Phys. Rev. Lett. **77**, 1683 (1996); ibid **81**, 1562 (1998)
3. S. Weinberg: Phys. Rev. **166**, 1568 (1968)
4. J. Grasser and H. Leutwyler: Nucl. Phys. B **250**, 465 (1985)
5. W. Fetscher, H.-J. Gerber and K.F. Johnson: Phys. Lett. B **173**, 102 (1986),
6. P. Herczeg: Z. Phys. C **56**, 129 (1992)
7. W. Buchmüller and D. Wyler: Nucl. Phys. B **268**, 621 (1986)
8. H.A. Weldon and A. Zee: Nucl. Phys. B **173**, 269 (1980)
9. K.S. Babu and S. Pakvasa: arXiv:hep-ph/0204236
10. C. Athanassopoulos et al: Phys. Rev. C **54**, 2685 (1996); ibid. **58**, 2489 (1998)
11. K. Hagiwara et al: Phys. Rev. D **66**, 010001 (2002)
12. F. Navarria et al: ETHZ-IPP-PR-98-04, `http://www.sno.phy.queensu.ca`, `http://www.triumpf.ca`, `http://nectar.nd.rl.ac.uk/~rikenral`
13. P. Vilain: Phys. Lett. B **364**, 121 (1995)
14. L.A. Ahrens et al: Phys. Rev. D **41**, 3297 (1990)
15. R.C. Allen et al: Phys. Rev. D **47**, 11 (1993)
16. R. Akers et al: Z. Phys. C **67**, 555 (1995)
17. M.L. Brooks et al: Phys. Rev. Lett. **83**, 1521 (1999)
18. U. Bellgardt et al: Nucl. Phys. B **299**, 1 (1988)
19. A. Bauer and H. Spiesberger: [in preparation]

Jets in Deep Inelastic Scattering and High Energy Photoproduction at HERA

Gerd W. Buschhorn

Max-Planck-Institut für Physik (Werner-Heisenberg-Institut), Föhringer Ring 6, 80805 München
gwb@mppmu.mpg.de

1 Introduction

The hard partonic interactions at the center of photon-induced hadronic processes in deep inelastic electron/positron-proton-scattering are charaterised by jets. The virtuality Q^2 of the exchanged photon can be varied from high values in deep inelastic scattering (DIS) to zero in photoproduction. According to the factorization theorem the hadronic process can be factorized into a pertubatively treated hard part, which includes the evolution of the parton entering the hard process, and a soft part, which describes the parton distribution of the target nucleon; the characteristic energy for this separation is given by the factorization scale. In DIS processes involving light quarks only, an energy scale is provided by Q^2 or the transverse energy of the emerging hadronic jet (or jet system), whereas in photoproduction the jet energy is the only available energy scale; in heavy quark production the quark mass may provide an additional energy scale. In jet studies details of the hard QCD calculations and their matching to the evolution of the target partons are probed; the investigations may also provide information on the parton distribution functions (pdfs) of the proton and photon and can serve to measure α_s and its scale dependence.

In this report, recent results from the H1- and ZEUS-collaboration on the production of standard jets i.e. jets without heavy quark tagging in neutral current DIS processes and photoproduction are discussed; jets from diffractive processes and jet shapes are not included.

For jet studies in DIS the Breit frame is preferred, which in the quark parton model (QPM) is defined as the reference frame in which the struck parton emerges collinear with the incoming purely time-like virtual photon $\mathbf{q} = (0, 0, -Q)$ such that $2x\mathbf{p} + \mathbf{q} = 0$. Working in the Breit frame suppresses the QPM background and provides maximum separation between fragments from hard scattering QCD processes and beam remnants. Since the Breit frame is related to the photon-hadron-cms by a boost in proton beam direction (defined as $+z$-direction), jet variables are convenient, which are boost-invariant – like the transverse energy E_T, the azimuthal angle ϕ and the pseudorapidity $\eta = -\ln\tan(\theta/2)$.

The jet finding algorithm has to be stable against infrared radiation and collinear splitting and has to take into account the fact that at HERA not all final state particles are detected. In recent years, the algorithm preferred in jet studies by the HERA-collaborations has become the inclusive k_T-cluster algorithm [1]. The algorithms proceeds by calculating for each track or energy cluster i the quantity $d_i = (E_{Ti})^2$ and for each pair i, j the quantity $d_{ij} = \min(E_{Ti}^2, E_{Tj}^2)[(\eta_i - \eta_j)^2 + (\phi_i - \phi_j)^2]$. If of all resulting $\{d_i, d_{ij}\}$ the smallest one belongs to $\{d_i\}$ it is kept as a jet and not treated further, if it belongs to $\{d_{ij}\}$ the corresponding tracks or clusters are combined to a single object; the procedure is repeated until all tracks or clusters are assigned to jets. The algorithm is suited to handle overlapping jets, facilitates the separation between beam remnants and hard scattering fragments and yields smaller higher order QCD and fragmentation corrections than e.g. the cone algorithm.

Jet cross sections are calculated by convoluting the matrix element of the specific hard scattering process with the pdfs of the target proton. Examples of LO hard scattering processes are boson-gluon-fusion and QCD-Compton scattering (Fig. 1). The parton evolution requires the summation of $\ln Q^2$- and $\ln 1/x$-terms, which is equivalent to a summing of ladder graphs pictured in Fig. 1. In the DGLAP-approach [2] only leading powers of $\ln Q^2$-terms are summed implying a strongly ordered increase of the virtuality of the exchanged partons towards the hard scattering. At small x where at HERA Q^2 is constrained to not too large values, the BFKL treatment of the parton evolution [3] is more appropriate, which resums $\ln 1/x$-terms to all orders but retains the full Q^2 dependence; the parton chain is strongly ordered in x i.e. $x \ll x_g \ll \ldots \ll 1$, implying k_T-unintegrated (gluon) pdfs and off-shell matrix elements for the hard scattering process. In the CCFM ansatz [4] colour coherence in the evolution ladder is taken into account imposing an angular-ordering of the gluon emission with the maximum angle determined by the hard scattering subprocess; the cross section factorizes into pdfs unintegrated both in k_T and the gluon angle and an off-shell matrix element. At asymptotic energies CCFM evolution approaches BFKL at small x and DGLAP at large x and Q^2.

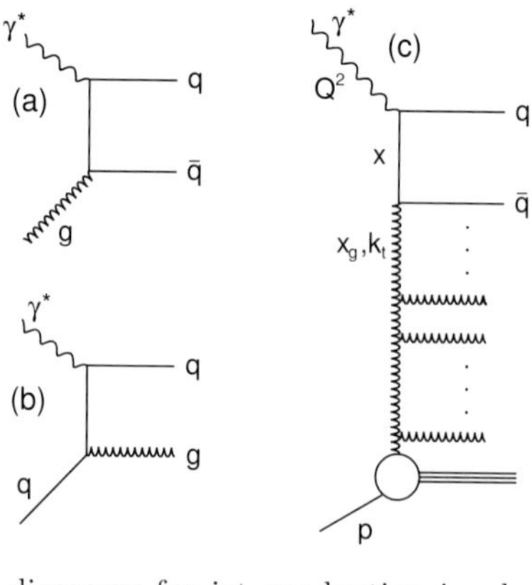

Fig. 1. Leading order diagrams for jet production in photon-proton-interaction: (**a**) photon-gluon-fusion; (**b**) QCD-Compton process; (**c**) parton evolution

While at high Q^2 the virtual photon behaves as a pointlike particle, at low Q^2 and in particular in photoproduction with $Q^2 = 0$, its hadronic structure has to be taken into account. The momentum transfer from the scattered electron to the parton cascade then takes place via a parton of the "resolved photon", which itself may evolve in a parton cascade towards the hard scattering process. Such resolved processes contribute a background to hard scattering processes but also can serve as a tool to investigate the photon structure function.

The response of the detectors to jet events is studied with Monte Carlo generated events, which are based on the measured pdfs of the proton and photon and on simulations of the fragmentation and hadronization of the partons. In RAPGAP the QCD matrix element is calculated in LO with NLO corrections evaluated by perturbative generation of parton showers; the hadronization of the partons is performed via JETSET, which is based on the Lund-string model. In ARIADNE, the parton cascade is generated using the colour-dipole-model (CDM) while in CASCADE it is calculated via CCFM evolution.

2 Jets in DIS

The inclusive jet cross section σ_j results from the convolution of the measured pdfs of the proton $f_a(x_a, \mu_F)$ with the pertubatively calculated hard partonic cross section $\mathrm{d}\sigma_a(x_a, \alpha_s, \mu_R, \mu_F)$,

$$\sigma_j = \sum_a \int \mathrm{d}x f_a(x, \mu_F^2) \mathrm{d}\sigma_a \left(x, \alpha_s, \mu_R^2, \mu_F^2\right) (1 + \delta_{\mathrm{hadr}})$$

where the sum is to be taken over all partons a of the proton; σ_j has to be folded with the bremsstrahlungsspectrum. μ_R is the renormalization scale and μ_F the factorization scale. $(1+\delta_{\mathrm{hadr}})$ represents the nonperturbative correction accounting for the hadronization of the partons; it is estimated from MC generators.

(i) Single jet cross sections have been measured by H1 and ZEUS in the Breit frame using the k_{T}-cluster algorithm. The kinematical parameters for the H1 data [5] are $150 < Q^2 < 5000\,\mathrm{GeV}^2$, $7 < E_T < 50\,\mathrm{GeV}$, $-1 < \eta_{\mathrm{lab}} < 2.5$, while for the ZEUS data [6] they are $Q^2 > 125\,\mathrm{GeV}^2$, $E_T > 8\,\mathrm{GeV}$, $-2 < \eta_{\mathrm{lab}} < 1.8$. The good agreement of both measurements with NLO QCD calculations (DISENT) at high Q^2 values (Fig. 2) has enabled precision determinations of α_s and its scale dependence from these data (see Sect. 5).

The ZEUS data have been used to study the azimuthal distribution of jets in the Breit frame [7]. The measured ϕ-distribution agrees well with the NLO QCD predictions from DISENT with either $\mu_R = E_T$ or Q, providing an independent test of QCD.

H1 has extended the inclusive jet measurements [8] to low Q^2 values of $5 < Q^2 < 100\,\mathrm{GeV}^2$, $-1 < \eta_{\mathrm{lab}} < 2.8$, for $E_T > 5\,\mathrm{GeV}$ and $0.2 < y < 0.7$. The good agreement of the differential cross section in E_T with NLO QCD (setting $\mu_R = E_T$) in the backward and central region worsens towards the forward direction i.e. $\eta_{\mathrm{lab}} > 1.5$ and for $E_T < 20\,\mathrm{GeV}$ (Fig. 3). In this region, NLO corrections are large and theoretical uncertainties due to the uncertainty in the renormalization scale are larger than the experimental errors. Further studies have shown the discrepancies to NLO QCD to develop at $Q^2 < 20\,\mathrm{GeV}^2$.

(ii) Data for inclusive jet production near the forward direction i.e. $7^\circ < \theta_{j\mathrm{lab}} < 20^\circ$ have been taken by H1 [9] for $0.5 < Q^2/E_T^2 < 2$. The jet analysis using the k_{T}-cluster algorithm has been performed in the lab frame with cuts corresponding to $5 < Q^2 < 75\,\mathrm{GeV}^2$ and $0.1 < y < 0.7$. In Fig. 4 the data are compared with different QCD models. Reasonable agreement with DGLAP is achieved only if resolved photon processes are included, but also the colour dipole model describes the data reasonably well.

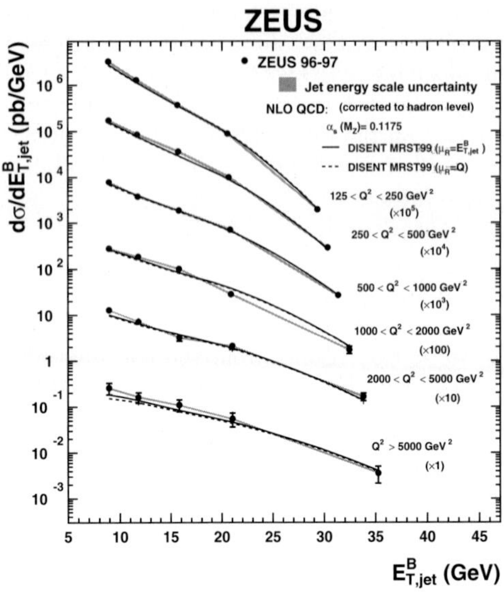

Fig. 2. Inclusive jet cross section in DIS in the Breit system (Figure from [6])

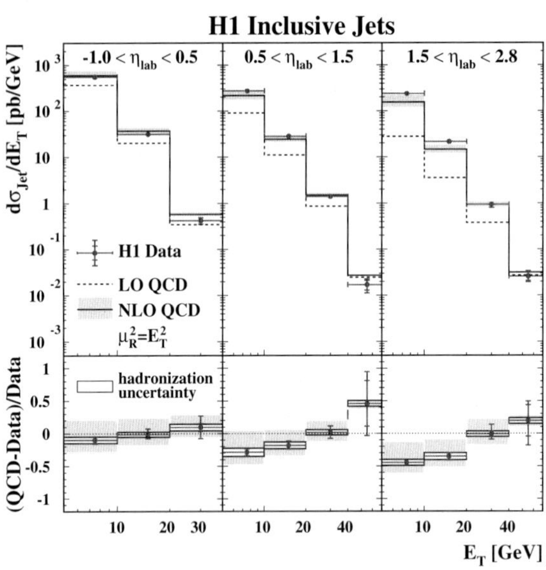

Fig. 3. Inclusive jet cross section in DIS in the Breit system integrated over $5 < Q^2 < 100\,\mathrm{GeV}^2$ and $0.2 < \eta_{\mathrm{lab}} < 0.6$, NLO QCD: DISENT without hadronization correction; shaded band: change of renorm. scale by 1/2 and 2 (Figure from [18])

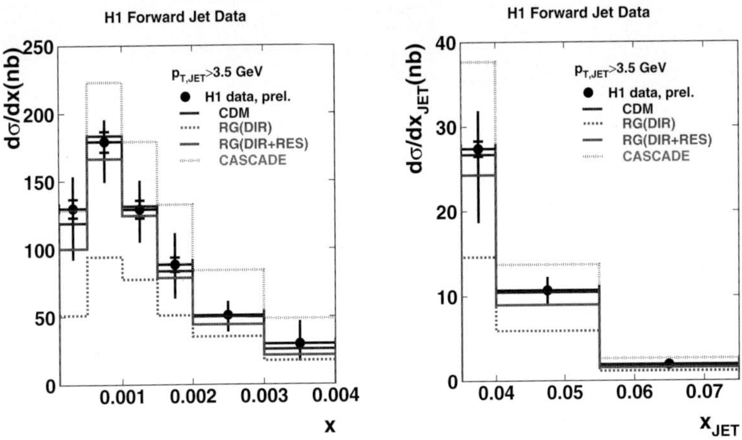

Fig. 4. Inclusive jet cross section in DIS in the forward direction i.e. $7° < \theta_{j\text{lab}} < 20°$; CDM: Color-Dipole-Model; RG: RAPGAP; CASCADE: CCFM; (preliminary data, from [9])

A process related to forward jet production is forward π°-production, which has been studied by H1 [10]; here the forward parton is tagged by a single energetic fragmentation product i.e. the π°. While in principle smaller angles than in jet production can be reached, the cross sections are lower and the hadronization uncertainties are higher. The data can be described by DGLAP models only after including resolved photon processes, but also BFKL models provide a reasonable description of the data.

(iii) LO contributions to dijet production are the QCD-Compton effect (QCDC) and boson gluon fusion (BGF) (see Fig. 1). In the high Q^2 region where QCDC is dominating and the pdfs are well constrained by fits to F_2 data, dijet measurements using α_s as input can serve to test pQCD; alternatively, by comparing the data with pQCD, α_s (and its running) can be determined. At low Q^2, where the cross section is dominated by BGF, the comparison of data with assumptions on the pdfs can constrain the gluon density distribution.

Dijet data have been taken by H1 [1] and ZEUS [11] in similar kinematical regions and analyzed by similar methods. Asymmetric cuts have been applied in E_T in order to avoid infrared sensitive regions of the phase space. At not too low Q^2 and E_T i.e. $Q^2 > 10\,\text{GeV}^2$ and $E_T > 5\,\text{GeV}$, the data are reproduced (Fig. 5) within 10% by NLO QCD (DISENT). For the α_s analysis of these data see Sect. 5.

Recent dijet-data from H1 [12] in the low-x region i.e. $10^{-4} < x < 10^{-2}$ for $10 < Q^2 < 100\,\text{GeV}^2$ agree within errors with DGLAP based NLO QCD predictions for inclusive cross sections. Serious disagreement is observed, however, in the ratio S of the number of dijet events with azimuthal separation $\phi < 2\pi/3$ to the total number of dijet events. While the LO (DISENT) prediction is much to low, the agreement is improved except in the low-x and low-Q^2 region if a third hard jet is included in the NLO (NLOJET) calculation (Fig. 6). A comparison with RAPGAP shows that the inclusion of resolved processes improves the agreement except in

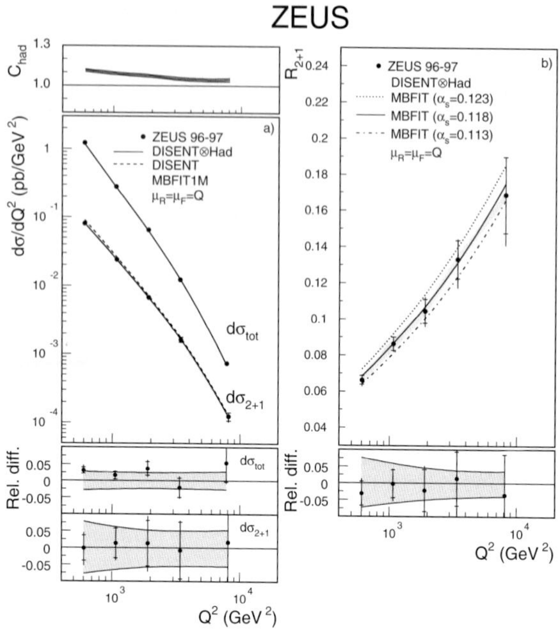

Fig. 5. Inclusive jet ($d\sigma_{tot}$) and dijet ($d\sigma_{2+1}$) cross sections with dijet fraction R_{2+1} in DIS; shaded band: uncertainty in absolute energy scale of jets (Figure from [11])

the very low-x and low-Q^2 region. From a comparison of the data with the CCFM based CASCADE MC a strong sensitivity to the assumptions on the unintegrated gluon distribution can be inferred, which may be used to constrain the gluon pdf. ARIADNE, which is based on CDM, describes the low-x and -Q^2 region reasonably well, fails, however, at larger Q^2 values.

The jet studies in deep inelastic scattering have been extended to trijets [13], [14] as well as to subjets [15].

3 Jets in Photoproduction

The description of photoproduction involves the convolution of the hard partonic scattering cross section with the pdfs of both the proton and the photon. In standard jet production i.e. jets from light quarks only, for the energy scale $\mu_R = \mu_F = \mu = E_T$ is taken. Apart from the folding with the electron bremsstrahlung spectrum, the jet cross section is written as

$$\sigma_j = \sum_{a,b} \int \int \mathrm{d}x_\gamma \mathrm{d}x_p f_{pa}(x_p, \mu^2) f_{\gamma b}(x_\gamma, \mu^2) \mathrm{d}\sigma_{ab}(x_p, x_\gamma, \alpha_s, \mu^2)(1 + \delta_{\mathrm{hadr}}),$$

where the sum is to be taken over all partons a, b from the proton and photon.

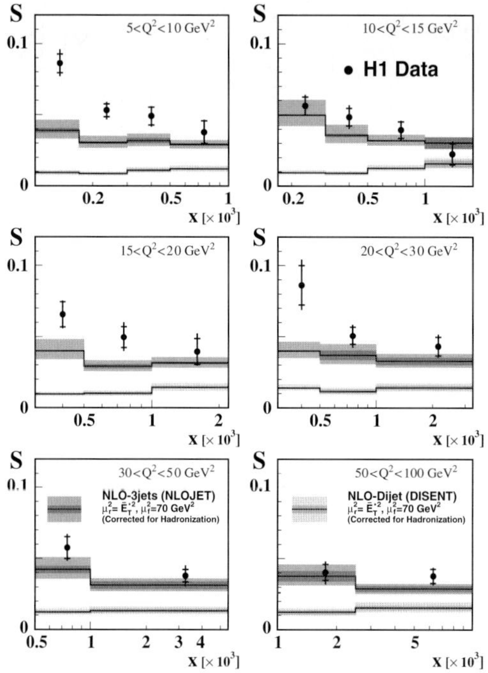

Fig. 6. Ratio S of number of dijet events in DIS with azimuthal separation $\phi < 2\pi/3$ between the two highest transverse momentum jets to the total number of dijet events compared with LO and NLO QCD predictions for dijet and trijet production; light shaded band: quadr. sum of hadroniz. error (*dark shaded band*) and renorm. error; (Figure from [12])

(i) In single jet production a wider kinematic range is accessible than in dijet production, cross sections are higher and systematic errors due to the treatment of infrared unsafe regions are avoided; on the other hand, the reaction is less sensitive to details of the hard scattering process.

In recent measurements at $Q^2 < 1\,\mathrm{GeV}^2$ of differential cross sections by ZEUS [16] and H1 [17] using the k_T-cluster algorithm in the lab frame, the kinematical range of $E_T > 5\,\mathrm{GeV}$, $-1 < \eta_\mathrm{lab} < 2.5$ and the photon-proton-cms energy W of $95 < W < 293\,\mathrm{GeV}$ has been covered. While LO QCD calculations fail to reproduce the data in their normalization, good agreement is found with the corresponding NLO predictions (Fig. 7, left). A subset of data taken by H1 at $Q^2 < 0.01\,\mathrm{GeV}^2$ for $5 < E_T < 12\,\mathrm{GeV}$ indicate an excess over predictions towards increasing η values. The precision of the experimental data and of the predictions is not yet sufficient to discriminate between different proton and photon pdfs. For the determination of $\alpha_s(M_Z)$ and its scale dependence from the data of [12] see Sect. 5.

In the ZEUS experiment [16] also the scaled invariant jet cross section $E_T^4 E d^3\sigma/\mathrm{d}p_x \mathrm{d}p_y \mathrm{d}p_z$ has been measured as a function of the dimensionless variable $x_\mathrm{T} = 2E_T/W_{\gamma p}$ at $\langle W_{\gamma p}\rangle = 180$ and $255\,\mathrm{GeV}$. The cross section ratio as function of x_T violates scaling as predicted from QCD (Fig. 7, right).

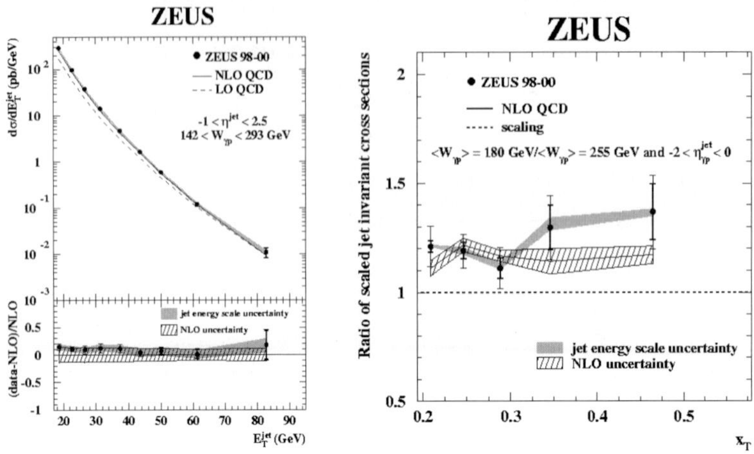

Fig. 7. Inclusive jet production cross section in photoproduction. Ratio of scaled invariant cross sections (av. over $-2 < \eta_{\rm lab} < 0$) for the two shown $\langle W_{\gamma p}\rangle$ compared with NLO QCD; (Figure from [16])

(ii) The dijet photoproduction cross section is dependent on the dijet angle θ^* in the parton-parton-cms, which is sensitive to the matrix element of the hard scattering subprocess. Direct photon processes are dominated by BGF with the angular dependence of the spin 1/2-progator $\sim$ $(1 - \cos\theta^*)^{-1}$ whereas in resolved processes gluon-exchange with the steeper angular dependence of the spin 1-progator $\sim$ $(1 - \cos\theta^*)^{-2}$ is dominating. The dijet events can be enriched with direct or resolved processes by cutting on the fraction x_γ^{obs} of the photon-momentum transferred to the two jets of highest transverse energy, which is given by $x_\gamma^{obs} = (E_{T1}\mathrm{e}^{-\eta_1} + E_{T2}\mathrm{e}^{-\eta_2})/2yE_e$ with y the fractional electron energy carried by the photon in the p-rest frame.

Dijet photoproduction has been studied by ZEUS [18] and H1 [19] under similar kinematical conditions using the $k_{\rm T}$-cluster algorithm. The differential cross sections in the dijet invariant mass, E_T and η agree well with NLO QCD expectations and in particular the measured dijet angular distributions (Fig. 8) shows the expected difference for direct ($x_\gamma^{obs} > .75$) and resolved processes ($x_\gamma^{obs} < .75$). The sensitivity of the data to the parametrization of the photon-pdfs appears to be affected by the choice for the E_T-cuts (ZEUS: $E_{T1} > 14\,\mathrm{GeV}$, $E_{T2} > 11\,\mathrm{GeV}$; H1: $E_{T1} > 25\,\mathrm{GeV}$, $E_{T2} > 15\,\mathrm{GeV}$).

4 α_s-Determinations

The good agreement of jet cross sections with QCD predictions has enabled determinations of α_s and its scale dependence.

The H1 inclusive jet cross sections [5] as a function of E_T have been fitted to the QCD prediction for four regions of Q^2 using the proton-pdfs and μ_R and μ_F as

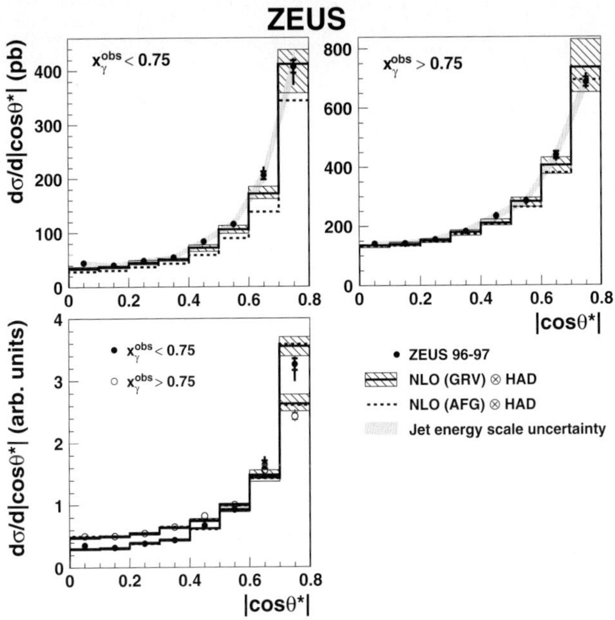

Fig. 8. Inclusive dijet photoproduction as a function of the dijet angle θ^* in the parton-parton-cms for different cuts on the fractional photon momentum x_γ^{obs} transferred to the dijet system. Comparison with NLO QCD for different photon-pdfs; (Figure from [18])

input. The resulting α_s has been shown to be stable against variations of the jet algorithm. The combined fit to all Q^2-data evolved to m_Z is shown in the summary Fig. 9).

For the analysis of the ZEUS data [6] the following procedure has been applied: The cross sections $\mathrm{d}\sigma/\mathrm{d}A$ with $A = Q^2$, E_T have been calculated in NLO QCD for the same pdf-set with three α_s-values. These calculations were used to parametrize the α_s-dependence of the binned cross sections $(\mathrm{d}\sigma/\mathrm{d}A)_i$ for each bin i according to $(\mathrm{d}\sigma(\alpha_s)/\mathrm{d}A)_i = C_{1i}.\ \alpha_s^1 + C_{2i}.\ \alpha_s^2$ with $\alpha_s = \alpha_s(M_Z)$. From a χ^2-fit of this ansatz to the measured $\mathrm{d}\sigma/\mathrm{d}A$, α_s was obtained for the chosen regions of A. The best fit was obtained for $Q^2 > 500\,\mathrm{GeV}^2$ (Fig. 9). The same procedure has been applied to determine the scale dependence of α_s. With E_T as energy scale, the α_s dependence of $\mathrm{d}\sigma/\mathrm{d}E_T$ was parametrized in terms of $\alpha_s(\langle E_T \rangle)$ where $\langle E_T \rangle$ is the mean value of E_T in bin i. The result is shown in Fig. 10.

The same method has been applied to other jet results from ZEUS i.e. the dijet fraction [12], subjet multiplicities ([15]) and jets from photoproduction [15]. A summary of the results is shown in Figs. 9 and 10.

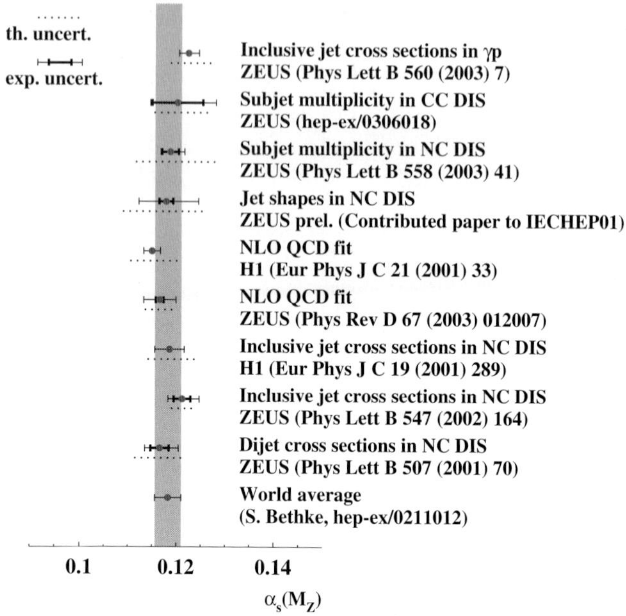

Fig. 9. $\alpha_s(M_Z)$ from recent H1 and ZEUS measurements and the world average

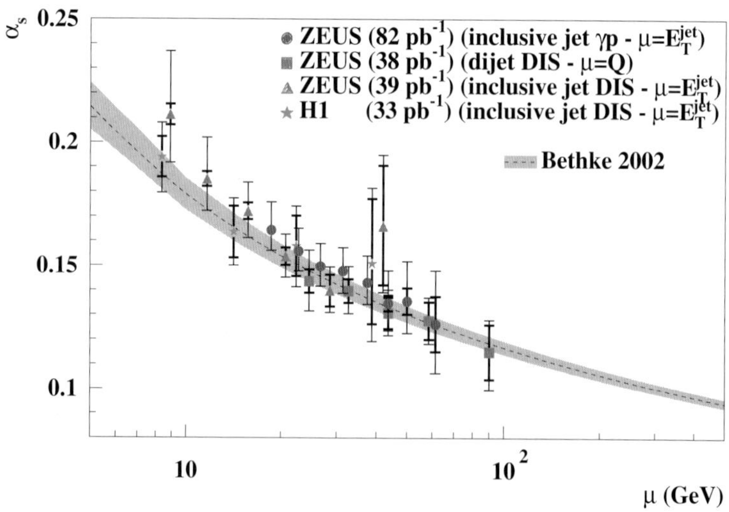

Fig. 10. Scale dependence of α_s from recent H1 and ZEUS measurements and QCD predictions for $\alpha_s(M_Z) = 0.1183$ (world average)

5 Summary

In the analysis of jet production in DIS and photoproduction at HERA the longitudinally invariant k_T-cluster algorithm in its inclusive version has become the standard jet finding algorithm.

In DIS, the inclusive jet cross sections at higher values of Q^2 and E_T are well described by NLO QCD; in this region higher order and hadronization corrections are small; this holds as well for photoproduction.

In the forward region and/or at smaller Q^2 resp. x values, the situation is less satisfactory. DGLAP based calculations are expected to become less reliable in this region, the mentioned corrections are sizable and the hadronic structure of the photon i.e. resolved processes have to be taken into account. Calculations based on BFKL or CCFM evolution only partly show better agreement. An increase in experimental statistics and efforts to reduce the theoretical uncertainties are highly desirable for a better understanding of this challenging region.

The measurements of α_s and its running from jet data have yielded results which are of perhaps unexpected precision; they can well compete with other precision measurements e.g. from e^+e^--annihilation and are in good agreement with the world average.

Acknowledgements

I thank G. Grindhammer for valuable discussions and comments on the paper.

References

1. S.D. Ellis and D.E. Soper, Phys. Rev. D48, 3160 (1993) [hep-ph/9305266], S. Catani et al., Nucl. Phys. B406, 187 (1993)
2. V. Gribov and L.N. Lipatov, Sov. J. Nucl. Phys. 15, 438, 675 (1972); L.N. Lipatov, Sov. J. Nucl. Phys. 20, 94 (1975); G. Altarelli and G. Parisi, Nucl. Phys. B126, 298 (1977); Y.L. Dokshitzer, Sov. Phys. JETP 46, 641 (1977)
3. V.S. Fadin, E.A. Kuraev and L.N. Lipatov, Sov. Phys. JETP 44, 443 (1976); V.S. Fadin, E.A. Kuraev and L.N. Lipatov, Sov Phys. JETP 45, 199 (1977); Y. Balitsky and L.N. Lipatov, Sov. J. Nucl. Phys. 28, 822 (1978)
4. M. Ciafolini, Nucl. Phys. B296, 49 (1988); S. Catani, F. Fiorani and G. Marchesini, Phys. Lett. B234, 339 (1990); S. Catani, F. Fiorani and G. Marchesini, Nucl. Phys. B336, 18 (1990); G. Marchesini, Nucl. Phys. B445, 49 (1995) [hep-ph/9412327]
5. C. Adloff et al., Europ. Phys. J. C19, 289 (2001) [DESY 00-145; hep-ex/0010054]
6. S. Chekanov et al., Physics Letters B547, 164 (2002) [DESY 02-112; hep-ex/0208037]
7. S. Chekanov et al., Physics Letters B551, 226 (2003) [DESY 02-171; hep-ex/0210064]
8. C. Adloff et al., Physics Letters B542, 193 (2002), [DESY 02-079; hep-ex/0206029]

9. J. Jung: H1 Collaboration, ICHEP 02, Amsterdam (2002)
10. L. Goerlich: H1 Collaboration, DIS 2003, St. Petersburg (2003)
11. J. Breitweg et al., Physics Letters B507, 70 (2001) [DESY 01-018; hep-ex/0102042]; S. Chekanov et al., Europ. Phs. J. C23, 13 (2002) [DESY 01-127; hep-ex/0109029]
12. A. Aktas et al.: H1-collaboration DESY 03-160; hep-ex/0310019
13. C. Adloff et al., Physics Letters B515, 17 (2001) [DESY 01-073; hep-ex/0106078]
14. N. Krumnack: ZEUS Collaboration, DIS 2003, St. Petersburg (2003)
15. S. Chekanov et al., Physics Letters B558, 41 (2003) [DESY 02-217; hep-ex/0212030]
16. S. Chekanov et al., Physics Letters B560, 7 (2003) [DESY 02-228; hep-ex/0212064]
17. C. Adloff et al., submitted to Europ. Phys. J. C [DESY 02-225; hep-ex/0302034]
18. S. Chekanov et al., Europ. Phys. J. C23, 615 (2002) [DESY 01-220; hep-ex/0112029]
19. C. Adloff et al., Europ. Phys. J. C25, 13 (2002) [DESY 01-225; hep-ex/0201006]

CP Violation from Orbifold: From Examples to Unification Structures

Nicolas Cosme[1]

Service de Physique Théorique, CP225 Université Libre de Bruxelles, Bld du Triomphe, 1050 Brussels, Belgium
ncosme@ulb.ac.be

1 Introduction

Higher dimensional theories have recently been explored as alternative to the "usual" scalar spontaneous symmetry breaking for gauge symmetry, supersymmetry and discrete symmetries. Along that line, we present here examples of gauge theories in which CP violation is introduced with the gauge breaking through the dimensional reduction process.

We illustrate first the purpose with a toy model [1] and then give the minimal gauge structure for a realistic realisation. The last section present the first Grand Unified group in this context which turns out to be $SO(11)$ instead of the usual $SO(10)$ in four dimensions [2].

1.1 A Toy Model

The reduction of five dimensional gauge theories introduces in four dimensions, together with a tower of Kaluza-Klein states for each particle, a scalar degree of freedom that is the extra component of the gauge field.

Its coupling to fermions through the gauge invariant line integral along the extra dimension $\int \mathrm{d}y A_y = \langle A_y \rangle$ may then induce an effective complex mass:

$$\bar{\psi}(M + \mathrm{i}\gamma_5 \langle A_y \rangle)\psi \ , \tag{1}$$

since the Clifford algebra is extended to $\gamma_B = (\gamma_\mu, i\gamma_5)$ for $4+1$ dimensions ($B = 0, 1, \cdots, (4 = y)$).

While, in the simplest $U(1)$ case, we can remove this phase by a redefinition of spinors, it is easily seen that the extension to a non-abelian group, e.g. $SU(2)$, gives rise to a physically observable phase, and hence to CP violation.

Indeed, let us consider a massive doublet $\Psi = (\psi_1, \psi_2)$ and let the extra component of the gauge field get an expectation value $\langle W_y \rangle = \int \mathrm{d}y\ W_y = w\ \sigma^3$. This breaks the group to vectorlike effective interactions in $3+1$ dimsensions:

$$\left(\bar{\psi}_1 \ \bar{\psi}_2 \right) \mathrm{i}(\partial^\mu - \mathrm{i}W_a^\mu \tau^a)\gamma_\mu \begin{pmatrix} \psi_1 \\ \psi_2 \end{pmatrix} \ , \tag{2}$$

with two massive $W^\pm$ and one massless W^3.

On the other hand, the fermion mass matrix receives a complex contribution from the Wilson loop, that is:

$$\begin{pmatrix} \bar{\psi}_1 & \bar{\psi}_2 \end{pmatrix} \begin{pmatrix} M + iw\gamma_5 & \\ & M - iw\gamma_5 \end{pmatrix} \begin{pmatrix} \psi_1 \\ \psi_2 \end{pmatrix} . \tag{3}$$

The mass matrix can be made real by a bi-unitary transformation, $\mathcal{M}' = U_R^\dagger \mathcal{M} U_L$, but then, fermion coupling to charged gauge bosons is no longer purely vectorial but includes a phase between the L and R parts. It results in a W^3-dipole moment at one loop level.

CP violation is introduced here in a fundamentally CP symmetrical framework where all initial couplings are real.

2 Realistic Models

We turn now to the realisation of realistic models. We first point out, from the previous toy model, the required features to match this goal:

1. Obviously, the model must contain fermions in the extra dimension and requires the reduction of these.
2. Since chirality does not exist in $4+1d$, $4+1d$ "vectorlike" couplings must be broken in the reduction to get realistic weak interactions in $3+1d$. This will be done through orbifold boundary conditions [3].
3. Nevertheless, one should keep some left and right fermionic components, in order to form mass terms through the gauge Wilson loop. We will therefore consider groups that contain the minimal left-right extension of weak interactions $SU(2)_L \times SU(2)_R \times U(1)_{(B-L)}$.

Let now consider an $SU(4)$ gauge group in five dimensions where the extra dimension is an S^1/Z_2 orbifold. We specify the Z_2-parity on the gauge group to be $P_{\mathcal{G}} = diag(1,1,-1,-1)$ on the fundamental. We verify easily that $P_{\mathcal{G}}$ commutes with an $SU(2) \times SU(2) \times U(1)$ group. This implies that the zero modes for gauge fields are [4]:

$$\begin{aligned} A_\mu^{\bar{a},0} &\rightarrow (1,1)(0) + (3,1)(0) + (1,3)(0) , \\ A_y^{\hat{a},0} &\rightarrow (2,2)(2) + (2,2)(-2) . \end{aligned}$$

For fermions in the 4 representation we get the following zero modes:

$$\begin{pmatrix} u_L^0 & d_L^0 & u_R^0 & d_R^0 \end{pmatrix} \leftrightarrow (2,1)(1) + (1,2)(-1).$$

An adjoint scalar Φ coupled to fermions gets its zero modes in the same representation as A_y.

We get indeed interresting features for a left-right symmetric model [that is a complex scalar bidoublet and left and right doublets for fermions] but here the abelian part distinguishes between left and right fermions. A solution is to consider the direct product $Sp(4) \times U(1)_{B-L}$ with the field content shown in Table 1. This constitutes the minimal realistic structure which induces CP violation through the mechanism considered here.

Table 1. Field content for $Sp(4) \times U(1)_{B-L}$

$Sp(4) \times U(1)_{B-L}$		$SU(2)_L \times SU(2)_R \times U(1)_{B-L}$
B_μ	1_0	$(1,1)_0$
A_μ	10_0	$(3,1)_0 + (1,3)_0$
A_y	10_0	$(2,2)_0$
Ψ	$4_{(B-L)}$	$(2,1)_{(B-L)} + (1,2)_{(B-L)}$
Φ	10_0	$(2,2)_0$
$\chi_{L,R}$	10_2	$(3,1)_2 + (1,3)_2$

3 Unification

If we now consider the embedding of this minimal model in the usual Grand Unified structure $SO(10)$, we will face a problem for the introduction of chirality from extra dimensional fermions.

Actually this can be easily seen since the usual unification for $SO(10)$ contains only left-handed fermions, e.g. $((u_L, d_L) \ \cdots \ (u_L^c, d_L^c))$. The alternative here is to use $SO(11)$.

Indeed, $SO(11)$ contains $SO(5) \times SO(6)$ (i.e. up to an isomorphism $Sp(4) \times SU(4)$) which can be broken to $SU(2)_L \times SU(2)_R \times SU(3)_c \times U(1)_{B-L}$. The first breaking could be done through a twist in the periodic conditions : we just add a gauge transformation

$$T_{\mathcal{G}} = \begin{pmatrix} I_{5\times 5} & 0 \\ 0 & -I_{6\times 6} \end{pmatrix}$$

in the fundamental of $SO(11)$.

Fermions are in the spin 1/2 representation of $SO(11)$, i.e. the 32, which breaks to $(4,4) + (4,\bar{4})$ under $Sp(4) \times SU(4)$. We can verify that $\bar{4}$ is a faithfull representation under $-I_{6\times 6}$ of $SO(6)$ and thus takes a minus; while 4 remains unchanged. As a consequence, the only $(4,4)$ provides zero modes for fermions.

The second breaking (a Z_2 symmetry for $Sp(4)$ and a usual scalar breaking for $SU(4)$) gives rise to the following zero modes:

$$(4,4) \to (2,1,1)_3 + (1,2,1)_3 + (2,1,3)_{-1} + (1,2,3)_{-1},$$

which is an entire fermion family with respectively: left and right doublets of leptons and left and right doublets of quarks.

This achieved the embedding of the proposed mechanism in a Grand Unified theory which takes care explicitly of extra dimensional fermions. Of course, three generations still need to be introduced but we have shown a mechanism for breaking $SO(11)$ to the standard model and generating the CP violating part of the Yukawa couplings.

Acknowledgements

This work was done in collaboration with J.-M. Frère and is supported in part by IISN, la Communauté Francaise de Belgique (ARC), and the Belgian federal governement (IUAP).

References

1. N. Cosme, J. M. Frere and L. Lopez Honorez, Phys. Rev. D **68** (2003) 096001 [arXiv:hep-ph/0207024].
2. N. Cosme and J. M. Frere, arXiv:hep-ph/0303037.
3. Y. Kawamura, Prog. Theor. Phys. **103** (2000) 613 [arXiv:hep-ph/9902423]. R. Barbieri, L. J. Hall and Y. Nomura, Phys. Rev. D **63** (2001) 105007 [arXiv:hep-ph/0011311]. A. Hebecker and J. March-Russell, Nucl. Phys. B **625** (2002) 128 [arXiv:hep-ph/0107039]. Q. Shafi and Z. Tavartkiladze, Phys. Rev. D **66** (2002) 115002.
4. R. Slansky, Phys. Rept. **79** (1981) 1.

Doubly Projected Functions in Out of Equilibrium Thermal Field Theories

Ivan Dadić

Ruder Bošković Institute, Zagreb, Croatia
dadic@thphys.irb.hr

1 Introduction

Out of equilibrium thermal field theory [1] has recently attracted considerable interest. In many applications, especially in heavy ion collisions [2], one is interested for short time scale features in the systems far from equilibrium. For such processes one usually assumes that the initial system, described by some single-particle density operator , starts its evolution at some finite time ($t_i = 0$) and ends the evolution at the (finite) time ($t_f = T$). We find that the natural framework for description of such systems is based on two-point functions projected [3] to finite time interval $[0, T]$ i.e. on doubly projected functions.

2 Doubly Projected Functions

Let us start with the two-point function $G(x, y)$ where $0 < x_0, y_0 < T$. We define the Wigner variables as usual $X = x + y/2,\ s = x - y$. The limits imposed on x_0, y_0 imply $0 < X_0, \ -2\bar{X}_0 < s_0 < 2\bar{X}_0$, where $\bar{X}_0 = min[X_0, T - X_0]$. Using Wigner variables we can write

$$G(x,y) = \Theta(X_0)\Theta(T-X_0)\Theta(2\bar{X}_0 - s_0)\Theta(2\bar{X}_0 + s_0)\bar{G}\left(X + \frac{s}{2}, X - \frac{s}{2}\right) . \quad (1)$$

Note here that the function $\bar{G}$ defined by (1) in general depends on X_0. The two-point function can be expressed in terms of the Wigner transform (i.e. Fourier transform with respect to s_0, s_i):

$$G\left(X + \frac{s}{2}, X - \frac{s}{2}\right) = (2\pi)^{-4} \int \mathrm{d}^4 p e^{-\mathrm{i}(p_0 s_0 - \boldsymbol{ps})} G(p_0, \boldsymbol{p}; X) , \quad (2)$$

$$\begin{aligned} G(p_0, \boldsymbol{p}; X) &= \int_{-2\bar{X}_0}^{2\bar{X}_0} \mathrm{d}s_0 \int \mathrm{d}^3 s e^{\mathrm{i}(p_0 s_0 - \boldsymbol{ps})} G\left(X + \frac{s}{2}, X - \frac{s}{2}\right) = \\ &= \int_{-\infty}^{\infty} \mathrm{d}s_0 \int D^3 s e^{\mathrm{i}(p_0 s_0 - \boldsymbol{ps})} \Theta(X_0)\Theta(T - X_0)\Theta(2\bar{X}_0 - s_0)\Theta(2\bar{X}_0 + s_0) \\ &\times \bar{G}\left(X + \frac{s}{2}, X - \frac{s}{2}\right) . \end{aligned} \quad (3)$$

The product of Θ functions is a projection operator with simple properties:

$$P_{X_0,T}(p_0,p_0') = \frac{\Theta(X_0)\Theta(T-X_0)}{2\pi}\int_{-2\bar{X}_0}^{2\bar{X}_0} \mathrm{d}s_0 \mathrm{e}^{\mathrm{i}s_0(p_0-p_0')}$$
$$= \frac{\Theta(X_0)\Theta(T-X_0)}{\pi}\frac{\sin\left(2\bar{X}_0(p_0-p_0')\right)}{p_0-p_0'} , \quad (4)$$

$$\mathrm{e}^{-\mathrm{i}s_0p_0'}\Theta(X_0)\Theta(T-X_0)\ \ \Theta(2\bar{X}_0+s_0)\Theta(2\bar{X}_0-s_0) =$$
$$= \int \mathrm{d}p_0 \mathrm{e}^{-\mathrm{i}s_0p_0} P_{X_0,T}(p_0,p_0') , \quad (5)$$

$$\lim_{X_0\to\infty}\lim_{T\to\infty} P_{X_0,T}(p_0,p_0') = \lim_{X_0\to\infty}\lim_{T\to\infty}\frac{\Theta(X_0)\Theta(T-X_0)\sin\left(2\bar{X}_0(p_0-p_0')\right)}{\pi(p_0-p_0')}$$
$$= \Theta(X_0)\Theta(T-X_0)\delta(p_0-p_0') . \quad (6)$$

$$P_{X_{0,M},T}(p_0,p"_0) = \int \mathrm{d}p_0' P_{X_0,T}(p_0,p_0')P_{X_0',T}(p_0',p"_0),$$
$$X_{0,M} = min(\bar{X}_0,\bar{X}_0'), \quad (7)$$

$$P_{X_0,\bar{T}}(p_0,p"_0) = \int \mathrm{d}p_0' P_{X_0,T}(p_0,p_0')P_{X_0,T'}(p_0',p"_0), \ \ \bar{T} = min(T,T') . \quad (8)$$

In this paper, the doubly projected function is a very special two-point function $F(T,x,y) = F(T,X+s/2,X-s/2)$: it does not depend on $\boldsymbol{X}$, it is a function of $(s_0,\boldsymbol{s})$ within the interval $-2\bar{X}_0 < s_0 < 2\bar{X}_0$ and identical to zero outside:

$$F\left(T,X+\frac{s}{2},X-\frac{s}{2}\right) = \Theta(X_0)\Theta(T-X_0)\Theta(2\bar{X}_0-s_0)\Theta(2\bar{X}_0+s_0)$$
$$\times\bar{F}_\infty(s_0,\boldsymbol{s}) , \quad (9)$$

$$\lim_{X_0\to\infty}\lim_{T\to\infty} F\left(T,X+\frac{s}{2},X-\frac{s}{2}\right) = \bar{F}_\infty(s_0,\boldsymbol{s}) . \quad (10)$$

The whole X_0 dependence is introduced by the projection operator

$$F_{X_0,T}(p_0,\mathbf{p}) = [P_{X_0,T}F_\infty](p_0,\mathbf{p}) = \int_{-\infty}^{\infty} \mathrm{d}p_0' P_{X_0,T}(p_0,p_0')F_\infty(p_0',\mathbf{p}) . \quad (11)$$

Important examples of projected functions are retarded, advanced, and Keldysh components of free propagators. Further examples emerge in the process of convoluting the functions and calculating the self-energies. For further analysis, the analytic properties of the $X_0 \to \infty$ limit of the WTDPF (Wigner transform of doubly projected function) as a function of complex energy are very important. We define the following properties: (1) the function of p_0 is analytic above (below) the real axis, (2) the function goes to zero as $|p_o|$ approaches infinity in the upper (lower) semiplane. The choice above (below) and upper (lower) refers to R(A) components.

2.1 Propagators as Doubly Projected Functions

The usual propagators defined in the R/A basis are, with an analytic (but not unique) definition of a sign function, identical to our propagators at $X_0 \to \infty$

$$G_R(p) = -G_{1,1} + G_{1,2} = G_{R,\infty} = \frac{-i}{p^2 - m^2 + 2i\epsilon p_0} , \tag{12}$$

$$G_K(p) = G_{1,1} + G_{2,2} = G_{K,\infty}(p) = -(1 \pm 2f(\omega_p))\times$$

$$\mathrm{sign}(p_0, \omega_p)\left(G_{R,\infty}(p) - G_{A,\infty}(p)\right), \mathrm{sign}(p_0,\omega_p) = (\frac{p_0}{\omega_p})^{2n+1} , \tag{13}$$

where n is freely choosen integer. The finite Wigner transform ($0 < x_0 < T$, $0 < y_0 < T$, $\bar{X}_0 = min[X_0, T]$) is obtained by smearing

$$\begin{aligned} G_{T,R,X_0}(p) &= [P_{X_0,T}G_{R,\infty}](p) = -G^*_{A,X_0,T}(p),\ G_{T,K,X_0}(p) = \\ &= [P_{X_0,T}G_{K,\infty}](p) . \end{aligned} \tag{14}$$

It is important to note that, at any finite X_0, the above expression is not singular at $p_0 = \pm\omega_p$.

3 Convolution Product of Two Two-Point Functions

Let us now consider the convolution product [4] of two Green functions:

$$C = A * B \Leftrightarrow C(x,y) = \int \mathrm{d}z A(x,z)B(z,y) . \tag{15}$$

In terms of Wigner transforms (we assume translational invariance):

$$\begin{aligned} C(p_0, \boldsymbol{p}; X, T) &= \int_{-2\bar{X}_0}^{2\bar{X}_0} \mathrm{d}s_0 \int \mathrm{d}^3 s \int \mathrm{d}^4 z \mathrm{e}^{\mathrm{i}(p_0 s_0 - \boldsymbol{p}\ \boldsymbol{s})} \\ &\times \frac{1}{(2\pi)^4} \int \mathrm{d}^4 p_1 \mathrm{e}^{-\mathrm{i}(p_{01}s_{01} - \boldsymbol{p}_1 \boldsymbol{s}_1)} A(p_{01}, \boldsymbol{p}_1; X_1, T) \\ &\times \frac{1}{(2\pi)^4} \int \mathrm{d}^4 p_2 \mathrm{e}^{-\mathrm{i}(p_{02}s_{02} - \boldsymbol{p}_2 \boldsymbol{s}_2)} B(p_{02}, \boldsymbol{p}_2; X_2, T) , \end{aligned}$$

$$X_1 = X + \frac{s_2}{2},\quad X_2 = X - \frac{s_1}{2},\ s_1 = x - z,\ s_2 = z - y ,$$

$$\bar{X}_0 = min[X_0, T - X_0] . \tag{16}$$

We can write the final expression as

$$C_{X_0,T}(p_0, \boldsymbol{p}) = \int \mathrm{d}p_{01}\mathrm{d}p_{02} P_{X_0,T}\left(p_0, \frac{p_{01}+p_{02}}{2}\right)$$

$$\frac{i}{2\pi}\frac{\mathrm{e}^{-\mathrm{i}X_0(p_{01}-p_{02})} - \mathrm{e}^{-\mathrm{i}(X_0-T)(p_{01}-p_{02})}}{p_{01}-p_{02}} A_\infty(p_{01}, \boldsymbol{p}) B_\infty(p_{02}, \boldsymbol{p}) . \tag{17}$$

Expression (17) is the key for finite-time thermal field theory. If both A and B are operators satisfying assumptions (1) and (2) for advanced (or both retarded) components, the expression simplifies:

$$C_{X_0,T}(p_0,\boldsymbol{p}) = \int \mathrm{d}p_{01} P_{X_0,T}(p_0,p_{01}) A_\infty(p_{01},\boldsymbol{p}) B_\infty(p_{01},\boldsymbol{p}) \,. \tag{18}$$

This is an extraordinary result: the convolution product of two WTDPF's is a WTDPF under conditions (1) and (2). As expected, in the $T \to \infty$, $X_0 = \infty$, $T > 2X_0$ limit (18) becomes a simple product

$$\lim_{X_0\to\infty} \lim_{T\to\infty} C_{X_0,T}(p_0,\boldsymbol{p}) = A_\infty(p_0,\boldsymbol{p}) B_\infty(p_0,\boldsymbol{p}) \,. \tag{19}$$

At finite X_0 (18) exhibits a smearing of energy (as much as it is necessary to preserve the uncertainty relations).

4 Modifications of the Feynman Rules

The calculations performed so far already contain all of the modifications of the Feynman rules required by the finite-t_i assumption. In coordinate space, the only modification is that the bare propagators [(12) and (13)] are limited by $0 < x_0 < T$ and $0 < y_0 < T$; thus they are doubly projected functions. In energy-momentum space, the above change reflects in the use of finite T, X_0 of propagators, vertices, and the overall factor. At the vertices the usual energy-conserving $\delta(\sum_i p_{0i})$ is substituted by (we have arranged the orientation of all the momenta to be incoming) $-i[1 - \mathrm{e}^{-\mathrm{i}T\sum_i p_{0i}}][2\pi(\sum_i p_{0i})]^{-1}$. Under the momentum integrals there is a left-over factor at the vertices j_A (by subscript A we indicate that j_A are vertices with amputated legs): $\mathrm{e}^{-\mathrm{i}\sum_{j_A} x_{j_A}(\sum_{i_{j_A}} \lambda_{i_{j_A}} p_{i_{j_A}})}$, where $\lambda = \pm$ depends on whether the corresponding momentum is outgoing of or incoming to the vertex j_A, and i_{j_A} is running through the nonamputated lines. The overall factor takes care of uncertainty relations: the larger the closer of distances ($\bar{X}_0 = min[X_0, T - X_0]$) from the moment of "turning on" (X_0) and "turning off" ($T - X_0$), the smaller the energy smearing.

In the vertex factor the energy is not explicitly conserved. This energy nonconservation is, through the uncertainty relations, related to the finiteness of X_0. In the limit of infinite T, X_0, energy conservation is recovered. For finite T there are two cases in which the energy conservation at the vertex cannot be achieved: 1) all incoming propagators are retarded and all outgoing propagators are advanced, 2) all incoming propagators are advanced and and all outgoing propagators are retarded. In [3] both cases are shown to occure in the perturbation expansion.

References

1. J. Schwinger: J. Math. Phys. 2, 407 (1961); L. V. Keldysh: ZH. Eksp. Teor. Fiz.47, 1515 (1964) [Sov. Phys.-JETP 20, 1018 (1965)].

2. D. Boyanovsky, H. J. de Vega, and S. -Y. Wang: Phys. Rev. D 61, 065006 (2000); S. -Z. Wang and D. Boyanovsky: Phys. Rev. D 63, 051702 (2001); I. Dadić: preprint hep-ph/0103025.
3. I. Dadić: Phys. Rev. D 63, 25011 (2001); I. Dadić: Phys. Rev. D66, 049903 (2002); I. Dadić: Nucl. Phys. A702 (2002) 356C-360C.
4. D. C. Langreth and J. W. Wilkins: Phys. Rev. B 6, 3189 (1972).

Nonfactorizable Contributions in $\overline{B^0} \to D_s^+ D_s^-$ and $\overline{B_s^0} \to D^+ D^-$ Decays

Jan O. Eeg[1], Svjetlana Fajfer[2] and Aksel Hiorth[1]

[1] Department of Physics, University of Oslo, P.O. Box 1048 Blindern, 0316 Oslo, Norway
j.o.eeg@fys.uio.no; aksel.hiorth@fys.uio.no

[2] Physics Department, University of Ljubljana and J. Stefan Institute, 1000 Ljubljana, Slovenia
svjetlana.fajfer@ijs.si

We are living in a fruitful era of B meson physics. A lot of experimental data are coming from BaBar, Belle and Tevatron on B meson decays stimulate many studies of their decay mechanism. In the treatment of decay amplitudes, usually the factorization assumption has been used. Recently, it has been shown [1] that some B meson decay modes exhibit *QCD factorization*. This means that their amplitudes might factorize into the product of two matrix elements of weak currents. Typically, the decay amplitudes which factorize in this sense are $B \to \pi\pi$ and $B \to K\pi$ where the energy release is big compared to the light meson masses. For decays where the energy release is of order 1 GeV, QCD factorization is not expected to hold. Here we discuss the leading contributions for $\overline{B^0} \to D_s^+ D_s^-$ and $\overline{B_s^0} \to D^+ D^-$ [2]. At quark level these decays occur through the annihilation mechanism $b\bar{s} \to c\bar{c}$ and $b\bar{d} \to c\bar{c}$, respectively. However, within the factorized limit the annihilation mechanism will give a vanishing amplitude due to current conservation [3]. These contributions are proportional to the numerically small Wilson coefficient C_1, which we will neglect in our analyzis. In contrast, the typical factorized decay modes which proceed through the spectator mechanism, say $\overline{B^0} \to D^+ D_s^-$, are proportional to the numerically larger Wilson coefficient C_2. In the approach of [2], the nonfactorizable contributions are coming from from the chiral loops and the gluon condensates. The chiral loops result from the K^0-exchanges as given in Fig. 1. There are also nonfactorizable contributions due to soft gluon emission. Such contributions can be calculated in terms of the (lowest dimension) gluon condensate within a recently developed Heavy Light Chiral Quark Model (HLχQM) [4], which is based on Heavy Quark Effective Theory (HQEFT) [5]. This model has been applied to processes with B-mesons in [6, 7]. The gluon condensate contributions is also proportional to the favorable Wilson coefficient C_2. We have followed the standard approach [8] for non-leptonic decays where one constructs an effective lagrangian $\mathcal{L}_W$ in terms of quark operators multiplied with Wilson coefficients containing all information of the short distance (SD) loop effects above a renormalization scale μ of order m_b. Within Heavy Quark Effective Theory (HQEFT) [5], the effective nonleptonic Lagrangian $\mathcal{L}_W$ can be further evolved down to the scale $\mu \sim \Lambda_\chi \sim$1 GeV [9, 10].

The use of factorization is illustrated in the $\overline{B^0} \to D^+ D_s^-$ decay:

$$\langle D_s^- D^+ | \mathcal{L}_W | \overline{B^0} \rangle_F = - \left(C_2 + \frac{1}{N_c} C_1 \right) \langle D_s^- | \bar{s} \gamma_\mu \gamma_5 c | 0 \rangle \langle D^+ | \bar{c} \gamma_\mu b | \overline{B^0} \rangle \, , \tag{1}$$

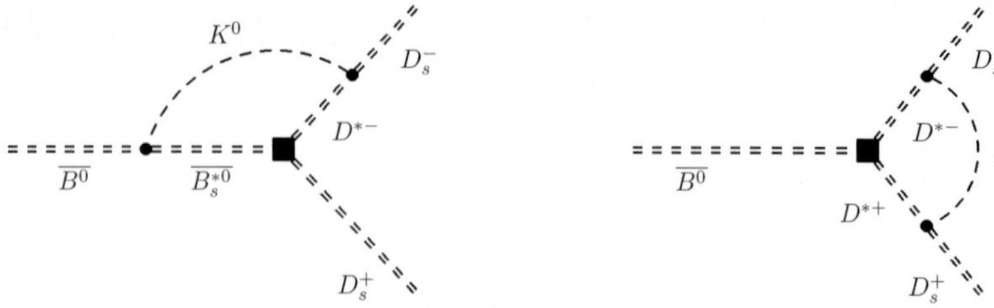

Fig. 1. Non-factorizable chiral loops for $\overline{B^0} \to D_s^+ D_s^-$

The coefficients $C_{1,2}$ are Wilson coefficients for the operators containing the product of two left-handed currents. In our notation $C_i = -\frac{G_F}{\sqrt{2}} V_{cb} V_{cs}^* a_i$, where the a_i are dimensionless, G_F is the Fermi coupling, V_{ij} are CKM parameters. Numerically, $a_1 \sim 10^{-1}$ and $a_2 \sim 1$ at the scale $\mu = m_b$, and $|a_1| \simeq 0.4$ and $|a_2| \simeq 1.4$ at $\mu \sim \Lambda_\chi \sim 1\,\mathrm{GeV}$ [9, 10]. Penguin operators may also contribute, but have rather small Wilson coefficients.

The factorized amplitude for $\overline{B^0} \to D_s^+ D_s^-$ is given by

$$\langle D_s^- D_s^+ | \mathcal{L}_W | \overline{B^0} \rangle_F = 4 \left(C_1 + \frac{1}{N_c} C_2 \right) \langle D_s^- D_s^+ | \overline{c_L} \gamma_\mu c_L | 0 \rangle \langle 0 | \overline{d_L} \gamma^\mu b_L | \overline{B^0} \rangle \,. \quad (2)$$

Unless one or both of the D-mesons in the final state are vector mesons, this matrix element is zero due to current conservation:

$$\langle D_s^+ D_s^- | \bar{c} \gamma_\mu c | 0 \rangle \langle 0 | \bar{d} \gamma^\mu \gamma_5 b | \overline{B^0} \rangle \sim f_B (p_D + p_{\bar{D}})^\mu \, \langle D_s^+ D_s^- | \bar{c} \gamma_\mu c | 0 \rangle = 0 \,. \quad (3)$$

Our approach is based on the use of bosinized currents [2] and by using them we first write down the amplitude for $\overline{B^0} \to D^+ D_s^-$. To calculate the chiral loop amplitudes we need the factorized amplitudes for $\overline{B_s^{*0}} \to D_s^+ D^{*-}$ and $\overline{B^0} \to D^{*+} D^{*-}$. After use of bosonised currents [2] we obtain the following chiral loop amplitude from the Fig. 1:

$$A(\overline{B^0} \to D_s^+ D_s^-)_\chi = (V_{cd}^* / V_{cs}^*) \; A(\overline{B_d^0} \to D_d^+ D_s^-)_F \cdot R^\chi \,, \quad (4)$$

where the $A(\overline{B_d^0} \to D_d^+ D_s^-)_F$ stands for the factorized amplitude for the process $\overline{B^0} \to D^+ D_s^-$ given in [2] and the quantity R^χ is a sum of contributions from the left and right part of Fig. 1 respectively. In the $\overline{MS}$ scheme we obtained

$$R^\chi = \frac{m_K^2}{(4\pi f)^2} g_A^2 \left[\left\{ \frac{(\omega + 1)}{(\omega + \lambda)} \left[r(-\omega) + r(-\lambda) \right] - 1 \right\} \ln \left(\frac{m_K^2}{\Lambda_\chi^2} \right) - 1 \right] \,. \quad (5)$$

with $\omega = M_B/(2M_D)$, $\lambda = [M_B^2/(2M_D^2) - 1]$ and f being the π decay constantwhile g_A stands for the coupling between H, H^* and M $(= \pi, K, \eta)$. The function $r(x)$ is given in [2]. Numerically, we find [2]:

$$R^\chi \simeq 0.12 - 0.26i \,. \quad (6)$$

The genuine nonfactorizable part for $\overline{B^0} \to D_s^+ D_s^-$ can, by means of Fierz transformations and identities for the product of two colour matrices, be written in terms of coulored currents

$$\langle D_s^- D_s^+ | \mathcal{L}_W | \overline{B^0} \rangle_{NF} = 8\, C_2 \, \langle D_s^- D_s^+ | (\overline{d}_L \gamma^\alpha t^a b_L) \, (\overline{c}_L \gamma_\alpha t^a c_L) \, | \overline{B^0} \rangle \, . \tag{7}$$

Within our approach, this amplitude is written in a quasi-factorized way in terms of matrix elements of colored currents:

$$\langle D_s^+ D_s^- | \mathcal{L}_W | \overline{B^0} \rangle_{NF}^G = 8\, C_2 \, \langle D_s^+ D_s^- | \overline{c}_L \gamma_\mu t^a c_L | G \rangle \langle G | \overline{d}_L \gamma^\mu t^a b_L | \overline{B^0} \rangle \, , \tag{8}$$

where a G in the bra-kets symbolizes emision of one gluon (from each current) as vizualized in Fig. 2.

In order to calculate the matrix elements in (8), we have used [2] the Heavy Light Chiral Quark Model (HLχQM) recently developed in [4], which incorporates emision of soft gluons modelled by a gluon condensate. Then we defined a quantity R^G for the gluon condensate amplitude analogously to R^χ in (4) and (5) for chiral loops. Numerically, we determine that the ratio between the two amplitudes is [2]

$$R_G \simeq 0.055 + 0.16i \, , \tag{9}$$

which is of order one third of the chiral loop contribution in (5).

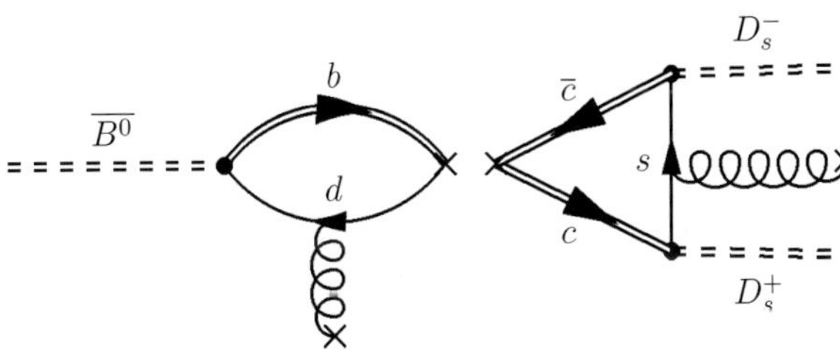

Fig. 2. Non-factorizable contribution for $\overline{B^0} \to D_s^+ D_s^-$ through the annihilation mechanism with additional soft gluon emision. The wavy lines represent soft gluons ending in vacuum to make gluon condensates

Adding the amplitudes R_χ and R_G we found that the amplitude for $\overline{B^0} \to D_s^+ D_s^-$ is of order 15–20% of the factorizable amplitude for $\overline{B^0} \to D^+ D_s^-$, before the different CKM-factors are taken into account. Multiplying also with the Wilson coefficient [9, 10] $a_2 \simeq 1.33 + 0.2i$, we obtained the branching ratios [2]

$$BR(\overline{B_d^0} \to D_s^+ D_s^-) \simeq 7 \times 10^{-5} \, ; \quad BR(\overline{B_s^0} \to D_s^+ D_s^-) \simeq 1 \times 10^{-3} \, . \tag{10}$$

We hope that current searches at BaBar and Belle might soon give the limit on the rate $\overline{B^0} \to D_s^+ D_s^-$, while the detection of the $\overline{B_s^0}$ mode seems to be more difficult due to trouble with $\overline{B_s^0}$ identification. Thus the mode $\overline{B^0} \to D_s^+ D_s^-$ will only be accessable at Tevtron and later at LHC.

Talk given by S. Fajfer

References

1. M. Beneke, G. Buchalla, M. Neubert, C.T. Sachrajda: Phys. Rev. Lett. **83**, 1914 (1999)
2. J.O. Eeg, S. Fajfer, A. Hiorth: Phys. Lett. B **570**, 46 (2003) Phys. Rev. D 64 (2001) 034010.
3. J.O. Eeg, S. Fajfer, J. Zupan: Phys. Rev. D **64**, 034010 (2001)
4. A. Hiorth and J. O. Eeg: Phys. Rev. D **66**, 074001 (2002)
5. See for example A.V. Manohar, M.B. Wise: *Heavy Quark Physics*, (Cambridge Monogr. Part. Phys. Nucl. Phys. Cosmol. 10) (2000)
6. J. O. Eeg, A. Hiorth, A. D. Polosa: Phys. Rev. D **65**, 054030 (2002)
7. A. Hiorth and J. O. Eeg: Eur. Phys. J. direct C **30**, 006 (2003)
8. See for example: G. Buchalla, A. Buras, and M. Lautenbacher: Rev. Mod. Phys. **68**, 1125 (1996)
9. B. Grinstein, W. Kilian, T. Mannel, and M.B. Wise: Nucl. Phys. B **363**, 19 (1991)
10. R. Fleischer: Nucl. Phys. B **412**, 201 (1994)

On the Geometry of Gauge Field Theories

Helmuth Hüffel and Gerald Kelnhofer

Institut für Theoretische Physik, Universität Wien
Boltzmanngasse 5, 1090 Vienna, Austria
helmuth.hueffel@univie.ac.at

1 Introduction

One of the interesting aspects of the stochastic quantization scheme [1] (for reviews see [2, 3]) lies in its rather unconventional treatment of gauge field theories. We recall that originally it was formulated by Parisi and Wu [1] without the introduction of gauge fixing terms and without the usual Faddeev-Popov ghost fields; later on a modified approach named stochastic gauge fixing was given by Zwanziger [4]; further generalizations and a globally valid path integral for Yang-Mills theory – overcoming the Gribov obstruction – were advocated in [5, 6].

In this talk we discuss new modifications [7] of the original Parisi-Wu stochastic process of scalar QED, showing that the standard – gauge fixed – QED path integral density can be identified with the equilibrium limit of the underlying Fokker–Planck probability distribution.

2 The Geometrical Setting of QED

Let $P \rightarrow M$ be a principal $U(1)$-bundle over the n-dimensional boundaryless connected, simply connected and compact Euclidean manifold M. The photon fields A are regarded as elements of the affine space $\mathcal{A}$ of all connections on P. The action of the gauge group $\mathcal{G}$ on $A \in \mathcal{A}$ is defined by

$$A \rightarrow A^g = A + g^{-1}\mathrm{d}g \; . \tag{1}$$

Let us define the subgroup $\mathcal{G}_0 \subset \mathcal{G}$ where $\mathcal{G}_0 = \mathcal{G}/U(1)$ denotes the group of all gauge transformations reduced by the constant ones. Since $\mathcal{G}_0$ acts freely on $\mathcal{A}$ we can prove that $\mathcal{A} \rightarrow \mathcal{M} = \mathcal{A}/\mathcal{G}_0$ is trivializable, a global section being given by $\hat{\sigma}([A]) = A^{\omega(A)^{-1}}$. Here $\omega(A) \in \mathcal{G}_0$ is defined by

$$\omega(A) = \exp[\triangle^{-1}d^*(A - A_0)] \tag{2}$$

and $\triangle$ denotes the invertible Laplacian; $A_0 \in \mathcal{A}$ is a chosen fixed background connection.

In order to discuss scalar matter fields ϕ we chose a representation ρ of $g \in \mathcal{G}_0$ on the vector space $V = \mathbf{C}$; $\rho(g)$ simply denotes multiplication with g. We consider the associated vector bundles $E = P \times_\rho V$ on M. Scalar fields are described by appropriately chosen sections of E; we denote by $\mathcal{F}$ the space of scalar fields. The action of $\mathcal{G}_0$ on $\Phi^i := (A, \phi) \in \mathcal{A} \times \mathcal{F}$ is given by

$$\Phi^i = (A,\phi) \longrightarrow (\Phi^g)^i = (A^g,\phi^g) = (A + g^{-1}dg, g^{-1}\phi) . \tag{3}$$

Using the previous construction of $\omega(A)$ we obtain a global section

$$\sigma([A,\phi]) = (\hat{\sigma}([A]), \phi^{\omega(A)^{-1}}) = (A^{\omega(A)^{-1}}, \phi^{\omega(A)^{-1}}) . \tag{4}$$

A $\mathcal{G}_0$-invariant Riemannian metric $h_{(A,\phi)}$ can be defined by

$$h_{(A,\phi)}\left((\tau_A^1, v_\phi^1),(\tau_A^2, v_\phi^2)\right) = \langle \tau_A^1, \tau_A^2\rangle + \langle v_\phi^1, v_\phi^2\rangle \tag{5}$$

where

$$\langle \alpha,\beta\rangle = \frac{1}{2}\int_M (\bar{\alpha}\wedge *\beta + \alpha\wedge *\bar{\beta}) \tag{6}$$

and $\alpha,\beta \in T\mathcal{A}$, or $T\mathcal{F}$, respectively; $*$ is the Hodge operator on M, $\bar{\alpha}$ denotes complex conjugation of α.

Let the globally defined gauge fixing surface Σ in $\mathcal{A}\times\mathcal{F}$ be given by

$$\Sigma = im\,\sigma = \{(B,\psi)\in\mathcal{A}\times\mathcal{F} | (B,\psi) = (A^{\omega(A)^{-1}}, \phi^{\omega(A)^{-1}})\} . \tag{7}$$

We note that B and ψ are invariant under the action of $\mathcal{G}_0$; B satisfies the gauge fixing condition

$$d^*(B - A_0) = 0 . \tag{8}$$

We define the adapted coordinates $\Psi^\mu = \{B,\,\psi,\,g\}$ via the bundle maps $\chi : \Sigma\times\mathcal{G}_0 \to \mathcal{A}\times\mathcal{F}$ and $\chi^{-1} : \mathcal{A}\times\mathcal{F} \to \Sigma\times\mathcal{G}_0$, where

$$\chi(B,\psi,g) = (B^g,\psi^g) \tag{9}$$

and

$$\chi^{-1}(A,\phi) = (\sigma([A,\phi]),\omega(A)) = (A^{\omega(A)^{-1}}, \phi^{\omega(A)^{-1}}, \omega(A)) . \tag{10}$$

The differentials $T\chi$ and $T\chi^{-1}$ are calculated straightforwardly and the vielbeins $\mathrm{e}^i{}_\mu = \frac{\delta\Phi^i}{\delta\Psi^\mu}$ and their inverses $E^\mu{}_i = \frac{\delta\Psi^\mu}{\delta\Phi^i}$ corresponding to the change of variables $\Psi^\mu = \{B,\psi,g\} \leftrightarrow \Phi^i = \{A,\phi\}$ are obtained by inspection. The $\mathcal{G}_0$ invariant metric h can be expressed in adapted coordinates as the pullback $G = \chi^* h$. In matrix notation we have $G = \mathrm{e}^\dagger\, e$ and $G^{-1} = E\,E^\dagger$. The determinant of G is given by $det\, G = \triangle$.

3 Parisi-Wu Stochastic Quantization

For scalar QED we have

$$S_{inv} = \langle D_A\varphi, D_A\varphi\rangle + m^2\langle\varphi,\varphi\rangle + \frac{1}{2}\langle F,F\rangle \tag{11}$$

where $D_A\varphi = (d-A)\varphi$ and $F = \mathrm{d}A$. The Parisi-Wu Langevin equations are given by

$$\mathrm{d}\Phi = -\frac{\delta S_{inv}}{\delta\Phi}\mathrm{d}s + \mathrm{d}\xi, \quad \mathrm{d}\xi\,\mathrm{d}\xi^\dagger = 2\cdot\mathbf{1}\,\mathrm{d}s\ , \tag{12}$$

where $\Phi^i = (A,\phi)$. Using Ito calculus we transform the Parisi-Wu Langevin equations into adapted coordinates

$$\mathrm{d}\Psi = \left[-G^{-1}\frac{\delta S_{inv}}{\delta\Psi} + \frac{\delta G^{-1}}{\delta\Psi}\right]\mathrm{d}s + \mathrm{d}\zeta\ , \quad \text{where} \quad \mathrm{d}\zeta\mathrm{d}\zeta^\dagger = 2\,G^{-1}\mathrm{d}s\ . \tag{13}$$

4 Generalized Stochastic Quantization

Our equivalence proof relies on specific modifications of the metric on the field space which governs the stochastic process. These modifications keeping expectation values of gauge invariant observables unchanged can be achieved in such a way that the associated Fokker–Planck operator has a positive kernel and is annihilated on its *right* by the standard gauge fixed QED path integral density:

First a damping force along the gauge orbit is introduced [4] in order to maintain the probabilistic interpretation of the Fokker–Planck formulation. However, due to its presence the standard gauge fixed QED action will not annihilate the Fokker–Planck operator on its right side: We recall that the bundle metric $h_{(A,\phi)}$ is giving rise to a natural connection γ, whose horizontal subbundle $\mathcal{H}$ is orthogonal to the fiber. Explicitly we can prove that its curvature is nonvanishing which implies that any vector field along the gauge group cannot be written as a gradient with respect to the metric $h_{(A,\phi)}$.

We are considering now a larger class of stochastic processes by modifying the Wiener increments $d\xi$ as well; extra so called Ito-terms have to be added correspondingly. This induces a metric on the space $\mathcal{A} \times \mathcal{F}$ of gauge and matter fields implying a specific connection with a potentially analogous obstruction as discussed above. A necessary requirement is therefore that the corresponding curvature has to vanish. Indeed, there exists a flat connection in our bundle which is the pull-back of the Maurer–Cartan form on $\mathcal{G}_0$ via the global trivialization χ^{-1}. In the adapted coordinates the induced field metric $\widetilde{G}$ has to be chosen by defining

$$E = \begin{pmatrix} E^{\Sigma} \\ \\ e_{\mathcal{G}}^{\dagger} \end{pmatrix} \quad \text{where} \quad \widetilde{G}^{-1} = \widetilde{E}\,\widetilde{E}^{\dagger} \,. \tag{14}$$

Finally we have to specify the vertical drift term: it is related to the gradient of an extra action $S_{\mathcal{G}}$, which we are allowed to choose as $S_{\mathcal{G}}[g] = \frac{1}{2}\langle d^* g^* \theta^{U(1)}, d^* g^* \theta^{U(1)}\rangle$, $\theta^{U(1)}$ is the Maurer Cartan form on $U(1)$. Note that in the original variables $S_{\mathcal{G}}$ is just the standard QED background-gauge fixing term $S_{\mathcal{G}}(\omega(A)) = \frac{1}{2}\langle d^*(A - A_0), d^*(A - A_0)\rangle$. Summarizing we have

$$\mathrm{d}\Psi = \left[-\widetilde{G}\,\frac{\delta S_{tot}}{\delta \Psi} + \frac{\delta \widetilde{G}}{\delta \Psi} \right] \mathrm{d}s + \mathrm{d}\widetilde{\zeta} \tag{15}$$

where

$$S_{tot} = S_{inv} + S_{\mathcal{G}} \,, \quad \text{and} \quad \mathrm{d}\widetilde{\zeta}\,\mathrm{d}\widetilde{\zeta}^{\dagger} = 2\widetilde{G}^{-1}\mathrm{d}s \,. \tag{16}$$

It is easy now to prove the equivalence of the stochastic quantization scheme with the path integral quantization. For the formulation in terms of the adapted coordinates $\Psi = \{B, \psi, g\}$ the associated Fokker–Planck equation is derived in straightforward manner

$$\frac{\partial \rho[\Psi, s]}{\partial s} = \frac{\delta}{\delta \Psi}\,\widetilde{G}^{-1} \left[\frac{\delta S_{tot}[\Psi]}{\delta \Psi} + \frac{\delta}{\delta \Psi} \right] \rho[\Psi, s] \,, \tag{17}$$

where the Fokker–Planck operator is appearing in just factorized form. Due to the positivity of $\widetilde{G}$ the fluctuation dissipation theorem applies and the equilibrium Fokker–Planck distribution $\rho^{\mathrm{eq}}[\Psi]$ obtains by direct inspection as

$$\rho^{\mathrm{eq}}[\Psi] = \frac{\mathrm{e}^{-S_{inv}[B,\psi]}\,\mathrm{e}^{-S_{\mathcal{G}}[g]}}{\int_{\mathbf{\Sigma}} \mathrm{d}B\mathrm{d}\psi\,\mathrm{e}^{-S_{inv}[B,\psi]}\int_{\mathcal{G}_0}\mathrm{d}g\,\mathrm{e}^{-S_{\mathcal{G}}[g]}}\,. \tag{18}$$

This result is completely equivalent to the standard background-gauge fixed QED path integral prescription. The additional *finite* contributions of the gauge degrees of freedom always cancel out when evaluated on gauge invariant observables.

Similarly, in terms of the original variables $\Phi = \{A, \phi\}$ the Fokker–Planck equilibrium distribution is given by the standard background-gauge fixed path integral density

$$\rho^{\mathrm{eq}}[\Phi] = \frac{\mathrm{e}^{-S_{inv}[A,\phi]-S_{\mathcal{G}}(\omega(A))}}{\int_{\mathbf{A}} \mathrm{d}A\mathrm{d}\phi\,\mathrm{e}^{-S_{inv}[A,\phi]-S_{\mathcal{G}}(\omega(A))}}\,. \tag{19}$$

References

1. G. Parisi and Wu Yongshi, Sci. Sin. **24**, 483 (1981)
2. P. Damgaard and H. Hüffel, Phys. Rep. **152**, 227 (1987)
3. M. Namiki: *Stochastic Quantization* (Springer, Heidelberg 1992)
4. D. Zwanziger, Nucl. Phys. B **192**, 259 (1981)
5. H. Hüffel and G. Kelnhofer, Ann. of Phys. **270**, 231 (1998)
6. H. Hüffel and G. Kelnhofer, Phys. Lett. B **472**, 101 (2000)
7. H. Hüffel and G. Kelnhofer, hep-th/0312315

On the Singlet Penguin in $B \to K\eta'$ Decay

Jan Olav Eeg[1], Krešimir Kumerički[2], and Ivica Picek[2]

[1] Department of Physics, University of Oslo, 0316 Oslo, Norway
[2] Department of Physics, Faculty of Science, University of Zagreb, P.O.B. 331, 10002 Zagreb, Croatia

Recent measurements by CLEO, BaBar and Belle collaborations [1, 2, 3, 4, 5] of two-body charmless hadronic B meson decays with η' meson in final state [6]

$$\begin{aligned} Br(B^+ \to K^+\eta') &= (77.6 \pm 4.6) \cdot 10^{-6} \ , \\ Br(B^0 \to K^0\eta') &= (65.2 \pm 6.0) \cdot 10^{-6} \ , \end{aligned} \qquad (1)$$

indicate an enhancement by a factor of 5–6 when compared to the corresponding decays to the pion instead of η'. In an attempt to explain a dynamical origin of such an enhancement, one can start from the observation that some exceptional properties of η' particle are related to the QCD anomaly. Namely, the puzzle of the unexpectedly large η' *mass* (famous U(1) problem) was resolved by taking into account the QCD axial anomaly and the pure gluonic component of the η' quantum state

$$|\eta'\rangle = \cdots + |\mathrm{gg}\rangle + \cdots \ . \qquad (2)$$

Thus, it was no surprise that in various dynamical mechanisms invoked to explain the enhancement (1) [7, 8, 9, 10, 11, 12, 13, 14, 15, 16, 17, 18] this gluonic component often played a prominent role.

On the other hand, more phenomenological analysis based on the SU(3) flavour symmetry and the corresponding diagrammatical formalism [19], suggests that besides the usual penguin (P) diagrams (Fig. 1a) dominating the $B \to K\pi$ rates, the singlet penguin (SP) diagram (Fig. 1b) should also contribute substantially to the processes of the $B \to K\eta'$ type. Since the singlet penguin diagram can naturally be realized via creation of η' particle by the pure gluonic intermediate state (as already suggested on Fig. 1b) this points to such gluonic mechanisms as good candidates for explaining the enhancement for η' modes (1).

In our recent paper [20] we identified a well-defined short-distance (SD) contribution to the singlet penguin amplitude, generating $b \to s\eta'$ transition displayed on Fig. 2. This SD amplitude corresponds to the hard gluons in the intermediate state producing a final η' particle via a quark triangle loop (represented by the blob N in Fig. 2), thus corresponding to a sort of a remnant of the QCD anomaly (dubbed the "anomaly tail" in [20]).

Now we supplement this analysis by investigating further contributions to the bi-gluonic $b \to s\eta'$ amplitude on Fig. 2 with vertex N represented by the expression

$$N_{\mu\nu}^{aa'}(\bar{Q}^2, \omega) = -\mathrm{i} F_{\eta' g^* g^*}(\bar{Q}^2, \omega) \epsilon_{\mu\nu k_1 k_2} \delta^{aa'} \ , \qquad (3)$$

where $\bar{Q}^2 = -(k_1^2 + k_2^2)/2$ is average virtuality of gluons and $\omega = (k_1^2 - k_2^2)/q^2$ is asymmetry parameter. The form-factor $F_{\eta' g^* g^*}(\bar{Q}^2, \omega)$ for the symmetric case $\omega = 0$, and in the asymptotic limit valid for large $\bar{Q}^2$ is given by [21, 22]

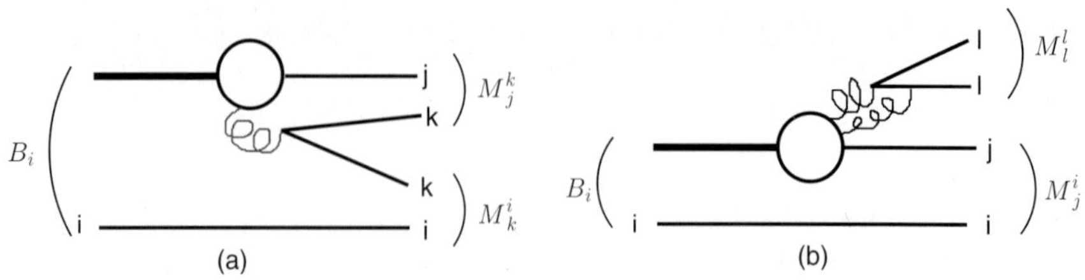

Fig. 1. Two flavour topologies contributing to $B \to K\eta'$ decay in SU(3) symmetry approach: penguin (a) and singlet-penguin (b). B_i is B meson triplet and $M^j{}_i$ is a pseudoscalar meson nonet

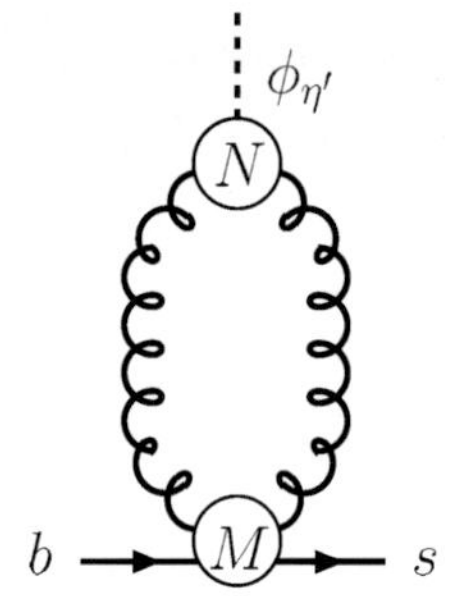

Fig. 2. The hard gluon loop contribution to the $b \to s\eta'$ transition

$$F_{\eta' g^* g^*}(\bar{Q}^2, 0) = 4\pi\alpha_s(\bar{Q}^2) \frac{f^1_{\eta'}}{\sqrt{3}\bar{Q}^2} , \tag{4}$$

with $f^1_{\eta'} \approx 0.15\,\mathrm{GeV}$. Combining this with the amplitude $A_i(-Q^2)$ given in [20] for the flavour-changing $b \to sg^*g^*$ transition (the blob M in Fig. 2), we obtain the $b \to s\eta'$ amplitude

$$\mathcal{A}(b \to s\eta') = \frac{G_F f^1_{\eta'}}{2\sqrt{6}\pi^2} (\phi_{\eta'} \bar{s}\gamma \cdot KLb) \sum_{i=u,c,t} \lambda_i \int \mathrm{d}Q^2 \, \frac{\alpha_s^2(Q^2)}{Q^2} \, A_i(-Q^2) \, . \tag{5}$$

Whereas the form-factor (4) takes into account only transition to the quark $|q\bar{q}\rangle$ quantum state of η', it is interesting to study the influence of gluonic $|gg\rangle$ component of η' on this result. It turns out [22] that in the symmetric case of two gluons having similar momenta ($\omega \approx 0$) the effect of the $|gg\rangle$ component can be included by multiplying the $\eta' g^* g^*$ form-factor (4) by a factor

$$1 - \frac{1}{12} B^g_2 \, .$$

Here, the range of allowed values for the Gegenbauer coefficient B^g_2 (obtained by a fit to the η' transition form-factor) can be found in [22]. The error in B^g_2 is large, and for the most of the allowed region the gluon contribution will interfere destructively, because of the minus sign in the above factor. Accordingly, the amplitude will be smaller than in the pure $|q\bar{q}\rangle$ case by an average of 30 percent. This is displayed

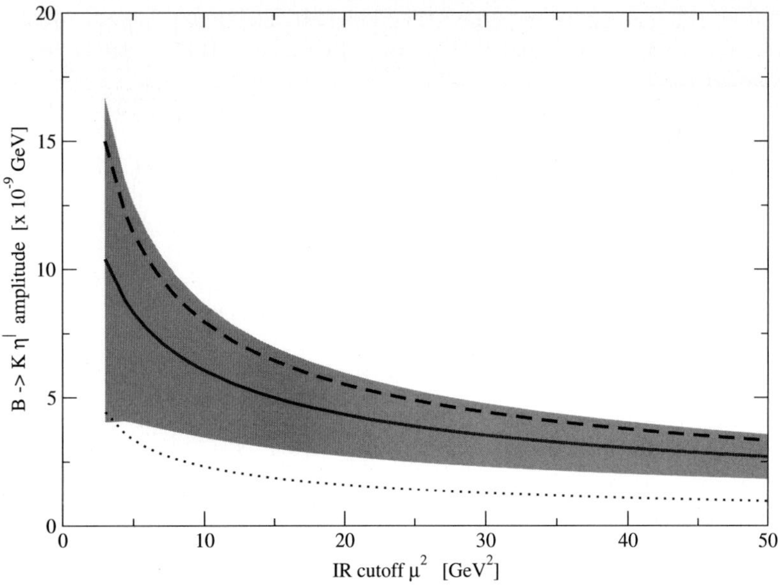

Fig. 3. Short-distance hard gluon contribution to $B \to K\eta'$ amplitude. Dashed line corresponds to pure quark content of η' (4), while shaded area corresponds to allowed region when also gluonic content of η' is taken into account. Dotted line is "anomaly tail" from [20]

on Fig. 3 where the quark transition amplitude (5) is combined with the spectator quark in order to produce the physical $B \to K\eta'$ amplitude.

Comparison to results of [20] shows that the SD contributions considered here are substantially larger then the "anomaly tail" part. Still, they cannot explain the observed η' enhancement (1) by themselves. Apart from some attempts to invoke new physics beyond the Standard Model [16], another mechanism incorporating long distance aspects of the QCD anomaly [23] and/or the one of the penguin interference [24, 17] seems to be needed to complete the picture.

References

1. CLEO, D. Cronin-Hennessy et al., Phys. Rev. Lett. **85**, 515 (2000).
2. CLEO, S. J. Richichi et al., Phys. Rev. Lett. **85**, 520 (2000), hep-ex/9912059.
3. BABAR, B. Aubert et al., Phys. Rev. Lett. **87**, 151802 (2001), hep-ex/0105061.
4. BELLE, K. Abe et al., Phys. Rev. Lett. **87**, 101801 (2001), hep-ex/0104030.
5. BELLE, K. Abe et al., Phys. Lett. **B517**, 309 (2001), hep-ex/0108010.
6. Heavy flavour averaging group, http://www.slac.stanford.edu/xorg/hfag/ .
7. D. Atwood and A. Soni, Phys. Lett. **B405**, 150 (1997), hep-ph/9704357.
8. I. E. Halperin and A. Zhitnitsky, Phys. Rev. **D56**, 7247 (1997), hep-ph/9704412.
9. A. L. Kagan and A. A. Petrov, (1997), hep-ph/9707354.

10. W.-S. Hou and B. Tseng, Phys. Rev. Lett. **80**, 434 (1998), hep-ph/9705304.
11. A. Datta, X. G. He, and S. Pakvasa, Phys. Lett. **B419**, 369 (1998), hep-ph/9707259.
12. D.-s. Du, C. S. Kim, and Y.-d. Yang, Phys. Lett. **B426**, 133 (1998), hep-ph/9711428.
13. M. R. Ahmady, E. Kou, and A. Sugamoto, Phys. Rev. **D58**, 014015 (1998), hep-ph/9710509.
14. A. Ali, J. Chay, C. Greub, and P. Ko, Phys. Lett. **B424**, 161 (1998), hep-ph/9712372.
15. E. Kou and A. I. Sanda, Phys. Lett. **B525**, 240 (2002), hep-ph/0106159.
16. Z.-j. Xiao, K.-T. Chao, and C. S. Li, Phys. Rev. **D65**, 114021 (2002), hep-ph/0204346.
17. M. Beneke and M. Neubert, Nucl. Phys. **B651**, 225 (2003), hep-ph/0210085.
18. H. Fritzsch and Y.-F. Zhou, (2003), hep-ph/0301038.
19. C.-W. Chiang, M. Gronau, and J. L. Rosner, Phys. Rev. **D68**, 074012 (2003), hep-ph/0306021.
20. J. O. Eeg, K. Kumerički, and I. Picek, Phys. Lett. **B563**, 87 (2003), hep-ph/0304274.
21. A. Ali and A. Y. Parkhomenko, Phys. Rev. **D65**, 074020 (2002), hep-ph/0012212.
22. P. Kroll and K. Passek-Kumerički, Phys. Rev. **D67**, 054017 (2003), hep-ph/0210045.
23. J. O. Eeg, K. Kumerički, and I. Picek, in preparation
24. H. J. Lipkin, Phys. Rev. Lett. **46**, 1307 (1981).

Bjorken-Like Limit versus Fermi-Watson Approximation in High Energy Hadron Diffraction

Andrzej R. Malecki

KEN Pedagogical University, 30-084 Krakow, Poland,
amalecki@ultra.ap.krakow.pl

1 Fermi-Watson Approximation

The amplitude $T_{fi} \equiv \langle f|T|i\rangle$ of transition from the initial state $|i\rangle$ to the final state $|f\rangle$ does not change under a simultaneous *unitary transformation* of the physical states $|\, j\rangle \rightarrow|\, Uj\rangle$ and of the transition operator $T \rightarrow T_0 = U^\dagger TU$. The amplitude can thus be expressed through the matrix elements of the transformation operator U and of the transition operators T or T_0: $T_{fi} = \sum_{|j\rangle,|k\rangle} U_{fk}\, U^\star_{ij}\, t_{kj}$ where $t_{kj} \equiv \langle Uk \mid T \mid Uj\rangle = \langle k \mid T_0 \mid j\rangle$. The unitary transformation operator may alternatively be written as: $U \equiv \mathrm{e}^{\mathrm{i}M} \equiv 1 - \Lambda$ where the operator M is *hermitian* while the *normal* operator Λ satisfies the relation: $\Lambda\Lambda^\dagger = \Lambda^\dagger\Lambda = \Lambda + \Lambda^\dagger$.

In frequent cases of physical interest the part of the T-matrix, corresponding to a given entrance channel, is *nearly diagonal*: $T = T_0 + i\epsilon$ where T_0 is *diagonal* and the matrix elements of the operator ϵ are *all small.* Such a situation is typical for reactions at relatively *low energy* as compared to the thresholds for the most important inelastic channels; e.g. the above Ansatz was used by Fermi [1] in the case of pion production by nucleons. The decomposition of the transition operator in “large” (or “hard”) and “small” (or “soft”) parts can also be viewed as the result of a suitable unitary transformation which would *diagonalize* operator T subject to condition the transforming operators M or Λ were “soft”, i.e. their matrix elements were small. While the diagonalization of the operator T through a unitary transformation is always possible, the satisfaction of *both* unitarity and softness (which means $M^2 = \Lambda^2 = 0$) conditions imposes a severe constraint on the transforming operators: $\Lambda = -iM$ and $\Lambda = -\Lambda^\dagger$ instead of the less stringent relation of normality.

In the case $U = \mathrm{e}^{\mathrm{i}M} \approx 1 + iM$ we obtain the following “soft” limit of the Haussdorf expansion in terms of multiple commutators of the operators M and T_0: $T = T_0 + \sum_{n=1}^{\infty}[M,\ldots,[M,T_0],\ldots,]\frac{i^n}{n!} \approx T_0 + i[M,T_0]$. Alternatively, when $U = 1 - \Lambda$ we have : $T = T_0 - \Lambda T_0 - T_0\Lambda^\dagger + \Lambda T_0 \Lambda^\dagger \approx T_0 - [\Lambda, T_0]$. Thus the above “soft” non-diagonal operator reads: $\epsilon = [M,T_0] = i[\Lambda,T_0]$. The antisymmetry of the commutator together with symmetry under time reversal means that the matrix elements of both M and Λ are *antisymmetric* which implies $ReM_{jk} = Im\Lambda_{jk} = 0$. From the above two equations, on the account of diagonality of T_0, one obtains the amplitude of transition:

$$T_{fi} = T_{ii}\delta_{fi} - iM_{fi}(T_{ff} - T_{ii}) = T_{ii}\delta_{fi} + \Lambda_{fi}(T_{ff} - T_{ii}) \qquad (1)$$

where $T_{jj} \approx t_j$ are the amplitudes of elastic scattering, approximately equal to the eigenvalues of T_0 in a state $|j\rangle$. Equation (1) is referred to as the *Fermi-Watson theorem* [1] which allowed e.g. to relate the phases of pion photoproduction on nucleons to the phase-shifts of the elastic pion-nucleon scattering.

2 Eigenstates of Diffraction

The Fermi-Watson theorem established around the year 1953 was rediscovered 20 years later, in a quite different context of high energy diffraction of hadrons. The representation of the inelastic diffractive amplitude [2] as a *difference* between elastic amplitudes for the entrance and exit channels became a part of the folklore of high energy physics [3]. It is most easily understood in terms of the "eigenstates of diffraction" [4] which do not mix with each other, undergoing only the elastic scattering caused by their absorption at the expense of inelastic channels. The diagonalization in all states $T_0 \mid j\rangle = t_j \mid j\rangle$ can be replaced with a weaker assumption of diagonalization of T in a particular class of states [D] only: $T_0 \mid j\rangle = t_j \mid j\rangle + \sum_{|k\rangle\in[\sim D]} t_{kj} \mid k\rangle$ where the states belonging to [D] are called diffractive states and those from its orthogonal complement $[\sim D]$ are referred to as non-diffractive. The states of the diffractive sector are thus subject only to elastic scattering which arises from absorption related to the production of non-diffractive states.

If the transition takes place between two diffractive states, one has then

$$T_{f\,i} = t_i\delta_{fi} + \frac{1}{2}(\Lambda_{fi} - \Lambda^\star_{if})\,(t_f - t_i) + \sum_{|j\rangle\in[D]} \Lambda_{fj}\,\Lambda^\star_{ij}\left[t_j - \frac{1}{2}(t_f + t_i)\right] \tag{2}$$

while the elastic scattering amplitude reads: $T_{i\,i} = t_i + \sum_{|j\rangle\in[D]} |\Lambda_{ij}|^2(t_j - t_i)$. If all Λ_{ij} were small then retaining only terms linear (violating thus unitarity) in $\Lambda = 1-U$ one would yield the elastic scattering amplitude equal to the eigenvalue of T_0 in the initial state. Instead, the inelastic diffraction amplitude would be proportional to the *difference* of the scattering amplitudes in the initial and final states [2], which is just what states the Fermi-Watson theorem.

The above assumption regarding partial diagonalization of T may still be put in doubt by invoking the very sense of diffraction as a feed-back process coupled to the inelastic channels. Therefore, apart of respecting unitarity by keeping quadratic terms in Λ, one should also consider the most general expression which includes contributions of inelastic diffraction to elastic scattering: $T_0 \mid j\rangle = \sum_{|k\rangle\in[D]} t_{kj} \mid k\rangle + \sum_{|k\rangle\in[\sim D]} t_{kj} \mid k\rangle$.

3 Diffractive (Bjorken-Like) Limit

In terms of the operator Λ the amplitude of diffractive transtions reads: $T_{f\,i} = t_{f\,i}\delta_{f\,i} - \sum_{|k\rangle\in[D]} \Lambda_{fk}\, t_{ki} - \sum_{|j\rangle\in[D]} t_{fj}\Lambda^\star_{ij} + \sum_{|j\rangle,|k\rangle\in[D]} \Lambda_{f\,k} t_{kj}\Lambda^\star_{ij}$. This can be rewritten as $T_{f\,i} = t_i\delta_{f\,i} - N_{fi}(T_0)\Lambda_{fi}t_i - \Lambda^\star_{if}t_f N^\star_{if}(T_0^\dagger) + \sum_{|j\rangle\in[D]} N_{fj}(T_0)\Lambda_{fj}t_j\Lambda^\star_{ij}$ where the undimensional quantities N_{kj} are defined [6] as follows : $N_{kj}(T_0) \equiv (\sum_{|l\rangle\in[D]} \Lambda_{kl}t_{lj})(\Lambda_{kj}t_{jj})^{-1}$.

If the subspace [D] contains a very large number of diffractive states then $N_{kj} \equiv N \to \infty$ for any pair of states $|k\rangle$ and $|j\rangle$. In fact, since Λ is a non-singular operator

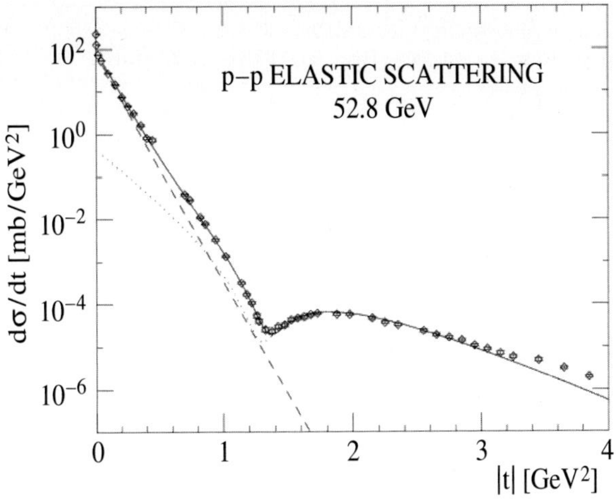

Fig. 1. The p-p elastic differential cross-section at c.m. energy $\sqrt{s} = 52.8\,\text{GeV}$ in the function of the squared momentum transfer $|t|$. The experimental data [7] are compared with the results of our approach [6] (*solid curve*). The first term in (4) which is mostly feed by the shadow of non-diffractive transitions (*dashed curve*) is important at small momentum transfers and negligible at the dip. By contrast, the diffractive term (*dotted curve*) is dominating around and above the dip

its matrix elements vary smoothly under the change of diffractive states. This leads to a great simplification in the limit $N \to \infty$:

$$T_{fi} = t_i \delta_{fi} - N(\Lambda_{fi} t_i + \Lambda^{\star}_{if} t_f - \sum_{|j\rangle \in [D]} \Lambda_{fj} t_j \Lambda^{\star}_{ij}). \tag{3}$$

In general, the effect of non-diagonal transitions inside the diffractive subspace [D] gets factorized. e.g., in the case of elastic scattering one has:

$$T_{ii} = t_i + N \sum_{|j\rangle \in [D]} |\Lambda_{ij}|^2 (t_j - t_i) = t_i + g_i N (t_{av} - t_i) \tag{4}$$

where $g_i = \sum_{|j\rangle} |\Lambda_{ij}|^2 = 2Re(\Lambda_{ii})$ and $t^{(i)}_{av} = (g_i)^{-1} \sum_{|j\rangle} |\Lambda_{ij}|^2 t_j$ is the average value of the diagonal matrix elements t_j.

The second term in (4) can be referred to as the diffractive contribution to elastic scattering since it originates from the action of the operator Λ which filters as intermediate states only those equivalent to the initial state. It is built as an *infinite sum* of the *infinitesimal* contributions from all intermediate states belonging to [D]. The expressions of the form $N\Delta t$ where Δt represents diversity of t_j over the diffractive subspace [D] are to be considered in the Bjorken-like *diffractive limit* [5, 6]: $N \to \infty$, $\Delta t \to 0$ such that $N\Delta t$ is finite.

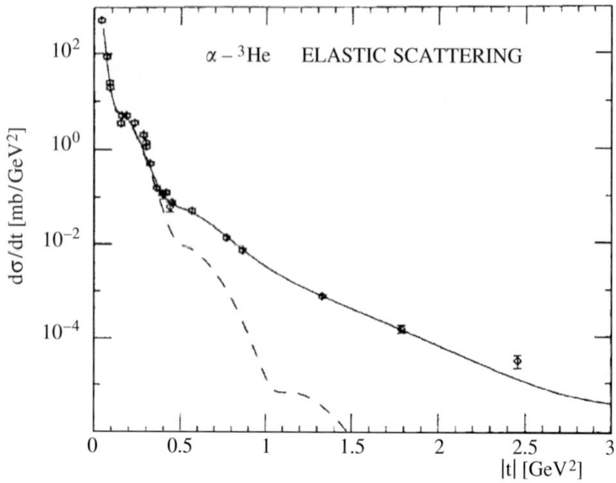

Fig. 2. The ^{4}He–^{3}He elastic cross-section in the function of the squared momentum transfer $|t|$. The experimental data [8] are compared with the results of our approach (*solid curve*). The non-diffractive contribution to the cross-section is shown separately (*dashed curve*)

References

1. E. Fermi, Nuovo Cimento Suppl. **2**, 17 (1955), K. M. Watson, Phys. Rev. **95**, 228 (1954)
2. A. Bialas, W. Czyz and A. Kotanski, Ann. of Phys. **73**, 439 (1972), W. Czyz, Phys. Rev. D **8**, 3219 (1973)
3. J. V. Noble, Phys. Rev. C **8**, 2508 (1973)
4. M. L. Good and W. D. Walker, Phys. Rev. **120**, 1857 (1960)
5. E. Etim, A. Malecki and L. Satta, Phys. Lett. B **184**, 99 (1987)
6. A. Malecki, Phys. Rev. D **54**, 3180 (1996)
7. K.R. Schubert: Tables of nucleon-nucleon scattering. In: Landolt-Börnstein, Numerical data and functional relationship in science and technology, New Series, Vol.1/9a (1979)
8. L. Satta et al., Phys. Lett. B **139** 263 (1984)

Some Aspects of Radiative Corrections and Non-Decoupling Effects of Heavy Higgs Bosons in Two Higgs Doublet Model

Michal Malinský

Institute of Particle and Nuclear Physics, Charles University, Prague;
S.I.S.S.A., Trieste
malinsky@ipnp.troja.mff.cuni.cz

1 Introduction

Although the two-Higgs-doublet extension of the standard model (THDM) was invented about 30 years ago [1], it still belongs among viable candidates for a theory beyond the electroweak standard model (SM). Despite its simplicity it is quite popular, namely because of its capability to include various aspects of "new physics" like for example the additional sources of CP violation (see e.g. [2], [3]). Moreover, its two Higgs doublet structure mimics many features of the Higgs sector of perhaps the most popular SM extension, the minimal supersymmetric standard model (MSSM). On the other hand, since the Higgs sector of THDM is less constrained, it can lead to various effects which are not present in MSSM, in particular to the non-decoupling behaviour of the heavy Higgs boson contributions in the electroweak scattering amplitudes.

As in the MSSM, the presence of the additional doublet leads to five physical Higgs states: 2 CP even Higgs scalars h^0 and H^0, a CP odd pseudoscalar A^0 and a charged pair of $H^{\pm}$. The lightest scalar h^0 is quite similar to the SM Higgs boson η i.e. the mass of h^0 should be close to the weak scale. On the other hand the typical mass scale of the other Higgses (M_H) is not so constrained in general, the unitarity bounds [4] permit M_H around one TeV (if there is no new physics in the game at this scale). Therefore a natural question arises as to whether these additional Higgs bosons tend to decouple from the weak-scale amplitudes. As we shall see in the next section, the answer is "not in general".

2 Non-Decoupling of Heavy Higgs Bosons in THDM

The reason why the heavy Higgs bosons *need not* decouple from the weak-scale physics in the THDM, but they do so within MSSM [5] is roughly the following. Since the Higgs self-couplings are driven by SUSY, the only way to make the four additional Higgs bosons (H^0, A^0 and $H^{\pm}$) sufficiently heavy in MSSM is to adjust the $SU(2)_L \otimes U(1)_Y$ singlet mass parameters in the Higgs potential; in such case these masses have to decouple in accord with the famous Appelquist-Carazzone theorem [6]. In the THDM case one can do the job also by a convenient choice of the Higgs couplings λ_i and the SSB parameter $\tan\beta$, keeping at the same time the singlet mass parameters small. Notice that even the violation of the simple unitarity bounds could be fully compatible with the requirement of perturbativity of the

Higgs sector ($\lambda_i < 1$) provided one can choose a sufficiently large value of $\tan\beta$. As an illustration, consider the following tree-level THDM Higgs mass relations:

$$\begin{aligned}
m^2_{h^0} &= \frac{1}{2}(1-\kappa)M^2 + \left[B_2\sin^2\beta - A_1\cos^2\beta + \frac{1}{4}C(1+\cos 2\alpha\cos 2\beta)\right]\frac{v^2}{\cos 2\alpha} \\
m^2_{H^0} &= \frac{1}{2}(1+\kappa)M^2 + \left[A_2\cos^2\beta - B_1\sin^2\beta - \frac{1}{4}C(1-\cos 2\alpha\cos 2\beta)\right]\frac{v^2}{\cos 2\alpha} \\
m^2_{A^0} &= M^2 - \frac{1}{2}\left[2\lambda_5^R + \lambda_6^R\cot\beta + \lambda_7^R\tan\beta\right]v^2 \qquad (1)\\
m^2_{H^\pm} &= M^2 - \frac{1}{2}\left[\lambda_4 + \lambda_5^R + \lambda_6^R\cot\beta + \lambda_7^R\tan\beta\right]v^2
\end{aligned}$$

where (using the superscript R to denote the real part of a quantity)

$$\begin{aligned}
M^2 &\equiv \frac{m_{12}^{2\ R}}{\sin\beta\cos\beta} & \kappa &\equiv -\frac{\cos 2\beta}{\cos 2\alpha} \\
A_1 &\equiv \lambda_1\sin^2\alpha - \lambda_7^R\tan\beta\cos^2\alpha & B_1 &\equiv \lambda_2\sin^2\alpha - \lambda_6^R\cot\beta\cos^2\alpha \\
A_2 &\equiv \lambda_1\cos^2\alpha - \lambda_7^R\tan\beta\sin^2\alpha & B_2 &\equiv \lambda_2\cos^2\alpha - \lambda_6^R\cot\beta\sin^2\alpha \\
C &\equiv \lambda_7^R\tan\beta - \lambda_6^R\cot\beta & D &\equiv \lambda_7^R\tan\beta + \lambda_6^R\cot\beta
\end{aligned}$$

The model and notation are those used in [7]. Notice that in the case $\lambda_6 = \lambda_7 = 0$ one recovers the relations obtained in [8]. Moreover, using

$$\cos^2(\alpha-\beta) = \frac{m^2_{h^0} - m^2_L}{m^2_{H^0} - m^2_{h^0}} \quad \text{and} \quad m^2_L \equiv \lambda_1\cos^4\beta + \lambda_2\sin^4\beta + \frac{1}{2}\left(\lambda + 2D\sin^2\beta\right)v^2 \qquad (2)$$

it is easy to se that if the weak-scale contributions (square-brackets in (1)) are small compared to M^2, the requirement of having h^0 light and the others much heavier forces the heavy multiplet to be almost degenerate with masses proportional to M, which is the signature of the so-called decoupling regime [9]. Therefore, it is the *distortion of the heavy Higgs spectrum* which matters concerning the possible nondecoupling effects of the additional Higgses in THDM.

In this work I would like to demonstrate these issues at the particular case of the amplitude of the proces $e^+e^- \to W^+W^-$ at one-loop level in THDM in comparison with the well-known one-loop SM result [10]. Note that there are already earlier papers on this topic in the literature [11], [12] but these usually make use of some specific approximations (in particular, the equivalence theorem for longitudinal vector bosons [13]) which we would like to avoid.

3 The Process $e^+e^- \to W^+W^-$

For the considered process, the central quantity of our interest is the deviation of the differential cross-section, calculated within THDM, from its SM value; this is defined by

$$\delta \equiv \mathrm{d}\sigma^{THDM}/\mathrm{d}\sigma^{SM} - 1 \qquad (3)$$

Expanding the THDM amplitude around the SM value and keeping just the leading terms, one gets [14]

$$\delta \doteq 2\mathrm{Re}\frac{\Delta\mathcal{M}[\Delta\Gamma^{TGV}]_{1-loop}}{\mathcal{M}^{SM}_{tree}} \tag{4}$$

Here $\Delta\mathcal{M}_{1-loop}$ stands for the difference of the THDM and SM 1-loop amplitudes, which descends primarily (the leading term) from the differences of the triple gauge vertex corrections $\Delta\Gamma^{TGV}$:

$\Delta\mathcal{M}[\Delta\Gamma^{TGV}]_{1-loop} =$

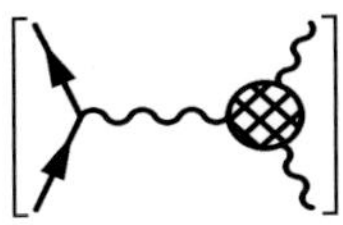

$_{THDM}$ $-$

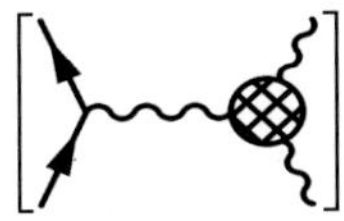

$_{SM}$

Since most of the technical aspects of the calculation are covered in [7],[14] let us emphasize only several salient points. *i*) We have chosen to work in the on-shell renormalization scheme. There are two main reasons for that: the overall number of diagrams to be calculated is reduced with respect to other schemes and the mass-parameters we are playing with are the true physical masses. The only disadvantage is the need of treating carefully the finite parts of the counterterms which must be computed by means of Ward identities. On the other hand, the cancellation of UV-divergences provides a non-trivial consistency check. *ii*) There is also a simple consistency check for the finite parts of $\Delta\Gamma^{TGV}$: they should tend to vanish in the decoupling regime, i.e. in the case where the masses of heavy Higgs bosons are large and almost degenerate.

4 Summary of Results and Conclusion

Due to the large number of diagrams contributing to $\Delta\Gamma^{TGV}$ it is hard to get an analytic expression even for the leading terms in $\Delta\mathcal{M}_{1-loop}$. The numerical analysis shows that the formfactors $\Delta\Pi^{\gamma WW}$ and $\Delta\Pi^{ZWW}$ defined in [7] behave in accordance with the consistency conditions mentioned above. For example, let us look at $|\Delta\Pi_1^{\gamma WW}|$ as a function of the mass of the A^0, (Fig. 1): since the other Higgs masses are kept close to the weak scale, the heavy Higgs spectrum distortion grows with m_{A^0} and the non-decoupling effect in the formfactor as well. Concerning δ, one naturally expects a similar behaviour because it is linear in the formfactors (at the leading order, see [7]). Let us take the particular case: $e^+_L e^-_R \to W^+_L W^-_L$ (in this setup the leading term turns out to be $\cos\theta^*$-independent which allows us to draw simpler pictures). As can be seen in Fig. 2, for large distortions of the heavy Higgs spectrum one can get an effect of several percent. At least in principle, the nondecoupling effects of relatively heavy additional Higgs degrees of freedom can be used in an indirect exploration of the EW Higgs sector at future colliders.

Acknowledgements

The work was partially supported by "Centre for Particle Physics", project No. LN00A006 of the Czech Ministry of Education. I would like to thank Prof. Jiří Hořejší for useful discussions. I am grateful to S.I.S.S.A. for the financial support and the organizers for the possibility to take part at this nice event.

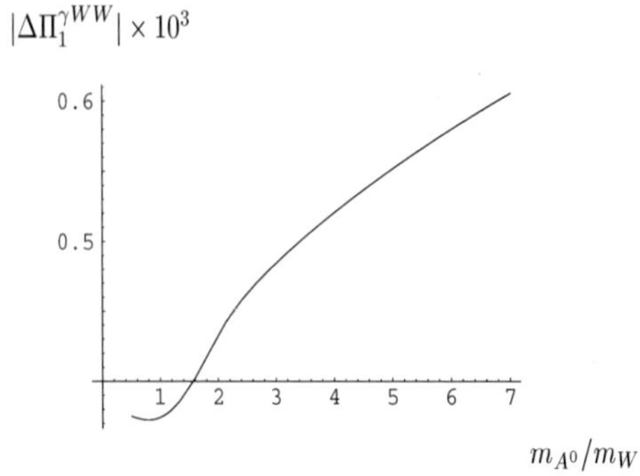

Fig. 1. $|\Delta\Pi_1^{\gamma WW}|$ as a function of m_{A^0}. The other masses are: $m_\eta = 105\,\text{GeV}$, $m_{h^0} = 125\,\text{GeV}$, $m_{H^0} = 145\,\text{GeV}$, $m_{H^\pm} = 180\,\text{GeV}$ and we take $\sqrt{s} = 250\,\text{GeV}$

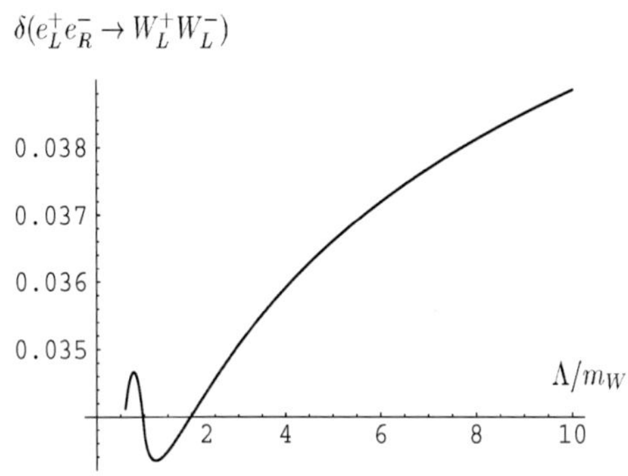

Fig. 2. δ as a function of $m_H^0 = 20\,\Lambda$, $m_{A^0} = 10\,\Lambda$, $m_{H^\pm} = 2\,\Lambda$. $\sqrt{s} = 320\,\text{GeV}$. However, for large Λ the unitarity bounds can be violated

References

1. T. D. Lee, Phys. Rev. D **8**, (1973) 1226.
2. M. Krawczyk, Acta Phys. Polon. B **33**, (2002) 2621. [arXiv:hep-ph/0208076].
3. E. O. Iltan, Phys. Rev. D **65**, (2002) 036003. [arXiv:hep-ph/0108230].
4. A. G. Akeroyd, A. Arhrib and E. M. Naimi, Phys. Lett. B **490**, 119 (2000)
5. A. Dobado, M. J. Herrero and S. Penaranda, Eur. Phys. J. C **17**, (2000) 487.
6. T. Appelquist and J. Carazzone, Phys. Rev. D **11**, (1975) 2856.
7. M. Malinsky and J. Horejsi, arXiv:hep-ph/0308247.
8. S. Kanemura, T. Kasai and Y. Okada, Phys. Lett. B **471**, 182 (1999)
9. J. F. Gunion and H. E. Haber, Phys. Rev. D **67**, (2003) 075019.
10. M. Bohm et al., Nucl. Phys. B **304** (1988) 463.
11. S. Kanemura and H. A. Tohyama, Phys. Rev. D **57**, (1998) 2949.
12. M. Malinsky, Acta Phys. Slov. **52**, 259 (2002) [arXiv:hep-ph/0207066].
13. J. M. Cornwall, D. N. Levin and G. Tiktopoulos, Phys. Rev. D **10**, (1974) 1145.
14. M. Malinský and J. Hořejší, $e^+e^- \to W^+W^-$ in THDM, in preparation

Towards a NNLO Calculation in Hadronic Heavy Hadron Production

Jürgen G. Körner[1], Zakaria Merebashvili[2] and Mikhail Rogal[3]

[1] Institut für Physik, Johannes Gutenberg-Universität, 55099 – Mainz, Germany
koerner@thep.physik.uni-mainz.de
[2] High Energy Physics Institute, Tbilisi State University, 380086 Tbilisi, Georgia
zaza@thep.physik.uni-mainz.de
[3] Institut für Physik, Johannes Gutenberg-Universität, 55099 – Mainz, Germany
rogal@thep.physik.uni-mainz.de

1 Introduction

The full next-to-leading order (NLO) corrections to hadroproduction of heavy flavors have been completed in 1988 [1, 2]. They have raised the leading order (LO) estimates [3] but were still below the experimental results (see e.g. [4]). In a recent analysis theory moved closer to experiment [4]. A large uncertainty in the NLO calculation results from the freedom in the choice of the renormalization and factorization scales. The dependence on the factorization and renormalization scales is expected to be greatly reduced at next-to-next-to-leading order(NNLO). This reduces the theoretical uncertainty. Furthermore, one may hope that there is yet better agreement between theory and experiment at NNLO.

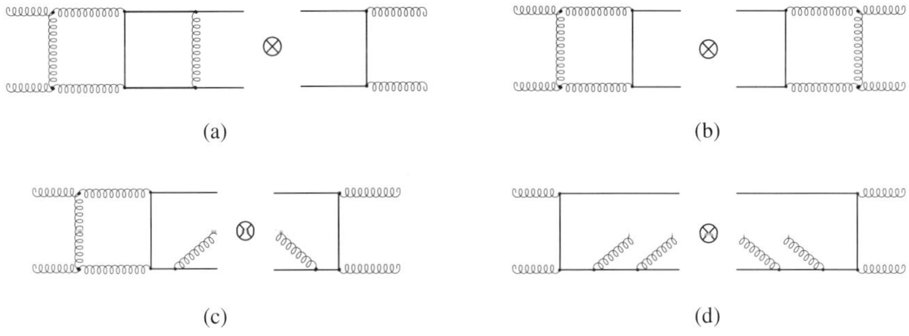

Fig. 1. Exemplary diagrams for the NNLO calculation of heavy hadron production

In Fig. 1 I show one generic diagram each for the four classes of contributions that need to be calculated for the NNLO corrections to hadroproduction of heavy flavors. They involve the two-loop contribution (Fig. 1a), the loop-by-loop contribution (Fig. 1b), the one-loop gluon emission contribution (Fig. 1c) and, finally, the two gluon emission contribution (Fig. 1d). An interesting subclass of the diagrams in Fig. 1c are those diagrams where the outgoing gluon is attached directly to the loop. One then has a five-point function which has to be calculated up to $\mathcal{O}(\varepsilon^2)$.

In our work we have concentrated on the loop-by-loop contributions exemplified by Fig. 1b. Specifically, working in the framework of dimensional regularization, we are in the process of calculating $\mathcal{O}(\varepsilon^2)$ results for all scalar one-loop one-, two-, three- and four-point integrals that are needed in the calculation of hadronic heavy flavour production. This is generally done by writing down the Feynman parameter representation for the corresponding integrals, integrating over Feynman parameters up to the last remaining integral, expanding the integrand of the last remaining parametric integral in terms of ε and doing the last parametric integration on the coefficients of the expansion. Because the one-loop integrals exhibit infrared (IR)/collinear (M) singularities up to $\mathcal{O}(\varepsilon^{-2})$ one needs to know the one-loop integrals up to $\mathcal{O}(\varepsilon^2)$ because the one-loop contributions appear in product form in the loop-by-loop contributions. It is clear that the spin algebra and the calculation of tensor integrals in the one-loop contributions also have to be done up to $\mathcal{O}(\varepsilon^2)$. This task will be left for the future.

Due to lack of space I can only present a few exemplary results. Since the four-point functions are the most difficult I concentrate on them.

2 Four-Point Functions

As a sample calculation I discuss the $(0, 0, m, 0)$-box with one massive propagator depicted in Fig. 2. As explained before one needs to calculate each one-loop integral up to $\mathcal{O}(\varepsilon^2)$ in order to obtain the finite terms in the loop-by-loop contributions. The box integral Fig. 2 is represented by the integral

$$D(-p_2, p_4, p_3, 0, 0, m, 0) = \mu^{2\varepsilon} \int \frac{\mathrm{d}^n q}{(2\pi)^n} \frac{1}{(q^2)(q-p_2)^2[(q-p_2+p_4)^2 - m^2](q-p_2+p_4+p_3)^2} ,$$

where p_1, p_2, p_3 and p_4 are external momenta with $p_1^2 = p_2^2 = 0$, $p_3^2 = p_4^2 = m^2$ and $n = 4 - 2\varepsilon$ is the dimension.

The ε^{-2}, ε^{-1} and ε^0 coefficients have been known for some time [1, 2] and will not be listed here. We define Mandelstam-type variables by $s \equiv (p_1 + p_2)^2$, $t \equiv (p_1 - p_3)^2 - m^2$ and $\beta = (1 - 4m^2/s)^{1/2}$, $x = (1 - \beta)/(1 + \beta)$.

For the real part of the $\mathcal{O}(\varepsilon)$ term we obtain:

$$\frac{iC_\varepsilon(m^2)}{12st}\varepsilon\Big[6 \ln^3 \tfrac{s}{m^2} + 20 \ln^3 \tfrac{-t}{m^2} + \ln^3 x + 6 \ln^2 x \ln \tfrac{-(s+2t-s\beta)}{2m^2}$$
$$- 12 \ln x \ln^2 \tfrac{-(s+2t-s\beta)}{2m^2}) + 8 \ln^3 \tfrac{-(s+2t-s\beta)}{2m^2}$$
$$- 12 \ln^2 \tfrac{-t}{m^2} \left(2 \ln(-1 - \tfrac{t}{m^2}) + 3 \ln x - \ln \tfrac{-(s+2t-s\beta)}{2m^2} + 4 \ln \tfrac{s+2t+s\beta}{2m^2}\right)$$
$$-3 \ln^2 \tfrac{s}{m^2} \left(6 \ln \tfrac{-t}{m^2} + 3 \ln x + 2 \ln \tfrac{-(s+2t-s\beta)}{2m^2} + 4 \ln \tfrac{s+2t+s\beta}{2m^2}\right)$$

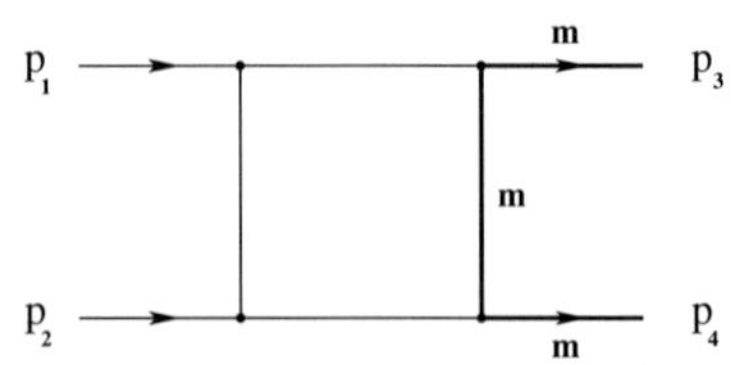

Fig. 2. One-loop box with one internal massive propagator

$+12\ \ln\frac{s}{m^2}+24\ \ln\frac{-t}{m^2}+24\ \ln\frac{s}{m^2}\ \ln\frac{-t}{m^2}+48\,Li_3(\frac{m^2+t}{t})+48\,Li_3(1+\frac{t}{m^2})+24\,Li_3(-x)-$
$24\,Li_3(\frac{-2\,m^2+t\,(-1+\beta)}{t\,(1+\beta)})-48\,Li_3(\frac{2\,m^2+t-t\,\beta}{2\,m^2})+24\,Li_3(\frac{2\,m^2+t-t\,\beta}{m^2\,(1+\beta)})$
$-24\,Li_3(\frac{2\,m^2+t-t\,\beta}{t-t\,\beta})-48\,Li_3(\frac{2\,m^2}{2\,m^2+t+t\,\beta})+24\,Li_3(-\frac{m^2\,(-1+\beta)}{2\,m^2+t+t\,\beta})$
$-24\,Li_3(\frac{t\,(-1+\beta)}{2\,m^2+t+t\,\beta})-24\,Li_3(\frac{2\,m^2+t+t\,\beta}{t+t\,\beta})-6\ \ln\frac{-t}{m^2}\Big(4+2\ \ln^2(-1-\frac{t}{m^2})-\ln^2 x-$
$4\ln x\ \ln\frac{-(s+2\,t-s\,\beta)}{2\,m^2}+6\ \ln^2\frac{-(s+2\,t-s\,\beta)}{2\,m^2}$
$-4\ \ln(-1-\frac{t}{m^2})\Big(\ln x+\ln\frac{-(s+2\,t-s\,\beta)}{2\,m^2}+\ln\frac{s+2\,t+s\,\beta}{2\,m^2}\Big)+8\,Li_2(1+\frac{t}{m^2})$
$+8\,Li_2(\frac{t\,(-1+\beta)}{2\,m^2})+4\,Li_2(\frac{m^2\,(-1+\beta)}{(m^2+t)\,(1+\beta)})-4\,Li_2(\frac{(m^2+t)\,(-1+\beta)}{m^2\,(1+\ \beta)})$
$+\ 8\,Li_2(\frac{2\,m^2}{2\,m^2+t+t\,\beta})\ -\ 12\zeta(2)\Big)\ -\ 12\ \ln x\,\zeta(2)\ +\ 48\ \ln\frac{-(s+2\,t-s\,\beta)}{2\,m^2}\,\zeta(2)\ -\ 6\ \ln\frac{s}{m^2}$
$\Big(2+4\ \ln^2\frac{-t}{m^2}-\ln^2 x-2\ \ln x\ \ln\frac{s+2\,t+s\,\beta}{2\,m^2}-4\ \ln\frac{-t}{m^2}\,\big(-1+\ln x+\ln\frac{-(s+2\,t-s\,\beta)}{2\,m^2}+$
$2\ \ln\frac{s+2\,t+s\,\beta}{2\,m^2}\big)-4\,Li_2(\frac{t\,(-1+\beta)}{2\,m^2})-4\,Li_2(\frac{2\,m^2}{2\,m^2+t+t\,\beta})+2\,\zeta(2)\Big)-36\,\zeta(3)\Big],$

where $C_\varepsilon(m^2)\equiv\frac{\Gamma(1+\varepsilon)}{(4\pi)^2}\left(\frac{4\pi\mu^2}{m^2}\right)^\varepsilon$.

One also needs the imaginary part of the $(0,0,m,0)$-box since the total contributions from the loop-by-loop contribution contains also imaginary parts via $|A|^2=(ReA)^2+(ImA)^2$. Note , however, that the imaginary part is only needed up to $\mathcal{O}(\varepsilon)$ since the IR/M singularities in the imaginary parts of the one-loop contributions are of $\mathcal{O}(\frac{1}{\varepsilon})$ only. For the $\mathcal{O}(\varepsilon)$ absorptive (imaginary) part we obtain:

$-\frac{iC_\varepsilon(m^2)}{4st}\varepsilon\pi\Big[3\ \ln^2\frac{s}{m^2}\ -\ 4\ \ln\frac{s}{m^2}\ \ln\frac{-t}{m^2}\ -\ 8\ \ln^2\frac{-t}{m^2}\ -\ 2\ \ln\frac{s}{m^2}\ \ln x\ +\ 4\ \ln\frac{-t}{m^2}\ \ln x\ +$
$3\ \ln^2 x-4\ \ln\frac{s}{m^2}\ \ln\frac{-(s+2\,t-s\,\beta)}{2\,m^2}\ -\ 4\ \ln x\ \ln\frac{-(s+2\,t-s\,\beta)}{2\,m^2}+4\ \ln^2\frac{-(s+2\,t-s\,\beta)}{2\,m^2}$
$-\ 4\ \ln\frac{s}{m^2}\ \ln\frac{s+2\,t+s\,\beta}{2\,m^2}+8\ \ln\frac{-t}{m^2}\ \ln\frac{s+2\,t+s\,\beta}{2\,m^2}+4\ \ln x\ \ln\frac{s+2\,t+s\,\beta}{2\,m^2}+8\,Li_2(\frac{t\,(-1+\beta)}{2\,m^2})+$
$8\,Li_2(\frac{2\,m^2}{2\,m^2+t+t\,\beta})+8\,\zeta(2)\Big].$

The ε^2–results for the $(0,0,m,0)$-box are too lengthy to fit into this report. They will be presented in a forthcoming publication [5]. A new feature of the ε^2–contributions is that the result can no longer be expressed in terms of logarithms and polylogarithms . They involve more general functions – the multiple polylogarithms introduced by Goncharov in 1998 [6]. A multiple polylogarithm is represented by

$$Li_{m_k,\ldots,m_1}(x_k,\ldots,x_1)=\int\limits_0^{x_1x_2\ldots x_k}\left(\frac{\mathrm{d}t}{t}\circ\right)^{m_1-1}\frac{\mathrm{d}t}{x_2x_3\ldots x_k-t}\circ$$
$$\left(\frac{\mathrm{d}t}{t}\circ\right)^{m_2-1}\frac{\mathrm{d}t}{x_3\ldots x_k-t}\circ\ldots\circ\left(\frac{\mathrm{d}t}{t}\circ\right)^{m_k-1}\frac{\mathrm{d}t}{1-t}\,,$$

where the iterated integrals are defined by

$$\int\limits_0^\lambda\frac{\mathrm{d}t}{a_n-t}\circ\ldots\circ\frac{\mathrm{d}t}{a_1-t}=\int\limits_0^\lambda\frac{\mathrm{d}t_n}{a_n-t_n}\int\limits_0^{t_n}\frac{\mathrm{d}t_n}{a_{n-1}-t_{n-1}}\times\ldots\times\int\limits_0^{t_2}\frac{\mathrm{d}t}{a_1-t_1}\,.$$

Besides the scalar $(0,0,m,0)$-box one also needs to calculate the scalar $(0,0,m,m)$ and $(m,m,0,m)$-boxes. Work is in progress on the calculation of these boxes.

3 Summary

I have reported on the results of an ongoing calculation of the ε-expansion of the scalar one-loop integrals up to $\mathcal{O}(\varepsilon^2)$ that are needed for the NNLO calculation of hadronic heavy hadron production. In order to arrive at the full amplitude structure of the one-loop contributions one still has to include the ε-dependence resulting from the Passarino-Veltman decomposition of the tensor integrals, and the ε-dependence of the spin algebra calculation. Putting all these pieces together one might optimistically say, when considering the four classes of diagrams Fig. 1, that the present calculation constitutes one-fourth of the full NNLO calculation of hadronic heavy hadron production.

References

1. P. Nason, S. Dawson and R. K. Ellis, Nucl. Phys. **B303** (1988) 607.
2. W. Beenakker, H. Kuijf, W. L. van Neerven and J. Smith, Phys. Rev. D **40**, 54 (1989).
3. M. Glück, J.F. Owens and E. Reya, Phys. Rev. D **17**, 2324 (1978);
   B. L. Combridge, Nucl. Phys. **B151** (1979) 429;
   J. Babcock, D. Sivers and S. Wolfram, Phys. Rev. D **18**, 162 (1978);
   K. Hagiwara and T. Yoshino, Phys. Lett. **80B**, 282 (1979);
   L. M. Jones and H. Wyld, Phys. Rev. D **17**, 782 (1978);
   H. Georgi et al., Ann. Phys. (N.Y.) **114**, 273 (1978).
4. M. Cacciari, S. Frixione, M. L. Mangano, P. Nason, G. Ridolfi, ArXiv: hep-ph/0312132.
5. J.G. Körner, Z. Merebashvili and M. Rogal, to be published.
6. A.B. Goncharov, Math. Res. Lett. **5**, 497 (1998), available at http://www.math.uiuc.edu/K-theory/0297.

Jet Physics at CDF

Sally Seidel[1]

Department of Physics and Astronomy, University of New Mexico, Albuquerque, NM 87131, USA
seidel@phys.unm.edu

1 Jets at CDF

Jets have been studied by the CDF Collaboration [1] as a means of searching for new particles and interactions, testing a variety of perturbative QCD predictions, and providing input for the global parton distribution function (PDF) fits. Unless otherwise indicated below, the jets were reconstructed using a cone algorithm [2] with cone radius $R = 0.7$ from data taken at the Fermilab Tevatron collider in Run 2, 2001–2003, with $\sqrt{s} = 1.96\,\text{TeV}$. Central jets, in the pseudorapidity range relative to fixed detector coordinates $0.1 < |\eta| < 0.7$, are used.

2 The Inclusive Jet Cross Section versus Transverse Energy

A measurement has been made of the inclusive jet cross section using 177 pb^{-1} of data. This cross section, which probes a distance scale below 10^{-17} cm, stimulated interest in obtaining improved precision on PDFs when initial Run 1 measurements [3] showed an excess of data over theoretical expectations at high transverse energy (E_T). The analysis uncouples the systematic shift in the cross section associated with the combined effects of energy mismeasurement and resolution limitation of the detector from the statistical uncertainty on the data. The data span an E_T range of 41–550 GeV, extending the upper limit from Run 1 by almost 150 GeV. The data are compared to next-to-leading (NLO) QCD predictions using the CTEQ6.1 PDF set [4] and found to be in good agreement. Figure 1 shows the ratio of measured and predicted cross sections as a function of E_T.

3 The Dijet Mass Spectrum

The dijet mass spectrum has been examined for evidence of new particles. A general search has been made for narrow resonances, and a direct search has been made for several particle types. The dijet masses were fitted to a smooth background function plus a mass resonance to obtain 95% confidence level upper limits on the cross section for production of new particles, as a function of mass M. The upper limit is compared to cross section predictions for axigluons [5], flavor universal colorons [6], excited quarks [7], and E_6 diquarks [8]. At 95% CL, the search excludes a model of

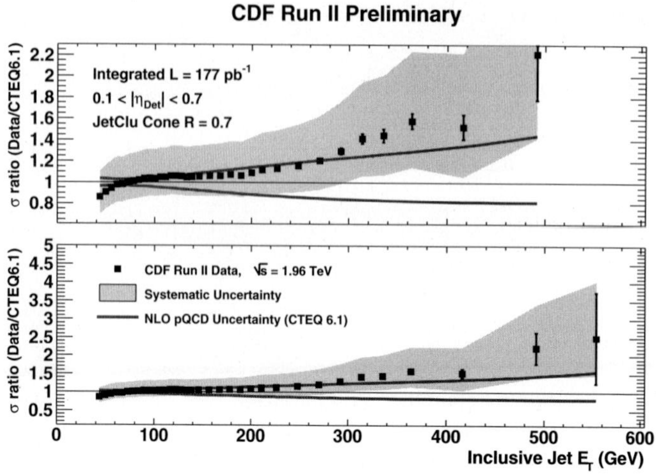

Fig. 1. The ratio of measured and predicted cross sections as a function of E_{T}. The gray band indicates the combined experimental systematic uncertainty. The *solid lines* represent the uncertainty associated with choice of PDF

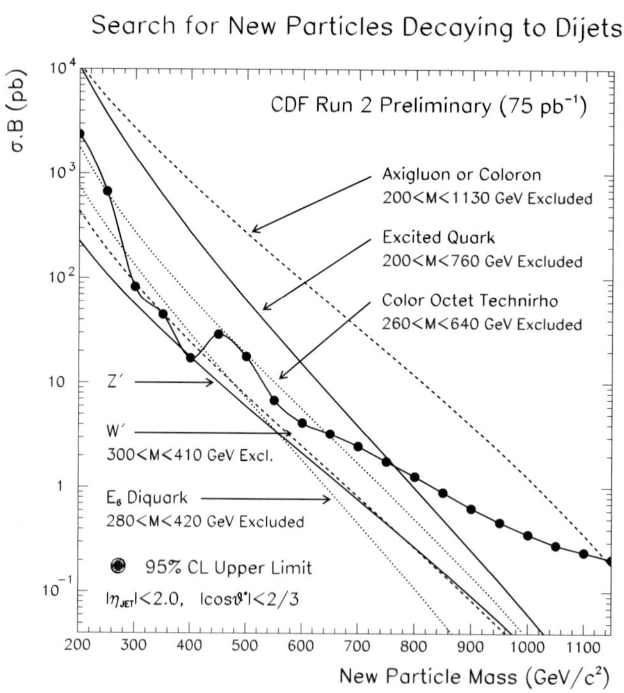

Fig. 2. The production cross section times branching ratio upper limits for new particles decaying to dijets, for $75\,\mathrm{pb}^{-1}$ of data

axigluons and colorons in the range $200 < M < 1130\,\mathrm{GeV/c^2}$, a model of excited quarks in the range $200 < M < 760\,\mathrm{GeV/c^2}$, and a model for E_6 diquarks in the range $280 < M < 420\,\mathrm{GeV/c^2}$. Figure 2 shows the production cross section times branching ratio upper limits for $75\,\mathrm{pb}^{-1}$ of data.

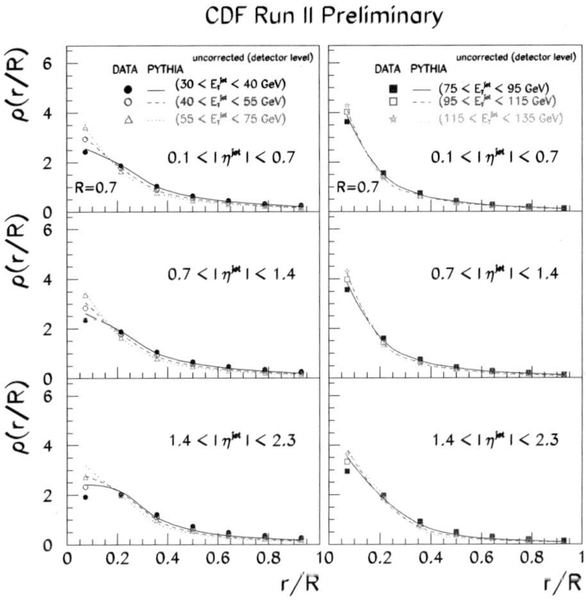

Fig. 3. The measured uncorrected differential jet shapes, $\rho(r)$, as computed using calorimetric information, in different regions of jet E_T and η, compared to Monte Carlo predictions

4 Jet Shapes

Jets shapes may be characterized in differential and integrated form as $\rho(r)$ and $\Psi(r)$, respectively, where r is a track's radial distance from the jet axis. The differential and integrated jet shapes are described by the average fraction of the jet's transverse energy that lies inside an annulus and a cone, respectively, concentric with the jet cone axis in the plane defined by pseudorapidity (η) and azimuthal angle (ϕ) relative to the detector. For an annulus of thickness Δr and a cone of radius R, we define the differential distribution of a jet containing N_jets jets as: $\rho(r) = \frac{1}{N_\mathrm{jet}} \frac{1}{\Delta r} \sum_\mathrm{jets} E_\mathrm{T}(r - \Delta r/2, r + \Delta r/2)/E_\mathrm{T}(0, R)$. We further define the integrated distribution by: $\Psi(r) = \frac{1}{N_\mathrm{jet}} \sum_\mathrm{jets} E_\mathrm{T}(0, r)/E_\mathrm{T}(0, R)$. A total of 75 pb^{-1} of data from calorimeter towers and from tracks in the central tracking chamber was examined, and the results were compared to leading order Monte Carlo predictions. Figures 3 and 4 show typical results for the differential and integrated distributions,

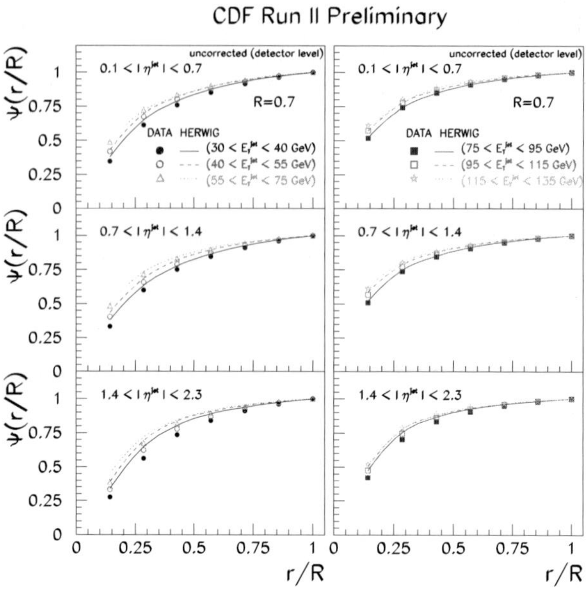

Fig. 4. The measured uncorrected integrated jet shapes, $\Psi(r)$, as computed using calorimetric information, in different regions of jet E_{T} and η, compared to Monte Carlo predictions

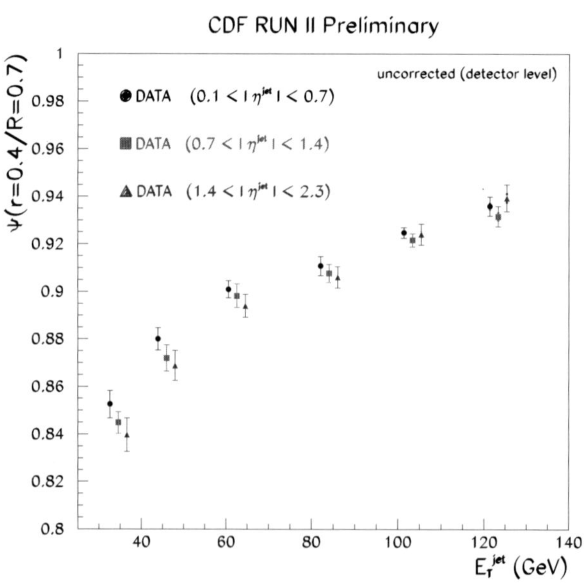

Fig. 5. Measured uncorrected jet shapes, $\Psi(r = 0.4)$, as computed using calorimetric information, in different regions of jet E_{T} and η. Inner error bars indicate statistical uncertainties, while outer error bars indicate the quadrature sum of statistical and systematic errors

respectively. One sees that HERWIG produces jets that are narrower than data, especially in the forward regions, but that the jet description improves with E_{T}. PYTHIA describes jet shapes fairly well but produces jets narrower than the data in some kinematic regions, in particular at low E_{T}. Figure 5 demonstrates, with an integrated jet shape measurement applied in three pseudorapidity regions for fixed cone opening angle $R = 0.4$, that jets narrow as jet transverse energy increases.

5 Jet Algorithms

The cone and k_{T} [9] algorithms have been compared. The k_{T} algorithm, which successively merges pairs of nearby objects in order of decreasing E_T, uses a parameter D to control the end of merging. The cone algorithm combines tracks into jets on the basis of their location relative to a cone of radius R in η-ϕ space. For the purpose of comparing the algorithms, we set $\Delta R \equiv \sqrt{(\eta_{cone} - \eta_{k_T})^2 + (\phi_{cone} - \phi_{k_T})^2} < 0.1$. With these definitions, Fig. 6 shows that the k_{T} algorithm typically captures less E_{T} than the cone. We also find that the difference in E_{T} assignment depends upon the E_{T} of the cone jet, and that the relative E_T captured by the two algorithms depends strongly on the value of D.

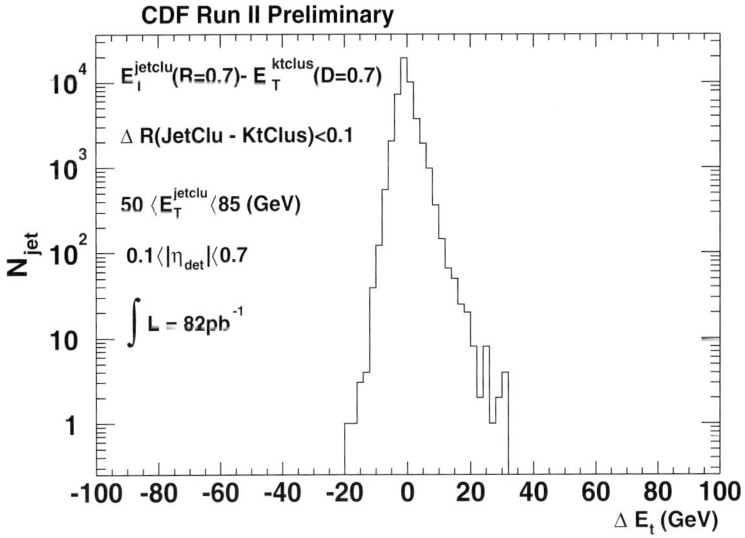

Fig. 6. The difference between the E_{T} distributions of jets selected with the cone and k_{T} algorithms, for R and D both equal to 0.7 and $\Delta R < 0.1$

References

1. A. Sill, et al: Nucl. Instr. and Meth. A 447, 1 (2000); T. Affolder, et al., FERMILAB-PUB-03/355-E; G. Ascoli, et al., Nucl. Instr. and Meth. A 268, 33 (1988); T. Dorigo, Nucl. Instr. and Meth. A 461, 560 (2001).
2. F. Abe, et al.: Phys. Rev. D 45, 1448 (1992).
3. F. Abe, et al.: Phys. Rev. Lett. 77, 438 (1996).
4. J. Pumplin, et al.: JHEP 0207, 012 (2002).
5. J. Bagger, et al.: Phys. Rev. D 37, 1188 (1988).
6. E. Eichten and K. Lane: Phys. Lett. B 327, 129 (1994).
7. U. Baur, et al.: Int. J. Mod. Phys. A2, 1285 (1987); U. Baur, et al.: Phys. Rev. D 42, 815 (1990).
8. G. Katsilieris, et al.: Phys. Lett. B 288, 221 (1992); J. Hewitt and T. Rizzo: Phys. Rep. 183, 193 (1989); V. Angelopoulos, et al.: Nucl. Phys. B 292, 59 (1987).
9. S.D. Ellis and D.E. Soper: Phys. Rev. D 48, 3160 (1993).

About the Meeting

The 9th Adriatic Meeting was organized by:
Theoretical Physics Division of the Rudjer Boskovic Institute, Zagreb; Theoretical Physics Department of the University of Zagreb; University of Split; Ludwig-Maximilians-Universitaet, Munich; Max-Planck-Institute for Physics, Munich and WIGV.

The Meeting was to large extent supported by the VolksWagen Stiftung.
We would also like to thank for support: Ministry of Science and Technology, Republic of Croatia, Croatian Academy of Sciences and Arts, Rudjer Boskovic Institute, The Abdus Salam International Centre for Theoretical Physics.

International Advisory Committee

D. Amati, L. Álvarez-Gaumé, K. Berkelman, M. Blagojević, L. Bonora, S. J. Brodsky, M. Burić, G. W. Buschhorn, D. Chang, D. Denegri, S. Fajfer, H. Grosse, B. Guberina, J. Hořejší, R. Jackiw, T. Jacobson, P. Jenni, B. Jurčo, A. Linde, A. Ljubičić, A. Patkós, I. Picek, M. Praszalowicz, G. Raffelt, H. Rubinstein, A. I. Sanda, H. Satz, P. Schupp G. Senjanović, M. Shifman, K. Suruliz, P. J. Steinhardt, R. D. Viollier.

National Advisory Committee

M. Dželalija, M. Furić, R. Horvat, K. Kadija, D. Klabučar, M. Martinis, K. Pisk, N. Zovko.

Organizing Committee

I. Andrić, Ž. Antunović, N. Bilić, D. Horvatić, L. Jonke, L. Moeller, H. Nikolić, S. Pallua, H. Štefančić, D. Tadić - Chairman, J. Trampetić - Chairman, J. Wess - Chairman.

List of Participants

Hrvoje ABRAHAM
Rudjer Boskovic Institute, Zagreb
ahrvoje@thphys.irb.hr

Guido ALTARELLI
CERN, Geneva
guido.altarelli@cern.ch

Luis ÁLVAREZ-GAUMÉ
CERN, Geneva
alvarez@nxth04.cern.ch

Ivan ANDRIĆ
Rudjer Boskovic Institute, Zagreb
iandric@rudjer.irb.hr

Željko ANTUNOVIĆ
University of Split
Zeljko.Antunovic@pmfst.hr

Paolo ASCHIERI
Ludwig-Maximilians-Universitaet, Munich
aschieri@theorie.physik.uni-muenchen.de

Ana BABIĆ
Rudjer Boskovic Institute, Zagreb
ababic@thphys.irb.hr

Antun BALAŽ
Institute of Physics, Zemun
antun@phy.bg.ac.yu

Astrid BAUER
ThEP - Theoretical Elementary Particle Physics, Mainz
bauea002@thep.physik.uni-mainz.de

Igor BERTOVIĆ
University of Zagreb
bertovic@plan.phy.hr

Neven BILIĆ
Rudjer Boskovic Institute, Zagreb
bilic@thphys.irb.hr

Ines-Ana BILJAN
Rudjer Boskovic Institute, Zagreb
inesbiljan@hotmail.com

Loriano BONORA
SISSA, Trieste
bonora@sissa.it

Damir BOSNAR
University of Zagreb
bosnar@phy.hr

Friedemann BRANDT
MPI for Mathematics in the Sciences, Leipzig
fbrandt@mis.mpg.de

Andrzej BURAS
Technical University Munich, Garching
aburas@ph.tum.de

Gerd W. BUSCHHORN
Max-Planck-Institut, Muenchen
gwb@mppmu.mpg.de

Martin CEDERWALL
Theoretical Physics, Goteborg
martin.cederwall@fy.chalmers.se

Nicolas COSME
Université Libre de Bruxelles, Brussels
ncosme@ulb.ac.be

Maro CVITAN
University of Zagreb
mcvitan@phy.hr

Ivan DADIĆ
Rudjer Boskovic Institute, Zagreb
dadic@thphys.irb.hr

Klaus DESCH
University of Hamburg
Klaus.Desch@desy.de

Nilendra DESHPANDE
Institute of Theoretical Science, Eugene
desh@oregon.uoregon.edu

Marija DIMITRIJEVIĆ
University of Belgrade
dmarija@theorie.physik.uni-muenchen.de

Goran DUPLANČIĆ
Rudjer Boskovic Institute, Zagreb
gorand@thphys.irb.hr

Mile DŽELALIJA
University of Split
mile@pmfst.hr

Svjetlana FAJFER
Fakultet za matematiko in fiziko, Ljubljana
svjetlana.fajfer@ijs.si

Daniel FERENC
University of California, Davis
daniel.ferenc@cern.ch

Michael FINGER
JINR, Dubna, Moscow region
michael.finger@cern.ch

Ephraim FISCHBACH
Purdue University, W. Lafayette
ephraim@physics.purdue.edu

Fabiola GIANOTTI
CERN, Geneva
fabiola.gianotti@cern.ch

Merab GOGBERASHVILI
Institute of Physics, Tbilisi
gogber@ICTP.Trieste.it

Ariel GOOBAR
Stockholm University
ariel@physto.se

Harald GROSSE
University of Vienna
harald.grosse@univie.ac.at

Branko GUBERINA
Rudjer Boskovic Institute, Zagreb
guberina@thphys.irb.hr

Wolfgang HOLLIK
Max-Planck-Institut fuer Physik, Munich
hollik@mppmu.mpg.de

Raul HORVAT
Rudjer Boskovic Institute, Zagreb
horvat@lei3.irb.hr

Davor HORVATIĆ
University of Zagreb
davorh@phy.hr

Dario HRUPEC
University of Zagreb
dario.hrupec@fer.hr

Tristan HÜBSCH
Howard University, Washington
thubsch@howard.edu

Helmuth HUEFFEL
Univ. of Vienna
helmuth.hueffel@univie.ac.at

Ted JACOBSON
University of Maryland, College Park
jacobson@physics.umd.edu

Antal JEVICKI
Brown University, Providence
antal@het.brown.edu

Larisa JONKE
Rudjer Boskovic Institute, Zagreb
larisa@irb.hr

Branislav JURČO
Universitaet Munich
jurco@lmu.de

Danijel JURMAN
Rudjer Boskovic Institute, Zagreb
djurman@irb.hr

Dubravko KLABUČAR
University of Zagreb
klabucar@oberon.phy.hr

Andrea KNEŽEVIĆ
Rudjer Boskovic Institute, Zagreb
andrea@lumen.irb.hr

Florian KOCH
Sektion Physik der Universitaet Muenchen
florian.koch@physik.uni-muenchen.de

Boris KOSYAKOV
Russian Federal Nuclear Center-VNIIEF, Sarov
kosyakov@vniief.ru

Milica KRČMAR
Rudjer Boskovic Institute, Zagreb
Milica.Krcmar@irb.hr

Vitaly KUDRYAVTSEV
University of Sheffield
v.kudryavtsev@sheffield.ac.uk

Krešimir KUMERIČKI
University of Zagreb
kkumer@phy.hr

Duško LATAS
University of Belgrade
latas@phy.bg.ac.yu

Ernest MA
University of California, Riverside
ma@phyun8.ucr.edu

Andrzej R. MALECKI
KEN Pedagogical University, Kraków
amalecki@ap.krakow.pl

Michal MALINSKY
Charles University, Praha
malinsky@ipnp.troja.mff.cuni.cz

Thomas MANNEL
University of Karlsruhe
thomas.mannel@physik.uni-karlsruhe.de

Donald MAROLF
Syracuse University, Syracuse
marolf@physics.syr.edu

Mladen MARTINIS
Rudjer Boskovic Institute, Zagreb
martinis@rudjer.irb.hr

Katarina MATIĆ
University of Belgrade
mkatarin@tesla.rcub.bg.ac.yu

Blaženka MELIĆ
Rudjer Boskovic Institute, Zagreb
melic@thphys.irb.hr

Frank MEYER
Max-Planck-Institute for Physics, Munich
Frank.Meyer@physik.uni-muenchen.de

Vesna MIKUTA-MARTINIS
Rudjer Boskovic Institute, Zagreb
vmikuta@rudjer.irb.hr

Marijan MILEKOVIĆ
University of Zagreb
marijan@phy.hr

Predrag MILENOVIĆ
"Vinca" Institute of Nuclear Sciences, Beograd
predragm@vin.bg.ac.yu

Peter MINKOWSKI
University of Bern
mink@itp.unibe.ch

Lutz MOELLER
Max-Planck-Institut fuer Physik, Muenchen
lmoeller@theorie.physik.
uni-muenchen.de

Hrvoje NIKOLIĆ
Rudjer Boskovic Institute, Zagreb
hrvoje@thphys.irb.hr

Bene NIŽIĆ
Rudjer Boskovic Institute, Zagreb
nizic@thphys.irb.hr

Igor NOVIKOV
Theoretical astrophysics center, Copenhagen
novikov@tac.dk

Silvio PALLUA
University of Zagreb
Pallua@phy.hr

Kornelija PASSEK-KUMERIČKI
Rudjer Boskovic Institute, Zagreb
passek@thphys.irb.hr

Ivica PICEK
University of Zagreb
picek@phy.hr

Anita PRAPOTNIK
Institute Jožef Štefan, Ljubljana
anita.prapotnik@ijs.si

Predrag PRESTER
University of Zagreb
pprester@phy.hr

Ivica PULJAK
Technical University of Split
Ivica.Puljak@fesb.hr

Voja RADOVANOVIĆ
Faculty of Physics Belgrade
rvoja@ff.bg.ac.yu

Georg RAFFELT
MPI Physik, Munich
raffelt@mppmu.mpg.de

Mikhail ROGAL
Mainz University
rogal@thep.physik.uni-mainz.de

Graham ROSS
Oxford University
g.ross1@physics.ox.ac.uk

Reinhold RUECKL
University of Würzburg
rueckl@physik.uni-wuerzburg.de

Peter SCHUPP
International University Bremen
p.schupp@iu-bremen.de

Sally SEIDEL
University of New Mexico, Albuquerque
seidel@phys.unm.edu

Mikhail SHIFMAN
TPI, Minneapolis
shifman@umn.edu

Joan SOLÀ
University of Barcelona
sola@ifae.es

Vikram SONI
National Physical Laboratory, New Delhi
v_soni@vsnl.com

Hrvoje ŠTEFANČIĆ
Rudjer Boskovic Institute, Zagreb
shrvoje@thphys.irb.hr

Harold STEINACKER
Ludwig-Maximilians Universitaet Muenchen
hsteinac@theorie.physik.uni-muenchen.de

Gordana TADIĆ
Insiel s.p.a., Trieste
gordana.tadic@insiel.it

Josip TRAMPETIĆ
Rudjer Boskovic Institute, ZAGREB
josipt@rex.irb.hr

Anca TUREANU
University of Helsinki
atureanu@pcu.helsinki.fi

Raoul VIOLLIER
Institute of Theoretical Physics and Astrophysics, Cape Town
viollier@physci.uct.ac.za

Ana WEBER
Rudjer Boskovic Institute, Zagreb
aweber@rudjer.irb.hr

Julius WESS
MPI fuer Physik, Muenchen
wess@theorie.physik.uni-muenchen.de

Michael WOHLGENANNT
University of Munich
miw@theorie.physik.uni-muenchen.de

Danilo ZAVRTANIK
Nova Gorica Polytechnic
danilo.zavrtanik@p-ng.si

Nikola ZOVKO
Rudjer Boskovic Institute, Zagreb
zovko@thphys.irb.hr

George ZUPANOS
National Technical University, Athens
george.zoupanos@cern.ch

SPRINGER PROCEEDINGS IN PHYSICS

46 **Cellular Automata and Modeling of Complex Physical Systems**
Editors: P. Manneville, N. Boccara, G.Y. Vichniac, and R. Bidaux

47 **Number Theory and Physics**
Editors: J.-M. Luck, P. Moussa, and M. Waldschmidt

48 **Many-Atom Interactions in Solids**
Editors: R.M. Nieminen, M.J. Puska, and M.J. Manninen

49 **Ultrafast Phenomena in Spectroscopy**
Editors: E. Klose and B. Wilhelmi

50 **Magnetic Properties of Low-Dimensional Systems II**
New Developments
Editors: L.M. Falicov, F. Mejía-Lira, and J.L. Morán-López

51 **The Physics and Chemistry of Organic Superconductors**
Editors: G. Saito and S. Kagoshima

52 **Dynamics and Patterns in Complex Fluids**
New Aspects of the Physics–Chemistry Interface
Editors: A. Onuki and K. Kawasaki

53 **Computer Simulation Studies in Condensed-Matter Physics III**
Editors: D.P. Landau, K.K. Mon, and H.-B. Schüttler

54 **Polycrystalline Semiconductors II**
Editors: J.H. Werner and H.P. Strunk

55 **Nonlinear Dynamics and Quantum Phenomena in Optical Systems**
Editors: R. Vilaseca and R. Corbalán

56 **Amorphous and Crystalline Silicon Carbide III, and Other Group IV–IV Materials**
Editors: G.L. Harris, M.G. Spencer, and C.Y. Yang

57 **Evolutionaly Trends in the Physical Sciences**
Editors: M. Suzuki and R. Kubo

58 **New Trends in Nuclear Collective Dynamics**
Editors: Y. Abe, H. Horiuchi, and K. Matsuyanagi

59 **Exotic Atoms in Condensed Matter**
Editors: G. Benedek and H. Schneuwly

60 **The Physics and Chemistry of Oxide Superconductors**
Editors: Y. Iye and H. Yasuoka

61 **Surface X-Ray and Neutron Scattering**
Editors: H. Zabel and I.K. Robinson

62 **Surface Science**
Lectures on Basic Concepts and Applications
Editors: F.A. Ponce and M. Cardona

63 **Coherent Raman Spectroscopy**
Recent Advances
Editors: G. Marowsky and V.V. Smirnov

64 **Superconducting Devices and Their Applications**
Editors: H. Koch and H. Lübbing

65 **Present and Future of High-Energy Physics**
Editors: K.-I. Aoki and M. Kobayashi

66 **The Structure and Conformation of Amphiphilic Membranes**
Editors: R. Lipowsky, D. Richter, and K. Kremer

67 **Nonlinearity with Disorder**
Editors: F. Abdullaev, A.R. Bishop, and S. Pnevmatikos

68 **Time-Resolved Vibrational Spectroscopy V**
Editor: H. Takahashi

69 **Evolution of Dynamical Structures in Complex Systems**
Editors: R. Friedrich and A. Wunderlin

70 **Computational Approaches in Condensed-Matter Physics**
Editors: S. Miyashita, M. Imada, and H. Takayama

71 **Amorphous and Crystalline Silicon Carbide IV**
Editors: C.Y. Yang, M.M. Rahman, and G.L. Harris

72 **Computer Simulation Studies in Condensed-Matter Physics IV**
Editors: D.P. Landau, K.K. Mon, and H.-B. Schüttler

73 **Surface Science**
Principles and Applications
Editors: R.F. Howe, R.N: Lamb, and K. Wandelt

Printing: Saladruck, Berlin
Binding: Stein+Lehmann, Berlin